Symmetry and Structural Properties of Condensed Matter: Proceedings of the Eighth International School on Theoretical Physics

(SSPCM 2005)

Journal of Physics: Conference Series Volume 30

Myczkowce, Poland
31 August - 7 September 2005

ISBN: 978-1-60560-973-7

Printed from e-media with permission by:

Curran Associates, Inc.
57 Morehouse Lane
Red Hook, NY 12571
www.proceedings.com

Some format issues inherent in the e-media version may also appear in this print version.

Printed by Curran Associates, Inc. (2009)

Institute of Physics
Dirac House, Temple Back
Bristol BS1 6BE UK

Phone: 44 1 17 929 7481
Fax: 44 1 17 920 0979

techtracking@iop.org

TABLE OF CONTENTS

Preface

This volume encloses the Proceedings of the Eight Summer School on Theoretical Physics under the banner title 'Symmetry and Structural Properties of Condensed Matter' (SSPCM 2005). The School, organised by Rzeszów University of Technology, Poland, together with Laboratory of Physical Foundation of Information Processing, Poland (LFPPI), was held between 31 August and 7 September 2005 in Myczkowce.

The main goal of the whole series of biannual SSPCM schools (since 1990) is promotion of advanced mathematical methods within condensed matter physics, directed towards its symmetry and structural properties.

This SSPCM 05 School focused on the following three main subjects:

A. decoherence and quantum computers;

B. the role of combinatorics in classification of solutions for exactly solvable models;

C. geometric aspects in nanophysics.

The Proceedings are divided into three parts accordingly, but particular topics can overlap between main subjects, and be related to problems pursued in previous SSPCM schools. In this way, the present school is concentrated on various aspects of theory and technology of quantum informatics, with the inclusion of exactly solvable models of statistical physics of condensed matter in low dimensions, as some natural theoretical prototypes of a quantum computer.

The last SSPCM 05 was devoted to the memory of Professor Brian G Wybourne, a great Inspirer, inestimable Patron and Lecturer of the whole series of these schools.

The School gathered together more than 50 participants, both advanced researchers in physics and mathematics, as well as their young collaborators and students, representing altogether 10 countries from all over the world.

The Organising Committee would like to express their gratitude to all members of the International Advisory Committee for their opinions and support and to all invited lecturers and contributors for their talks and preparation of their manuscripts. Special thanks are addressed to all participants and everyone who attended for creating such a stimulating and friendly atmosphere during our meeting, and for several valuable discussions. We thank all chairmen for their polite but efficient leading of sessions. Many thanks are due to the referees who improved significantly the quality of papers presented in this volume. The organisers address special thanks to The Nicholas C. Metropolis Mathematics Foundation (USA) for a substantial initiating support, and to the Polish State Committee for Scientific Research. The hospitality of the whole team of the hotel 'Energetyk' Myczkowce is also appreciated.

On behalf of the organizers of SSPCM 2005
Tadeusz Lulek, **Andrzej Wal** and **Barbara Lulek**

Conference committees

International advisory committee
Willem J Caspers, Enschede
Rainer Dirl, Wien
Jean-Pierre Gazeau, Paris
Jacek Karwowski, Toruń
Adalbert Kerber, Bayreuth
Maurice Kibler, Lyon
Ronald C King, Southampton
James D Louck, Los Alomos
Janusz A Morkowski, Poznań
Jan Mozrzymas, Wrocław
Yoshio Ohnuki, Nagoya
Viacheslav I Yukalov, Dubna
Apostel Vourdas, Bradford
Brian G Wybourne, Toruń

Local organizing committee
Tadeusz Lulek (chairman)
Tadeusz Paszkiewicz
Wiesława Sikora
Lucjan Pytlik
Barbara Lulek
Dorota Jakubczyk
Sławomir Wolski
Andrzej Wal (secretary)

Sponsors
Nicholas C Metropolis Mathematics Foundation (USA)
Polish State Committee for Scientific Research

Brian G Wybourne: Innovator

James D Louck
Los Alamos National Laboratory, Theoretical Division
Los Alamos, NM 87545 USA
E-mail: `jimlouck@aol.com`

We will hear much during this conference on Brian Wybourne's contributions to science. I wish to bring attention to another aspect of his role in science and science statesmanship by relating a personal story based on the discovery and development of the properties of quaternions during the 1844–45 period by Hamilton and Cayley. Brian Wybourne will enter in the middle of my presentation.

In the collection of reprints *Theory of Angular Momentum*, Ed., L.C. Biedenharn and H. Van Dam, Academic Press, 1965, there appears two paragraphs on the geometrical objects called 'turns.' It is described there in succinct descriptive terms how this concept is the generalization of the addition of geometrical vectors (directed line segments) in the plane, using the usual 'head-to-tail' rule and the equality of geometrical vectors under parallel transport, the generalization being the addition of directed line segments belonging to great circles of the unit sphere with parallel transport being the equivalence of two such 'directed arcs' of the same length belonging to the same great circle. It is remarked that for infinitesimal turns the commutation rules $\mathbf{J}\times\mathbf{J} = i\mathbf{J}$ are implied. These are, of course, the Lie algebra relations for the quantum theory of angular momentum, discovered some fifty years later by mathematicians (Poincaré, Cartan, Lie) in a purely mathematical context, and some ninety years later by physicists in the context of quantum theory. Many of us, including Brian, have spent years dedicated to the application of these relations and their generalizations to other symmetries, and the consequences for physical theory.

I first visited Larry Biedenharn at Duke University in 1968, and he gave me a copy of his book with Van Dam. It was a very valuable asset for me, and I was caught-up by the concept of turns. It was such a nice problem to visualize that I worked out the principal properties for my own enjoyment. (In fact, the concept does not appear directly in the work of Hamilton and Cayley.) Larry liked it very much, and tuned it greatly. I invite each of you to undertake the same enjoyable task before consulting the very detailed treatment in L. C. Biedenharn and J. D. Louck, *Angular Momentum in Quantum Physics,* in: Encyl. Math. and Its Applications (Ed. G.-C. Rota) Cambridge Univ., 1981, where the full apparatus of the group representation theory of angular momentum through the use of left and right translations, etc. is developed.

The essence of the treatment is presented [1] in Figures 1–3 which illustrate, respectively, the definition of a turn as a directed arc length on a great circle, the rule of addition for turns belonging to different great

[1] Thanks are expressed to Cambridge University Press for permission to reprint Figures 1–3.

circles and the noncommutivity of this addition rule, and finally the associativity of this addition rule.

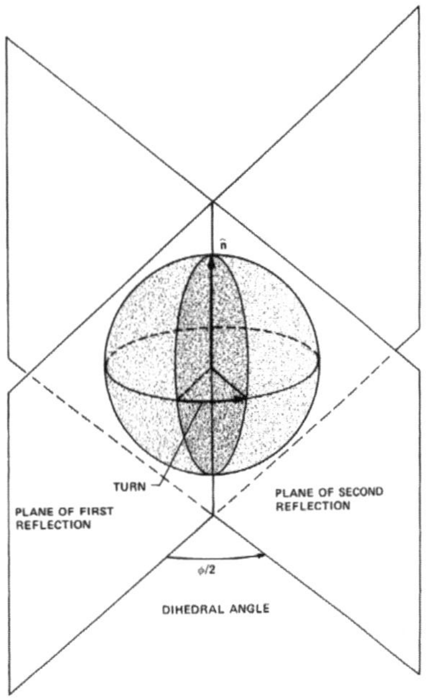

Figure 1. A turn is an ordered pair of points shown here, geometrically, as a directed arc length of a great circle. Each turn defines a rotation by specifying two intersecting planes.

The seeds of Lie algebra and group representation theory, going beyond the substitution (symmetric) group of Galois, were already present in the work of Hamilton and Cayley. Such developments in physics were to await the foundational work of Weyl, Wigner, and Racah. But we must also include here Brian Wybourne for his capacity and talent for bringing these concepts into forms suitable for real calculations in his books, publications, and computer programs.

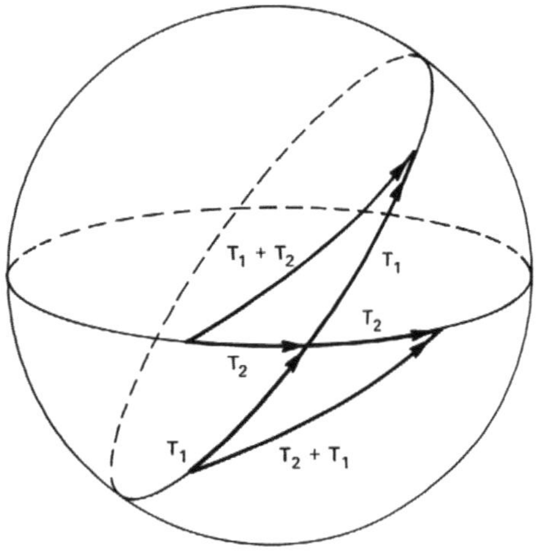

Figure 2. Addition of turns is noncommutative.

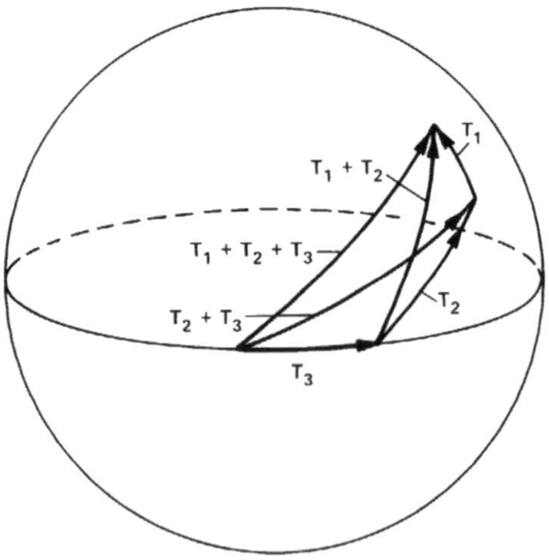

Figure 3. Addition of turns is associative.

It was but a few years later that Brian Wybourne entered my career in an indirect, but memorable way. Brian Wybourne was principally, if not solely, responsible for a series of lectures given at the University of Canterbury on basic advances in the spectroscopy of complex systems, using methods of symmetry groups. Brian Judd, Johns Hopkins University, and J. P. Elliott, University of Sussex, were the first Visiting Erskine Fellows at the University of Canterbury, where they delivered some very famous lectures published in: Topics in Atomic and Nuclear Physics, University of Canterbury, 1969. I expect that he met Larry Biedenharn at about the same time, perhaps, somewhat earlier, at the famous Moshinsky conferences in Mexico. In any case, he invited Larry to Canterbury to give the next Erskine Lecture in 1974. While Larry was already well-known for his independent discovery of the Biedenharn-Elliott identities for Racah coefficients, he chose to speak on the theory of turns. The lectures were never published, but the event is documented in the Biedenharn-Louck volumes with my name included. I was, of course, very pleased and honored to have had a small part in those lectures. (I also learned just today that Ron King was an Erskine Fellow, which initiated a long and fruitful collaboration that you will hear about soon.)

This personal story is intended to show Brian in his broader role in science as the communicator and innovator of ideas by bringing colleagues together; by his intense interest and insights into the foundations of a subject; by his extraordinary insight into what is important; and by the inspiring conversations he generated at informal meetings with colleagues and friends. And not least, the confidence he instilled in others such as me to do research. It has been a great privilege to share in this at the meetings of the Lulek SSPCM conferences, extending over fifteen years, and continuing today in his honor.

Qubits and invariant theory

Ronald C King and Trevor A Welsh

School of Mathematics, University of Southampton, Southampton SO17 1BJ, England

E-mail: R.C.King@soton.ac.uk and taw@maths.soton.ac.uk

Abstract. The invariants of a mixed two-qubit system are discussed. These are polynomials in the elements of the corresponding density matrix. They are counted by means of group-theoretic branching rules and the Molien function is determined. The fundamental invariants are then explicitly constructed and the relations between them are found in the form of syzygies. In this way, complete sets of primary and secondary invariants are identified: there are 10 of the former and 15 of the latter.

1. Introduction

The entanglement of a two-qubit system is a non-local property so that measures of entanglement should be independent of all local transformations of the two qubits seperately. Since a mixed two-qubit system is described by its density matrix, its non-local entangling properties must be described by local invariants of the density matrix. A complete set of local invariants is provided by local invariants that are homogeneous polynomials, that is polynomials of fixed total degree in the elements of the density matrix [1, 2].

It follows that there is a hierarchy of invariant theory problems relevant to a mixed two-qubit system. The first problem is to count the number of invariants n_m of a given degree m and to determine the corresponding generating function, that is to identify the Molien function $M(q)$ whose power series expansion takes the form $\sum_{m=0}^{\infty} n_m\, q^m$. The second problem is to construct the invariants explicitly, identifying a set of *fundamental* invariants, from which all others may be constructed as sums of products. By virtue of Hilbert's Theorem [3], the cardinality of any set of fundamental invariants is known to be finite. The third problem is to determine the structure of the ring of invariants, which is necessarily *Cohen-Macaulay* [4]. To this end it is necessary to identify a set of *primary* invariants and a set of *secondary* invariants.

To state the problem more precisely, it should be noted that a single qubit is described by a quantum state ψ_i with two degrees of freedom specified by $i = 0, 1$. A two-qubit system is then described by a linear superposition of direct product states $\Psi_{a,i}(A,B) = \psi_a(A)\psi_i(B)$ with $a, i = 0, 1$. For a mixed system the quantity of interest is the density matrix which takes the form

$$\rho = \sum_{a,i,b,j=0}^{1} \rho_{a,i}^{b,j}\ \Psi_{a,i}(A)\ \overline{\Psi}^{b,j}(B). \tag{1}$$

Under arbitrary and independent non-singular local transformations of the two qubits defined by

$$g:\psi_i(A)\;\rightarrow\;\sum_{j=0}^{1}\psi_j(A)\,g_{ji}\quad\text{and}\quad h:\psi_a(B)\;\rightarrow\;\sum_{b=0}^{1}\psi_b(B)\,h_{ba} \tag{2}$$

with $g\times h\in GL_A(2)\times GL_B(2)$, the density matrix elements transform as

$$g\times h:\rho_{a,i}^{b,j}\;\rightarrow\;\sum_{c,k,d,l} g_{ac}\,h_{ik}\,\rho_{c,k}^{d,l}\,g_{db}^{-1}\,h_{lj}^{-1}\,. \tag{3}$$

With this notation, the task is first to count and then to construct all polynomial functions of the 16 components of ρ that are invariant under the local transformations $g\times h$ of $GL_A(2)\times GL_B(2)$, viewed as a subgroup of $GL_{AB}(16)$.

2. Counting invariants

The 16 components, $\rho_{a,i}^{b,j}$, of the mixed 2-qubit density matrix form the basis of the defining irreducible representation V of $GL_{AB}(16)$. The homogeneous polynomials of degree m form the basis of the mth fold symmetrised power irreducible representation $V^{\{m\}}$ of $GL_{AB}(16)$. Let n_m be the number of linearly independent homogeneous polynomials that are invariant under the action of the local transformations of $GL_A(2)\times GL_B(2)$. Then n_m is just the number of times the trivial, one-dimensional irreducible representation of $GL_A(2)\times GL_B(2)$ appears in the restriction of $V^{\{m\}}$ from $GL_{AB}(16)$ to this subgroup.

Quite generally [5, 6], the polynomial irreducible representations of $GL(n)$ have characters $\{\lambda\}$, specified by partitions λ of length $\ell(\lambda)\leq n$, which are nothing other than Schur functions $s_\lambda(x_1,\ldots,x_n)$ of the eigenvalues x_i of the group elements of $GL(n)$. In addition there exist corresponding contragredient irreducible representations with characters $\{\overline{\lambda}\}$, which are Schur functions $s_\lambda(\overline{x}_1,\ldots,\overline{x}_n)$ of the eigenvalues $\overline{x}_i=x_i^{-1}$ of the inverse group elements.

We consider the following two group-subgroup chains from $GL_{AB}(16)$ to $GL_A(2)\times GL_B(2)$:

$$\begin{aligned} GL_{AB}(16) &\supset GL_A(4)\times GL_B(4)\\ &\supset (GL_A(2)\times\overline{GL_A(2)})\times(GL_B(2)\times\overline{GL_B(2)})\\ &\supset GL_A(2)\times GL_B(2); \end{aligned} \tag{4}$$

$$\begin{aligned} GL_{AB}(16) &\supset GL_{AB}(4)\times\overline{GL_{AB}(4)}\\ &\supset (GL_A(2)\times GL_B(2))\times(\overline{GL_A(2)}\times\overline{GL_B(2)})\\ &\supset GL_A(2)\times GL_B(2). \end{aligned} \tag{5}$$

The corresponding branching rules take the form [7]:

$$\begin{aligned} \{m\} &\rightarrow \sum_{\lambda}\{\lambda\}\times\{\lambda\}\\ &\rightarrow \sum_{\lambda,\mu,\nu,\sigma,\tau}\left(k_{\mu\nu}^{\lambda}\,\{\mu\}\times\{\overline{\nu}\}\right)\times\left(k_{\sigma\tau}^{\lambda}\,\{\sigma\}\times\{\overline{\tau}\}\right)\\ &\rightarrow \sum_{\lambda,\mu,\nu,\sigma,\tau} k_{\mu\nu}^{\lambda}\,k_{\sigma\tau}^{\lambda}\,(+\cdots+\delta_{\mu\nu}\{0\})\times(+\cdots+\delta_{\sigma\tau}\{0\}) \end{aligned} \tag{6}$$

and

$$
\begin{aligned}
\{m\} &\rightarrow \sum_{\lambda} \{\lambda\} \times \{\overline{\lambda}\} \\
&\rightarrow \sum_{\lambda,\mu,\nu,\sigma,\tau} \left(k^{\lambda}_{\mu\nu}\ \{\mu\} \times \{\nu\}\right) \times \left(k^{\lambda}_{\sigma\tau}\ \{\overline{\sigma}\} \times \{\overline{\tau}\}\right) \\
&\rightarrow \sum_{\lambda,\mu,\nu,\sigma,\tau} k^{\lambda}_{\mu\nu}\ k^{\lambda}_{\sigma\tau}\ (+\cdots+\delta_{\mu\sigma}\{0\}) \times (+\cdots+\delta_{\nu\tau}\{0\})\ , \qquad (7)
\end{aligned}
$$

where the coefficients $k^{\lambda}_{\mu\nu}$ are defined by the inner product rule for characters of irreducible representations of the symmetric group S_m, which takes the form:

$$
\chi^{\mu}_{\kappa}\,\chi^{\nu}_{\kappa} = \sum_{\lambda} k^{\lambda}_{\mu\nu}\,\chi^{\lambda}_{\kappa}\,. \qquad (8)
$$

In each of the two branching rules, $\lambda, \mu, \nu, \sigma, \tau$ are partitions of m, restricted by the conditions $\ell(\lambda) \leq 4$ and $\ell(\mu), \ell(\nu), \ell(\sigma), \ell(\tau) \leq 2$. These branching rules give rise to the following two formulae for the required multiplicities:

$$
\begin{aligned}
n_m &= \sum_{\lambda,\mu,\nu,\sigma,\tau} k^{\lambda}_{\mu\nu}\ k^{\lambda}_{\sigma\tau}\ \delta_{\mu\nu}\ \delta_{\sigma\tau} = \sum_{\lambda\vdash m;\ell(\lambda)\leq 4} \left(\sum_{\mu\vdash m;\ell(\mu)\leq 2} k^{\lambda}_{\mu\mu} \right)^2 \qquad (9) \\
&= \sum_{\lambda,\mu,\nu,\sigma,\tau} k^{\lambda}_{\mu\nu}\ k^{\lambda}_{\sigma\tau}\ \delta_{\mu\sigma}\ \delta_{\nu\tau} = \sum_{\lambda\vdash m;\ell(\lambda)\leq 4} \ \sum_{\mu,\nu\vdash m;\ell(\mu),\ell(\nu)\leq 2} (k^{\lambda}_{\mu\nu})^2. \qquad (10)
\end{aligned}
$$

For example, in the case $m = 4$, we find that λ is to be summed over $\{4\}$, $\{31\}$, $\{22\}$, $\{211\}$, $\{1111\}$. For each of these we have the branchings:

$$
\begin{aligned}
\{4\} &\rightarrow \{4\} \times \{4\} + \{31\} \times \{31\} + \{22\} \times \{22\}\,; \\
\{31\} &\rightarrow \{4\} \times \{31\} + \{31\} \times \{4\} + \{31\} \times \{31\} \\
&\qquad + \{31\} \times \{22\} + \{22\} \times \{31\}\,; \\
\{22\} &\rightarrow \{4\} \times \{22\} + \{31\} \times \{31\} + \{22\} \times \{4\} + \{22\} \times \{22\}\,; \\
\{211\} &\rightarrow \{31\} \times \{31\} + \{31\} \times \{22\} + \{22\} \times \{31\}\,; \\
\{1111\} &\rightarrow \{22\} \times \{22\}\,.
\end{aligned}
$$

In the case of the first formula, for each λ, we sum the coefficients of terms of the form $\{\mu\}\times\{\mu\}$ and then sum the squares of these numbers to give $n_4 = 3^2 + 1^2 + 2^2 + 1^2 + 1^2 = 16$, while in the case of the second formula we sum the squares of all coefficients of $\{\mu\} \times \{\nu\}$ to give $n_4 = 16 \times 1^2 = 16$.

Stimulated by the work of Davis *et al* [8, 9], Wybourne used his software package SCHUR [10] to carry out calculations of the above type and thereby to build up the corresponding Molien series term by term [11]:

$$
\begin{aligned}
M(q) &= \sum_{m=0}^{\infty} n_m\ q^m \\
&= 1 + q + 4q^2 + 6q^3 + 16q^4 + 23q^5 + 52q^6 + 77q^7 \\
&\quad +150q^8 + 224q^9 + 396q^{10} + 583q^{11} + 964q^{12} + 1395q^{13} + 2180q^{14} \\
&\quad +3100q^{15} + 4639q^{16} + 6466q^{17} + 9344q^{18} + 12785q^{19} + 17936q^{20} \\
&\quad +24121q^{21} + 33008q^{22} + 43674q^{23} + 58512q^{24} + 76277q^{25} + 100312q^{26} \\
&\quad +129009q^{27} + 166932q^{28} + 212022q^{29} + 270448q^{30} + O(q^{31}). \qquad (11)
\end{aligned}
$$

Rather than extending this series by proceeding term by term, it is preferable to derive the corresponding generating function. To this end we can use Molien's Theorem [3]. This states that if G is a compact continuous group, with elements g and Haar measure $d\mu(g)$, then

$$M(q) = \int_{g\in G} \frac{d\mu(g)}{\det(I - q\,g)}\,. \tag{12}$$

In our case G is in principle the non-compact group $GL_A(2)\times GL_B(2)$. However, the number of invariants is unchanged if we restrict from $GL_A(2)\times GL_B(2)$ to the compact subgroup $SU_A(2)\times SU_B(2)$. Moreover the invariants only arise from tensor products of pairs of irreducible representations and their contragredients of the form $\{\mu\}\times\{\overline{\mu}\}$ and $\{\sigma\}\times\{\overline{\sigma}\}$. All the irreducible representations that appear in these products are faithful irreducible representations not of $SU_A(2)\times SU_B(2)$ but of $SO_A(3)\times SO_B(3)$. This allows us to take $G = SO_A(3)\times SO_B(3)$ with the group element g corresponding to the 16-dimensional density matrix realised as the tensor product of two 4×4 matrices. This can be diagonalised to yield eigenvalues $\{1,e^{i\theta},1,e^{-i\theta}\}\times\{1,e^{i\phi},1,e^{-i\phi}\}$. The corresponding measure can be written in the form $d\mu(g) = (1-\cos\theta)(1-\cos\phi)\;d\theta\,d\phi$ [6]. Setting $z = e^{i\theta}$ and $w = e^{i\phi}$ then leads to integrations around $|z| = 1$ and $|w| = 1$. Molien's Theorem then gives

$$\begin{aligned} M(q) \;=\; & \oint_{|z|=1}\oint_{|w|=1} \frac{(1-z)^2(1-w)^2 z^{-2}w^{-2}}{(1-q)^4(1-qz)^2(1-q/z)^2(1-qw)^2(1-q/w)^2} \\ & \frac{dw\,dz}{(1-qzw)(1-qz/w)(1-qw/z)(1-q/zw)}\,. \end{aligned} \tag{13}$$

The repeated use of Cauchy's residue theorem then yields

$$\begin{aligned} M(q) &= \frac{1-q^2-q^3+2q^4+2q^5+2q^6-q^7-q^8+q^{10}}{(1-q)^{10}(1+q)^6(1+q^2)^2(1+q+q^2)^3} \\ &= \frac{1+q^4+q^5+3q^6+2q^7+2q^8+3q^9+q^{10}+q^{11}+q^{15}}{(1-q)(1-q^2)^3(1-q^3)^2(1-q^4)^3(1-q^6)}\,. \end{aligned} \tag{14}$$

where the first form is due to Grassl *et al* [1], while the second emphasises the fact that, for $|q| < 1$, the expansion as a power series in q yields a series in which all the coefficients n_m are non-negative integers, as required.

3. Construction of the invariants

First we note some general properties of the ring $\mathcal{R}^G$ of invariants. Hilbert's Theorem states that $\mathcal{R}^G$ is generated by a finite set of *fundamental* homogeneous invariants $\{I_1, I_2, \ldots, I_N\}$. The Cohen-Macaulay condition applies [4]. This states that there exists a set of *primary* invariants K_m with $m = 1, 2, \ldots, n$ for some n that are algebraically independent, and a set of secondary invariants invariants J_k with $k = 1, 2, \ldots, r$ for some r that are linearly independent, such that

$$\mathcal{R}^G = \oplus_{k=0}^{r} J_k\;\mathbb{C}[K_1, K_2, \ldots, K_n], \tag{15}$$

with $J_0 = 1$. This implies that an *arbitrary* invariant I can always be expressed in the form $I = \sum_{k=0}^{r} J_k\, P_k(K_1, K_2, \ldots, K_n)$ with each P_k polynomial in the K_m.

This general theory implies that the Molien function for $\mathcal{R}^G$ takes the form:

$$M(q) = \frac{\sum_{k=0}^{r} q^{\deg J_k}}{\prod_{m=1}^{n}(1-q^{\deg K_m})}. \tag{16}$$

Comparison with (14) then suggests, but does not prove, that in the mixed qubit case there are 10 primary invariants of degrees 1,2,2,2,3,3,4,4,4,6, and 15 secondary invariants of degrees 4,5,6,6,6,7,7,8,8,9,9,9,10,11,15.

To illustrate the fact that a knowledge of the Molien series generating function does not always determine the numbers and degrees of the primary and secondary invariants, consider the following example [12]. Let G be the finite group generated by the matrices:

$$\begin{pmatrix} -1 & 0 & 0 \\ 0 & -1 & 0 \\ 0 & 0 & 1 \end{pmatrix} \qquad \begin{pmatrix} 1 & 0 & 0 \\ 0 & 1 & 0 \\ 0 & 0 & i \end{pmatrix}$$

The Molien function of G in this representation is given by:

$$M(q) = \frac{1}{(1-q^2)^3}.$$

This suggests the existence of just 3 degree 2 primary invariants and no secondary invariants. However, if the space V on which G acts is spanned by vectors $v = (x, y, z)$, then x^2, y^2, xy, z^4 are fundamental invariants. Every homogeneous invariant is polynomial in these four. However, $(xy)^2 = x^2y^2$, so that xy is secondary and not primary. In fact the Molien function can be re-written in the form:

$$M(q) = \frac{1}{(1-q^2)^3} = \frac{1+q^2}{(1-q^2)^2(1-q^4)},$$

corresponding to the fact that there are 3 primary invariants x^2, y^2, z^4 and 1 secondary invariant xy.

Our task is to identify the invariants and to confirm or otherwise the consistency of the Molien series (14) obtained by direct integration with the form (16) exhibiting the Cohen-Macaulay structure of the ring of invariants $\mathcal{R}^G$.

In order to exploit our knowledge of $SO(3)$ invariants, it is convenient to take advantage of the $SO_A(3) \times SO_B(3)$ structure of our problem by writing the two-qubit density matrix in the form [2]:

$$\rho = t\,\hat{I} \otimes \hat{I} + s_a\,\hat{\sigma}_a \otimes \hat{I} + p_i\,\hat{I} \otimes \hat{\sigma}_i + b_{ai}\,\hat{\sigma}_a \otimes \hat{\sigma}_i, \tag{17}$$

where $\hat{I}$ is the 2×2 unit matrix and $\hat{\sigma}_k$ for $k = 1, 2, 3$ are the Pauli matrices, and the repeated indices a, i are summed over $1, 2, 3$.

Then the aim is to construct $SO_A(3) \times SO_B(3)$ invariants as polynomials in the components of t, $s = (s_1, s_2, s_3)$, $p = (p_1, p_2, p_3)$, $b = (b_{ai})_{1\leq a,i\leq 3}$ and $b^T = (b_{ia})_{1\leq a,i\leq 3}$. In doing this, use may be made of the symmetric tensors δ_{ab}, δ_{ij}, and the antisymmetric tensors ϵ_{abc} and ϵ_{ijk}. The property characterising an invariant is that following contraction there must remain no free indices of either type, a or i. A graphical approach to this problem is described elsewhere [13]. Here, following Makhlin [2], we summarise the outcome algebraically.

In accordance with the conjecture of Grassl *et al* [1], it is confirmed that there are just 21 fundamental invariants. Those involving no ϵs are:

$$\begin{array}{lll}
K_1 := t & K_6 := s_a b_{ai} p_i & U_2 := s_a b_{ai} b^T_{ib} b_{bj} p_j \\
K_2 := b_{ai} b^T_{ia} & K_7 := b_{ai} b^T_{ib} b_{bj} b^T_{ja} & X_1 := s_a b_{ai} b^T_{ib} b_{bj} b^T_{jc} s_c \\
K_3 := s_a s_a & K_8 := s_a b_{ai} b^T_{ib} s_b & X_2 := p_i b^T_{ia} b_{aj} b^T_{jb} b_{bk} p_k \\
K_4 := p_i p_i & K_9 := p_i b^T_{ia} b_{aj} p_j\,. &
\end{array}$$

Those involving one ϵ are:

$$\begin{array}{ll}
W_1 := \epsilon_{abc}\; s_a\; b_{bi} p_i\; b_{cj} b^T_{jd} s_d & V_1 := \epsilon_{ijk}\; p_i\; b^T_{ja} s_a\; b^T_{kb} b_{bl} p_l \\
W_2 := \epsilon_{abc}\; s_a\; b_{bj} p_j\; b_{ck} b^T_{kd} b_{dl} p_l & V_2 := \epsilon_{ijk}\; p_i\; b^T_{ja} s_a\; b^T_{kb} b_{bl} b^T_{lc} s_c \\
W_3 := \epsilon_{ijk}\; p_i\; b^T_{ja} b_{al} p_l\; b^T_{kb} b_{bm} b^T_{mc} s_c & V_3 := \epsilon_{abc}\; s_a\; b_{bi} b^T_{id} s_d\; b_{cj} b^T_{je} b_{ek} p_k \\
W_4 := \epsilon_{ijk}\; p_i\; b^T_{ja} b_{al} p_l\; b^T_{kb} b_{bm} b^T_{mc} b_{cn} p_n & V_4 := \epsilon_{abc}\; s_a\; b_{bi} b^T_{id} s_d\; b_{cj} b^T_{je} b_{ek} b^T_{kf} s_f\,.
\end{array}$$

and those involving both ϵs are

$$K_5 := \det b = \tfrac{1}{6}\epsilon_{abc}\,\epsilon_{ijk}\,b_{ai}\,b_{bj}\,b_{ck} \quad U_3 := \epsilon_{abc}\,\epsilon_{ijk}\,s_a\,b_{bj}\,b_{ck}\,p_i\,.$$

Thus the fundamental invariants and their degrees are

K_1	deg 1	X_1, X_2, V_1, W_1	deg 6
$K_2, K_3, K_4,$	deg 2	V_2, W_2	deg 7
K_5, K_6	deg 3	V_3, W_3	deg 8
K_7, K_8, K_9, U_3	deg 4	V_4, W_4	deg 9
U_2	deg 5		

The fact that there are no more fundamental invariants is discussed in detail elsewhere [13], but comes about through the existence of identities expressing every other homogeneous invariant as a polynomial in the above 21.

4. Identification of primary and secondary invariants

In order to identify and distinguish between primary and secondary invariants, the first step is to consider powers and products of the fundamental invariants and the relations between them, known as syzygies. A computer search reveals that there are 63 independent syzygies of the first kind. The invariants K_m with $m = 1, 2, \ldots, 9$ are found to be algebraically independent and primary, while U_i, V_j and W_k, with $i = 2, 3$ and $j, k = 1, 2, 3, 4$ are linearly independent but secondary by virtue of syzygies exemplified by:

$$\begin{aligned}&2\,U_2^2 - 2\,X_1\,K_9 - 2\,K_8\,X_2 + 2\,K_8\,K_9\,K_2 + K_6^2\,K_7 - K_6^2\,K_2^2\\&\quad -2\,K_3\,K_4\,K_5^2 + 2\,U_3\,K_5\,K_6 = 0\,;\end{aligned}$$

$$\begin{aligned}&U_3^2 + 8\,K_6\,U_2 - 4\,K_6^2\,K_2 - 4\,K_8\,K_9 + 4\,K_3\,K_9\,K_2 + 4\,K_8\,K_4\,K_2\\&\quad +2\,K_3\,K_4\,K_7 - 2\,K_3\,K_4\,K_2^2 - 4\,X_1\,K_4 - 4\,K_3\,X_2 = 0\,.\end{aligned}$$

There exists no similar expansion involving a term $U_2\,U_3$, however $U_i\,V_j$ and $U_i\,W_j$ can be eliminated by means of syzygies such as:

$$\begin{aligned}&2\,U_2\,V_3 - W_1\,K_6\,K_7 + W_1\,K_6\,K_2^2 - 2\,V_4\,K_9\\&\quad -2\,K_5\,V_1\,K_8 + 4\,K_5\,V_2\,K_6 - 2\,K_5\,W_3\,K_3 = 0\,;\end{aligned}$$

$$U_3\,W_2 + 2\,V_1\,K_6\,K_2 - 2\,V_2\,K_9 + 4\,W_3\,K_6 - 2\,W_4\,K_3 - 2\,K_5\,W_1\,K_4 = 0\,.$$

All V_i^2, W_j^2 and $V_i\,W_j$ may be eliminated in the same way, as can all $W_k\,X_1$ and $V_k\,X_2$ by means of syzygies such as:

$$\begin{aligned}&W_1\,X_1 - W_1\,K_8\,K_2 - V_3\,K_8 + V_4\,K_6 - K_5\,V_2\,K_3 = 0\,;\\&V_1\,X_2 - V_1\,K_9\,K_2 - W_3\,K_9 + W_4\,K_6 - K_5\,W_2\,K_4 = 0\,.\end{aligned}$$

The same is not true of any $W_k\,X_2$ or $V_k\,X_1$, nor of X_1^2 or X_2^2; none of which can be eliminated in this way. However, the elimination of $X_1\,X_2$ proceeds by way of the syzygy:

$$\begin{aligned}&2\,X_1\,X_2 - 2\,X_1\,K_9\,K_2 - 2\,K_8\,X_2\,K_2 + K_8\,K_9\,K_7\\&+K_8\,K_9\,K_2^2 - 2\,U_2\,K_6\,K_7 + 2\,U_2\,K_6\,K_2^2 + K_6^2\,K_7\,K_2\\&-K_6^2\,K_2^3 - 2\,U_4\,K_5 + 2\,U_3\,K_5\,K_6\,K_2 - 2\,K_6^2\,K_5^2\\&+2\,K_8\,K_4\,K_5^2 + 2\,K_3\,K_9\,K_5^2 - 2\,K_3\,K_4\,K_5^2\,K_2 = 0,\end{aligned} \qquad (18)$$

where it has been convenient to introduce $U_4 := U_2 U_3$.

Now let

$$D := \prod_{m=1}^{9} (1 - K_m),$$

$$U := 1 + \sum_{i=2}^{4} U_i, \quad V := \sum_{i=1}^{4} V_i, \quad W := \sum_{i=1}^{4} W_i.$$

Then, taking all syzygies into account, the generating function for polynomial invariants takes the form:

$$F := \frac{1}{D}\left(U + \frac{U X_1}{1 - X_1} + \frac{U X_2}{1 - X_2} + \frac{V}{1 - X_1} + \frac{W}{1 - X_2}\right), \tag{19}$$

in the sense that every linearly independent homogeneous invariant appears once and once only in the formal expansion of F.

As a check on the validity of this generating function, it is found that replacing each invariant I in (19) by $q^{\deg(I)}$ leads directly to the formula (14) for the Molien function $M(q)$, as required.

The fact that both X_1 and X_2 cannot be algebraically independent of one another and of K_m for $m = 1, 2, \ldots, 9$ comes about from the existence of a syzygy linking just X_1, X_2 and all these K_m. It is about 10 pages long and is of degree 48. Its dependence on X_1 and X_2 is illustrated by setting $K_m = z^{\deg(K_m)}$ for $m = 1, 2, \ldots, 9$ which gives:

$$\begin{aligned}
&16\, X_1^4\, X_2^4 + 8832\, z^{36}\, X_1\, X_2 - 7040\, X_1^2\, X_2\, z^{30} \\
&-2112\, z^{30}\, X_2^3 + 144\, z^{24}\, X_1^4 - 2112\, z^{30}\, X_1^3 - 1088\, z^{42}\, X_1 \\
&+1984\, X_1^2\, X_2^2\, z^{24} + 128\, z^{24}\, X_2^3\, X_1 - 1088\, z^{42}\, X_2 \\
&+144\, z^{24}\, X_2^4 + 896\, X_1^3\, z^{18}\, X_2^2 + 6496\, z^{36}\, X_2^2 \\
&-7040\, z^{30}\, X_2^2\, X_1 - 32\, z^{12}\, X_1^4\, X_2^2 - 64\, X_1^4\, X_2^3\, z^6 \\
&+192\, z^{18}\, X_1^4\, X_2 + 128\, X_1^3\, X_2\, z^{24} - 32\, z^{12}\, X_2^4\, X_1^2 \\
&+896\, X_1^2\, X_2^3\, z^{18} - 6512\, z^{48} + 192\, z^{18}\, X_2^4\, X_1 \\
&-384\, X_1^3\, X_2^3\, z^{12} + 6496\, z^{36}\, X_1^2 - 64\, z^6\, X_2^4\, X_1^3 = 0\,.
\end{aligned}$$

This implies that X_1 and X_2 cannot both be primary invariants. However, in our expansion of F in (19) it is clear that arbitrarily large powers of both X_1 and X_2 occur. The way out of this apparent impasse is to define $K_{10} := X_1 + X_2$ and $J_3 := X_1 - X_2$. Then K_m for $m = 1, 2, \ldots, 10$ are algebraically independent. Moreover $J_3^2 = K_{10}^2 - 4\, X_1\, X_2$ with $X_1\, X_2$ known to be linear in X_1, X_2, U_2, U_3, U_4 from (18). This is sufficient to conclude that J_3^2 is linear in J_3, U_2, U_3, U_4, with coefficients polynomial in $K_1, K_2, \ldots, K_{10}$, and can be eliminated.

It follows that our primary and secondary invariants and their degrees can be identified as follows: the 10 primary invariants:

K_1	deg 1
K_2, K_3, K_4	deg 2
K_5, K_6	deg 3
K_7, K_8, K_9	deg 4
$K_{10} = X_1 + X_2$	deg 6

and the 15 secondary invariants:

$J_1 := U_3$	deg 4	$J_8, J_9 := V_1, W_1$	deg 6
$J_2 := U_2$	deg 5	$J_{10}, J_{11} := V_2, W_2$	deg 7
$J_3 := X_1 - X_2$	deg 6	$J_{12}, J_{13} := V_3, W_4$	deg 8
$J_4 := J_1\, J_2$	deg 9	$J_{14}, J_{15} := V_4, W_4$	deg 9
$J_5 := J_1\, J_3$	deg 10		
$J_6 := J_2\, J_3$	deg 11		
$J_7 := J_1\, J_2\, J_3$	deg 15		

5. Conclusion

Recalling that $J_0 = 1$, the ring $\mathcal{R}^G$ of invariants takes the Cohen-Macaulay form:

$$\mathcal{R}^G = \oplus_{k=0}^{15} J_k\, \mathbb{C}[K_1, K_2, \ldots, K_{10}].$$

The Molien series for $\mathcal{R}^G$ is given by

$$M(q) = \frac{\sum_{k=0}^{15} q^{\deg J_k}}{\prod_{m=1}^{10} (1 - q^{\deg K_m})}$$

with 10 primary invariants K_m of degrees $1, 2, 2, 2, 3, 3, 4, 4, 4, 6$, and 15 secondary invariants J_k of degrees $4, 5, 6, 6, 6, 7, 7, 8, 8, 9, 9, 9, 10, 11, 15$.

Acknowledgments

It is a pleasure to thank Professor C J Cummins of Concordia University, Montreal for some helpful sugggestions given to one of us (RCK) regarding the search for syzygies using Maple. One of us (RCK) is grateful for the award of a Leverhulme Emeritus Fellowship which enabled him to offer partial support to the other (TAW).

References

[1] Grassl M, Rotteler M and Beth T 1998 Computing local invariants of qubit quantum systems *Phys. Rev.* A **58** 1833–1839

[2] Makhlin Y 2002 Nonlocal properties of two-qubit gates and mixed states and optimization of quantum computations *Quantum Info. Proc.* **1** 243–252

[3] Sturmfels B 1993 *Algorithms in Invariant Theory* (Wien: Springer)

[4] Hochster M and Roberts J 1974 Rings of invariants of reductive groups acting on regular rings are Cohen-Macaulay *Adv. Math.* **13** 115–175

[5] Weyl H 1946 *The Classical Groups* (Princeton, NJ: Princeton Univ. Press)

[6] Littlewood D E 1950 *The Theory of Group Characters* 2nd Ed. (Oxford: Clarendon Press)

[7] King R C 1975 Branching rules for classical Lie groups using tensor and spinor methods *J. Phys.* A **8** 429–449

[8] Davis R I A, Delbourgo R and Jarvis P D 2000 Covariance, correlation and entanglement *J. Phys.* A **33** 1895–1914

[9] Davis R I A, Delbourgo R and Jarvis P D 2000 Integrity bases for local invariants of composite quantum systems *J. Phys.* A **33** 3723–3725

[10] Wybourne B G 1990 *SCHUR: An Interactive Programme for Calculating Properties of Lie Groups* http://smc.vnet.net/schur.html

[11] Wybourne B G 2000 *Private communication*

[12] Stanley R P 1979 Invariants of finite groups and their applications to combinatorics *Bull. Amer. Math. Soc.* **1** 475–511

[13] Welsh T A, Jarvis P D and King R C *to appear* Invariants of mixed qubit systems, in *Proceedings of the Brian G Wybourne Commemorative Meeting* Eds: King R C, Karwowski J and Bylicki M (Toruń, Poland: Nicolaus Copernicus University Press)

Geometric Aspects of Quantum Mechanics and Quantum Entanglement

Dariusz Chruściński

Institute of Physics, Nicolaus Copernicus University, ul. Grudziądzka 5/7, 87-100 Toruń, Poland

E-mail: `darch@phys.uni.torun.pl`

Abstract. It is shown that the standard non-relativistic Quantum Mechanics gives rise to elegant and rich geometrical structures. The space of quantum states is endowed with nontrivial Fubini-Study metric which is responsible for the "peculiarities" of the quantum world. We show that there is also intricate connection between geometrical structures and quantum entanglement.

1. Introduction

It is well known that most theories of classical physics may be formulated in geometric language. The leading examples are Hamiltonian mechanics based on symplectic geometry, General Relativity based on Riemannian geometry and classical Yang-Mills theory which uses elegant theory of fibre bundles [1]. On the other hand Quantum Mechanics is intimately connected with linear operators in the Hilbert space and uses algebraic or functional analytic methods. We show that that the standard non-relativistic Quantum Mechanics may be formulated in the elegant geometric language. Moreover, it turns out that geometry enters Quantum Mechanics on the very fundamental level – the space of quantum states is endowed with rich and beautiful geometric structures.

Actually, geometry is present in quantum physics from the very beginning. Let us just mention celebrated Dirac magnetic pole. Dirac showed [2] that Quantum Mechanics leads to the celebrated charge quantization

$$e = \frac{n}{2}\frac{\hbar}{g}, \qquad n = 1, 2, 3, \ldots \tag{1}$$

where 'e' and 'g' stands for electric and magnetic charges, respectively. As wee shall see this condition is closely related to the geometrical properties of the qubit. Another well known example is provided by the Quantum Hall Effect [3]: the quantization of Hall conductance

$$\sigma_{\rm H} = n\,\frac{e^2}{h}\,, \qquad n = 1, 2, 3, \ldots \tag{2}$$

may be explained in topological terms. Geometry and topology are basic tools for Chern-Simons theory, Witten's topological quantum field theory and finally for string theory [4].

These examples shows that geometrical methods in quantum physics are not new. In what follows we shall concentrate on elementary Quantum Mechanics. In the next section we describe the geometrical structure of the space of quantum states – this space is endowed with so called *Fubini-Study* metric which enables one to measure a distance between quantum states. Actually, it turns out that the nontrivial geometry of the Fubini-Study metric is responsible for the "peculiarities" of the quantum world. Section 3 shows how the metric properties are connected with the quantum measurement and probabilistic interpretation of Quantum Mechanics. Section 4 relates the geometric properties of quantum states with elegant mathematical construction known as Hopf fibrations. Finally we show that there is also intricate connection between geometrical structures and quantum entanglement.

Geometric approach to Quantum Mechanics was originated in the work of Kibble [5] who showed how quantum theory could be formulated in the language of Hamiltonian phase- space dynamics. For recent investigation see [6, 7, 8] (see also [9]).

2. Space of quantum states

Any two vectors $\psi, \phi \in \mathcal{H}$ differing by a complex number $c \in \mathbb{C}$, i.e. $\psi = c\phi$, are physically equivalent $\psi \sim \phi$, that is, they define the same physical state. Therefore, the proper phase space of a quantum system is not the original Hilbert space $\mathcal{H}$ but rather the space of rays in $\mathcal{H}$:[1]

$$\mathcal{P}(\mathcal{H}) := \mathcal{H}/\sim\ , \tag{3}$$

called a *projective Hilbert space.* Points in $\mathcal{P}(\mathcal{H})$ are 1-dimensional rays in $\mathcal{H}$ or equivalently 1-dimensional projectors:

$$\psi \longrightarrow P_\psi := \frac{|\psi\rangle\langle\psi|}{\langle\psi|\psi\rangle} . \tag{4}$$

It is evident that if $\psi \sim \phi$, then $P_\psi = P_\phi$.

In this paper we shall consider only finite dimensional case, i.e. we assume that $\mathcal{H} \cong \mathbb{C}^{n+1}$. Normalized vectors in $\mathbb{C}^{n+1}$ define a $2n+1$–dimensional sphere

$$S^{2n+1} = \left\{ \psi \in \mathbb{C}^{n+1} \,\middle|\, \langle\psi|\psi\rangle = 1 \right\} . \tag{5}$$

Now, two points ψ and ϕ in S^{2n+1} define the same quantum state iff $\phi = e^{i\alpha}\psi$. Hence, the corresponding projective Hilbert space reads

$$\mathcal{P}(\mathcal{H}) = S^{2n+1}/U(1) =: \mathbb{C}P^n , \tag{6}$$

and it is usually called *complex projective space.*

A well known example is $\mathbb{C}P^1$, i.e. space of quantum states of a 2-level system – a qubit – $\mathbb{C}P^1 = S^3/U(1) \cong S^2$. This 2D sphere, called usually a *Bloch sphere*, defines a *true* space of quantum states, cf. Fig. 1.

The shape of $\mathbb{C}P^n$ for $n > 1$ is not known (see however interesting analysis in [10]). It turns out that $\mathbb{C}P^n$ defines n-dimensional complex space (n is a complex dimension!). Moreover, it is equipped with rich geometrical structures: *Fubini-Study* metric $g_{\alpha\beta}$ and symplectic form $\omega_{\alpha\beta}$.[2] These structures are inherited from $\mathcal{H}$. Observe that the scalar product in $\mathcal{H}$ may be decomposed as follows:

$$\langle\psi|\phi\rangle = G(\psi,\phi) + i\,\Omega(\psi,\phi) , \tag{7}$$

[1] Throughout this paper we shall consider only pure states.

[2] Symplectic structure is equivalent to a Poisson bracket known from the Hamiltonian mechanics [11, 12].

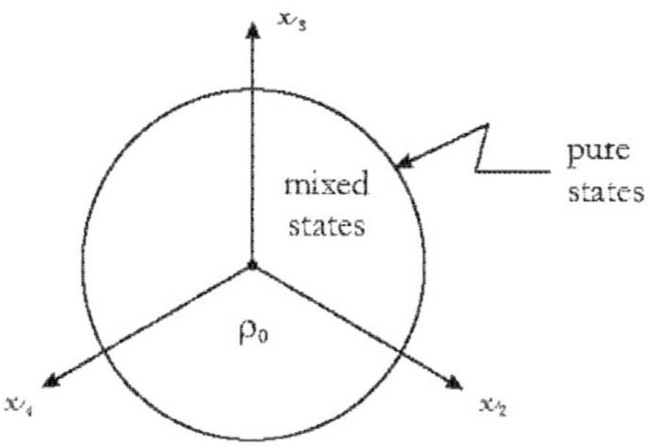

Figure 1. Bloch ball. Pure qubit states lie on the boundary $\mathbb{C}P^1 \cong S^2$. The maximally mixed state $\rho_0 = \frac{1}{2}\mathbb{1}$ lies in the center.

where $G(\psi,\phi) = \mathrm{Re}\,\langle\psi|\phi\rangle$, and $\Omega(\psi,\phi) = \mathrm{Im}\,\langle\psi|\phi\rangle$. These objects enjoy the following properties:

$$G(\psi,\phi) = G(\phi,\psi)\,, \qquad \Omega(\psi,\phi) = -\Omega(\phi,\psi)\,, \qquad \Omega(\psi,\phi) = G(\psi,i\,\phi)\,. \tag{8}$$

It shows that both $\mathcal{H}$ and $\mathcal{P}(\mathcal{H})$ define so called *Kähler spaces.*

Now, if e_α defines an orthonormal base in $\mathbb{C}^{n+1}$, i.e. any ψ may be decomposed as follows $\psi = \sum_\alpha \psi_\alpha e_\alpha$, then the Fubini-Study metric $g_{\alpha\beta}$ is given by the following formula

$$g_{\alpha\beta} = \frac{\langle\psi|\psi\rangle\delta_{\alpha\beta} - \psi_{(\alpha}\overline{\psi}_{\beta)}}{\langle\psi|\psi\rangle^2}\,, \tag{9}$$

that is, the infinitesimal distance is defined by $ds^2 = \sum_{\alpha,\beta} g_{\alpha\beta}\, d\psi_\alpha\, d\overline{\psi}_\beta$. For $n = 1$ (qubit) the Fubini-Study metric reproduces the standard *round* metric on S^2. It turns out that the nontrivial geometry of the Fubini-Study metric is responsible for the "peculiarities" of the quantum world.

3. Quantum measurement

In the standard formulation of Quantum Mechanics an observable is represented by a self-adjoint operator $\hat{F}$. Measuring $\hat{F}$ one obtains one of its eigenvalues:

$$\hat{F}\psi_k = \lambda_k\psi_k\,. \tag{10}$$

If the original state is represented by ψ then one measures λ_k with probability $p_k = |\langle\psi|\psi_k\rangle|^2$ and just after the measurement the state of the system jumps to ψ_k.[3]

Now, we show how to describe this process on the projective space $\mathcal{P}(\mathcal{H})$. Observe that $\hat{F}$ gives rise to the function f on $\mathcal{P}(\mathcal{H})$ defined by:

$$f(P) = \mathrm{Tr}(P\hat{F})\,, \tag{11}$$

where P is a quantum state in $\mathcal{P}(\mathcal{H})$ represented by a 1-dimensional projector (density matrix). Clearly, not all functions on $\mathcal{P}(\mathcal{H})$ have the above form. If the function 'f' may be represented by (11) we call it a *quantum observable.* This is one of the peculiarity of the quantum world: in classical mechanics all function on the corresponding phase-space are *classical observables.* In Quantum Mechanics it is no longer true. The natural question is therefore how to check whether a function 'f' is a quantum observable. It turns out that the appropriate test is provided by

[3] For simplicity we assume that λ_k are not degenerate.

the Fubini-Study metric. Let Δ be a Laplace operator on the projective space.[4] Now, 'f' is a quantum observable iff

$$\Delta f = 0 \ , \qquad \text{or} \qquad \Delta f = -2nf \ , \tag{12}$$

that is, 'f' is either a zero mode of Δ or is a eigenvector of Δ corresponding to the first non-vanishing eigenvalue '$-2n$'.

To see how this test works let us consider the simplest case $n = 1$. The corresponding projective space is given by the Bloch sphere S^2 and the eigenvalue problem $\Delta f = \lambda f$ is well known from the theory of angular momentum. One has

$$\Delta Y_{lm} = -l(l+1)Y_{lm} \ , \tag{13}$$

where Y_{lm} are spherical harmonics and the integer m runs from $-l$ to l. Note, that (12) implies that $l = 0$ or $l = 1$. In the first case Y_{00} defines a constant function on S^2 whereas in the second case we have three independent dipole functions

$$Y_{11} = x \ , \quad Y_{1-1} = y \ , \quad Y_{10} = z \ . \tag{14}$$

A constant function corresponds via (11) to the identity operator $\mathbb{1}$. The reader easily checks that dipole functions correspond to Pauli matrices: σ_x, σ_y and σ_z. This reproduces the well known fact that for 2-level system one has 4 independent observables.

Having described the structure of quantum observables let us see how to obtain a quantum spectrum $\{\lambda_1, \lambda_2, \ldots\}$ from (10). It turns out that the stationary states P_k are critical points of 'f', i.e. $f'(P_k) = 0$. Moreover, P_k correspond to projectors onto the rays generated by ψ_k, i.e. $P_k = |\psi_k\rangle\langle\psi_k|$. Now, the 'spectrum' of 'f' immediately follows:

$$\lambda_k = f(P_k) \ , \tag{15}$$

i.e. possible outcomes in measuring 'f' are values of 'f' in its critical points P_k.

Finally, let us see how the probability enters the game. Suppose that just before the measurement of 'f' the system was in the state P. Denote by $\gamma(P, P_k)$ a distance between P and P_k defined in terms of the Fubini-Study metric, i.e. it is the length of a geodesic connecting P and P_k. The corresponding probability p_k is given by the following formula:

$$p_k = \cos^2\{\gamma(P, P_k)\} \ , \tag{16}$$

i.e. a transition probability is determined by the distance between corresponding states. The system is more likely to collapse to a nearby state than to a distant one. The miracle of the Fubini-Study metric implies that these probabilities sum to 1!

4. Hopf fibrations

The map defining projection from S^{2n+1} to $\mathbb{C}P^n$ defines so called *Hopf fibration* or equivalently a *Hopf bundle*. The simplest and the most popular corresponds to $n = 1$. In this case $\mathbb{C}P^1 \cong S^2$ and hence it gives rise to a map $S^3 \longrightarrow S^2$. An example of such a map is easy to construct: let ψ be a normalized state vector of a qubit

$$\psi = \alpha|0\rangle + \beta|1\rangle \ ,$$

[4] Δ depends upon the metric. If x^k are local coordinates, then the formula for Δ reads as follows:

$$\Delta f = \frac{1}{\sqrt{g}} \frac{\partial}{\partial x^k} \left(\sqrt{g}\, g^{kl} \frac{\partial f}{\partial x^l} \right)$$

where $g = |\det g_{kl}|$.

with $|\alpha|^2 + |\beta|^2 = 1$. Let us define

$$x_1 = \langle\psi|\sigma_1|\psi\rangle\ , \quad x_2 = \langle\psi|\sigma_2|\psi\rangle\ , \quad x_3 = \langle\psi|\sigma_3|\psi\rangle\ . \tag{17}$$

It is easy to see that $x_1^2 + x_2^2 + x_3^2 = 1$, and hence the above assignment establishes a map from S^3 to S^2. To visualize the Hopf fibration $S^3 \longrightarrow S^2$ one may perform a stereographic projection from S^3 to $\mathbb{R}^3$. Each circular fibre of S^3 is mapped onto a circle in $\mathbb{R}^3$, cf. Figure 2.

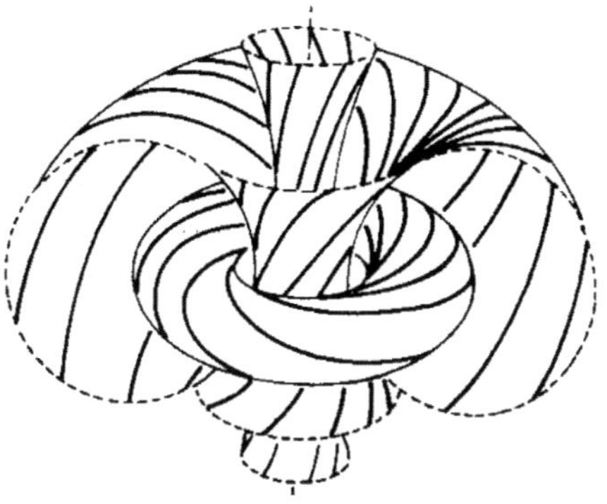

Figure 2. $S^3 \longrightarrow S^2$ Hopf fibration after a stereographic projection to $\mathbb{R}^3$. Circular fibres are grouped into a continuous family of nested tori.

Maps $S^3 \longrightarrow S^2$ were studied by Hopf [13] who found that any such map belongs to a class uniquely characterized by an integer number m – so called *Hopf number* of the map:

$$f\ :\ S^3 \longrightarrow S^2 \quad \Longrightarrow \quad \mathrm{Hopf}(f) = m\ ,$$

with $m \in \mathbb{Z}$. Any two maps f_1 and f_2 which can be continuously deformed one into another have the same Hopf numbers, i.e. $\mathrm{Hopf}(f_1) = \mathrm{Hopf}(f_2)$. One calls such maps homotopically equivalent. It turns out that the map defined by the qubit state $\psi \longrightarrow \langle\psi|\vec{\sigma}|\psi\rangle$ satisfies $\mathrm{Hopf}(f) = 1$, i.e. it belongs to the simplest nontrivial class of maps.

Actually, the Hopf map $S^3 \longrightarrow S^2$ is closely related to the mechanism of charge quantization developed by Dirac [2], see e.g. [4].

Now, let us recall that only three spheres S^n can be equipped with the group structures:

(i) S^1 which corresponds to normalized complex numbers,

(ii) S^3 which corresponds to normalized quaternions; $S^3 \cong SU(2)$,

(iii) S^7 which corresponds to normalized octonions.

Now, the Hopf map $S^3 \longrightarrow S^2$ corresponds to the coset space S^3/S^1. Another coset space S^7/S^3 gives rise to the following family of Hopf fibrations

$$f\ :\ S^7 \longrightarrow S^4\ .$$

It turns out that this family finds interesting applications in Yang-Mills theory (it describes so called instanton configuration of the classical Yang-Mills field) and in quantum entanglement.

5. Quantum entanglement

Consider two quantum systems A and B and let $\mathcal{H}_A$ and $\mathcal{H}_B$ denote the corresponding Hilbert spaces.[5] Suppose now that we are interested in the composite system AB made up of A and B. The composite system Hilbert space $\mathcal{H}_{AB}$ is a tensor product of $\mathcal{H}_A$ and $\mathcal{H}_B$:

$$\mathcal{H}_{AB} = \mathcal{H}_A \otimes \mathcal{H}_B\ , \tag{18}$$

[5] In quantum information theory one usually speaks about Alice and Bob systems, see [14].

which means that if $|A\rangle \in \mathcal{H}_A$ and $|B\rangle \in \mathcal{H}_B$ denote state vectors of A and B, respectively, then the joint system is in the state

$$|AB\rangle \equiv |A\rangle \otimes |B\rangle \in \mathcal{H}_{AB} \ . \tag{19}$$

Note however, that a general vector from $\mathcal{H}_{AB}$ can not be written this way. We call an element $\psi \in \mathcal{H}_{AB}$ a *separable state* if there are $\psi_A \in \mathcal{H}_A$ and $\psi_B \in \mathcal{H}_B$ such that $\psi = \psi_A \otimes \psi_B$. If this is not the case we call ψ an *entangled state* or simply nonseparable state.

Recall, that if $(e_1, \ldots, e_N)$ denote the basis in $\mathcal{H}_A$ and $(f_1, \ldots, f_M)$ the basis in $\mathcal{H}_B$, then

$$e_\mu \otimes f_\nu \ , \qquad \mu = 1, \ldots, N, \ \nu = 1, \ldots, M \ , \tag{20}$$

define a basis in $N \cdot M$ dimensional space $\mathcal{H}_{AB}$. It means that an arbitrary vector $\psi \in \mathcal{H}_{AB}$ may be represented as follows:

$$\psi = \sum_{\mu=1}^{N} \sum_{\nu=1}^{M} \psi_{\mu\nu} \, e_\mu \otimes f_\nu \ , \tag{21}$$

with $\psi_{\mu\nu} \in \mathbb{C}$. Now, given a state $\psi \in \mathcal{H}_{AB}$ is it separable or entangled? To answer this question let us observe that for every ψ in $\mathcal{H}_{AB}$ there exist orthonormal basis $\{\tilde{e}_\mu\}_{\mu=1}^N$ and $\{\tilde{f}_\mu\}_{\nu=1}^M$ such that

$$\psi = \sum_{\alpha=1}^{K} a_\alpha \, \tilde{e}_\alpha \otimes \tilde{f}_\alpha \ , \tag{22}$$

where $a_\alpha > 0$ with $\sum_{\alpha=1}^K a_\alpha^2 = 1$, and $K \leq \min\{N, M\}$. Formula (22) defines *Schmidt decomposition* of ψ.

The state $\psi \in \mathcal{H}_{AB}$ is separable if and only if its Schmidt decomposition contains only one terms, i.e. $K = 1$. As an example consider two qubits, that is $\mathcal{H}_A = \mathcal{H}_B = \mathbb{C}^2$. Denoting by $|0\rangle$ and $|1\rangle$ the standard orthonormal basis in $\mathbb{C}^2$ one introduces so called *Bell states*

$$|\psi^\pm\rangle := \frac{1}{\sqrt{2}} \left(|01\rangle \pm |10\rangle\right) \ , \tag{23}$$

and

$$|\phi^\pm\rangle := \frac{1}{\sqrt{2}} \left(|00\rangle \pm |11\rangle\right) \ , \tag{24}$$

where $|01\rangle := |0\rangle \otimes |1\rangle$, etc. These four vectors define an orthonormal basis in $\mathbb{C}^2 \otimes \mathbb{C}^2 \cong \mathbb{C}^4$. Moreover, they are all entangled states.[6]

6. Entanglement and geometry

Consider now a composite 2-qubit system. Any normalized 2-qubit state may be written as follows

$$\psi = \alpha \, |00\rangle + \beta \, |01\rangle + \gamma \, |10\rangle + \delta \, |11\rangle \ , \tag{25}$$

[6] The notion of separability and entanglement may be easily generalized to mixed states of the composed system [15]. A mixed state represented by a density matrix ρ in $\mathcal{H}_{AB}$ is called separable if and only if it can be represented as

$$\rho = \sum_{\alpha=1}^{K} p_\alpha \, \rho_\alpha \otimes \sigma_\alpha \ ,$$

where $p_\alpha > 0$ with $\sum_{\alpha=1}^K p_\alpha = 1$, and ρ_α and σ_α are mixed states of A and B, respectively. Surprisingly, contrary to the case of pure states the criterion of separability is still unknown for composite mixed states, see [16, 17] for more details.

where $\alpha, \beta, \gamma, \delta \in \mathbb{C}$ together with

$$|\alpha|^2 + |\beta|^2 + |\gamma|^2 + |\delta|^2 = 1 \ . \tag{26}$$

The above formula defines a unit 7D sphere S^7. A 2-qubit state ψ is separable if it is a tensor product of two 1-qubit states $\psi = \psi_1 \otimes \psi_2$. Representing ψ_k in $(|0\rangle, |1\rangle)$ basis

$$\psi_1 = a\,|0\rangle + b\,|1\rangle \ , \qquad \psi_2 = c\,|0\rangle + d\,|1\rangle \ , \tag{27}$$

one finds for separable state

$$\psi = ac\,|00\rangle + ad\,|01\rangle + bc\,|10\rangle + bd\,|11\rangle \ . \tag{28}$$

Hence, a 2-qubit state represented by (25) is separable if and only if

$$\alpha\delta = \beta\gamma \ . \tag{29}$$

Let us introduce $\mathbf{E} : \mathbb{C}^2 \otimes \mathbb{C}^2 \longrightarrow \mathbb{C}^2 \otimes \mathbb{C}^2$ by

$$\mathbf{E}\,\psi = (\sigma_y \otimes \sigma_y)\,\psi^* \ , \tag{30}$$

and following [18] define

$$x_0 = \langle\psi|\,\sigma_z \otimes \mathbb{1}\,|\psi\rangle \ , \quad x_1 = \langle\psi|\,\sigma_x \otimes \mathbb{1}\,|\psi\rangle \ , \quad x_2 = \langle\psi|\,\sigma_y \otimes \mathbb{1}\,|\psi\rangle \ ,$$

and

$$x_3 = \mathrm{Re}\,\langle\psi|\,\mathbf{E}\,|\psi\rangle \ , \quad x_4 = \mathrm{Im}\,\langle\psi|\,\mathbf{E}\,|\psi\rangle \ .$$

Using (25) one shows that $x_0^2 + x_1^2 + x_2^2 + x_3^2 + x_4^2 = 1$ and hence the above map $\psi \longrightarrow (x_0, x_1, x_2, x_3, x_4)$ defines Hopf fibration $S^7 \longrightarrow S^4$. Actually,

$$C(\psi) = |\langle\psi|\,\mathbf{E}\,|\psi\rangle| \ , \tag{31}$$

defines a Wootters concurence [19] which defines an entanglement measure for 2 qubit case.

Note that

$$x_3 = 2\mathrm{Re}\,(\alpha\delta - \beta\gamma) \ , \quad x_4 = 2\mathrm{Im}\,(\alpha\delta - \beta\gamma) \ ,$$

which shows that for separable states $x_3 = x_4 = 0$. Therefore, the Hopf map $S^7 \longrightarrow S^4$ is entanglement sensitive and may be used to detect quantum entanglement in a 2-qubit system: a state is separable iff its image under the Hopf map belongs to the equatorial 2D sphere defined by the intersection of S^4 with the 3D plane $x_3 = x_4 = 0$.

For other studies on geometric approach to quantum entanglement see e.g. [7, 10, 20].

Acknowledgments

This work was partially supported by the Polish State Committee for Scientific Research Grant *Informatyka i inżynieria kwantowa* No PBZ-Min-008/P03/03.

References

[1] Nash N and Sen S 1983 *Topology and Geometry for Physicists* (London: Academic)
Nakahara M 1990 *Geometry, Topology and Physics* (Bristol: Adam Hilger)
Eguchi T Gilkey P G and Hanson J 1980 *Phys. Rep.* **66** 213

[2] Dirac P A M 1931 *Proc. Roy. Soc. London* **A 133** 60
Dirac P A M 1948 *Phys. Rev.* **74** 817

[3] Prange R E and Girvin S M 187 *The Quantum Hall Effect* (New York: Springer)
Morandi G 1988 *Quantum Hall Effect* (Naples: Bibliopolis)

[4] Morandi G 1992 *The Role of Topology in Classical and Quantum Physics* Lecture Notes in Physics m7 (Berlin: Springer)

[5] Kibble T W B 1978 *Commun. Math. Phys.* **64** 73
Kibble T W B 1979 *Commun. Math. Phys.* **65** 189

[6] Ashtekar A and Schilling T A 1998 *On Einsteins Path* (Berlin: Springer) ed Harvey A (*Preprint* gr-qc/9706069)

[7] Brody D C and Hughston L P 2001 *J. Geom. Phys.* **38** (Berlin: Springer)

[8] Cirelli R Mania A and Pizzocchero L 1990 *J. Math. Phys.* **31** 2891

[9] Chruściński D and Jamiołkowski A 2004 *Geometric Phases in Classical and Quantum Mechanics* (Boston: Birkhäuser)

[10] Bengtsson I Brännlund J and Życzkowski K 2001 *CP^n, or, Entanglement Illustrated* Preprint quant-ph/0108064

[11] Abraham R and Marsden J E 1978 *Foundations of Mechanics* (New York: Addison-Wesley)

[12] Arnold V I 1989 *Mathematical Methods of Classical Mechanics* (Berlin: Springer)

[13] Hopf H 1931 *Mat. Annalen.* **104** 637

[14] Nielsen M A and Chuang I L 2000 *Quantum Computation and Quantum Information* (Cambridge: Cambridge University Press)

[15] Werner R F 1989 *Phys. Rev.* **A 40** 4277

[16] Peres A 1995 *Quantum Theory: Concepts and Methods,* (Dordrecht: Cluwer Academic Publisher)

[17] Horodecki M Horodecki P and Horodecki R 2001 *Mixed-state entanglement and quantum communication* in *Quantum Information: An Introduction to Basic Theoretical Concepts and Experiments* eds Alber G Beth T Horodecki M Horodecki P Horodecki R Rotteler M Weinfurter H Werner R and Zeilinger A (Berlin: Springer)

[18] Mosseri R and Dandaloff R 2001 *J. Math. Phys. A: Math. Gen.* **34** 10243

[19] Wootters K W 1998 *Phys. Rev. Lett.* **80** 2245

[20] Kuś M and Życzkowski K 2001 *Phys. Rev.* **A 63** 032307

Rapid Magnon Relaxation

J Morkowski

Institute of Molecular Physics, Polish Academy of Sciences, ul. Smoluchowskiego 17, 60-179 Pozna , Poland

E-mail: jmorkows@ifmpan.poznan.pl

Abstract. The standard procedure for calculating life-times of magnons is to construct equations of motion for magnon occupancy numbers and to solve them for a state of a weak departure from equilibrium. In the last decade a number of experiments used pulse techniques to study dynamics of magnetization after very short pulses bringing magnetic systems out of equilibrium. Therefore a relevant question is how magnons relax after a short strong pulse of magnetic field. The three- and four-magnon relaxation processes are discussed. It is shown that the inverse relaxation time in a short interval after a strong perturbation is enhanced by a temperature independent contribution, depending on the strength of the perturbation.

1. Introduction

Magnon relaxation in ferromagnets was studied intensively some 50 years ago in connection with investigations of the ferromagnetic resonance and related phenomena of absorption of microwaves. The reason of returning to the problem of magnon relaxation is the recent development of new experimental techniques which allow studying magnon relaxation more directly than in the old resonance experiments. The new methods [1-10] are based upon so called pulse-probe techniques in which a short laser pulse disturbs the system studied. After a well defined and very short delay time another short laser pulse ("probe") examines the momentary state of the system. Thus by changing the delay time it is possible to study the time evolution of the system from a disturbance to equilibrium. The technique is very powerful, and can be applied to various phenomena, notably magnetic relaxation. The technique is steadily improved giving the possibilities of shortening the durations of the pulses and the delay times thus enabling to study faster and faster processes. As an example of the progress an experiment [11] can be mentioned in which direct visualization of electron movement in an atom (of krypton, in this case) was possible operating in the time scale of the order of 100 attoseconds (1 attosec = 10^{-18} sec).

The magnon relaxation theory started by the paper [12] was excellently summarized in the article [13] and the book [14]. The theory is based upon the Boltzmann type equations describing time evolution of magnon numbers of a particular wave vector. Assuming small deviations from equilibrium the Boltzmann equations are linearized with respect to these deviations, easily solved giving exponential decay in time of the deviations, governed by the relaxation time. This simple theoretical scheme is complicated by the fact that there are numerous scattering mechanisms, e.g. involving three, four or even more magnons, interaction of magnons with conduction electrons, lattice phonons and structural or magnetic defects of different kinds contributing to a total relaxation rate (i.e. the sum of the inverse relaxation times of processes involved).

Only a few out of many experiments over the past years were successful in reliably confirming predictions of the theory for some processes (see e.g. [14]). These best experiments were done on a single material, the yttrium-iron garnet (YIG), which is a localized spin, non-conducting ferrimagnet and can crystallize with negligible amount of defects. By clever experiments particular relaxation processes were found dominant making possible verification of the theory (see [14] for more information).

2. Magnon interaction

For the range of magnon wave vectors accessible to experiments based on excitations of magnon modes by high frequency magnetic fields the important ingredient is the energy of magnetic dipolar interactions of electrons responsible for ferromagnetism. Although the magnetic forces are by far weaker than the exchange interactions of an electrostatic origin, because of their long range they strongly affect energy and interactions of magnons in the experimentally important range of magnon wave vectors small as compared with the inverse lattice constant.

For the description of magnons we shall use the simple localized spin models although the results are in fact model independent, since also for itinerant electron magnetism the spin wave theory can be cast into the traditional formalism [15]. The system considered here is defined by the spin hamiltonian

$$H = -J\sum_{<i,j>}\vec{S}_i\cdot\vec{S}_j + \sum_{i,j}\frac{(2\mu_B)^2}{\left|\vec{R}_i-\vec{R}_j\right|^3}\cdot\left\{\vec{S}_i\cdot\vec{S}_j - 3\frac{\left[\vec{S}_i\cdot(\vec{R}_i-\vec{R}_j)\right]\cdot\left[\vec{S}_j\cdot(\vec{R}_i-\vec{R}_j)\right]}{\left|\vec{R}_i-\vec{R}_j\right|^2}\right\} - 2\mu_B H\sum_j S_j^z \tag{1}$$

in the standard notation (e.g. [13]). As usually the spin operators are approximately expressed in terms of the harmonic oscillator operators [16, 17] describing flips of spin moments and those are expanded into their Fourier components. Formally $S_j^x + iS_j^y = N^{-1/2}\sum_k e^{i\vec{k}_i\vec{R}_j} a_k^+ + \cdots$, where **k** are the magnon wave vectors and $n_k \overset{def}{=} a_k^+ a_k$ is the (operator of) magnon occupations, attaining 0, 1, 2, etc. as its eigenvalues.

In the representation of boson operators the hamiltonian (1) takes the familiar form

$$\begin{aligned} H = &\sum_k \varepsilon_k a_k^+ a_k + \sum_{kq}(W_{kq}^{(3)} a_k^+ a_q^+ a_{k+q} + h.c.) \\ &+ \sum_{kqq'}(\Omega_{k,q,q'} a_k^+ a_q^+ a_{q'} a_{k+q+q'} + h.c.) \\ &+ \sum_{kqq'}(T_{kqq'} a_k^+ a_q^+ a_{q'}^+ a_{k+q+q'} + h.c.) \end{aligned} \tag{2}$$

Due to the presence of the dipolar interactions in order to diagonalising the harmonic part of the hamiltonian (1) expressed in terms of the Fourier components of the spin flip operators the standard canonical transformation [16] was used to get (2). The magnon energy $_k$ in (2) for long wavelength magnons (e.g. for wave vector **k** much smaller than the inverse lattice constant but also much larger than the inverse dimension of the ferromagnetic crystal considered) is

$$\varepsilon_k = \left\{(2\mu_B H + Dk^2)(2\mu_B H + 8\pi\mu_B M\sin^2\vartheta_k + Dk^2)\right\}^{1/2} \tag{3a}$$

With small numerical errors, not important in the present problem, the magnon energy ε_k can be approximated by a simpler expression

$$\varepsilon k \cong 2\mu BH + 4\pi\mu BM\sin 2\vartheta_k + Dk2 \tag{3b}$$

where $H = H_z - 4\pi N_z M$, $\sin^2\vartheta_k = (k_x^2 + k_y^2)/k^2$. Here z-axis is taken along the applied magnetic field, N_z is the demagnetizing factor for that direction.

The procedure leading to eq. (2) is well known (an excellent reference is [13]), so the description here is very brief. But now a comment is needed: the transformation diagonalising the dipolar terms leads to the interaction terms (of higher than the second in *a*-s) in eq. (2) containing coefficients W..., ..., T... renormalized by the so-called ellipticity factors. If $H \gg M$ the resulting corrections are small and usually are neglected in discussing magnon relaxation (for the leading relaxation processes they were shown [18] to produce errors of a few percent).

3. Time evolution of average magnon occupation numbers

The basic variable for describing magnon relaxation is the time dependent average magnon occupation number $n_k(t)=\langle a_k^+ a_k\rangle$ averaged over a non equilibrium statistical ensemble. The time evolution of $n_k(t)$ is calculated from a balance of scattering processes. The probability $P_{i\to f}$ of transition in a unit time from an initial state $|\Psi_i\rangle$ to a final one $|\Psi_f\rangle$ is calculated from the well-known Fermi golden rule, $P_{i\to f}=(2\pi/\hbar)|\langle\Psi_i|H_I|\Psi_f\rangle|^2\cdot\delta(E_i-E_f)$ where H_I is the interaction energy causing the transition, and E_i, E_f are the energies of the initial and final states, respectively.

For clarity we shall discuss separately relaxation due to various scattering processes allowed by the interaction part of the hamiltonian (2), the 3-magnon and 4-magnon ones. In the standard linearized theory the inverse relaxation times for these various processes are additive (see e.g. [13]), the same holds true for the initial rapid relaxation discussed in the present lecture.

The lowest order are the 3-magnon processes, which can be of two types: the so called magnon confluence processes and magnon splitting ones [13]. In the confluence processes a given magnon, say **k** is annihilated together with another magnon **q** to creating a magnon of wave vector $(\mathbf{k}+\mathbf{q})$. This is the most effective process for long-wavelengths magnons **k** since it is easy to conserve energy in the scattering process for experimentally accessible conditions. For the splitting processes in which a given magnon **k** splitts into two other, say **q** and $(\mathbf{k}-\mathbf{q})$ the energy conservation can be fulfilled only for large enough $|\mathbf{k}|$.

The number of transitions $(\mathbf{k})+(\mathbf{q})\to(\mathbf{k}+\mathbf{q})$ in a unit time, determined by the second term in (2), is (cf. [13])

$$P_{k,q}=W_{kq}n_k n_q(n_{k+q}+1) \tag{4}$$

where

$$W_{kq}=\frac{1}{\tau}N^{-1}|f_k+f_q|^2\,\delta(\varepsilon_k+\varepsilon_q-\varepsilon_{k+q}) \tag{5}$$

and

$$f_k=k_z(k_x+ik_y)/k^2\ . \tag{6}$$

For convenience the magnon energy in (5) in dimensionless, measured in units of $k_B(\times 1°\mathrm{K})$, and $\tau\cong(\mu_B M)^2/(\hbar k_B)$ ($\tau\cong 1.7\times10^{-12}$ sec for $M=140$ gauss, as for the yttrium-iron garnet at room temperature), and wave vectors are is measured in units of the inverse lattice constant.

The transition rate for reverse process is $W_{kq}(n_k+1)(n_q+1)n_{k+q}$. The resulting total change of $n_k(t)$ is thus

$$\frac{dn_k}{dt}=-\sum_q W_{kq}n_k n_q(n_{k+q}+1)+\sum_q W_{kq}(n_k+1)(n_q+1)n_{k+q}\ . \tag{7}$$

In the standard theory this Boltzmann type equation is solved under the assumption of small deviations $\zeta_k(t)$ of $n_k(t)$ from their thermal averages $n_k^0=(\mathrm{e}^{\varepsilon_k/k_BT}-1)^{-1}$,

$$n_k(t)=n_k^0+\zeta_k(t)\,;\,|\zeta_k(t)|\ll n_k^0\quad .$$

The equation (7) linearized with respect to $\zeta_k(t)$ takes the form

$$\frac{d\zeta_k(t)}{dt} \cong -\frac{1}{\tau_k}\zeta_k(t) + \cdots \tag{8}$$

which has thc obvious solution $\zeta_k(t) = \zeta_k(0)e^{-t/\tau_k}$. Here

$$1/\tau_k = \sum_q W_{kq}(n_q^0 - n_{k+q}^0) \tag{9}$$

defines the relaxation time τ_k for 3-magnon confluence processes.

The system of equations (7) will be now solved for a strong pulse-like disturbance from equilibrium. The equations (7) can be cast into the form

$$\frac{d\zeta_k(t)}{dt} = -\left\{\frac{1}{\tau_k} + \Psi_k(t)\right\}\zeta_k(t) + \Psi_{1k}(t), \tag{10}$$

where

$$\Psi_k(t) = \sum_q (W_{kq} - W_{k,q-k})\zeta_q \tag{11}$$

and

$$\Psi_{1k}(t) = -n_k^0\Psi_k(t) + \sum_q W_{kq}\left[(n_q^0 + 1)\zeta_{q+k} + n_{q+k}^0\zeta_q\right] + \sum_q W_{kq}\zeta_q\zeta_{q+k}. \tag{12}$$

The system of equations (10) is formally equivalent to the following

$$\zeta_k(t) = \{\zeta_k(0) + \alpha_k(t)\}\,\mathrm{e}^{-t/\tau_k - \eta_k(t)} \tag{13}$$

where

$$\eta_k(t) = \int_0^t du\Psi_k(u) = \sum_q (W_{kq} - W_{k,q-k})\int_0^t du\zeta_q(u) \tag{14}$$

and

$$\alpha_k(t) = \int_0^t du\Psi_{1k}(u)\,\mathrm{e}^{u/\tau_k + \eta_k(u)}. \tag{15}$$

Now suppose one can generate a pulse-like excitation of magnons peaked at a particular wave vector, say **Q**. Thus we assume $\zeta_k(0) = \mathrm{f}\,\lambda_{kQ}$ where the distribution λ_{kQ} is sharp, e.g. $\lambda_{kQ} \sim \mathrm{e}^{-(\vec{k}-\vec{Q})^2/\Delta^2}$. An asymptotically exact result can be obtained assuming $\zeta_k(0) = \mathrm{f}\,\delta_{kQ}$ and f very large. Then, for $\mathbf{k} = \mathbf{Q}$ from (13), neglecting $\alpha_Q(t)$ as compared with $\mathrm{f} \gg 1$ it follows

$$\zeta_Q(t) \cong f\,\mathrm{e}^{-t/\tau_Q - \eta_Q(t)}. \tag{16}$$

From (14) we get the self-consistency condition, retaining in (14) only one leading term

$$\eta_Q(t) = W_{QQ} f \int_0^t \mathrm{d}u\ \mathrm{e}^{-u/\tau_Q - \eta_Q(u)} \tag{17}$$

(since, as can be seen from (5), $W_{Q,0} = 0$ for any Q).

The equation (17) can be solved to give (for $\mathrm{f} \gg 1$):

$$\eta_Q(t) = \ln\left[1 + \tau_Q f W_{QQ}(1 - \mathrm{e}^{-t/\tau_Q})\right]. \tag{18}$$

At the initial evolution of the system, in the limit $t \ll \tau_Q$, $\eta_Q(t) \cong \mathrm{f}\,W_{QQ}\,t$ whereas finally, for $t \gg \tau_Q$, $\eta_Q(t) \cong \ln(1 + \mathrm{f}\,W_{QQ})$ i.e. tends to the constant. Therefore, from (13) we get

$$\zeta_Q(t) \cong f\,\mathrm{e}^{-(1/\tau_Q + fW_{QQ})t} \quad \text{for} \quad t \ll \tau_Q \tag{19a}$$

$$\zeta_Q(t) \cong \frac{f}{1+fW_{QQ}\tau_Q} e^{-t/\tau_Q} \text{ for } \quad t \geq \tau_Q. \tag{19b}$$

The result (19a) tells us that in the initial stage of relaxation, for $t \ll \tau_Q$, where τ_Q is the standard relaxation time obtained from linearized Boltzmann equation, the effective relaxation rate $1/\tau_Q^f$ is

$$1/\tau_Q^f = 1/\tau_Q + \mathrm{f}\, W_{QQ} \tag{20}$$

This is an universal result holding also for higher order processes, as we shall show later on. The important observation is that the initial (i.e. for $t \ll \tau_Q$) relaxation rate is enhanced by a contribution $\mathrm{f}\, W_{QQ}$ which is directly proportional to the strength of the disturbance and is essentially temperature independent (apart from an indirect temperature dependence due to the magnetization M entering into the expression for W_{QQ}).

For $t \geq \tau_Q$ the standard relaxation time τ_Q determines the decay, eq. (19b).

Even as we consider very large initially excited number of magnons f, $\mathrm{f} \gg 1$, the value of the correction $\mathrm{f}\, W_{QQ}$ in eq. (20) can be detectable under narrowly defined experimental condition. The explicit expression for W_{QQ} is

$$W_{QQ} = \frac{1}{\tau} N^{-1} 4\cos^2\vartheta_Q \sin^2\vartheta_Q \cdot \delta(2\varepsilon_Q - \varepsilon_{2Q}) \tag{21}$$

where $\cos\vartheta_Q = Q_z/Q$, $\omega_0 = 2\mu_B(H - 4\pi M)/k_B$, $a = 4\pi\mu_B M/k_B$ and

$$2\varepsilon_Q - \varepsilon_{2Q} = \left\{ \left[\omega_0 (1 - \frac{a}{\omega_0}\cos^2\vartheta_Q) \right] - 2DQ^2 \right\}.$$

The crucial quantity in (21) is $\mathrm{f}\, N^{-1}$ which is equal to the relative change in magnetization $\Delta M/M$ due to initially exciting f magnons of a given wave vectors $\mathbf{Q} = Q\left(\sin\vartheta_Q, 0, \cos\vartheta_Q\right)$, $\mathrm{f}\, N^{-1} = \Delta M/M$. It is not an easy task to make such an experiment. Although it is experimentally feasible to excite well defined magnons in geometrically constrained ferromagnetic samples (e.g. in thin films), attaining large value of $\Delta M/M$ is difficult. In experiments rather a range of magnon wave vectors from an interval $(Q-\Delta, Q+\Delta)$ can be excited instead of a single well defined value Q. Thus a relevant quantity in place of W_{QQ} is the average $\bar{W}_{QQ}$

$$\bar{W}_{QQ} = \frac{1}{2\Delta} \int_{Q-\Delta}^{Q+\Delta} dQ' W_{Q'Q'};$$

explicitly

$$f\bar{W}_{QQ} = \frac{1}{2\Delta D} \cdot \frac{\cos^2\vartheta_Q \sin^2\vartheta_Q}{\sqrt{\frac{\omega_0}{2D}(1 - \frac{a}{\omega_0}\cos^2\vartheta_Q)}} \cdot \frac{1}{\tau} fN^{-1}. \tag{22}$$

An alternative procedure is to average W_{QQ} over directions $\mathbf{Q}$ with the result:

$$f\langle W_{QQ}\rangle = fN^{-1}\frac{1}{\tau}2(\frac{D}{a}Q^2 - \frac{\omega_0}{a})(1 + \frac{\omega_0}{a} - \frac{D}{a}Q^2)^{1/2} \tag{23}$$

for Q satisfying the condition $\frac{\omega_0}{a} < \frac{D}{a}Q^2 < 1 + \frac{\omega_0}{a}$.

To estimate numerical values of the averaged $\mathrm{f}W_{QQ}$ as given by eq. (22) and (23) we take the material parameters for YIG and assume a reasonable value for magnetic field, giving $\omega_0 \cong 2a$, and $Q = 3.5\times10^{-2}$ (within the limits imposed by eq. (23)), and $\Delta/Q \cong 10^{-3}$. For such choice of parameters the average of $\mathrm{f}W_{QQ}$ from the both approaches, (22) and (23) is of the order of magnitude $\langle \mathrm{f}W\rangle \cong 3\times10^{11}(\mathrm{f}/N)1/\mathrm{sec}$. The estimate of the standard, i.e. calculated from linearized Boltzmann equation [14], relaxation rate $1/\tau_Q$ in eq. (20) for the same set of parameters for YIG is roughly equal $1/\tau_Q \cong 4\times10^6(T/300°K)\ 1/\mathrm{sec}$. Therefore, even for not too large relative number (f/N) of excited magnons $\mathbf{Q}$, say already for $(\mathrm{f}/N)\geq10^{-5}$, both contributions, $1/\tau_Q$ and $\langle \mathrm{f}W_{QQ}\rangle$ to the initial relaxation rate $1/\tau_Q^f$ are of importance. The linear dependence of $1/\tau_Q$ on temperature and very weak temperature dependence of $\mathrm{f}\langle W_{QQ}\rangle$ can also be helpful in a possible verification of the present predictions by an experiment.

4. A two-mode model

The analysis so far is incomplete since first, it still lacks an estimate of the function $\alpha_Q(t)$ in eq. (13) for $\mathbf{k}=\mathbf{Q}$ to confirm that in fact $\alpha_Q(t)$ is much smaller that f and, second, the time evolution of modes $\mathbf{k}\neq\mathbf{Q}$ is not yet determined. Although the equations (10), (11) and (12) provide a complete set to determine $\zeta_k(t)$, $\alpha_k(t)$ and $\eta_k(t)$ a direct solution of the set is obviously out of question. For a qualitative analysis we shall consider a two-mode model with only two sets of magnons, those externally excited of the vectors $\mathbf{Q}$ and magnons $\mathbf{k}$, arbitrarily picked from the magnon manifold. We shall consider only short time intervals $t \ll \tau_k$ after the excitation. To simplify the otherwise lengthy formulae we shall assume low temperature so the thermal magnon occupations $n_k^0 = (e^{\varepsilon_k/k_BT}-1)^{-1} \cong e^{-\varepsilon_k/k_BT} \ll 1$ are very low and n_k^0 can be dropped out from calculations.

It can be shown that for $t \ll \tau_k$

$$\eta_k(t) \cong \mathrm{f}\left(W_{kQ}-W_{kQ-k}\right)t + O\left(t^2\right) \tag{24}$$

whereas $\eta_k(t)$ for large t approaches a constant, non-zero value. So properties of $\eta_k(t)$ are similar to those for $\mathbf{k}=\mathbf{Q}$. Also it follows that $\alpha_Q(t)\cong O\left(t^2\right)$ for small t and for $\mathbf{k}\neq\mathbf{Q}$

$$\alpha_k(t) \cong \mathrm{f}W_{k,Q-k}t + O\left(t^2\right) \tag{25}$$

Since $\alpha_Q(t) \sim O\left(t^2\right)$ so indeed $\alpha_Q(t)$ is negligible in comparison with f during the initial stage, $t \ll \tau_k$, of relaxation and the validity of eq. (16) is thus confirmed. For $\mathbf{k}\neq\mathbf{Q}$ the initial relaxation is given by

$$\zeta_k(t) \cong (fW_{k,Q-k}t)e^{-t/\tau_k^f} \tag{26}$$

where the enhanced effective relaxation rate is given by

$$1/\tau_k^f = 1/\tau_k + f(W_{kQ}-W_{k,Q-k}). \tag{27}$$

5. Other mechanisms of rapid relaxation

The rapid relaxation by 3-magnon confluence processes of magnetic dipolar origin was discussed in details since they are the dominant ones in prevailing experimental conditions. For the 3-magnon splitting processes the magnon energy conservation conditions restrict the possible enhancement of the relaxation rates to values of the magnetic field and magnon wave vectors and a result similar to eq. (19) is difficult to prove with any rigor.

Rapid relaxation can proceed through two types of 4-magnon processes two in-two out ("2→2") scattering processes of both exchange and magnetic dipolar origin and by three in-one out ("3→1") 4-magnon confluence processes of magnetic dipolar origin.

For the higher order processes qualitatively the result is analogous to eq. (20), namely the effective relaxation rate consists of the sum of inverses of the standard, Boltzmann type, relaxation times for all processes possible (2→1; 2→2; 3→1) and a term which is temperature independent and increases with the strengths f of the excitation. The only difference is that for some values of the excited magnons wave vector Q besides the linear term in f there appears also a term of the order f^2 due to 3→1 confluence processes [19].

6. Discussion

The main result of the theory presented here is the following: if we excite a large number f of magnons of a particular wave vector **Q** at a given time, say t=0, then the initial relaxation (i.e. for the time interval $t \ll \tau_k$) is determined by an effective relaxation time τ_Q^{ef}, eq. (20), which is shorter than the relaxation time τ_Q for small departures from equilibrium, and the effective relaxation rate $1/\tau_Q^{ef}$ is enhanced by the contribution which is (essentially) temperature independent and proportional to excitation strength f. This result was discussed in details for the usually dominant scattering mechanisms which are the 3-mangnon confluence processes of dipolar origin. Qualitatively the same behaviour can be shown for higher order magnon processes. The physical interpretation of eq. (20) is clear. At low temperatures the equilibrium magnons occupations n_k^0 are small, $n_k^0 \ll 1$, so only a small number of magnons is present in the system. If at $t = 0$ we excite a large number $n_Q(0) \cong \zeta_Q(0) = \mathrm{f}$ of magnons **Q** the dominant scattering processes will consist in combining two magnons **Q** into a third one having the wave vector (2**Q**). These processes have the relaxation rate $\mathrm{f}\,W_{QQ}$, essentially independent on the presence of a small number of thermal magnons so it is temperature independent (apart for a slight temperature dependence of M entering into W_{QQ}). The processes are allowed, i.e. $W_{QQ} \neq 0$, for a narrow but experimentally accessible interval of Q values.

The decaying of the primary magnons **Q** is also going through their interaction with the thermal magnons at the rate $1/\tau_Q$ in eq. (20). After a short time interval $t \ll \tau_Q$, τ_k the relaxation of the primary magnons **Q** as well as the secondary ones **k** is continued by interactions with thermal magnons and the decay rates are determined by the standard relaxation times τ_Q, τ_k calculated for small deviations from equilibrium, known in literature (see e.g. [13], [14]).

References

[1] Vaterlaus A, Beutler T and Meier F 1991 *Phys. Rev. Lett.* **67** 3314

[2] Beaurepaire E, Merle J C, Daunois A and Bigot YJ 1996 *Phys. Rev. Lett.* **76** 4250

[3] Aeschliemann M, Bauer M, Pawlik S, Weber W, Burgemeister R, Oberli D and Siegmann H C 1997 *Phys. Rev. Lett.* **79** 5158

[4] Güdde J, Conrad U, Jähnke V, Hohlfeld J and Matthias E 1999 *Phys. Rev.* **B59** R6608

[5] Hohlfeld J, Matthias E, Knorren R and Bennemann K H 1997 *Phys.Rev. Lett.* **78** 4861

[6] Scholl A, Baumgarten L, Jacquemin R and Eberhardt W 1997 *Phys. Rev. Lett.* **79** 5146

[7] Guo C, Rodriguez G and Taylor A J 2001 *Phys. Rev. Lett.* **86** 1638

[8] ChoiB C, Belov M, Hiebert W K, Ballentine G E and Freeman M R 2001 *Phys. Rev. Lett.* **86** 728

[9] Zhang G P and Hübner W 2000 *Phys. Rev. Lett.* **85** 3025

[10] Woltersdorf G, Buess M, Heinrich B and Back C H 2005 *Phys. Rev. Lett.* **95** 037401

[11] Drescher M et al 2002 *Nature* **419** 803
[12] Akhiezer A I 1946 *J. Phys.* (USSR) **10** 217
[13] Keffer F 1966 *Spin waves in: Encyclopedia of Physics* vol. XVIII/2 ed. S Flügge (Springer-Verlag) pp 1-340
[14] Sparks M 1964 *Ferromagnetic relaxation theory* (McGraw-Hill)
[15] Morkowski J 1973 *Acta Phys. Polon.* **A43** 809
[16] Holstein T and Primakoff H 1940 *Phys. Rev.* **58** 1098
[17] Dyson F J 1956 *Phys. Rev.* **102** 1217
Dyson F J 1956 *Phys. Rev.* **102** 1230
[18] Schloemann E 1961 *Phys. Rev.* **121** 1312
[19] Morkowski J 2003 *Molecular Physics Reports* **38** 74

Dynamical dephasing of optically controlled charge qubits in quantum dots

P Machnikowski, L Jacak

Institute of Physics, Wrocław University of Technology, 50-370 Wrocław, Poland

E-mail: Pawel.Machnikowski@pwr.wroc.pl

Abstract. We describe the mechanism of phonon-induced dynamical dephasing of an exciton confined in a quantum dot and discuss methods of reducing the dephasing effect by appropriately tailored driving.

Quantum dots (QDs) are promising candidates for implementing quantum information processing schemes due to their atomic-like discrete spectra, controllable properties, feasibility of building arrays of many coupled QDs (providing for scalability) and availability of fast optical control methods. Indeed, proposals for QD-based quantum computing [1, 2] were recently followed by an experimental demonstration of a controlled-NOT gate [3]. On the other hand, the progress of coherent control over confined charge states in QDs encounters serious obstacles resulting from strong dephasing effects due to the dynamics of the surrounding crystal environment. Experimentally, this dephasing manifests itself as a partial decay (on picosecond time scale) of the optically induced coherent polarization in QDs [4]. In self-assembled dots, this effect may be attributed to carrier-phonon interactions [5]. Another experiment where the effects of dephasing may be observed are pulse area dependent Rabi oscillations [6].

An interesting feature of these dephasing processes is that they appear in a structure with strong quantization of levels under resonant coherent excitation to the ground confined exciton state and at low temperatures so that real phonon-assisted transitions and scattering on free carriers may be excluded. It turns out that the decay of coherence may be interpreted in terms of the lattice response to a change of the carrier distribution in a QD. Due to carrier-lattice interactions, the equilibrium positions of lattice ions is different in the presence of a confined exciton. An excitation of the carrier subsystem much faster than the typical phonon response times (i.e., in the sub-picosecond range) is followed by spontaneous lattice relaxation to this new equilibrium [7]. A theoretical description formulated in the limit of infinitely short pulses agrees very well with the experimental data [8] on the decay of coherent polarization.

On the other hand, in order to explain the damping of Rabi oscillations observed in experiments [6] the dynamics under driving pulses of finite duration must be described. This is also necessary for a study of coherence properties of an *excitonic qubit* [2] implemented on a confined exciton with the quantum logical states $|0\rangle$ and $|1\rangle$ corresponding to the absence and presence of the exciton, respectively. Here, we present a method that allows one to perturbatively include phonon perturbation into the driven dynamics of the excitonic system for arbitrary control pulses. We discuss also a few applications of this formalism.

Under resonant driving conditions the system may be treated in a two-level approximation (ground system state and a single confined exciton) with the exciton state coupled to lattice displacement by various carrier-phonon interaction mechanisms (the deformation potential coupling to longitudinal acoustic phonons and the Fröhlich coupling to longitudinal optical phonons are most important in the InAs/GaAs system considered here). The two charge states are coupled to a pulsed laser beam which is used for driving. Thus, the system is described by the Hamiltonian (in the rotating wave approximation)

$$H = \epsilon|1\rangle\langle 1| + \frac{1}{2}\left[f(t)e^{i\omega t}|0\rangle\langle 1| + \text{H.c.}\right] + \sum_{\boldsymbol{k}} \hbar\omega_{\boldsymbol{k}} b_{\boldsymbol{k}}^{\dagger} b_{\boldsymbol{k}} + |1\rangle\langle 1| \sum_{\boldsymbol{k}} f(\boldsymbol{k})(b_{\boldsymbol{k}}^{\dagger} + b_{-\boldsymbol{k}}), \tag{1}$$

where ϵ is the exciton energy, $f(t)$ is the slowly varying envelope of the laser pulse, $\omega_{\boldsymbol{k}}$ is the frequency of phonons with wave vector $\boldsymbol{k}$, $b_{\boldsymbol{k}}, b_{\boldsymbol{k}}^{\dagger}$ are the corresponding phonon creation and annihilation operators (the branch index is included in $\boldsymbol{k}$), and $f(\boldsymbol{k})$ are carrier-phonon coupling constants (see [9] for explicit formulas corresponding to various coupling mechanisms). The exciton wave function (determining the coupling constants $f(\boldsymbol{k})$) is modelled as a product of two anisotropic Gaussians with lateral extension of l_{e} for the electron, and $l_{\mathrm{h}} \approx 0.8 l_{\mathrm{e}}$ for the hole (based on numerical diagonalization [7]) and with a much smaller width along the growth direction of the structure.

Our approach is based on the evolution equation for the density matrix of the total system in the second order (Born) approximation with respect to the carrier-phonon interaction

$$\varrho(t) = \varrho(s) + \frac{1}{i\hbar}\int_s^t d\tau [V(\tau), \varrho(s)] - \frac{1}{\hbar^2}\int_s^t d\tau \int_s^{\tau} d\tau' [V(\tau'), [V(\tau''), \varrho(s)]], \tag{2}$$

where $\varrho(t)$ and $V(t)$ are the density matrix and the carrier-phonon interaction (the last term in Eq. (1)) in the interaction picture with respect to the first three terms of Eq. (1). Note that the effect of the driving field is included non-perturbatively.

Upon tracing out the phonon degrees of freedom, the first (zeroth order) term in (2) gives the unperturbed evolution, the second term vanishes (since it contains the thermal average of an odd number of phonons) and the third (second order) term describes the leading-order phonon correction to the dynamics of the carrier subsystem. The latter may be represented in the form of the spectral integral (see [9] for details)

$$\rho^{(2)}(t) = \int d\omega \frac{R(\omega)}{\omega^2} \hat{S}(\omega), \tag{3}$$

where $\hat{S}(\omega)$ is a certain operator depending on the initial system state and on the driving field and

$$R(\omega) = \frac{1}{\hbar^2}|n_{\mathrm{B}}(\omega) + 1|\frac{1}{N}\sum_{\boldsymbol{k}} |f(\boldsymbol{k})|^2 \left[\delta(\omega - \omega_{\boldsymbol{k}}) + \delta(\omega + \omega_{\boldsymbol{k}})\right] \tag{4}$$

is the phonon spectral density. Physical quantities are linear functionals of the density matrix and may be expressed in the form identical to Eq. (3) with a certain scalar function $S(\omega)$.

As a first application of the presented method, let us considered the excitonic qubit driven with a Gaussian pulse by an angle $\pi/2$ in the pseudo-spin (Bloch sphere) picture (this corresponds e.g. to creating an equal superposition from the ground state). The quality of such an operation is quantified by the *fidelity* F, i.e., the probability that after the operation the qubit is found in the desired (unperturbed) state. Since, in general, the initial state of the qubit in a quantum computing process is not known, we calculate the error $\delta = 1 - F$ averaged over initial states, which is given by [10]

$$\delta = \int d\omega \frac{R(\omega)}{\omega^2} S(\omega), \quad S(\omega) = \frac{|F(\omega)|^2 + |F(-\omega)|^2}{12}, \quad F(\omega) = \int_{-\infty}^{\infty} e^{i\omega t} \exp\left[i\int_{-\infty}^{t} f(\tau)d\tau\right],$$

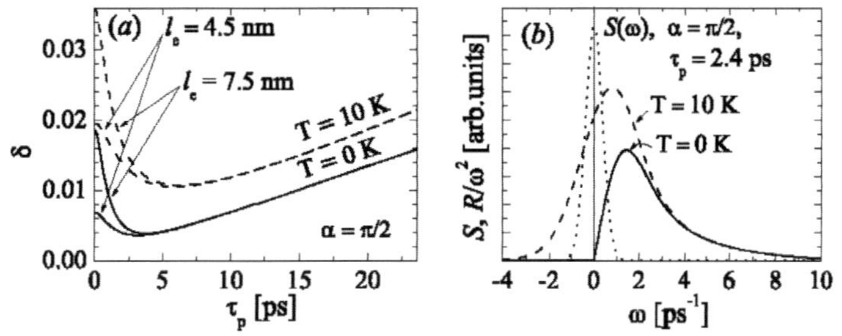

Figure 1. (a) The total error for a $\pi/2$ rotation on an excitonic qubit, resulting from pure dephasing and finite exciton lifetime [4] for two QD diameters (QD height is 20% of its diameter). (b) The spectral density of the phonon reservoir at $T = 0$ K and $T = 10$ K (solid and dashed) and the spectral function of the driving (dotted) corresponding to a $\pi/2$ rotation performed by a Gaussian pulse.

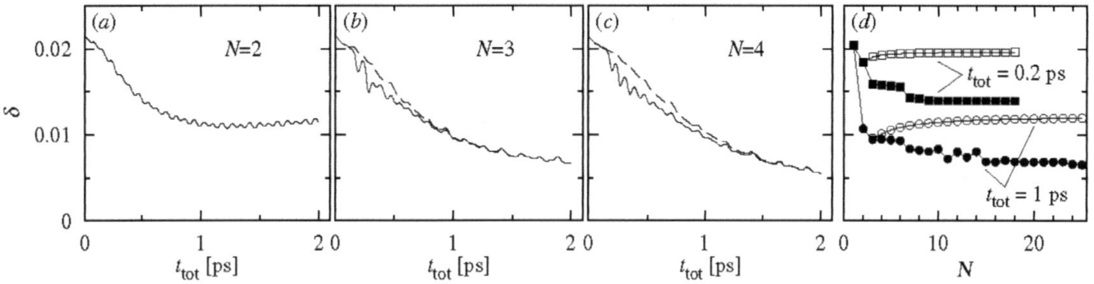

Figure 2. (a-c) The error resulting from performing a $\pi/2$ rotation on an excitonic qubit by a series of $N = 2, 3, 4$ pulses of duration $\tau_p = 100$ fs, either identical (dashed) or optimized (solid), with the total duration of the pulse sequence t_{tot}; the small oscillations come from the crossing between the spectral function of the driving and LO phonon lines which are included in this calculation. (d) The error as a function of the number of pulses for two durations of the pulse sequence. Here $l_e = 4.5$ nm.

where we assume that the initial state is defined before the driving pulse is switched on and the final state is taken after it has been switched off. For a family of pulses of the form $f(t) = \tilde{f}(t/\tau_p)$ ($\tilde{f}$ defines the pulse shape and τ_p is the pulse duration), the nonlinear spectral transform has the scaling property $F(\omega) = \tilde{F}(\tau_p \omega)$, where $\tilde{F}(x)$ is fixed for a given pulse shape. It turns out that the error induced by the carrier-phonon interaction decreases with growing pulse duration τ_p, as seen in Fig. 1a for short durations. This may be understood by recalling that the dephasing is due to spontaneous lattice relaxation after a fast change of carrier distribution. For slow driving, high-frequency modes follow the charge evolution adiabatically and reversibly so that only the decreasing fraction of slow, low-frequency modes contributes to dephasing. This is formally confirmed by Fig. 1b: for long pulses the scaling rule causes the function $S(\omega)$ to shrink around $\omega = 0$ and the overlap integral decreases. On the other hand, extending the switching time is restricted by other dephasing processes (e.g., related to the finite exciton lifetime) which accumulate over the driving time, as becomes clear in Fig. 1a for longer times. Thus, the contrary requirements imposed by the dynamical dephasing (slow dynamics) and by the accumulation decoherence resulting from finite lifetime (short operation) leads to a trade-off situation, reflected by a minimum of δ for a certain τ_p, and limits the possibility of reducing dephasing.

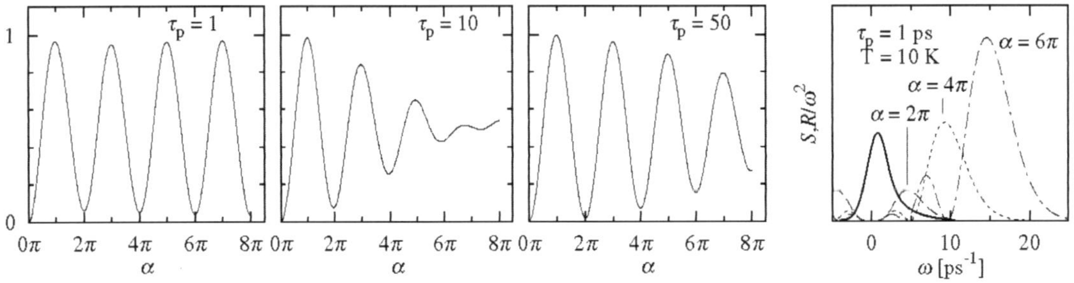

Figure 3. (a-c) Rabi oscillations (final exciton occupation, or projection on the state $|1\rangle$, as a function of the pulse area) performed on the excitonic qubit by a Gaussian pulse of length τ_p as shown. (d) The spectral density of the phonon reservoir at $T = 10$ K and the spectral functions of the driving for 1,2 and 3 full Rabi rotations. Here $l_e = 4.5$ nm.

In view of the limited possibility of reducing decoherence by using long pulses, it is interesting to try to improve coherence by driving the system with shaped pulses. Our approach is to perform the same $\pi/2$ qubit rotation using a series of equally spaced narrow Gaussian pulses with either equal or optimized amplitudes. In Fig. 2a-c we plot the error (calculated as above) for such pulsed driving as a function of the total duration of the pulse sequence t_{tot} (see [11] for details). As previously, the error drops down with growing duration of the control sequence. Although this reduction of decoherence is larger for a larger number of pulses most of the coherence gain is achieved already with 3-4 pulses, while further increase of the number of pulsespulse leads only to minor improvement. Moreover, it turns out that the overall duration of the pulse sequence is often shorter than the duration of a Gaussian pulse necessary for achieving the same level of coherence [11]. The fact that a considerable reduction of decoherence may be obtained with modest experimental resources (without the need to generate arbitrarily shaped pulses [12]) is important from the point of view of practical realization of the proposed control scheme.

Finally, let us discuss the effect of phonon-induced pure dephasing on pulse-area-dependent Rabi oscillations on a single QD [13]. In the experiment [6], a pulse with the area $\int f(t)dt = \alpha$ and duration $\tau_p \sim 1$ ps is applied to a single quantum dot placed in a photo diode structure. The photo-generated carriers then tunnel out of the dot, contributing to the photo-current, which is thus proportional to the average number of photo-excited excitons. This average number of excitons is then plotted as a function of the pulse area α and, in the ideal case, should show sinusoidal oscillations, corresponding to coherently driving the system from the ground state to the state with an exciton and then back to the ground state, when the pulse area exceeds π. In reality, one observes oscillations which are damped for larger pulse areas (note that the duration of the process is always the same). The perturbative method yields the final exciton occupation in the form

$$n = \sin^2\frac{\alpha}{2} + \int d\omega \frac{R(\omega)}{\omega^2} S(\omega), \quad S(\omega) = \frac{\cos\alpha}{16}|F(\omega) - F^*(-\omega)|^2 - \frac{\sin\alpha}{16}\mathrm{Im}[F^2(\omega) - F^{*2}(-\omega)],$$

where the first term in n is the unperturbed result and the second term describes the phonon perturbation. The resulting Rabi oscillations are plotted in Fig. 3a-c. The most striking effect is the improvement of the oscillations both for short and for long pulses. Again, this may be explained using the spectral representation (Fig. 3d). For large pulse areas (many pseudo-spin rotations) the spectral function $S(\omega)$ takes the form of a series of maxima. Taking into account the scaling behavior of this spectral function it is clear that a reduction of the overlap with the phonon spectral density $R(\omega)$ is indeed achieved in both limits, with the spectral function either stretched beyond the phonon frequencies or squeezed around $\omega = 0$. Thus, phonon damping of the Rabi oscillations has a resonant character and becomes very strong when the frequency of modulation of the charge distribution (corresponding to the strongest maximum of $S(\omega)$) lies in the region of strongly coupled phonon modes. This semiclassical idea of resonance provides a 'rule of thumb' for optimizing the driving conditions towards high-fidelity coherent driving.

To summarize, we have analyzed the physical nature of phonon-induced dephasing of excitons on QDs and proposed a perturbative method for the description of these dephasing effects under arbitrary driving. This allows us to explain the experimentally observable dephasing effects and to propose means to reduce the decoherence by using appropriate driving conditions.

Acknowledgments

Supported by the Polish Ministry of Scientific Research and Information Technology (Grant No. 2 P03B 02 424). P.M. is grateful to Alexander von Humboldt Foundation for support.

References

[1] Zanardi P and Rossi F 1998 *Phys. Rev. Lett.* **81** 4752

[2] Biolatti E, Iotti R. C, Zanardi P, and Rossi F 2000 *Phys. Rev. Lett.* **85** 5647

[3] Li X, Wu Y, Steel D, Gammon D, Stievater T, Katzer D, Park D, Piermarocchi C, and Sham L 2003 *Science* **301** 809

[4] Borri P, Langbein W, Schneider S, Woggon U, Sellin R. L, Ouyang D, and Bimberg D 2001 *Phys. Rev. Lett.* **87** 157401

[5] Vagov A, Axt V. M, and Kuhn T 2003 *Phys. Rev. B* **67** 115338

[6] Zrenner A, Beham E, Stufler S, Findeis F, Bichler M, and Abstreiter G 2002 *Nature* **418** 612

[7] Jacak L, Machnikowski P, Krasnyj J, and Zoller P 2003 *Eur. Phys. J. D* **22** 319

[8] Vagov A, Axt V. M, Kuhn T, Langbein W, Borri P, and Woggon U 2004 *Phys. Rev. B* **70** 201305(R)

[9] Grodecka A, Jacak L, Machnikowski P and Roszak K 2005 Phonon impact on the coherent control of quantum states in semiconductor quantum dots, in: Ling P. A, editor, *Quantum Dots: Research Developments* (New York: Nova Science Publishers), cond-mat/0404364 p 47

[10] Alicki R, Horodecki M, Horodecki P, Horodecki R, Jacak L, and Machnikowski P 2004 *Phys. Rev. A* **70** 010501(R)

[11] Axt V. M, Machnikowski P, and Kuhn T 2005 *Phys. Rev. B* **71** 155305

[12] Hohenester U and Stadler G 2004 *Phys. Rev. Lett.* **92** 196801

[13] Machnikowski P and Jacak L 2004 *Phys. Rev. B* **69** 193302

$\widehat{P}$-matrices in orbit spaces and invariant theory

V Talamini

INFN, Sezione di Padova, via Marzolo 8, I-35131 Padova, Italy

E-mail: vittorino.talamini@pd.infn.it

Abstract. In many physical problems or applications one has to study functions that are invariant under the action of a symmetry group G and this is best done in the orbit space of G if one knows the equations and inequalities defining the orbit space and its strata. It is reviewed how the $\widehat{P}$-matrix is defined in terms of an integrity basis and how it can be used to determine the equations and inequalities defining the orbit space and its strata. It is shown that the $\widehat{P}$-matrix is a useful tool of constructive invariant theory, in fact, when the integrity basis is only partially known, calculating the $\widehat{P}$-matrix elements, one is able to determine the integrity basis completely.

1. Introduction

In many physical applications one has to do with functions that are invariant under transformations of a compact symmetry group. These functions may be considered as functions defined on the orbit space of the group and some arguments, like the study of symmetry breaking or the study of phase transitions, are better understood if studied in the orbit space. To study invariant functions in the orbit space one has to know first the equations and inequalities that define the orbit space and its strata. A standard way to find these equations and inequalities is by using an integrity basis and the $\widehat{P}$-matrix. This method is reviewed in sections 2 and 4. The orbit spaces have not been widely employed in the study of invariant functions yet because the integrity basis is often unknown (the determination of an integrity basis of a general group is still an open problem of constructive invariant theory). Section 3 reviews some selected arguments of invariant theory and section 5 (the main section of this article) shows how the $\widehat{P}$-matrix can be used together with the Molien function to determine the integrity basis of a compact group.

2. Orbit Spaces

In this article G is a compact group acting effectively in a finite dimensional space. Then, in all generality G is a group of real orthogonal matrices acting on the real vector space $\mathbf{R}^n$.

The *orbit* through a point $x \in \mathbf{R}^n$ is the set: $\Omega(x) = \{g \cdot x, \forall g \in G\}$. Orbits do not intersect and define a partition of $\mathbf{R}^n$. The *isotropy subgroup* (or *stabilizer*) G_x of the point $x \in \mathbf{R}^n$ is the subgroup of G that leaves x unchanged: $G_x = \{g \in G \mid g \cdot x = x\}$. Isotropy subgroups of points in a same orbit are conjugated: $G_{g \cdot x} = g \cdot G_x \cdot g^{-1}$. The *orbit type* of the orbit $\Omega(x)$ is the conjugacy class $[G_x]$ of its isotropy subgroups. The orbit types define a partition of orbits (and of $\mathbf{R}^n$), each equivalence class $\Sigma_{[H]}$ is called a *stratum* (of $\mathbf{R}^n$): $\Sigma_{[H]} = \{x \in \mathbf{R}^n \mid G_x \in [H]\}$. The orbits and the strata can be partially ordered according to their orbit types. The orbit type $[H]$ is said to be *smaller* than the orbit type $[K]$: $[H] < [K]$, if $H' \subset K'$ for some $H' \in [H]$ and $K' \in [K]$. Then $[K]$ is *greater* than $[H]$. The greatest orbit type is $[G]$ and its stratum $\Sigma_{[G]}$

contains all fixed points of the G-action. $\Sigma_{[G]}$ coincides with the origin of $\mathbf{R}^n$ if the G-action is effective. Due to the compactness of G, the number of different orbit types is finite and there is a unique smallest orbit type $[G_p]$, called the *principal* orbit type. The stratum $\Sigma_{[G_p]}$ is called the *principal stratum*, all other strata are called *singular*.

The *orbit space* is the quotient space $\mathbf{R}^n/G$. The natural projection $\pi : \mathbf{R}^n \to \mathbf{R}^n/G$ maps orbits of $\mathbf{R}^n$ into single points of $\mathbf{R}^n/G$. Projections of strata of $\mathbf{R}^n$ define strata of $\mathbf{R}^n/G$. The principal stratum of $\mathbf{R}^n/G$ is always open connected and dense in $\mathbf{R}^n/G$. If $[K] > [H]$, then $\pi(\Sigma_{[K]})$ lies in the boundary of $\pi(\Sigma_{[H]})$ and the boundary of the principal stratum contains all singular strata.

A *G-invariant* function $f : \mathbf{R}^n \to \mathbf{R}$ is such that $f(gx) = f(x)$, $\forall g \in G$, $x \in \mathbf{R}^n$. f is then constant on the orbits and it is in fact a function defined on the orbit space.

Some symbols. $\mathbf{R}[\mathbf{R}^n]$ and $\mathbf{R}[\mathbf{R}^n]^G$ indicate the rings of real polynomial functions and of real G-invariant polynomial functions defined in $\mathbf{R}^n$, respectively; $\mathbf{R}[\mathbf{R}^n]_d$ and $\mathbf{R}[\mathbf{R}^n]^G_d$ indicate the rings of homogeneous polynomial functions of degree d in $\mathbf{R}[\mathbf{R}^n]$ and in $\mathbf{R}[\mathbf{R}^n]^G$, respectively; $\mathbf{R}[p_1, \ldots, p_q]$ indicates the ring of polynomials with real coefficients in the q indeterminates $p_1, \ldots, p_q$.

By Hilbert's theorem, $\mathbf{R}[\mathbf{R}^n]^G$ has a finite number q of generators [1]:

$$\forall\, p \in \mathbf{R}[\mathbf{R}^n]^G,\ \exists\, \widehat{p} \in \mathbf{R}[p_1, \ldots, p_q] \ \mid\ p(x) = \widehat{p}(p_1(x), \ldots, p_q(x)),\ \forall\, x \in \mathbf{R}^n.$$

The q generators for $\mathbf{R}[\mathbf{R}^n]^G$ are called *basic invariants* or *basic polynomials* and they form an *integrity basis* for G (and for $\mathbf{R}[\mathbf{R}^n]^G$). The integrity basis $p_1, \ldots, p_q$ is said to be *minimal* (abbreviated in MIB), if no p_a, $a = 1, \ldots, q$, is a polynomial in the other elements of the basis. The basic invariants $p_1, \ldots, p_q$ may be supposed homogeneous and ordered according to their degrees d_a, for example $d_a \leq d_{a+1}$. One may choose, in all generality, $p_1(x) = \sum_{i=1}^n x_i^2$. The choice of a MIB is not unique, but the numbers q and $d_a, a = 1, \ldots, q$ are uniquely determined by G.

A MIB can be used to represent the orbits as points of $\mathbf{R}^q$. In fact, given an orbit Ω, the vector map $p(x) = (p_1(x), p_2(x), \ldots, p_q(x))$ is constant on Ω, and, given two orbits $\Omega_1 \neq \Omega_2$, $\exists\ p_a \in$ MIB $\mid p_a(x_1) \neq p_a(x_2)$, $\forall x_1 \in \Omega_1$, $x_2 \in \Omega_2$. $p(x)$ then defines a point $p = p(x) \in \mathbf{R}^q$ that can be considered the image in $\mathbf{R}^q$ of Ω because no other orbit is represented in $\mathbf{R}^q$ by the same point.

The vector map: $p : \mathbf{R}^n \to \mathbf{R}^q : x \to (p_1(x), p_2(x), \ldots, p_q(x))$, is called the *orbit map* and maps $\mathbf{R}^n$ onto the subset $\mathcal{S} = p(\mathbf{R}^n) \subset \mathbf{R}^q$. p induces a one to one correspondence between $\mathbf{R}^n/G$ and $\mathcal{S}$ so that $\mathcal{S}$ can be concretely identified with the orbit space of the G-action. All the strata of $\mathcal{S}$ are images of the strata of $\mathbf{R}^n$ through the orbit map and if $[K] > [H]$, $p(\Sigma_{[K]})$ lies in the boundary of $p(\Sigma_{[H]})$. The interior of $\mathcal{S}$ coincides with the image of the principal stratum and the image of all singular strata lie in the bordering surface of $\mathcal{S}$. $p(\Sigma_{[G]})$ always coincides with the origin of $\mathbf{R}^q$.

Generally, the q basic invariants are algebraic dependent, that is, some polynomial $f \in \mathbf{R}[p_1, \ldots, p_q]$ exists such that $f(p_1(x), \ldots, p_q(x)) \equiv 0, \forall x \in \mathbf{R}^n$. In this case the polynomial $f(p_1, \ldots, p_q)$ is called a *syzygy (of the first kind)*. Given a syzygy f, the equation $f(p_1, \ldots, p_q) = 0$ defines a surface in $\mathbf{R}^q$ and $\mathcal{S}$ must be contained in that surface. So, $\mathcal{S} \subset \mathcal{Z}$, where $\mathcal{Z}$ is the intersection of all syzygy surfaces.

All real G-invariant C^∞-functions (not only polynomial functions) can be expressed as real C^∞-functions of the q basic invariants of a MIB [2], and hence define C^∞ functions on $\mathbf{R}^q$: $f(x) = \widehat{f}(p_1(x), \ldots, p_q(x)) \to \widehat{f}(p) = \widehat{f}(p_1, \ldots, p_q)$. The functions $\widehat{f}(p)$ are defined also in points $p \notin \mathcal{S}$ but only the restrictions $\widehat{f}(p)\mid_{p \in \mathcal{S}}$ have the same range as $f(x)$, $x \in \mathbf{R}^n$. All G-invariant C^∞ functions can then be studied in the orbit space $\mathcal{S}$ but one needs to know all equations and inequalities defining $\mathcal{S}$ and its strata.

Details and proofs of the statements here recalled may be found in [3, 4, 5, 6] and references therein.

3. Invariant Theory

Given a MIB $p_1, \ldots, p_q$, there are in general $s \leq q$ basic polynomials that are algebraic independent. If $s < q$ necessarily there are syzygies (of the first kind). One can determine a finite minimal set of r_1 linearly independent homogeneous polynomials, $f_1^{(1)}, \ldots, f_{r_1}^{(1)} \in \mathbf{R}[p_1, \ldots, p_q]$, of degrees $m_{11} \leq \ldots \leq m_{1r_1}$ with respect to the natural grading assigned to the p_a ($\deg(p_a) = \deg(p_a(x))$), that generate all the syzygies of the first kind as a module over $\mathbf{R}[p_1, \ldots, p_q]$. One can continue to define syzygies. Linear combinations of the r_1 variables $f_1^{(1)}, \ldots, f_{r_1}^{(1)}$, with coefficients in $\mathbf{R}[p_1, \ldots, p_q]$, that vanish identically by substituting the $f_i^{(1)}$ with their expressions in terms of the p_a, are called *syzygies of the second kind.* Let $f_1^{(2)}, \ldots, f_{r_2}^{(2)}$ be a linearly independent basis for the syzygies of second kind, made up by homogeneous polynomials (with respect to the natural grading) of degrees $m_{21} \leq \ldots \leq m_{2r_2}$. Linear combinations of the r_2 variables $f_1^{(2)}, \ldots, f_{r_2}^{(2)}$, with coefficients in $\mathbf{R}[p_1, \ldots, p_q]$, that vanish identically by substituting the $f_i^{(2)}$ with their expressions in terms of the $f_j^{(1)}$, are called *syzygies of the third kind*, and so on. There exist a number m such that one has syzygies of the m-th kind $f_1^{(m)}, \ldots, f_{r_m}^{(m)}$, but no syzygies of the $(m+1)$-th kind. Changing the MIB, one changes the syzygies too but does not change the numbers q, s, m, d_a, $\forall a = 1, \ldots, q$, r_i, m_{ij}, $\forall\, i = 1, \ldots, m$, $j = 1, \ldots, r_i$, that are characteristic of $\mathbf{R}[\mathbf{R}^n]^G$.

By a theorem of Hochster and Roberts, in $\mathbf{R}[\mathbf{R}^n]^G$ one can choose a maximal subset of s algebraically independent polynomials: $p_{h_1}, \ldots, p_{h_s}$, of degrees $d'_1, \ldots, d'_s$, called *primary invariants* (or *homogeneous system of parameters*), and a number t of other homogeneous polynomials: $f_1, \ldots, f_t$, of degrees $m_1, \ldots, m_t$, called *secondary invariants*, such that all $p \in \mathbf{R}[\mathbf{R}^n]^G$ can be written in a unique way in following form:

$$p = \sum_{k=0}^{t} q_k\, f_k, \tag{1}$$

where $q_k \in \mathbf{R}[p_{h_1}, \ldots, p_{h_s}]$, and, by definition, $f_0 = 1$ [7, 8]. In other words, all $p \in \mathbf{R}[\mathbf{R}^n]^G$ can be written in a unique way in terms of $p_{h_1}, \ldots, p_{h_s}$, $f_1, \ldots, f_t$, using the f_k at most linearly. The most convenient way to choose the p_{h_l} and the f_k is to take $p_{h_1}, \ldots, p_{h_s}$ and $f_1, \ldots, f_{q-s}$, equal to the q basic invariants in the MIB, with $p_{h_1} = p_1$, and $f_{q-s+1}, \ldots, f_t$, equal to some of the products $f_k f_l$, with $1 \leq k \leq l \leq q - s$, in such a way that all products $f_k f_l$ may be expressed in the form (1) (using in case the syzygies). This choice implies that the degrees d'_l of the p_{h_l} are equal to d_a, $a \in \{h_1, \ldots, h_s\}$, and the degrees m_k of the f_k are equal to $d_a, a = 1, \ldots, q$, $a \notin \{h_1, \ldots, h_s\}$, or to their sums. Then, the degrees m_{1l} of the syzygies of the first kind $f_l^{(1)}$ are $m_{1l} = m_i + m_j$, the degrees m_{2l} of the of the syzygies of the second kind $f_l^{(2)}$ are $m_{2l} = m_i + m_j + m_k$, and so on.

The main problem of constructive invariant theory is to determine a MIB for a given group G, and after that all the syzygies. It is then of great interest to find some algorithms that suggest what the numbers q, s, d_a, m, r_i, m_{ij}, t, m_k, are.

A key role is played by the *Molien function* $M(\eta)$ that may be calculated without the knowledge of the MIB in the following way:

$$M(\eta) = \frac{1}{|G|} \sum_{g \in G} \frac{1}{\det(e - \eta g)}, \quad \text{or} \quad M(\eta) = \int_G \frac{dg}{\det(e - \eta g)},$$

respectively for finite [7] or compact continuous [9] groups, where $\eta < 1$ is an abstract variable, e is the unit matrix of G, $|G|$ is the order of G, dg is the normalized Haar measure and the integration is over the compact group variety. The integral may be reduced to a sum of integrals over unit circles in the complex plane and calculated using the residues [10, 11]. An alternative algebraic method to compute the Molien function for compact connected groups is presented in [12].

When one calculates the Molien function from the formulas written above, one ends up with $M(\eta)$ expressed by a rational function.

The expansion of $M(\eta)$ in power series has the following interpretation:

$$M_1(\eta) = \sum_{d=0}^{\infty} \dim(\mathbf{R}[\mathbf{R}^n]_d^G)\eta^d,$$

where $\dim(\mathbf{R}[\mathbf{R}^n]_d^G)$ is the dimension of $\mathbf{R}[\mathbf{R}^n]_d^G$ as a linear space. Let's call $M_1(\eta)$ the *first form* of the Molien function. $M_1(\eta)$ gives a first information about the number and degrees of the basic invariants of small degrees. One has $\dim(\mathbf{R}[\mathbf{R}^n]_d^G) = n_d^{(0)} - n_d^{(1)} + n_d^{(2)} - \dots$, where $n_d^{(0)}$ is the number of linearly independent polynomials of degree d, $n_d^{(1)}$ is the number of linearly independent syzygies of the first kind of degree d, and so on. For sufficiently small d, one may find out the decomposition of $\dim(\mathbf{R}[\mathbf{R}^n]_d^G)$ in terms of $n_d^{(0)}, n_d^{(1)}, \dots$, but in general one is not able to deduce the numbers $n_d^{(i)}$.

Two rational forms of the Molien function are particularly interesting.

The rational function

$$M_2(\eta) = \frac{\sum_{k=0}^{t} \eta^{m_k}}{\prod_{l=1}^{s}(1-\eta^{d'_l})},$$

here called the *second form* of the Molien function, has the denominator that is the product of s factors of the kind $(1-\eta^{d'_l})$, one for each primary invariant p_{h_l}, and the numerator that is a reciprocal polynomial that gives the number t of the secondary invariants f_k, and their degrees m_k.

The rational function

$$M_3(\eta) = \frac{\sum_{i=0}^{m}(-1)^i \sum_{j=1}^{r_i} \eta^{m_{ij}}}{\prod_{a=1}^{q}(1-\eta^{d_a})},$$

here called the *third form* of the Molien function, has the denominator that is the product of q factors of the kind $(1-\eta^{d_a})$, one for each p_a in a MIB, and the numerator that is a polynomial that gives a partial information about the number r_i of the syzygies of i-th kind $f_j^{(i)}$, and their degrees m_{ij}.

When one calculates $M(\eta)$ one ends up with a rational function that is generally different from both $M_2(\eta)$ and $M_3(\eta)$. To get $M_2(\eta)$ and $M_3(\eta)$ from $M(\eta)$ one has to multiply numerator and denominator of $M(\eta)$ by convenient factors. Obviously, one has $M(\eta) = M_1(\eta) = M_2(\eta) = M_3(\eta)$.

If G is finite then $s = n$. If G is a compact Lie group then $s = n - \dim G + \dim G_p$.

The expressions for $M_2(\eta)$ and $M_3(\eta)$ coincide only if $q = s$ (and this implies $t = m = 0$). Many groups for which $q = s$ are classified in [13, 14, 15, 16].

In the general case $s < q$, $M_2(\eta)$ and $M_3(\eta)$ are unknown as long as one does not know the MIB completely. A possible expression for $M_2(\eta)$ can easily be found by requiring only positive coefficients in the numerator. At this point other possible expressions for $M_2(\eta)$ are obtained by multiplying the numerator and the denominator of $M_2(\eta)$ by factors like $(1+\eta^{d'_l})$, where d'_l appears in the exponents of the denominator, increasing in this way the number of secondary invariants required by the numerator. Generally, when one starts to calculate the

MIB, one supposes the simplest form for $M_2(\eta)$ as the correct one, but it is possible, during the calculation, that one has to change it. For a given expression of $M_2(\eta)$, one has to guess the correct expression of $M_3(\eta)$. One notes that:

(i) One obtains $M_3(\eta)$ by multiplying numerator and denominator of $M_2(\eta)$ by a certain (unknown) number $(q-s)$ of factors like $(1-\eta^{d_a})$, one for each basic non-primary invariant f_k, $k = 1, \ldots, q-s$ (and this implies the equality $m = q - s$).

(ii) The coefficients in the numerator of $M_3(\eta)$ corresponding to the syzygies of the first kind are negative.

(iii) An invariant f_k, whose degree m_k appears in the numerator of $M_2(\eta)$, must coincide or with a basic non-primary invariant, or with a product of them. This implies that: (i) for each f_k, whose degree m_k cannot be written as a sum of degrees of basic polynomials, there is a factor $(1-\eta^{m_k})$ in the denominator of $M_3(\eta)$; (ii) the degrees of the syzygies of the first kind are $m_{1j} \geq 2m_1$.

If one does not know the numbers q and d_a, there are many "possibly right" expressions for $M_3(\eta)$ that satisfy all the points written above, but only one is the correct one, corresponding to the right numbers q and d_a. Moreover, given the right expression for $M_3(\eta)$, the numbers m, r_i, m_{ij}, are not immediately readable from the numerator because of simplifications between similar terms. In any case, if there is only one kind of syzygies of degree d (for example if d is small), the coefficient of η^d in the numerator allows to know the number of independent syzygies of degree d.

Summarizing, the Molien function is a first powerful device to study $\mathbf{R}[\mathbf{R}^n]^G$ because it gives some information about the degrees d_a and allows, in any case, to determine correctly the number and degrees of the basic invariants of small degrees.

When one knows the degrees of some invariants, one may find them explicitly using known properties of the group action or by means of the classical procedure of averaging in the group. Given an arbitrary function $f(x)$, its average in the group is the invariant function $F(x)$, where:

$$F(x) = \frac{1}{|G|} \sum_{g \in G} f(gx), \quad \text{or} \quad F(x) = \int_G f(gx)\, \mathrm{d}g,$$

respectively for finite [17] or continuous compact [18] groups. Obviously, for the linearity of G, if $f(x) \in \mathbf{R}[\mathbf{R}^n]_d$, then $F(x) \in \mathbf{R}[\mathbf{R}^n]_d^G$, so to get an invariant of degree d one may take the average in the group of a monomial of degree d. If d and n are small, one has only a limited number of possible choices of monomials of degree d, but, if d or n are great, this averaging procedure becomes difficult to carry out, because of the many possible monomials of degree d one may choose to average and because most of the invariants that one finds out in this way are not independent from those that one already knows.

For a detailed presentation of invariant theory and for the proofs that are here omitted, the interested reader may read [1, 7, 10, 11, 19, 20, 21] and the references therein.

4. $\widehat{P}$-matrices and Orbit Spaces

Given a stratum $\Sigma \subset \mathbf{R}^n$, in a point $x \in \Sigma$, the number of linear independent gradients of the basic invariants is equal to the dimension of the stratum $p(\Sigma) \subset \mathcal{S}$ [22]. One may calculate the $q \times q$ Grammian matrix $P(x)$ with elements $P_{ab}(x)$ that are scalar products of the gradients of the basic invariants [23]:

$$P_{ab}(x) = \nabla p_a(x) \cdot \nabla p_b(x) = \sum_{i=1}^{n} \frac{\partial p_a(x)}{\partial x_i} \frac{\partial p_b(x)}{\partial x_i},$$

so, if $x \in \Sigma$, $\text{rank}(P(x)) = \dim(p(\Sigma))$.

From the covariance of the gradients of G-invariant functions ($\nabla f(g \cdot x) = g \cdot \nabla f(x)$) and the orthogonality of G, the matrix elements $P_{ab}(x)$ are G-invariant homogeneous polynomial functions of degree $d_a + d_b - 2$, so they may be expressed as polynomials of the basic invariants:

$$P_{ab}(x) = \widehat{P}_{ab}(p_1(x), \ldots, p_q(x)) = \widehat{P}_{ab}(p) \quad \forall x \in \mathbf{R}^n \text{ and } p = p(x).$$

One can define then a matrix $\widehat{P}(p)$ in $\mathbf{R}^q$, called a *$\widehat{P}$-matrix*, having $\widehat{P}_{ab}(p)$ for elements. At the point $p = p(x) \in \mathcal{S}$, image in $\mathbf{R}^q$ of the point $x \in \mathbf{R}^n$ through the orbit map, the matrix $\widehat{P}(p)$ is the same as the matrix $P(x)$, $\widehat{P}(p)$ is however defined in all $\mathbf{R}^q$, also outside $\mathcal{S}$, but only in $\mathcal{S}$ it reproduces $P(x)$, $\forall x \in \mathbf{R}^n$.

$\widehat{P}(p)$ is a real, symmetric $q \times q$ matrix and its matrix elements $\widehat{P}_{ab}(p)$ are homogeneous polynomial functions of degree $d_a + d_b - 2$ with respect to the natural grading, $\text{rank}(\widehat{P}(p))$ is equal to the the dimension of the stratum containing p, and $\mathcal{S}$ is the *only* region of $\mathcal{Z}$ where $\widehat{P}(p)$ is positive semidefinite [6, 24].

If $s \leq q$ basic invariants are algebraic independent, one has $\text{rank}(\widehat{P}(p)) = s$, $\forall p$ in the principal stratum, and $\mathcal{S}$ has dimension s.

The matrix $\widehat{P}(p)$ completely determines $\mathcal{S}$ and its stratification. Defining $\mathcal{S}_k$ the union of all k-dimensional strata of $\mathcal{S}$, $(k = 1, \ldots, s)$, one has:

$$\mathcal{S} = \{p \in \mathcal{Z} \mid \widehat{P}(p) \geq 0\}, \quad \mathcal{S}_k = \{p \in \mathcal{Z} \mid \widehat{P}(p) \geq 0, \ \text{rank}(\widehat{P}(p)) = k\}$$

To find out the defining equations and inequalities of k-dimensional strata, one may impose that all the principal minors of $\widehat{P}(p)$ of order greater than k are zero and that at least one of those of order k, and all those of order smaller than k are positive.

The $\widehat{P}$-matrix contains all information necessary to determine the geometric structure of the orbit space and its strata, so, if one wishes to classify geometrically the orbit spaces, it is sufficient to classify the $\widehat{P}$-matrices. This, in turn, gives a non-standard group classification. One discovers so that different groups, no matter if finite or continuous, share the same orbit space structure, despite the fact that the strata correspond to different orbit types (some examples are in Table X of [25]). They exist different MIB's (of different linear groups), with the same number of invariants and the same degrees, that determine: (i) the same $\widehat{P}$-matrix (some examples are in Table X of [25]); (ii) different $\widehat{P}$-matrices (an example is given by Entries 6 and 25 of Table V of [25]).

Some examples using the $\widehat{P}$-matrices to study orbit space stratifications, minima of invariant polynomials and phase transitions may be found in [26, 25, 27].

5. $\widehat{P}$-matrices and Invariant Theory

The *$\widehat{P}$-matrix method* to determine a MIB is summarized in the following items (i)-(iv) and applied in three examples.

Let $p_1, \ldots, p_q$ be a set of q independent G-invariant homogeneous polynomials.

(i) With $p_1, \ldots, p_q$ one tries to construct the $\widehat{P}$-matrix. By Euler's theorem on homogeneous functions and the standard choice for p_1, one has $\widehat{P}_{1a}(p) = \widehat{P}_{a1}(p) = 2d_a p_a$, $\forall a = 1, \ldots, q$. Taking into account the symmetry of $\widehat{P}(p)$, one has then to calculate only the matrix elements $\widehat{P}_{ab}(p)$, with $2 \leq a \leq b \leq q$.

(ii) If it is not possible to express $P_{ab}(x)$ as a polynomial in $p_1(x), \ldots, p_q(x)$, then $P_{ab}(x)$ is an invariant of degree $d_a + d_b - 2$ that is independent from the known ones. One has then to add a new basic invariant to the set $p_1, \ldots, p_q$. One may choose, for example, an irreducible invariant factor of $P_{ab}(x)$ independent from $p_1, \ldots, p_q$.

(iii) When one succeeds to express all the matrix elements $P_{ab}(x)$ in term of the basic invariants, one has found a "closed" $\widehat{P}$-matrix $\widehat{P}(p)$ and a set $p_1, \ldots, p_q$ that is a MIB (or part of it) for a group $\widehat{G}$ that contains G as a subgroup (see Example 3).

(iv) To check if $p_1, \ldots, p_q$ do form a complete MIB for G one has to look to the Molien function of G. If $M_2(\eta)$ and $M_3(\eta)$ agree with the MIB $p_1, \ldots, p_q$, then this is the MIB for G one was searching in. If instead $M_3(\eta)$ requires a different set of degrees for the basic polynomials, then one has to find out (by averaging in the group or in some other way) at least one of the missing basic polynomials of the lowest degree. If p_0 is this new invariant, first of all one has to check if the higher degree invariants in the set $p_1, \ldots, p_q$ are all independent from p_0, otherwise one has to discard the non independent ones. Then, return to item (i) above with p_0 in the set of the q invariant polynomials.

Some technical suggestions. With the $\widehat{P}$-matrix method, with only invariants of degree 2 one is not able to determine higher degrees invariants and with only even degree invariants one is not able to determine odd degree invariants. So it is convenient to start to build the $\widehat{P}$-matrix having determined in some way at least all invariants of degree 2, all invariants of the smallest odd degree and all the invariants of the smallest even degree greater than 2. One then starts to determine the $\widehat{P}$-matrix as indicated in items (i)-(iv) above, proceeding step after step to calculate the matrix elements of increasing degrees. During this calculation, every time one adds a basic invariant to the set of basic invariants $p_1, \ldots, p_q$, according to item (ii) above, it is convenient to look if this MIB is consistent with the guessed expression of $M_2(\eta)$ and $M_3(\eta)$. In case, one has to modify the expressions of $M_2(\eta)$ and/or $M_3(\eta)$ to include the new basic invariant. One has so a partial check of the correctness either of $M_2(\eta)$ and $M_3(\eta)$, either of the MIB. Every time one starts to determine the matrix elements $\widehat{P}_{ab}(p)$ of a given degree d, it is convenient to control if one knows explicitly all the basic invariants of degree lower than d that are predicted by $M_3(\eta)$. If one misses one of these invariants it is strongly recommended to find out this basic invariant first, otherwise, one does longer and harder calculations.

If a subset $p_{l_1}, \ldots, p_{l_k}$ of the MIB for G is a complete MIB for a group $\widehat{G} \subseteq O(n, \mathbf{R})$, containing G as a subgroup, then the principal minor of the $\widehat{P}$-matrix of G formed by the rows and columns $l_1, \ldots, l_k$ is expressed only in terms of $p_{l_1}, \ldots, p_{l_k}$.

Example 1. Let $G = O_h$, the symmetry group of the cube. The Molien function of G has the following expression:

$$M_2(\eta) = M_3(\eta) = \frac{1}{(1-\eta^2)(1-\eta^4)(1-\eta^6)}$$

and predicts a MIB with 3 algebraic independent polynomials of degree 2, 4 and 6 that have been calculated by many authors (see for ex. Table 4 of [5]). As an elementary example of the $\widehat{P}$-matrix method to determine a MIB, let us start with the basic invariants $p_1 = x^2 + y^2 + z^2$ and $p_2 = x^4 + y^4 + z^4$. The only non trivial element of the $\widehat{P}$-matrix formed with p_1 and p_2 is $\widehat{P}_{22}$. One finds easily $\nabla p_2 = (4x^3, 4y^3, 4z^3)$ and $P_{22} = \nabla p_2 \cdot \nabla p_2 = 16(x^6 + y^6 + z^6)$. As $\widehat{P}_{22}$ cannot be written in the form $c_1 p_1^3 + c_2 p_1 p_2$, the basis p_1 and p_2 is not complete and one may take $p_3 = x^6 + y^6 + z^6$ for a third element of the MIB. All elements of the $\widehat{P}$-matrix constructed with p_1, p_2 and p_3 can be written only in terms of p_1, p_2 and p_3, so the MIB is complete, it agrees with the Molien function and in fact is a possible MIB for G. This trivial Example shows that by constructing the $\widehat{P}$-matrix with only part of the MIB known, one succeeds to find out all the missing basic invariants.

Example 2. Let G be the group of order 192 generated by the matrices

$$g_1 = \frac{\sqrt{2}}{2}\begin{pmatrix} 1 & 1 & 0 & 0 \\ 1 & -1 & 0 & 0 \\ 0 & 0 & 1 & 1 \\ 0 & 0 & 1 & -1 \end{pmatrix}, \qquad g_2 = \begin{pmatrix} 1 & 0 & 0 & 0 \\ 0 & 0 & 0 & -1 \\ 0 & 0 & 1 & 0 \\ 0 & 1 & 0 & 0 \end{pmatrix}.$$

This group is the realification of the complex unitary group considered in [19], and one is referred to [19] for details about its structure.

The Molien function of G has the following first and second form:

$$\begin{aligned} M_1(\eta) = {} & 1 + \eta^2 + \eta^4 + \eta^6 + 4\eta^8 + 4\eta^{10} + 7\eta^{12} + 7\eta^{14} + 12\eta^{16} + \\ & +17\eta^{18} + 22\eta^{20} + 22\eta^{22} + 36\eta^{24} + 43\eta^{26} + \ldots, \\ M_2(\eta) = {} & \frac{1+2\eta^8+2\eta^{12}+2\eta^{16}+5\eta^{18}+5\eta^{24}+2\eta^{26}+2\eta^{30}+2\eta^{34}+\eta^{42}}{(1-\eta^2)(1-\eta^8)(1-\eta^{12})(1-\eta^{24})}. \end{aligned}$$

From $M_1(\eta)$ one gets the following kind of information. The term $4\eta^8$ suggests that there are 4 lin. indep. invariants of degree 8: p_1^4, p_2, p_3, p_4. The term $7\eta^{12}$ suggests that there are 7 lin. indep. invariants of degree 12: p_1^6, $p_1^2p_2$, $p_1^2p_3$, $p_1^2p_4$, p_5, p_6, p_7. The term $12\eta^{16}$ suggests that there are 12 lin. indep. invariants of degree 16: 13 lin. indep. invariants are the terms in the expansion of the polynomial $(p_2 + p_3 + p_4 + p_1^4)^2 + (p_5 + p_6 + p_7)p_1^2$ and it must exist one syzygy of the first kind of degree 16, $f_1^{(1)}$, so that $12 = 13 - 1$. The term $17\eta^{18}$ suggests that there are 17 lin. indep. invariants of degree 18: 12 independent invariants are found by multiplying by p_1 the 12 independent invariants of degree 16, and there must exist 5 new basic invariants of degree 18: p_8, p_9, p_{10}, p_{11}, p_{12}, so that $17 = 12 + 5$. One may continue this analysis and find out that there are 4 syzygies of the first kind of degree 20: $f_2^{(1)}, \ldots, f_5^{(1)}$ and no basic invariants and no syzygies of degree 22. The analysis becomes very uncertain for degree 24 because one might have either basic invariants, either syzygies of the first and of the second kind, and one only knows that their numbers $n_{24}^{(0)}$, $n_{24}^{(1)}$ and $n_{24}^{(2)}$ must satisfy the condition $n_{24}^{(0)} - n_{24}^{(1)} + n_{24}^{(2)} = 36$.

The information obtained so far from $M_1(\eta)$ and $M_2(\eta)$ suggest to obtain the third form of the Molien function by multiplying numerator and denominator of $M_2(\eta)$ by $(1-\eta^8)^2(1-\eta^{12})^2(1-\eta^{18})^5$. Note that the exponents 2, 2, 5 are the coefficients of η^8, η^{12}, η^{18} in the numerator of the second form of the Molien function. Doing this, one finds a third form of the Molien function that shows at the numerator the existence of the syzygies $f_1^{(1)}, \ldots, f_5^{(1)}$ but suggests, wrongly, that there is only one basic invariant of degree 24. At this point it is convenient to stop the (uncertain) analysis of the Molien function and start to build the $\widehat{P}$-matrix.

The Molien function suggests so far a MIB with 1 invariant of degree 2: p_1, 3 invariants of degree 8: p_2, p_3, p_4, 3 invariants of degree 12: p_5, p_6, p_7, 5 invariants of degree 18: $p_8, p_9, p_{10}, p_{11}, p_{12}$, and one invariant of degree 24: p_{13}. In this example the explicit calculation of all the invariants by the method of averaging in the group is not trivial, because of the high degree of the invariants, so, it is convenient to use the $\widehat{P}$-matrix method.

First of all, one defines: $p_1(x) = x_1^2 + x_2^2 + x_3^2 + x_4^2$ and finds out the three invariants of degree 8, for example by averaging in the group the monomials $x_1^2x_2^2x_3^2x_4^2$, $x_1^4x_3^4$ and $x_1^5x_3^3$. One then starts to build up the $\widehat{P}$-matrix. One calculates first the matrix elements of degree 14: $P_{2,2}(x), \ldots, P_{4,4}(x)$ and one finds out the 3 basic invariants of degree 12. One then calculates the matrix elements of degree 18: $P_{2,5}(x), \ldots, P_{4,7}(x)$, and one finds out the 5 basic invariants of degree 18. One then calculates the matrix elements of degree 22: $P_{5,5}(x), \ldots, P_{7,7}(x)$, with no need of new basic invariants. One then calculates the matrix elements of degree 24: $P_{2,8}(x), \ldots, P_{4,12}(x)$, and one finds out two basic invariants of degree 24: $p_{13}(x)$ and $p_{14}(x)$, not only one, as expected. Then the third form of the Molien function has to be corrected by multiplying its numerator and denominator by $(1-\eta^{24})$ and obtaining:

$$M_3(\eta) = \frac{1-\eta^{16}-4\eta^{20}-\eta^{24}-8\eta^{26}+4\eta^{28}-8\eta^{30}+8\eta^{34}-19\eta^{36}+\dots}{(1-\eta^2)(1-\eta^8)^3(1-\eta^{12})^3(1-\eta^{18})^5(1-\eta^{24})^2}\,.$$

Calculating all the remaining elements of the $\widehat{P}$-matrix one does not need any new basic invariant. As $p_1,\dots,p_{14}$ agrees with the third form of the Molien function, they form a complete MIB for G. It is important to note that the invariant p_{14} was not predicted from the Molien function, but it is required to express some of the elements of the matrix $P(x)$ in terms of the MIB.

This example shows that the Molien function is not sufficient to specify the number and degrees of all the basic invariants in a MIB, but its combined use with the construction procedure of the $\widehat{P}$-matrix allows one to do that. Moreover, the $\widehat{P}$-matrix method is constructive, in the sense that it allows to determine concretely the MIB.

In this case no proper subset of the MIB $p_1,\dots,p_{14}$, is a complete MIB for a group $\widehat{G}$ containing G as a subgroup, because no principal minor of the $\widehat{P}$-matrix can be written only in terms of the corresponding subset of basic invariants.

Having found the MIB, all the syzygies of any kind can be determined.

Example 3. Let G be the cyclic group of order 3, generated by the matrix

$$g = \begin{pmatrix} -\frac{1}{2} & -\frac{\sqrt{3}}{2} \\ \frac{\sqrt{3}}{2} & -\frac{1}{2} \end{pmatrix}.$$

The three forms of the Molien function are easily found. The second form is

$$M_2(\eta) = \frac{1+\eta^3}{(1-\eta^2)(1-\eta^3)}$$

and suggests a MIB of 3 elements of degree 2, 3, 3. Using the standard form of p_1 and averaging the monomials x_1^3 and x_2^3, one finds out the following MIB:

$$p_1 = x_1^2+x_2^2, \quad p_2 = x_1^3-3x_1x_2^2, \quad p_3 = x_2^3-3x_2x_1^2\,.$$

If one uses the $\widehat{P}$-matrix method in this trivial example starting with the basic invariants p_1 and p_2, one does not need any other invariant to write down the $\widehat{P}$-matrix and would not find p_3. The group that leaves p_1 and p_2 invariant however is G, that has 3 invariants, according to its Molien function. To use the $\widehat{P}$-matrix method it is then necessary to start with all invariants of the smaller degree greater than 2.

Let now H be the cyclic group of order 3, generated by the 4×4 matrix

$$h = \begin{pmatrix} g & o \\ o & g \end{pmatrix},$$

where o is the null matrix of order 2 and g is the matrix written above. The three forms of the Molien function are easily found. The second form is

$$M_2(\eta) = \frac{1+2\eta^2+6\eta^3+2\eta^4+\eta^6}{(1-\eta^2)^2(1-\eta^3)^2}\,,$$

and suggests a MIB of 12 elements: $p_1,\dots,p_4$ of degree 2, and $p_5,\dots,p_{12}$ of degree 3. All the basic invariants can be easily found by averaging in the group. For example, one may use the standard form of p_1 and the averages of the monomials x_1^2, x_1x_3, x_1x_4, x_1^3, x_2^3, x_3^3, x_4^3, $x_1x_2x_3$, $x_1x_2x_4$, $x_1x_3x_4$, $x_2x_3x_4$ for the invariants $p_2, \ldots, p_{12}$. One verifies easily that none of them can be expressed as a polynomial in the other ones, so they form a MIB for H.

In this case if one uses the $\widehat{P}$-matrix method starting with $p_1, \ldots, p_4$ and one arbitrary invariant of degree 3 one would also find out all the basic polynomials of a MIB (generally with more complicated expressions than those obtained from the averages).

If one uses the $\widehat{P}$-matrix method starting for example with the basic invariants $p_1 = x_1^2+x_2^2+x_3^2+x_4^2$, $p_2 = x_1^2+x_2^2$ and $p_5 = x_1^3-3x_1x_2^2$, one does not need any other invariant to write down the $\widehat{P}$-matrix. One knows from the Molien function that the polynomials p_1, p_2, p_5 do not form a MIB for H and one understands that one has to use more invariants at the beginning to find out the MIB with the $\widehat{P}$-matrix method.

The invariance group of p_1, p_2, p_5 is the group $\widehat{H}$, containing H as a subgroup, generated by an arbitrary rotation in the plane x_3, x_4 and by the matrix g acting in the plane x_1, x_2. We know from above, or from the Molien function of $\widehat{H}$:

$$M_2(\eta) = \frac{1+\eta^3}{(1-\eta^2)^2(1-\eta^3)}\,,$$

that a MIB for $\widehat{H}$ has 2 invariants of degree 2 and 2 of degree 3, so p_1, p_2, p_5 is not a complete MIB even for $\widehat{H}$.

This example suggests that it is convenient to start using the $\widehat{P}$-matrix method after having determined all invariants of degree 2 and all invariants of the smaller degree greater than 2 predicted by the Molien function. Moreover, it shows that the Molien function plays a fundamental role to decide if a given set of invariants do form a complete MIB for some group or not.

References

[1] Springer T A 1977 *Invariant Theory* (*Lecture Notes in Mathematics* vol 585) (Berlin: Springer-Verlag)

[2] Schwarz G W 1975 Smooth functions invariant under the action of a compact Lie group *Topology* **14** 63–8

[3] Bredon G 1972 *Introduction to Compact Transformation Groups* (New York: Academic Press)

[4] Schwarz G W 1980 Lifting smooth homotopies of orbit spaces *I.H.E.S. Publ. Math.* **51** 37–135

[5] Michel L and Zhilinskii B I 2001 Symmetry, invariants, topology. Basic tools *Phys. Rep.* **341** 11–84

[6] Abud M and Sartori G 1983 The geometry of spontaneous symmetry breaking *Ann. Phys., NY* **150** 307–72

[7] Stanley R P 1979 Invariants of finite groups and their applications to combinatorics *Bull. Amer. Math. Soc.* **1** 475–511

[8] Kempf G 1979 The Hochster-Roberts theorem of invariant theory *Michigan Math. J.* **26** 19–32

[9] Weyl H 1926 Zur Darstellungstheorie und Invariantenabzählung der projektiven, der Komplex und der Drehungsgruppe *Acta Mathematica* **48** 255–78, reprinted in 1968 *Ges. Abh. Bd. III* (Berlin: Springer-Verlag) pp 1–25

[10] Stanley R P 1979 Combinatorics and invariant theory *Proc. of Symp. in Pure Mathematics* vol 34 ed D K Ray-Chaudhuri (Providence RI: American Mathematical Society) pp 345–55

[11] Popov V L and Vinberg E B 1994 Invariant theory *Algebraic Geometry IV* (*Encyclopaedia of Mathematical Sciences* vol 55) ed A N Parshin and I R Shafarevich (Berlin: Springer-Verlag) pp 127–278

[12] Broer B 1994 A new method for calculating Hilbert series *J. Algebra* **168** 43–70

[13] Shephard G C Todd J A 1954 Finite unitary reflection groups *Canad. J. Math.* **6** 274–304

[14] Schwarz G W 1978 Representations of simple Lie groups with regular rings of invariants *Invent. Math.* **49** 167–91

[15] Littelmann P 1989 Koreguläre und äquidimensionale Darstellungen *J. Algebra* **123** 193–222

[16] Wehlau D L 1993 Equidimensional representations of 2-simple groups *J. Algebra* **154** 437–89

[17] Burnside W 1911 *Theory of Groups of Finite Order* second ed. (Cambridge: Cambridge University Press)

[18] Weyl H 1939 *The Classical Groups* (Princeton: Princeton University Press)

[19] Sloane N J A 1977 Error-correcting codes and invariant theory: new applications of a nineteenth-century technique *Amer. Math. Monthly* **84** 82–107

[20] Springer T A 1982 Séries de Poincaré dans la theorie des invariants *Séminaire d'Algèbre* (*Lecture Notes in Mathematics* vol 1029) (Berlin: Springer-Verlag) pp 37–54

[21] Benson D J 1993 *Polynomial Invariants of Finite Groups* (*London Mathematical Society Lecture Notes Series* vol 190) (Cambridge: Cambridge University Press)
[22] Sartori G 1983 A theorem on orbit structure (strata) of compact linear Lie groups *J. Math. Phys.* **24** 765–8
[23] Abud M and Sartori G 1981 The geometry of orbit space and natural minima of Higgs potentials *Phys. Lett.* **104 B** 147–52
[24] Procesi C and Schwarz G W 1985 Inequalities defining orbit spaces *Invent. Math.* **81** 539–54
[25] Sartori G and Talamini V 1998 Orbit spaces of compact coregular simple Lie groups with 2, 3 and 4 basic polynomial invariants: effective tools in the analysis of invariant potentials *J. Math. Phys.* **39** 2367–401
[26] Sartori G and Valente G 1996 Orbit spaces of reflection groups with 2, 3 and 4 basic polynomial invariants, *J. Phys. A: Math. Gen.* **29** 193–223
[27] Gufan Yu M, Popov Al V, Sartori G, Talamini V, Valente G and Vinberg E B 2001 Geometric invariant theory approach to the determination of ground states of D-wave condensates in isotropic space *J. Math. Phys.* **42** 1533–62

Reducing pure dephasing of quantum bits by collective encoding in quantum dot arrays

A Grodecka, P Machnikowski, L Jacak

Institute of Physics, Wrocław University of Technology, 50-370 Wrocław, Poland

E-mail: `Anna.Grodecka@pwr.wroc.pl`

Abstract. We show that phonon-induced pure dephasing of an excitonic (charge) quantum bit in a quantum dot (QD) may be reduced by collective encoding of logical qubits in QD arrays. We define the logical qubit on an array of 2, 4 and 8 QDs, connecting the logical $|0\rangle$ state with the presence of excitons in the appropriately chosen half of dots and the logical $|1\rangle$ state with the other half of the dots occupied. We give quantitative estimates of the resulting total error of a single qubit operation for an InAs/GaAs system.

The recent experimental demonstration of an optically controlled two-qubit conditional quantum gate implemented on a bi-exciton system confined in a quantum dot (QD) [1] shows that these semiconductor systems are promising candidates for implementation of quantum information processing schemes. The idea of encoding quantum logical values in charge states of confined carriers [2, 3] is motivated by the discrete, atomic-like spectrum of confined carrier states in QDs that allows one to single out a pair of optically active levels that may be assigned the logical values of 0 and 1. The most natural choice is the ground crystal state and a single confined exciton. The relatively large separation of levels in a self-assembled structure (at least several meV) seems to allow for ultrafast control of such a qubit with sub-picosecond optical pulses. Apparently, this should make it possible to perform of order of 10^4 operations within the lifetime of a confined exciton which is usually around 1 ns.

However, optical experiments [4] have demonstrated that the coherence of an exciton state is destroyed to a large extent within a few picoseconds after optically creating the state. This results from a pure dephasing process due to carrier-phonon interaction [5, 6]: spontaneous relaxation of the lattice after an optically induced change of a confined charge distribution is accompanied by emission of phonon wave packets [7, 8], which leaves a kind of *which way* trace in the crystal, distinguishing between the qubit states and thus correlating the subsystems. This short-time dephasing is a purely non-Markovian process resulting from the fact that an abrupt change of a charge distribution cannot be followed adiabatically by the lattice ion positions (deformation). If the operation is performed extremely slowly a local polaron-like deformation forms around the confined charge distribution in a reversible, adiabatic manner, without perturbing the remote parts of the system, thus without dephasing. To be more specific, for a given rate of switching the carrier state, the faster modes follow adiabatically while the slower ones stay behind and then relax spontaneously, contributing to dephasing. From this discussion it is clear that the degree of this dynamically-induced dephasing is reduced for slower driving (longer control pulses), as was indeed confirmed by a formal analysis [9]. Obviously, in a real system the duration of a control operation cannot be extended endlessly since for longer times the exponentially accumulating

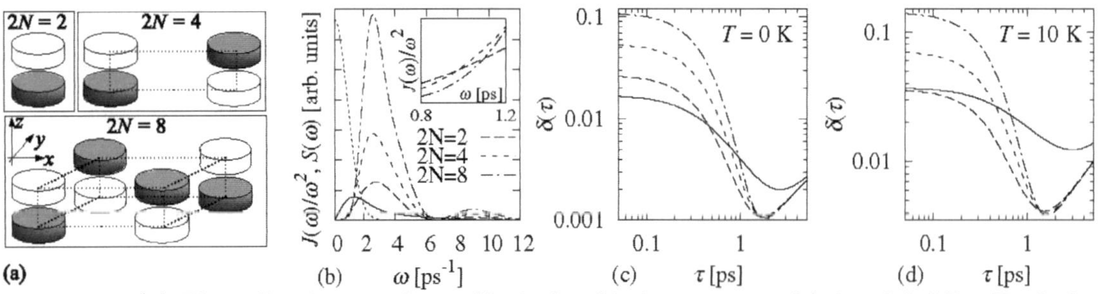

Figure 1. (a) The collective encoding of logical qubit in an array of 2,4 and 8 QDs. Dark dots are occupied in the $|0\rangle$ logical state and the white ones in the $|1\rangle$ state. In the calculations, the dot sizes were 1 nm along z and 5 nm in the xy plane, while the separation between the dots was, respectively, 5 nm and 20 nm. (b) The spectral function corresponding to performing a $\pi/2$ qubit rotation by a Gaussian pulse (dotted) and the spectral densities for the original physical qubit (1 dot, solid line) and for the logical qubit encoded in 2,4 and 8 dots (lines as shown). (c,d) The total (dynamical and Markovian) error $\delta = \delta_{\mathrm{d}} + \tau/\tau_{\mathrm{X}}$, with the exciton lifetime $\tau_{\mathrm{X}} = 2$ ns and 0.5 ns for $T = 0$ K and $T = 10$ K, respectively [13].

Markovian dephasing processes (e.g., related to a finite exciton lifetime) become important. This leads to a tradeoff situation where the optimum pulse duration must be sought against the two opposite dephasing mechanisms [9] (some further reduction of dephasing may also be achieved by pulse shaping [10]). As a result, maximum fidelity of a control operation on a given system is limited.

In this contribution we use a perturbative theory describing the open system evolution under arbitrary driving [9, 11, 12] to show that pure dephasing may be reduced by collective encoding of logical qubits in an array of QDs. The idea relies on the fact that, as discussed above, for sufficiently slow rotation in the qubit space, only a certain long wavelength (low frequency, i.e. slow modes) sector of the phonon reservoir (selected dynamically, independently of the system geometry) contributes to dephasing. The quantum logical $|0\rangle$ state is defined as a state of an array of $2N$ QDs in which appropriately selected N dots are occupied by excitons, while in the $|1\rangle$ state the other half of dots is occupied (see Fig. 1a). As we will show, the proper definition of logical qubits leads to destructive interference of low-frequency phonon excitations induced by qubit switching and, in consequence, to a reduction of the low-frequency part of the phonon spectral density, which leads to reduced dephasing for sufficiently slow operations. Although this scheme works only for relatively slow evolution and the resulting fidelity gain is restricted by finite lifetimes, experimentally measured exciton lifetimes allow one to reduce the overall error.

The QDs are approximated by two-level systems corresponding to the absence or presence of an exciton with a fixed polarization, with transition energies ϵ_n, $n = 1, \ldots 2N$. The wave functions of carriers confined in different dots do not overlap so that no phonon-assisted transfer is possible. The Hamiltonian of this system is

$$H_0 = \sum_n \epsilon_n a_n^\dagger a_n + \sum_k \hbar\omega_k b_k^\dagger b_k + \sum_{n,k} a_n^\dagger a_n f_n(\boldsymbol{k}) \left(b_k^\dagger + b_{-k} \right), \tag{1}$$

where $a_n^\dagger, a_n$ are fermionic creation and annihilation operators for the exciton in the nth dot, $b_k^\dagger, b_k$ are bosonic operators for phonon modes, ω_k are the corresponding frequencies and $f_n(\boldsymbol{k}) = f_n^*(-\boldsymbol{k})$ are the carrier-phonon coupling constants for an exciton in the nth dot. We restrict our discussion to the deformation potential coupling. Assuming, for simplicity, that the exciton state in the nth dot is described by a product of identical electron and hole wave functions $\psi_n(\boldsymbol{r} - \boldsymbol{r}_n)$, where $\boldsymbol{r}_n$ is the position of the dot, the coupling constants may be written as [7]

$$f_n(\boldsymbol{k}) = \sqrt{\frac{\hbar k}{2\rho V c}}(\sigma_{\rm e} - \sigma_{\rm h})e^{-i\boldsymbol{k}\cdot\boldsymbol{r}_n}\mathcal{F}_n(\boldsymbol{k}), \quad \mathcal{F}_n(\boldsymbol{k}) = \int d^3r|\psi_n(\boldsymbol{r})|^2 e^{-i\boldsymbol{k}\cdot\boldsymbol{r}}. \tag{2}$$

Here $\rho = 5360$ kg/m^3 denotes the crystal density, V is the system volume, $c = 5150$ m/s is the sound speed and $\sigma_{\rm e/h}$ ($\sigma_{\rm e} - \sigma_{\rm h} = 8$ eV) are the deformation potential constants for electrons and holes (the material parameters correspond to GaAs). Note that $\mathcal{F}(\boldsymbol{k}) = 1 + O(\boldsymbol{k}^2)$, so that for wavelengths much larger than the dot size the coupling is independent of size and shape and our simplifying assumptions do not play any role.

In our discussion we will assume that a control Hamiltonian of the form

$$H_{\rm C} = \frac{1}{2}f(t)e^{-i\Delta Et/\hbar}a^\dagger_{n_1}\ldots a^\dagger_{n_N}a_{n'_1}\ldots a_{n'_N} + \text{H.c.} \tag{3}$$

is available, that couples the relevant pair of states of the multi-QD register by transferring N excitons from one subset of dots to another without leaving the subspace spanned by the logical qubit states. For 2 QDs this can be achieved using the Förster coupling and the optical Stark effect [14, 15].

The Hamiltonian H_0 [Eq. (1)] has the structure of the independent boson model and may be diagonalized exactly [16]. To this end, one defines the operator

$$\mathbb{W} = \exp\left[\sum_{n,\boldsymbol{k}} a^\dagger_n a_n g^*_n(\boldsymbol{k}) b_{\boldsymbol{k}} - \text{H.c}\right],$$

where $g_n(\boldsymbol{k}) = f_n(\boldsymbol{k})/(\hbar\omega_{\boldsymbol{k}})$ and writes the Hamiltonian H_0 in terms of the operators $\alpha_n = \mathbb{W}^\dagger a_n \mathbb{W} = a_n W_n$ and $\beta_{\boldsymbol{k}} = \mathbb{W}^\dagger b_{\boldsymbol{k}} \mathbb{W} = b_{\boldsymbol{k}} + \sum_n \alpha^\dagger_n \alpha_n g_n(\boldsymbol{k})$, where $W_n = \exp[\sum_{\boldsymbol{k}} g^*_n(\boldsymbol{k}) b_{\boldsymbol{k}} - \text{H.c.}]$ (note that $[W_n, W_m] = 0$ for non-overlapping exciton states). The result is

$$H_0 = \sum_n E_n \alpha^\dagger_n \alpha_n + \sum_{\boldsymbol{k}} \hbar\omega_{\boldsymbol{k}} \beta^\dagger_{\boldsymbol{k}} \beta_{\boldsymbol{k}}, \tag{4}$$

where $E_n = \epsilon_n - \sum_{\boldsymbol{k}} |f_n(\boldsymbol{k})|^2/(\hbar\omega_{\boldsymbol{k}})$.

In terms of the redefined degrees of freedom, the control Hamiltonian (3) to the leading order in phonon coupling reads

$$H_{\rm C} = \frac{1}{2}f(t)e^{-i\Delta Et/\hbar}\alpha^\dagger_{n_1}\ldots \alpha^\dagger_{n_N}\alpha_{n'_1}\ldots \alpha_{n'_N}\left[1 + \sum_{\boldsymbol{k}} G(\boldsymbol{k})(\beta_{-\boldsymbol{k}} - \beta^\dagger_{\boldsymbol{k}})\right] + \text{H.c.}, \tag{5}$$

where $G(\boldsymbol{k}) = [g_{n'_1}(\boldsymbol{k}) + \ldots + g_{n'_N}(\boldsymbol{k})] - [g_{n_1}(\boldsymbol{k}) + \ldots + g_{n_N}(\boldsymbol{k})]$. The phonon perturbation in Eq. (5) does not drive the system out of the two-level subspace but leads to reservoir excitations accompanying any operation on the carrier subsystem. The effect of the phonon coupling on the driven dynamics is calculated using the second-order (Born) expansion of the evolution equation for the density matrix [17, 12], including the first term in Eq. (5) exactly and the second one as a perturbation. After tracing out the reservoir degrees of freedom one finds the phonon-induced correction $\Delta\tilde{\rho}$ to the reduced density matrix of the carrier subsystem in the interaction picture. For a pure initial state $|\psi_0\rangle$, the fidelity of an operation on the logical qubit is then $F = \sqrt{1-\delta}$, with the error $\delta = \langle\psi_0|\Delta\tilde{\rho}|\psi_0\rangle$. The dynamical error, averaged over initial states of the qubit, may be represented as [9]

$$\delta_{\rm d} = \int_0^\infty d\omega \frac{J(\omega)}{\omega^2}\coth\left(\frac{\hbar\omega}{2k_{\rm B}T}\right)S(\omega), \quad S(\omega) = \frac{|F(\omega)|^2 + |F(-\omega)|^2}{12}, \tag{6}$$

where $J(\omega) = \sum_k |G(\boldsymbol{k})|^2 \omega_k^2 \delta(\omega - \omega_k)$ is the spectral density of the phonon reservoir and the function $F(\omega)$ is a nonlinear spectral characteristics of the driving,

$$F(\omega) = \frac{1}{\hbar} \int_{-\infty}^{\infty} dt e^{i\omega t} f(t) e^{i\Phi(t)}, \quad \Phi(t) = \frac{1}{\hbar} \int_{-\infty}^{t} f(\tau) d\tau$$

(see Fig. 1b). For a family of control pulses $f(t) = (\hbar/\tau)\tilde{f}(t/\tau)$ ($\tilde{f}$, defines the pulse shape and τ is the pulse duration), the spectral function scales as $F(\omega) = \tilde{F}(\tau\omega)$, where $\tilde{F}(x)$ is fixed for a given pulse shape. Thus, for large τ, $S(\omega)$ is concentrated in the low frequency sector, in accordance with the qualitative discussion presented above.

For the coupling constants of Eq. (2) one finds the low-frequency behavior of the spectral densities $J(\omega) \sim \omega^5, \omega^7, \omega^9$ for $2N = 2, 4, 8$, respectively (the full dependence is shown in Fig. 1b). Thus, collective encoding indeed reduces the spectral function at low frequencies. However, it is clear from Fig. 1b that this reduction appears at lower and lower frequencies as the size of the logical qubit is increased. Hence, taking advantage of the collective encoding requires using longer and longer pulses which leads to growing contribution of the Markovian component. The resulting total error is plotted as a function of the pulse duration τ in Fig. 1c,d, using exciton lifetimes from the experimental data [13]. It is clear that for the system geometry assumed here considerable reduction of the error is possible, although not much is gained by extending the collective encoding beyond two QDs.

The proposed collective encoding is similar to the earlier idea of implementing logical qubits robust against real transitions on QD arrays [2] (noiseless encoding, closely related to the effect of sub-radiance). However, our scheme does not assume that the individual dots in the register are identical, which is essential for its possible implementation in artificial semiconductor structures.

Acknowledgments

Supported by the Polish Ministry of Scientific Research and Information Technology (Grant No. 1 P03B 09826). P.M. is grateful to the Alexander von Humboldt Foundation for support.

References

[1] Li X, Wu Y, Steel D, Gammon D, Stievater T, Katzer D, Park D, Piermarocchi C, and Sham L 2003 *Science* **301** 809
[2] Zanardi P and Rossi F 1998 *Phys. Rev. Lett.* **81** 4752
[3] Biolatti E, Iotti R. C, Zanardi P, and Rossi F 2000 *Phys. Rev. Lett.* **85** 5647
[4] Borri P, Langbein W, Schneider S, Woggon U, Sellin R. L, Ouyang D, and Bimberg D 2001 *Phys. Rev. Lett.* **87** 157401
[5] Krummheuer B, Axt V. M, and Kuhn T 2002 *Phys. Rev. B* **65** 195313
[6] Vagov A, Axt V. M, and Kuhn T 2003 *Phys. Rev. B* **67** 115338
[7] Jacak L, Machnikowski P, Krasnyj J, and Zoller P 2003 *Eur. Phys. J. D* **22** 319
[8] Vagov A, Axt V. M, and Kuhn T 2002 *Phys. Rev. B* **66** 165312
[9] Alicki R, Horodecki M, Horodecki P, Horodecki R, Jacak L, and Machnikowski P 2004 *Phys. Rev. A* **70** 010501(R), quant-ph/0302058
[10] Axt V. M, Machnikowski P, and Kuhn T 2005 *Phys. Rev. B* **71** 155305, cond-mat/0408322
[11] Grodecka A, Jacak L, Machnikowski P, and Roszak K. 2005 in: Ling P. A, editor, *Quantum Dots: Research Developments* (Nova Science, Hauppauge NY), cond-mat/0404364
[12] Roszak K, Grodecka A, Machnikowski P, and Kuhn T 2005 *Phys. Rev. B* **71** 195333
[13] Langbein W, Borri P, Woggon U, Stavarache V, Reuter D, and Wieck A. D 2004 *Phys. Rev. B* **70** 033301
[14] Lovett B. W, Reina J. H, Nazir A, Kothari B, and Briggs G. A. D 2003 *Phys. Lett. A* **315** 136
[15] Lovett B. W, Reina J. H, Nazir A, and Briggs G. A. D 2003 *Phys. Rev. B* **68** 205319
[16] Mahan G. D 2000 *Many-Particle Physics* (Kluwer, New York)
[17] Alicki R, Horodecki M, Horodecki P, and Horodecki R 2002 *Phys. Rev. A* **65** 062101, quant-ph/0105115

Quantum Monte Carlo simulations of interacting electrons

Ryszard Olchawa
Institute of Physics, Opole University, Opole, Poland

Abstract. A brief introduction to the Monte Carlo methods and the most common types of Quantum Monte Carlo, Variational Monte Carlo and Diffusion Monte Carlo are presented. The fermion sign problem in Diffusion Monte Carlo is outlined and a way of bypassing the fixed-node approximation is discussed.

1. The idea of Monte Carlo Methods

In general all Monte Carlo methods have in common the use of random sampling, and they are the most efficient way to integrate numerically functions in many dimensions.

The starting point for the Monte Carlo integration is the expression for the average value of a function f on the interval $[a, b]$

$$\bar{f} = \frac{1}{(b-a)} \int_a^b f(x)dx. \tag{1}$$

If we reverse this relation we get the expression for the value of the integral I

$$I = \int_a^b f(x)dx = (b-a)\ \bar{f}. \tag{2}$$

This means that if we have the value of $\bar{f}$ we also have the value of I. The aim of Monte Carlo simulation is to estimate the value of $\bar{f}$ by random sampling of the the function f on interval $[a, b]$. The value of the integral I can be approximated by expression

$$I \approx (b-a) \left(\frac{1}{N_S} \sum_{i=1}^{N_S} f(x_i) \right) \tag{3}$$

where x_i are random points uniformly distributed on $[a, b]$ and N_s is the number of this points. Such a method of sampling (uniform sampling) is rather ineffective, because in many practical cases, the contributions to the integral in some regions should be sampled with bigger density, in order to obtain more accurate results in reasonable time. We can do this is by use of the importance sampling method. The importance sampling is applied by introducing a Probability

Distribution Function (PDF): $\rho(x) > 0$ on $[a, b]$, $\rho(x) = 0$ outside. It is convenient to assume that this function is normalized $\int_a^b \rho(x)dx = 1$. Then we can write

$$I = \int_a^b \rho(x)\frac{f(x)}{\rho(x)}dx \approx (b-a)\left(\frac{1}{N_S}\sum_{i=1}^{N_S}\rho(x_i)\left[\frac{f(x_i)}{\rho(x_i)}\right]\right). \quad (4)$$

If the interval $[a, b]$ is sampled according to $\rho(x)$ then the above expression reduce to the final estimator of the integral

$$I \approx \frac{1}{N_S}\sum_{i=1}^{N_S}\left[\frac{f(x_i)}{\rho(x_i)}\right]. \quad (5)$$

The points x_i must be distributed now according to $\rho(x)$. Usually they are obtained as elements of a Markow chain generated by the Metropolis algorithm [1].

2. Quantum Monte Carlo Methods

The notion Quantum Monte Carlo denotes several different techniques based on random sampling. The simplest of these, Variational Monte Carlo (VMC), uses the Importance Sampling MC integration method to evaluate expectation values for a chosen trial wave function.

The other methods, Projector Monte Carlo are using a projection technique to obtain the ground-state component of a starting trial wave function. There are several Projector Monte Carlo methods depend on the projector operator they use, the most popular is Diffusion Monte Carlo (DMC) method, based on the imaginary time Schrodinger equation evolution operator.

Both VMC and DMC are used to calculate ground-state expectation values for correlated systems, they can also be used to specific excited states. For simulating systems at finite temperature the Path Integral Monte Carlo (PIMC) method, based on the path-integral formalism of quantum mechanics elaborated by Feyman is used. The advantage of the DMC and PIMC methods in application to fermion systems is tempered by the sign problem arising from the antisymmetry of meny-fermion wave function. One of the widely used techniques to overcome this problem is so-called "fixed-node approximation" in DMC and "restricted path approximation" in PIMC. The application of DMC to fermion system and an attempt to bypass the fixed-node approximation would be presented in more details further.

Despite of the sign problem Quantum Monte Carlo QMC methods are very powerful tools for calculating the electronic properties of interacting electrons in atoms, molecules and solids. The computational cost in QMC increases as the cube of the number of particles, so it is possible to study systems containing hundreds or even several thousands of electrons. This enables to model real condensed matter with surprising precision.

The general form of the Hamiltonian of interacting electrons in some external potential field V can be written as

$$\hat{H} = -\frac{1}{2}\sum_{i=1}^{N}\nabla_i^2 + \sum_{i=1}^{N}V(\mathbf{r}_i) + \sum_{1\leq i<j}^{N}\frac{1}{|\mathbf{r}_i - \mathbf{r}_j|}. \quad (6)$$

The potential V in the case of solid systems describes the Coulomb interaction between electrons and ionic cores.

3. Variational Monte Carlo (VMC)

In VMC [2] a trial many-electron functions containing a number of adjustable parameters is chosen and expectation values are evaluated

$$E_T = \frac{\left\langle \Psi_T \middle| \hat{H} \middle| \Psi_T \right\rangle}{\langle \Psi_T | \Psi_T \rangle}. \quad (7)$$

The values of these parameters are determined by minimizing the trial energy E_T, the ground-state energy $E_0 \le \min(E_T)$. In a system of N electrons the required integrals are 3N dimensional. For such problems the Monte Carlo integration is much more efficient than conventional numerical methods. In VMC as the importance sampling PDF function usually the physical density function is chosen

$$\rho(\mathbf{R}) = \Psi_T^*(\mathbf{R})\Psi_T(\mathbf{R}) \tag{8}$$

where $\mathbf{R} = \mathbf{r}_1, .., \mathbf{r}_N$.

If we define local energy $E_L(\mathbf{R})$

$$E_L(\mathbf{R}) = \frac{\hat{H}\Psi_T(\mathbf{R})}{\Psi_T(\mathbf{R})}, \tag{9}$$

then we can write

$$\frac{\left\langle \Psi_T \left| \hat{H} \right| \Psi_T \right\rangle}{\langle \Psi_T | \Psi_T \rangle} = \frac{\int \rho(\mathbf{R}) E_L(\mathbf{R}) d\mathbf{R}}{\int \rho(\mathbf{R}) d\mathbf{R}} \approx \frac{1}{N_S} \sum_{i=1}^{N_S} E_L(\mathbf{R}_i) \tag{10}$$

where $\mathbf{R}_i$ are random points in the underlying configuration space, distributed according to $\rho(x)$. The quality of the trial function can be estimated by the local energy variance $\sigma^2(E_L)$.

The main problem of VMC is that the accuracy of calculations depends strongly on the shape and parametrization of the trial wave function. Usually, the VMC calculations are used as a preliminary step for more advanced method DMC.

4. The form of the trial many-particle wave function

The quality of the trial wave function is crucial for the final accuracy of any VMC or DMC simulation. The time of evaluation of the trial wave function is also very essential feature. So, trial wave functions should be both accurate and easy to evaluate. QMC simulations of solids typically use trial wave functions of the Slater-Jastrow type [3], consisting of a single Slater determinant multiplied by a symmetric non-negative Jastrow correlation factor.

$$\Psi_T(\mathbf{x}_1, ..., \mathbf{x}_N) = \Psi_{AS}(\mathbf{x}_1, ..., \mathbf{x}_N) e^{-\sum_{1 \le i < j}^{N} u_{s_i s_j}(|\mathbf{r}_i - \mathbf{r}_j|)}, \tag{11}$$

where $\Psi_{AS}(\mathbf{x}_1, ..., \mathbf{x}_N)$ is the Slater determinant

$$\Psi_{AS}(\mathbf{x}_1, ..., \mathbf{x}_N) = \frac{1}{\sqrt{N!}} \begin{vmatrix} \psi_1(\mathbf{x}_1) & \psi_1(\mathbf{x}_2) & \cdots & \psi_1(\mathbf{x}_N) \\ \psi_2(\mathbf{x}_1) & \psi_2(\mathbf{x}_2) & \cdots & \psi_2(\mathbf{x}_N) \\ \vdots & \vdots & \cdots & \vdots \\ \psi_N(\mathbf{x}_1) & \psi_N(\mathbf{x}_2) & \cdots & \psi_N(\mathbf{x}_N) \end{vmatrix} \tag{12}$$

and $\mathbf{x}_i = (\mathbf{r}_i, \sigma_i)$, where $\sigma_i = \pm\frac{1}{2}$ are the spin states. The orbitals $\psi_i(\mathbf{x})$ in the Slater determinant are usually obtained from HF or LDA calculations. The terms $u_{ss'}(r)$ describe the electron-electron correlations

$$u_{ss'}(r) = \frac{A}{r}(1 - \mathrm{e}^{-\frac{r}{F}}) \tag{13}$$

The correlations between pairs of electrons depend on their mutual spins orientation, so the constants A and F are spin dependent and are related by the cusp conditions

$$\left.\frac{du_{ss'}}{dr}\right|_{r=0} = \begin{cases} -\frac{1}{2} & \text{opposite spins} \\ -\frac{1}{4} & \text{parallel spins} \end{cases} \tag{14}$$

Their values are adjusted by VMC calculations.

5. Diffusion Monte Carlo

DMC is a Monte Carlo method for solving of the imaginary-time Schrodinger equation

$$-\frac{\partial|\psi\rangle}{\partial\tau} = (\hat{H} - E_T)|\psi\rangle \tag{15}$$

where $\tau = it$. The formal solution of the equation can be written as

$$\psi(\mathbf{R},\tau) = e^{-\tau(\hat{H}-E_T)}\psi(\mathbf{R},0) \tag{16}$$

The action of the imaginary-time evolution operator on the function $\psi(\mathbf{R},0)$ can be easily presented if the function is expanded in eigenstates φ_n of Hamiltonian $\hat{H}$, ordered according to corresponding eigenvalues, so that $E_n \leq E_{n+1}$,

$$\psi(\mathbf{R},\tau) = \sum_{n=0}^{\infty} c_n e^{-\tau(E_n-E_T)}\varphi_n(\mathbf{R}) \tag{17}$$

$$\lim_{\tau\to\infty}\psi(\mathbf{R},\tau) = c_0 e^{-\tau(E_0-E_T)}\varphi_0(\mathbf{R}). \tag{18}$$

It follows from (17) that in the long time limit, the imaginary-time evolution operator projects out the component corresponding to minimal energy E_0. If the initial states $\psi(\mathbf{R},0)$ is not orthogonal to the ground state, then the obtained state is in fact the ground-state. In order to introduce the importance sampling, we define a new function [4, 9]

$$f(\mathbf{R},\tau) = \psi_G(\mathbf{R})\psi(\mathbf{R},\tau), \tag{19}$$

where $\psi_G(\mathbf{R})$ is the PDF function (named "guiding" function) and is the best known approximation of the exact ground-state. Substituting (19) into (19), lead to the equation for the new function $f(\mathbf{R},\tau)$, which is a type of Smoluchowski-Foker-Plank equation

$$-\frac{\partial f(\mathbf{R},\tau)}{\partial\tau} = -\frac{1}{2}\nabla^2 f(\mathbf{R},\tau) + \nabla\cdot[\mathbf{F}(\mathbf{R})f(\mathbf{R},\tau)] + (E_L(\mathbf{R}) - E_T)f(\mathbf{R},\tau) \tag{20}$$

where

$$\mathbf{F}(\mathbf{R}) = \frac{\nabla\psi_G(\mathbf{R})}{\psi_G(\mathbf{R})} \tag{21}$$

is a vector in the configuration space interpreted as quantum force. To use the Monte Carlo integration, the equation (20) is transformed to the integral form [5]

$$f(\mathbf{R}',\tau) = \int G(\mathbf{R}\to\mathbf{R}',\tau)f(\mathbf{R},0)d\mathbf{R} \tag{22}$$

Introducing the short time approximation, the Green's function $G(\mathbf{R}\to\mathbf{R}',\tau)$ can be written as

$$G(\mathbf{R}\to\mathbf{R}',\tau) \approx \underbrace{e^{-\tau\left(\frac{E_L(\mathbf{R}')+E_L(\mathbf{R})}{2}-E_T\right)}}_{\text{branching}}\ \underbrace{e^{-\frac{(\mathbf{R}'-\mathbf{R}-\tau\mathbf{F}(\mathbf{R}))^2}{2\tau}}}_{\text{diffusion}} \tag{23}$$

The two factors in (23) correspond to two processes diffusion and branching respectively. In Monte Carlo simulation the two processes are treated separately. In practice, the simulation is based on a set of walkers $\{\mathbf{R}_m\}$ (i.e set of points in the configuration space) initially distributed according to the guiding function. During the simulation, each walker moves through the configuration space in such a way that their distribution converges to the distribution of the ground-state. Each step of a walker splits into two stages: diffusion and branching. In the first stage a walker $\mathbf{R}_m$ is proposed to move to a new position $\mathbf{R}'_m$

$$\mathbf{R_m}' = \mathbf{R_m} + \tau \mathbf{F}(\mathbf{R}) + \eta \tag{24}$$

where η is a random vector in the configuration space and the move is accepted with probability

$$p_m = \min\left(1, \frac{|\psi_G(\mathbf{R_m}')|^2 G(\mathbf{R}_m \rightarrow \mathbf{R}'_m, \tau)}{|\psi_G(\mathbf{R_m})|^2 G(\mathbf{R}'_m \rightarrow \mathbf{R}_m, \tau)}\right), \tag{25}$$

according to Metropolis algorithm. In the next stage the branching process is realized, depending on the value of the branching factor

$$b_m = e^{-\tau\left(\frac{E_L(\mathbf{R}'_m)+E_L(\mathbf{R}_m)}{2} - E_T\right)} \tag{26}$$

the walker can be destroyed or new $\mathrm{int}(b_m + u)$ copies of the walker created, where u is a random number of $[0, 1]$. To keep the average number of workers fixed, the energy E_T should be properly adjusted.

6. The fermion problem in DMC

The presented algorithm so far is exact in a sense of numerical calculation, and leads to the ground-state of the system. For electron systems, the antisymmetry of the wave function is required. This restriction is realized by applying the antisymmetry (together with all other required symmetries) to the guiding function $\psi_G(\mathbf{R})$. However, antisymmetric function changes sign each time when the walker crosses nodes, defined as the solution of the equation

$$\psi_G(\mathbf{R}) = 0. \tag{27}$$

This leads to the well known fermion sign problem. The most popular solution of the problem is the fixed node approximation [6, 7]. Within the method all walker moves that cross a node are rejected. In fact, each walker is confined to some region (pocket) in the configuration space. This means that within the fixed node approximation, the final state has the same nodal structure as the guiding function, so accuracy of the calculations are limited by the accuracy of the nodes of the guiding function [8]. For electrons in one-dimensional system majority of nodes are determined by electron collisions $\mathbf{r}_i = \mathbf{r}_j$. Collision planes are $N - 1$ dimensional and devide the configuration space into separated parts. In Figure 1 two examples of the nodal structure for four (with the same spin orientation) electrons in 1d are presented. The picture was obtained for fixed positions of two electrons. The filled circle corresponds to fixed position (x_3, x_4) wheras the empty circle the transposition (x_4, x_3), the collision planes are represented by the straight lines. However, if the physical space is two(three)-dimensional then the collision condition is given by two (three) equations $x_i = x_j$, $y_i = y_j$ $(z_i = z_j)$, so the collision planes are $2N - 2$ $(3N - 3)$ – dimensional, whereas the nodal surfaces are $2N - 1$ $(3N - 1)$ – dimensional. This means, that the nodal surfaces entirely result from the shape of the guiding function. Some nodal structures for the simplest posible case of two electrons in 2d are shown in Figure 2. The picture was obtained for fixed position (x_2, y_2) one of the electrons (filled circle). It could be shown that for the linear combination of two antisymmetric functions

$$\Psi_{AS}(\mathbf{x}_1, ..., \mathbf{x}_N) + a\ \Phi_{AS}(\mathbf{x}_1, ..., \mathbf{x}_N) \tag{28}$$

continues changing of the parameter a leads to continues deformation of the nodal surfaces. In fact, the structures presented in Figure 1 were obtained for the same functions Ψ_{AS} and Φ_{AS}

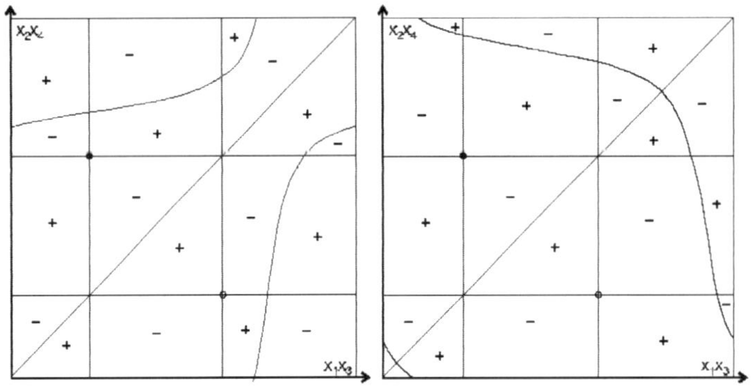

Figure 1. Nodal structures for 4 electrons in 1d, $[x_3, x_4]$ fixed (filled and empty circle).

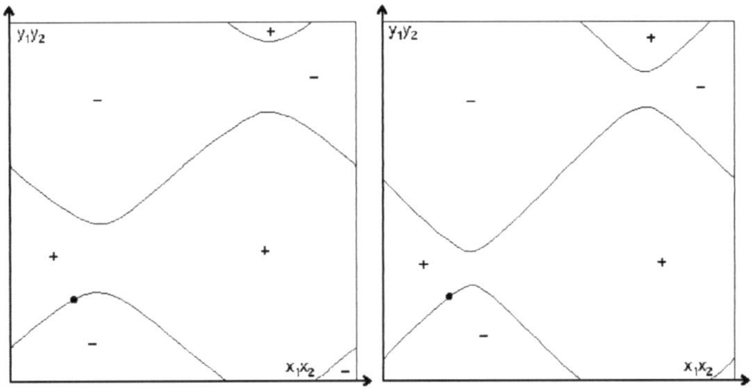

Figure 2. Nodal structures for 2 electrons in 2d, $[x_2, y_2]$ fixed (filled circle).

but for different values of a. Assuming, that the nodal structure is much more importing then the knowledge of the values of the guiding function, the technique of optimization of the guiding function based on (28) was investigated.

The guiding function Ψ_{AS} (which is the best known approximation of the exact ground state of the system) was modified by the substitution

$$\Psi'_{AS}(\mathbf{x}_1, ..., \mathbf{x}_N) = \Psi_{AS}(\mathbf{x}_1, ..., \mathbf{x}_N) + \sum_n a_n \Phi_{n_{AS}}(\mathbf{x}_1, ..., \mathbf{x}_N), \tag{29}$$

where $\Phi_{n_{AS}}$ are Slater determinants build of plane-wave functions. The coefficients a_n where adjusted during simulation in VMC manner using current states of all walkers. Additional restrictions for the coefficients resulting from the geometric symmetry of the system, were fulfilled during calculations. As the operation requires transfer of lot of data between computers in grid and is time consuming it was performed only a few dozen of times during simulation. To investigate the idea some testing calculations for the jellium model were performed. In all calculation noticeable lowering of the average total energy was observed. The typical evolution of the energy during calculating is shown in Figure 3.

7. Concluding remarks

The fixed node approximation for fermions system used in DMC calculations is the most crucial limitation of the accuracy of the calculations and requires very careful and advanced pre-

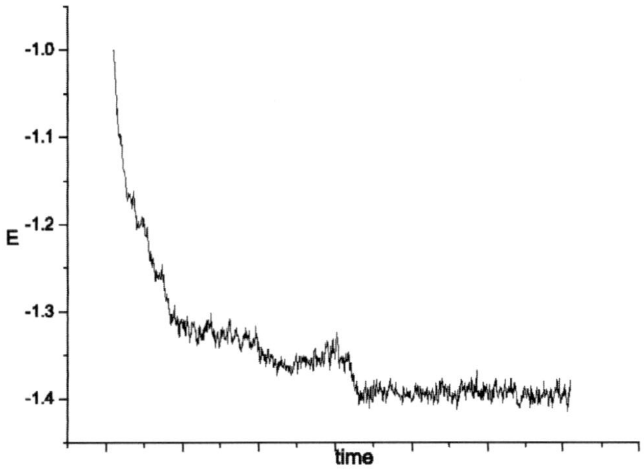

Figure 3. The local energy evolution during simulation.

calculations for guiding function. The presented here idea of dynamic changes of the nodal structure has occurred to be very promising for overcoming of the limitation. However, used in this test the linear combinations of the Slater determinants build of plane waves are very time consuming. It seems to be sensible to look for some other antisymmetric function easy to calculate and flexible in changing its nodal structure.

Acknowledgments
The computations were performed on the Polish grid Clusterix.

References

[1] Metropolis N, Rosenbluth A W, Rosenbluth M N, Teller A M, and Teller E 1953 *J. Chem. Phys.* **21** 1087
[2] Umrigar C J, Wilson K G and Wilkins J W 1988 *Phys. Rev. Lett.* **60** 1719
[3] Jastrow R J 1955 *Phys. Rev.* **98** 1479
[4] Reynolds P J, Ceperley D M, Alder B J and Lester W A 1982 *J. Chem. Phys.* **77** 5593
[5] Umrigar C J, Nightingale M P and Runge K J 1993 *J. Chem. Phys.* **99** 2865
[6] Anderson J 1976 *J. Chem. Phys.* **63** 130
[7] Reynolds P J, Ceperley D M, Alder B J, and Lester W A 1982 *J. Chem. Phys.* **77** 5593
[8] Ceperley D M 1991 *J. Stat. Phys.* **63**, 1237
[9] Foulkes W M C, Mitas L, Needs R J and Rajagopal G 2001 *Rev. of Modern Phys.* **73** 1

Which way interpretation of the dephasing of charge qubits in quantum dots

K Roszak, P Machnikowski, L Jacak

Institute of Physics, Wrocław University of Technology, 50-370 Wrocław, Poland

E-mail: Katarzyna.Roszak@pwr.wroc.pl

Abstract. We show that phonon-induced dephasing of exciton states confined in a quantum dot may be interpreted in terms of *which way* information transferred from the carrier system to the lattice environment. Using distinguishability of quantum states as a measure for this information transfer we cast this interpretation in the form of a quantitative complementarity relation.

The ability of a quantum system to remain in a superposition of states is the key feature that distinguishes it from a classical one and provides the unusual power for quantum computing schemes. However, the phase coherence of such a quantum superposition is affected by interaction with the surrounding world. The resulting effects of *decoherence* or *dephasing* not only belong to the most fundamental aspects of the quantum theory [1] but also present serious limitations to quantum computing schemes, where maintaining system coherence over many control operations is of primary importance [2]. Thus, not only quantitative estimates but also qualitative understanding of such dephasing processes is of great interest.

Contrary to the classical situation, where system states are perturbed due to influence of external noise, dephasing of a quantum state may be interpreted as a result of perturbation exerted *by the system on its environment* [3]. A state-dependent trace imprinted on the environment may allow one to infer the system state by observing only the environment. Because of the historical relation to the interference of two spatial paths in the Young experiment destroyed by the knowledge of the system state, the latter is customarily referred to as *which way* (or "*welcher Weg*") information. It should be stressed that dephasing is due to correlations between the two physical systems and is present no matter whether the *which way* information is really stored or extracted in a form useful to a human.

An exciton confined in a semiconductor quantum dot (QD) has been proposed as a natural implementation of an optically controlled quantum bit (qubit) with the two quantum logical values encoded as the presence ($|1\rangle$) or absence ($|0\rangle$) of the exciton in the QD [4]. Here, the dephasing effect is observed as a decay of the coherent polarization response from an optically excited QD (or an ensemble thereof) [5]. Another class of experiments where coherence properties of the system are manifested are time-domain interference experiments [6, 7, 8], where the final average number of excitons created in a QD with a phase-locked two-pulse sequence is measured as a function of the time delay between the pulses. This time delay is analogous to path length difference in the space-domain (Young) setup and produces a shift between the phase of the superposition state crated by the first pulse and the phase of the second pulse, leading to an oscillation of the final average exciton number between 0 and 1. The visibility of this exciton occupancy fringes is affected by the dephasing of the superposition state between the pulses.

In this paper we relate the dephasing of an exciton in a QD to the *which way* trace left in its lattice environment due to carrier–phonon interactions. By using a formal measure for information on the system state broadcast into the environment in the course of the dephasing process [9, 10] we will derive a *quantitative* relation between the degree of system coherence (as manifested by the amplitude of coherent radiation or visibility of exciton occupancy fringes) and the amount of *which way* information. In this way, we prove complementarity between the knowledge of the system state and the degree of coherence retained by the system.

The present discussion will be restricted to the ground state of an exciton confined in a QD, which is a very good approximation at low temperatures, under resonant excitation with an appropriately polarized laser beam. The observed decrease of coherence in such a system has been accounted for by a model [11] invoking interaction with acoustical phonons. The model reproduces experimental data very well [11, 12] and may serve as a reliable starting point to describe the evolution of the combined system of confined carriers and lattice modes. Under conditions assumed here, the most important effect is that of deformation potential coupling to longitudinal acoustic phonons.

The complete Hamiltonian of the system is

$$H = \epsilon|1\rangle\langle 1| + H_{\mathrm{ph}} + |1\rangle\langle 1| \sum_{\boldsymbol{k}} \hbar\omega_{\boldsymbol{k}}(g_{\boldsymbol{k}}^* b_{\boldsymbol{k}} + g_{\boldsymbol{k}} b_{\boldsymbol{k}}^\dagger), \tag{1}$$

where the first term describes the energy of the confined exciton (ϵ is the energy difference between the states without phonon corrections), $H_{\mathrm{ph}} = \sum_{\boldsymbol{k}} \hbar\omega_{\boldsymbol{k}} b_{\boldsymbol{k}}^\dagger b_{\boldsymbol{k}}$ is the Hamiltonian of the phonon subsystem ($\omega_{\boldsymbol{k}}$ is the frequency of a phonon with wave vector $\boldsymbol{k}$ and $b_{\boldsymbol{k}}^\dagger, b_{\boldsymbol{k}}$ are the corresponding creation and annihilation operators) and the third term describes the interaction, with

$$g_{\boldsymbol{k}} = (\sigma_{\mathrm{e}} - \sigma_{\mathrm{h}}) \sqrt{\frac{1}{2\varrho V_{\mathrm{N}} \hbar c^3 k}} \int_{-\infty}^{\infty} d^3\boldsymbol{r} \psi^*(\boldsymbol{r}) e^{-i\boldsymbol{k}\cdot\mathbf{r}} \psi(\boldsymbol{r}), \tag{2}$$

where $\sigma_{\mathrm{e,h}}$ are deformation potential constants for electrons and holes, V_{N} is the normalization volume of the phonon system, ϱ is the crystal density, c is the speed of sound, and $\psi(\boldsymbol{r})$ are single-particle wave functions which are assumed identical for both types of carriers. In our calculations we use typical parameters for a self-assembled InAs/GaAs structure: $\psi(\boldsymbol{r})$ modelled by Gaussians with 4 nm width in the xy plane and 1 nm along z, $\sigma_{\mathrm{e}} - \sigma_{\mathrm{h}} = 9.5$ eV, $\varrho = 5300$ kg/m^3, $c = 5150$ m/s.

The Hamiltonian (1) is diagonalized by the unitary transformation $W = |0\rangle\langle 0| \otimes \mathbb{I} + |1\rangle\langle 1| \otimes w$, where $w = \exp\left[\sum_{\boldsymbol{k}} \left(g_{\boldsymbol{k}} b_{\boldsymbol{k}}^\dagger - g_{\boldsymbol{k}} b_{\boldsymbol{k}}\right)\right]$, $\mathbb{I}$ is the identity operator, and the tensor product refers to the carrier subsystem (first component) and its phonon environment (second component). This allows us to find the system evolution exactly. Let us assume that an ultrashort optical pulse at $t = 0$ performed a $\pi/2$ rotation, i.e., prepared the system state $\rho_0 = (|\psi_0\rangle\langle\psi_0|) \otimes \rho_{\mathrm{E}}$, where ρ_{E} is the density matrix of the phonon subsystem (environment) at thermal equilibrium and $|\psi_0\rangle = (|0\rangle + |1\rangle)/\sqrt{2}$. The system state at time $t > 0$ may be then written in the form [13]

$$\rho(t) = \frac{1}{2} \begin{pmatrix} \rho_{\mathrm{E}} & e^{iEt/\hbar} \rho_{\mathrm{E}} w^\dagger(t) w \\ e^{-iEt/\hbar} w^\dagger w(t) \rho_{\mathrm{E}} & w^\dagger w(t) \rho_{\mathrm{E}} w^\dagger(t) w \end{pmatrix}, \tag{3}$$

where $E = \epsilon - \sum_{\boldsymbol{k}} \hbar\omega_{\boldsymbol{k}} |g_{\boldsymbol{k}}|^2$, $w(t) = e^{-iH_{\mathrm{ph}}t/\hbar} w e^{iH_{\mathrm{ph}}t/\hbar}$.

The density matrix for the carrier subsystem is obtained by tracing out the phonon degrees of freedom,

$$\rho_{\mathrm{S}}(t) = \mathrm{Tr}_{\mathrm{E}} \rho(t) = \frac{1}{2} \begin{pmatrix} 1 & e^{iEt/\hbar} \langle w^\dagger(t) w \rangle \\ e^{-iEt/\hbar} \langle w^\dagger w(t) \rangle & 1 \end{pmatrix}. \tag{4}$$

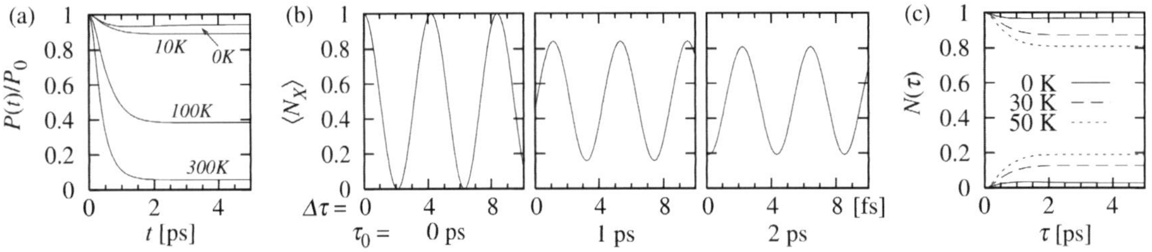

Figure 1. (a) Decay of the coherent radiation from a confined exciton at various temperatures, as shown. (b) Occupation interference fringes for different delay times $\tau = \tau_0 + \Delta\tau$ at 50K. The three panels show the dependence of the final exciton occupation on the delay time $\Delta\tau$ on femtosecond scale and correspond to three different time windows shifted by $\tau_0 = 0$, 1 and 2 picoseconds from the initial time (we assume $E/\hbar = 1.5\ \mathrm{fs}^{-1}$). (c) The envelope of the occupation fringes at three different temperatures as shown.

The first of the two classes of experiments mentioned above consists in measuring the amplitude of the coherent optical polarization $P(t)$ which is proportional to the non-diagonal element of the reduced density matrix (4),

$$|P(t)|^2 = P_0^2 |\langle w^\dagger(t) w\rangle|^2. \tag{5}$$

In the other experiment, at time $t = \tau > 0$ one applies a second pulse that performs a $\pi/2$ rotation, $U_{\pi/2} = [\mathbb{I} + i\sin\phi\sigma_x - i\cos\phi\sigma_y]/\sqrt{2}$, where $\phi = E\tau/\hbar$, and measures the average occupation of the state $|1\rangle$, i.e., the average exciton occupancy $N(\tau) = \langle 1|U_{\pi/2}\rho_{\mathrm{S}}(\tau)U_{\pi/2}^\dagger|1\rangle = \frac{1}{2}\left[1 + |\langle w^\dagger(\tau)w\rangle|\cos(\phi+\theta)\right]$, where we wrote $\langle w^\dagger(\tau)w\rangle = |\langle w^\dagger(\tau)w\rangle|e^{i\theta}$. The value of $N(\tau)$ shows oscillations (interference fringes) in function of ϕ, or of the time delay τ, on the time scale $\hbar/E \sim 1$ fs, modulated by a much more slowly changing function $|\langle w^\dagger(\tau)w\rangle|$, related to dephasing. The visibility of the fringes is

$$\mathcal{V}(\tau) = \frac{N_{\max} - N_{\min}}{N_{\max} + N_{\min}} = |\langle w^\dagger(\tau)w\rangle|. \tag{6}$$

It may be shown [13] that $|\langle w^\dagger(t)w\rangle| = \exp\left[\sum_{\boldsymbol{k}} |g_{\boldsymbol{k}}|^2(\cos\omega_{\boldsymbol{k}}t - 1)(2n_{\boldsymbol{k}} + 1)\right]$, where $n_{\boldsymbol{k}}$ are bosonic equilibrium occupation numbers. For sufficiently regular coupling constants (as in the case of all carrier-phonon coupling mechanisms in QDs) the above quantity decreases form the initial value of 1 to a certain finite value that depends on the system parameters and temperature. As a result, the coherent polarization response is partly reduced over the first few picoseconds of the system evolution (Fig. 1a; cf. Ref. [14]). The visibility of time-domain interference fringes undergoes the same reduction as a function of the delay time between the pulses (Fig. 1b,c).

Let us now derive a quantitative complementarity relation between the degree of coherence in the quantum dot system and the amount of *which way* information transferred to the environment [9]. This relation is analogous to the visibility–distinguishability relation in the double-slit setup [9, 10].

A measure of information on the carrier system contained in its phonon environment is defined as follows. One performs a measurement on the environment and uses its result to predict if an exciton will be found in the QD in a subsequent measurement. The probability of a correct prediction ranges from 1/2 (guessing at random in absence of any correlations) to 1 (knowing for sure, when the systems are maximally entangled). Quantitatively, an intrinsic measure is provided by the *distinguishability of states* [9, 10], $\mathcal{D} = 2(p - 1/2)$, where p is the probability for a correct prediction for the exciton state maximized over all possible measurements on the environment. In this way, guessing at random and knowing for sure correspond to $\mathcal{D}=0$ and $\mathcal{D}=1$, respectively.

For a general density matrix of the compound system $\rho=\sum_{i,j=0,1}|i\rangle\langle j|\otimes\rho_{ij}$, it may be shown [9, 13] that the distinguishability is given by $\mathcal{D}=(1/2)\mathrm{Tr}|\rho_{00}-\rho_{11}|$. Thus, for the carrier-phonon state of Eq. (3) one finds for the distinguishability of carrier states due to correlations with the phonon environment $\mathcal{D}(t)=(1/2)\mathrm{Tr}|\rho_E-w^\dagger w(t)\rho_E w^\dagger(t)w|$. One has, in general [15], $(1/4)\mathrm{Tr}^2|\rho-\rho'|\leq 1-\mathrm{Tr}^2\sqrt{\rho^{1/2}\rho'\rho^{1/2}}$. Hence

$$\mathcal{D}^2(t)\leq 1-\left[\mathrm{Tr}\sqrt{\rho_E^{1/2}w^\dagger w(t)\rho_E w^\dagger(t)w\rho_E^{1/2}}\right]^2=1-\left[\mathrm{Tr}\left|\rho_E^{1/2}w^\dagger w(t)\rho_E^{1/2}\right|\right]^2. \quad (7)$$

Since $\mathrm{Tr}|A|\geq|\mathrm{Tr}A|$ we may write

$$\mathrm{Tr}|\rho_E^{1/2}w^\dagger w(t)\rho_E^{1/2}|\geq|\mathrm{Tr}(\rho_E w^\dagger w(t))|=|\langle w^\dagger(t)w\rangle|. \quad (8)$$

Using Eqs. (5) and (6) and combining Eqs. (7) and (8) we find the inequalities

$$\left[\frac{P(t)}{P_0}\right]^2+\mathcal{D}^2(t)\leq 1,\quad \mathcal{V}^2(\tau)+\mathcal{D}^2(\tau)\leq 1, \quad (9)$$

which relate the degree of coherence to the *which way* knowledge of the system state.

Eqs. (9) show that extracting information on the system state must inevitably destroy coherence. They reflect the fundamental complementarity that lies at the foundation of quantum theory. While phase relations pertain to wave-like properties of the quantum system, an attempt to determine the system state refers to particle-like properties of the same system, like indivisibility of the relevant entity: when measured, an exciton in a QD may be either present or absent, with no intermediate states. Here we have shown that optical experiments with semiconductor quantum dots can contribute to these fundamental issues.

Acknowledgments

Supported by the Polish Ministry of Scientific Research and Information Technology (Grant No. 1 P03B 09826). P.M. is grateful to Alexander von Humboldt Foundation for support.

References

[1] Joos E, Zeh H. D, Kiefer C, Giulini D, Kupsch J and Stamatescu I.-O 2003 *Decoherence and the Appearance of a Classical World in Quantum Theory* (Springer, Berlin)
[2] DiVincenzo D. P and Loss D 1998 *Superlattices and Microstructures* **23** 419
[3] Zurek W. H 2003 *Rev. Mod. Phys.* **75** 715
[4] Biolatti E, Iotti R. C, Zanardi P and Rossi F 2000 *Phys. Rev. Lett.* **85** 5647
[5] Borri P, Langbein W, Schneider S, Woggon U, Sellin R. L, Ouyang D and Bimberg D 2001 *Phys. Rev. Lett.* **87** 157401
[6] Bonadeo N. N, Erland J, Gammon D, Katzer D. S, Park D and Steel D. G 1998 *Science* **282** 1473
[7] Kamada H, Gotoh H, Temmyo J, Takagahara T and Ando H 2001 *Phys. Rev. Lett.* **87** 246401
[8] Htoon H, Takagahara T, Kulik D, Baklenov O, Holmes Jr. A. L and Shih C. K 2002 *Phys. Rev. Lett.* **88** 087401
[9] Englert B.-G 1996 *Phys. Rev. Lett.* **77** 2154
[10] Jaeger G, Shimony A and Vaidman L 1995 *Phys. Rev. A* **51** 54
[11] Vagov A, Axt V. M and Kuhn T 2003 *Phys. Rev. B* **67** 115338
[12] Vagov A, Axt V. M, Kuhn T, Langbein W, Borri P and Woggon U 2004 *Phys. Rev. B* **70** 201305(R)
[13] Roszak K and Machnikowski P. 2005 "Which path" decoherence in quantum dot experiments, cond-mat/0504135
[14] Krummheuer B, Axt V. M and Kuhn T 2002 *Phys. Rev. B* **65** 195313
[15] Nielsen M. A and Chuang I. L 2000 *Quantum Computation and Quantum Information* (Cambridge University Press, Cambridge)

Quantum market games: implementing tactics via measurements

I Pakuła[1], E W Piotrowski[2] and J Sładkowski[1]

[1] Institute of Physics, University of Silesia, Uniwersytecka 4, Pl 40007 Katowice, Poland
[2] Institute of Theoretical Physics, University of Białystok, Lipowa 41, Pl 15424 Białystok, Poland

E-mail: ipakula@wp.pl, ep@alpha.uwb.edu.pl, sladk@us.edu.pl

Abstract. A major development in applying quantum mechanical formalism to various fields has been made during the last few years. Quantum counterparts of Game Theory, Economy, as well as diverse approaches to Quantum Information Theory have been found and currently are being explored. Using connections between Quantum Game Theory and Quantum Computations, an application of the universality of a measurement based computation in Quantum Market Theory is presented.

1. Introduction

In this paper we describe some applications of Quantum Computing methods in the field of Quantum Game Theory and to Quantum Market Games in particular. Quantum Computing gave us tools to overcome some of the most challenging computational problems. We can give as an example Shor's number factorisation algorithm, which works with a polynomial time instead of sub-exponential, but not polynomial one, achieved by the best classical algorithm so far (Number Field Sieve), or Grover's database search algorithm with quadratic advantage over the classical one. Together with the birth of Quantum Computation Theory there arose a question of a universal set of elementary quantum operations, which could accomplish any kind of quantum computation – an analogue of classical $NAND$ gate. Such a set is well known to consist of Hadamard gate H, controlled-not $CNot$ and a phase shift gate T [8] (actually, this set is only approximatively universal [9], that is it can simulate any unitary operation with any given accuracy). It can be further proven that any of the forementioned universal gates can be realised by projective measurement – either based on the teleportation scheme [10, 7, 11] or the state transfer [12, 9]. Of this two, the latter, developed by Jorrand and Perdrix, is more resource efficient, for it needs smaller number of auxiliary qubits than the former. We will show a direct application of the second approach into Quantum Market Game Theory – the existence of such universal primitives is necessary for successful implementation of quantum games.

2. Quantum Game Theory, Market Games and Measurements

Game Theory, a branch of mathematics widely used in Economy, Biology and social sciences [1, 2], has recently attracted attention of physicists. By adapting quantum mechanical language to well known game-theoretical problems new solutions have been found and a new field of research has been discovered [3, 4, 5, 6]. Many classical problems, when formulated in this new way, acquire better solutions and new optimal strategies – *vide* quantum versions of The Prisoner's Dilemma or The Battle of the Sexes. The former game is not only a pure theoretical concept, for it was verified experimentally by Du *et al.* [13] using the NMR method.

Quantum Market Games [14, 15, 16] are applications of Quantum Game Theory in Economy. In this approach, players' (traders') strategies are represented by qubits and tactics – choices of particular strategies – by unitary operators in the strategy Hilbert space (direct sum of Hilbert spaces of all players) acting on subspaces belonging to particular qubits. A transaction is a transition from the state of traders' strategies $|\psi_{in}\rangle$ to the state describing the capital flow $|\psi_{out}\rangle = \mathcal{T}\,|\psi_{in}\rangle$, where $\mathcal{T} := \sum_{k_d} |q\rangle_{k_d k_d}\langle q| + \sum_{k_s} |p\rangle_{k_s k_s}\langle p|$ is the projective operator given by the division of the set of traders to buyers and sellers in the round of transaction in question. An analogue of a measurement in this picture consists of all capital flows resulting from an ensemble of coincident transactions.

Thus Quantum Market Games are represented by sequences of unitary tactics acting on qubits (players' strategies). If we allow alliances between traders, realised by *CNot* gates, a quantum market game appears to be equivalent to a quantum computation, hence a practical realisation of a quantum market game can be done by applying measurement based computation. As will be shown, every quantum market game can be accomplished using projective measurements of the tactics $X = \sigma_x$, $G = \frac{\sigma_z+\sigma_y}{\sqrt{2}}$ and $X \otimes X' = \sigma_x \otimes \sigma_z$ in the way proposed by Jorrand and Perdrix.

We denote the corresponding measuring gates with a rounded off shape, according to the *Qcircuit.tex* standard macros [17]:

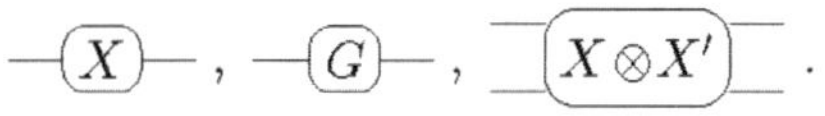

(1)

Although measurement of the tactics $X \otimes X'$ gives us the information whether the two strategies agree or disagree on the price, it reveals no data on the level of the price in question – to get this information we have to measure $X \otimes I$ or $I \otimes X'$. Simple calculations reveal that the measurement of X' can be accomplished by measurement of X and subsequently $X \otimes X'$, what is shown graphically by

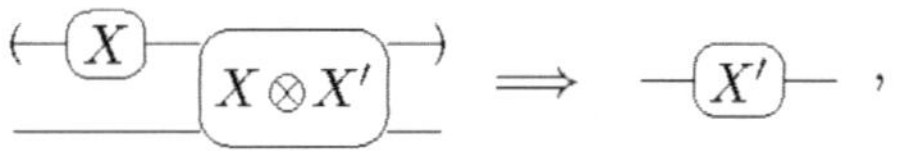

where the parentheses are used to denote auxiliary qubits.

Achieving Hadamard gate is done by a sequence of measurements given by the circuit below. As we can see, a strategy encoded in one qubit is transferred to another, and changed with a sequence of tactics σH, where σ is one of the Pauli matrices (including the identity matrix) depending on the measurement outcomes:

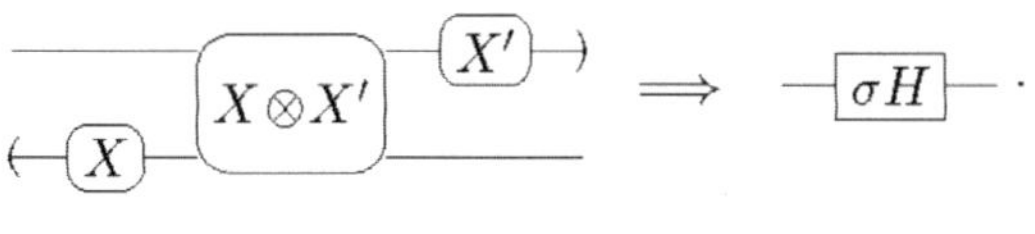

(2)

Random Pauli tactics can be further eliminated, because the gate (2) can be transformed in the following way:

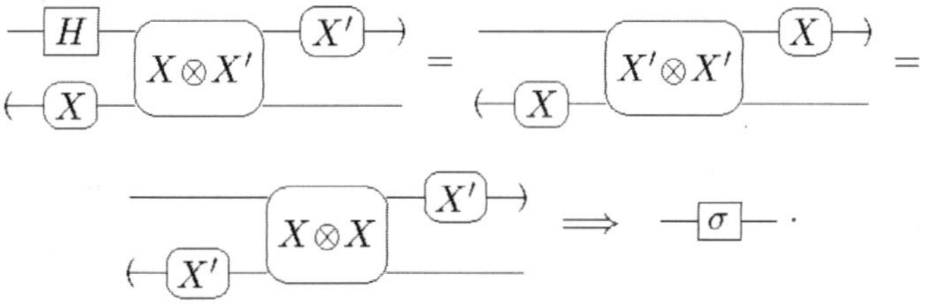

(3)

thus allowing us to wipe out all of the excess σ gates.

The gate (3) can be used to implement the phase-shift tactics:

$$T := \begin{pmatrix} 1 & 0 \\ 0 & \frac{1+\mathrm{i}}{\sqrt{2}} \end{pmatrix}.$$

As T commutes with X', we get:

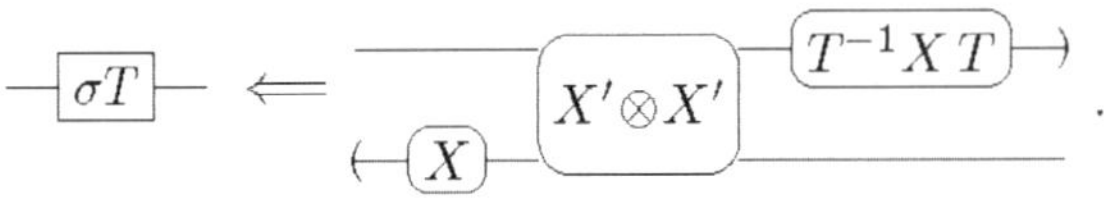

Simple calculations show that $T^{-1}XT = \frac{X-X''}{\sqrt{2}}$ and $H\frac{X-X''}{\sqrt{2}}H = G$, therefore:

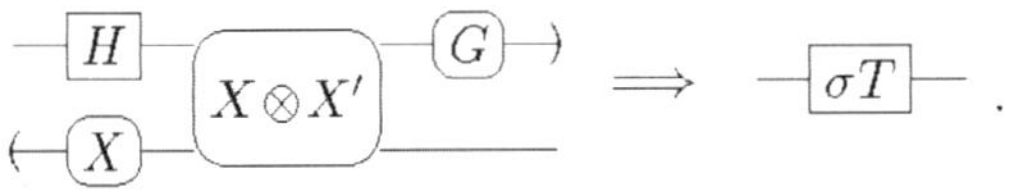

As we have already shown how to remove the superfluous Pauli operators (3), to finally prove the universality of the set of gates (1) we need to implement the alliance $CNot$ by measurements. This gate can be realised as the circuit (as before, the gate is constructed up to a Pauli tactics):

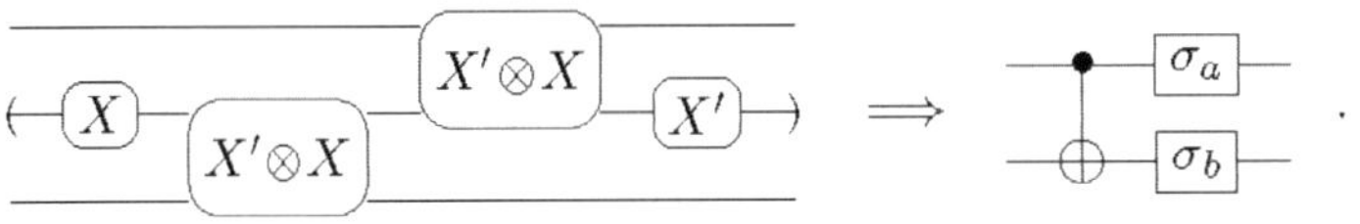

(4)

To eliminate interpretative problems caused by the Quantum Game Theory viewpoint, the measurement of the tactics G can be replaced by *controlled* $-$ H gate and the measurement of entanglement for another pair of conjugated bases $X' \otimes X''$ (where $X'' = \sigma_y$) in the set of universal primitives. Owing to the fact that $G = HGHGH$, the measurement of G can be done in the following way [7]:

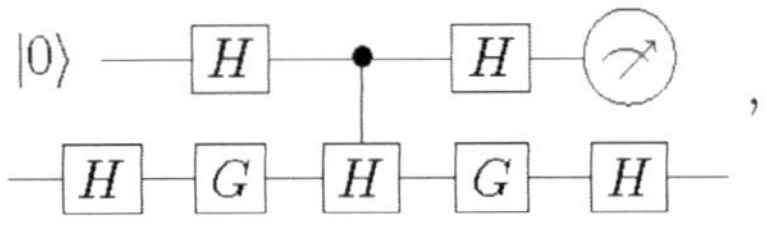

where the tactics G is obtained from (2) by the cyclic replacement $X \rightarrow X'$, $X' \rightarrow X''$ and $X'' \rightarrow X$:

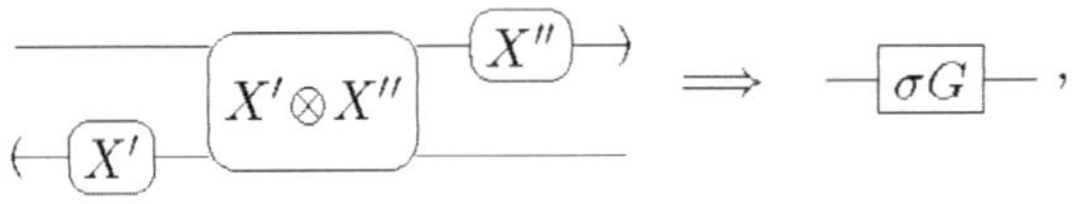

thus completing the set of gates needed to implement quantum market games.

3. Summary

We have shown that methods of measurement based quantum computing developed by Perdrix and Jorrand can also be used in order to realise quantum market games. This observation should be profitable in future direct applications of Quantum Market Theory, for it is potentially able to give traders new levels of effectiveness into their reach.

Acknowledgments

This paper has been supported by the **Polish Ministry of Scientific Research and Information Technology** under the (solicited) grant No **PBZ-MIN-008/P03/2003**.

References

[1] Makowski M, Wieczorek A and Sosnowska H 1997 *Konkurencja i Kooperacja. Teoria Gier w Ekonomii i Naukach Społecznych* (Warsaw: Polish Scientific Publishers PWN).

[2] Luce R D and Raiffa H 1957 *Games and Decisions* (New York: Wiley & Sons).

[3] Meyer D A Quantum strategies 1999 *Phys. Rev. Lett.* **82** 1052.

[4] Piotrowski E W and Sładkowski J 2004 *The next stage: Quantum Game Theory* (Mathematical Physics Frontiers) ed C V Benton (New York: Nova Science Publishers, Inc.) chapter 7 pp 213-34 (*Preprint* quant-ph/0308027).

[5] Piotrowski E W and Sładkowski J An invitation to quantum game theory 2003 *International Journal of Theoretical Physics* **42** 1089.

[6] Eisert J, Wilkens M and Lewenstein M Quantum games and quantum strategies 1999 *Phys. Rev. Lett.* **83** 3077.

[7] Nielsen M A and Chuang I L 2000 *Quantum Computation and Quantum Information* (Cambridge: Cambridge University Press).

[8] Barenco *et al* Elementary gates for quantum computation 1995 *Phys. Rev. A* **52** 3457.

[9] Jorrand P and Perdrix S Unifying quantum computation with projective measurements only and one-way quantum computation 2004 *Preprint* quant-ph/0404125.

[10] Nielsen M A Computable functions, quantum measurements, and quantum dynamics 1997 *Phys. Rev. Lett.* **79** 2915.

[11] Raussendorf R and Briegel H J Quantum computing via measuremets only 2000 *Phys. Rev. Lett.* **86** 5188.

[12] Perdrix S State transfer instead of teleportation in measurement-based quantum computation 2005 *International Journal of Quantum Information* **3** 219-23 (*Preprint* quant-ph/0402204).

[13] Du J it et al Experimental realisation of quantum games on a quantum computer 2002 it *Phys. Rev. Lett.* **88** 137902.

[14] Piotrowski E W and Sładkowski J Quantum market games 2002 *Physica A* **312** 208.

[15] Piotrowski E W and Sładkowski J Quantum computer: an appliance for playing market games 2004 *International Journal of Quantum Information* **2** 495.

[16] Piotrowski E W and Sładkowski J Quantum bargaining games 2002 *Physica A* **308** 391.

[17] Eastin B and Flammia S T Q-circuit tutorial 2004 *Preprint* quant-ph/0406003.

Beyond Lie algebras and group representations: combinatorics

James D Louck

Los Alamos National Laboratory, Theoretical Division, Los Alamos, NM 87545, USA

E-mail: jimlouck@aol.com

Abstract. While acknowledging the extraordinary influence and profound implications of symmetry methods in physics, we argue that these methods have a broader setting in enumerative combinatorics. This is illustrated in several ways by explicit results.

1. Introduction and Viewpoint

The role of symmetry in physics is still succinctly summarized by Wigner's [1] remark in the Preface that he had come to agree with M. von Laue's remark "that the recognition that almost all the rules of spectroscopy follow from the symmetry of the problem is the most remarkable result" contained in that volume. One now knows that this result extends to the term spectroscopy in the broadest sense: atomic, nuclear, particle, etc. The algebra of the operators corresponding to observables in quantum theory leads naturally to Lie algebra and group symmetry in physical law with a strong focus on representation theory and invariants, which arises naturally because of the action of operators in the Hilbert space of state vectors that describe physical systems, and the universality of the invariant subspaces that encode the patterns of symmetry.

Our friend and colleague, the late Brian Wybourne, recognized the importance of symmetry early in his career, and spent much of his time in developing the tools for implementing such techniques into complex atomic spectroscopy, as well as in other areas (Wybourne [2,3]), as we have heard in the opening remarks by Ron King. He was particular fascinated with the role of Schur functions and their plethysms. This paper is dedicated to his extraordinary enthusiasm for science and the hope that his contributions, insightful writings, and viewpoints will inspire others.

The role of symmetry in governing physical law cannot be overstated, and the viewpoints expressed in this paper are evolutions of that role to what I believe is the more general setting of combinatorics, much in the spirit of Brian's focus on Schur functions. Much of what I shall report has all been said before by my collaborators and me (see Refs.[4,5,11,12]). The purpose of this paper is to pull this all together in a compendium format that justifies the title of this presentation.

We contrast the present approach to the traditional methods of the representation theory of symmetry where the group and its Lie algebra are realized by their action in an abstract Hilbert space with suitable properties. An alternative approach is to use special Hilbert spaces that realize all the properties of the abstract postulates and perform calculations within that framework. The framework must be sufficiently rich in structure so as to apply to a manifold of physical situations. It is an approach that is particularly useful for revealing the combinatorial foundations of quantum angular momentum theory, and its generalizations to the general linear and unitary groups.

We shall introduce the subject in the simplest context through the concept of $SU(2)$ solid harmonics, which are the unitary group analogues of the familiar solid harmonics, or spherical harmonics, when restricted to the unit sphere of physical Cartesian $\mathbb{R}^3$-space, that we meet in elementary quantum theory.

2. $SO(3,\mathbb{R})$ and $SU(2)$ Solid Harmonics

It is often the case that when a mathematical entity is polynomial in, say, some constrained variables that define it, a useful procedure is to replace those variables by arbitrary indeterminates with no constraints, and then consider the properties of the new object. A simple example of this is provided by the binomial coefficients defined by

$$\binom{n}{k} = \frac{n!}{(n-k)!k!} = \frac{n(n-1)\cdots(n-k+1)}{k!},$$

which is polynomial in the nonnegative integer n. This suggests we consider the general polynomial $\binom{x}{k}$ for an arbitrary variable x defined by

$$\binom{x}{k} = \frac{x(x-1)\cdots(x-k+1)}{k!},$$

which is then a polynomial of degree k with roots $x = 0, 1, \ldots, k-1$, which reduces at $x = n \geq k$ to the binomial coefficient. These polynomials then have a host of properties generalizing those of the numerical binomial coefficients such as the sum rule

$$\binom{x+y}{k} = \sum_{k_1+k_2=k} \binom{x}{k_1}\binom{y}{k_2}.$$

The idea for obtaining the $SU(2)$ solid harmonics extends this simple example for binomial polynomials in the obvious way. The unitary irreducible representations of the unitary unimodular group $SU(2)$ can be written in terms of the elements of the 2 unitary matrix $U = (u_{ij})_{1\leq i,j\leq 2}, \det U = 1$, by the familiar expression

$$\begin{aligned} D^j_{m\,m'}(U) &= \sqrt{(j+m)!(j-m)!(j+m')!(j-m')!} \\ &\times \sum_k \frac{u_{11}^k u_{12}^{j+m-k} u_{21}^{j+m'-k} u_{22}^{k-m-m'}}{k!(j+m-k)!(j+m'-k)!(k-m-m')!}, \end{aligned} \tag{2.1}$$

where the angular momentum quantum number j is integral or half-odd integral, and each of the projection quantum numbers m and m' can assume values $j, j-1, \ldots, -j$. (See Biedenharn and Louck [5, p. 53] for this form).

Following the above advisement, we define the $SU(2)$ *solid harmonics* to be the polynomials homogeneous of degree $2j$ in four commuting indeterminates $Z = (z_{ij})_{1\leq i,j\leq 2}$ given by

$$D^j_{m\,m'}(Z) = \sqrt{\alpha!\,\beta!} \sum_{A\in\mathbb{M}_2(\alpha,\beta)} \frac{Z^A}{A!}, \tag{2.2}$$

in which

$$\alpha_1 = j+m, \alpha_2 = j-m, \beta_1 = j+m', \beta_2 = j-m', \tag{2.3}$$

and the nonnegative exponents $A = (a_{11}, a_{21}, a_{12}, a_{22})$ are subject to the constraints

$$\mathbb{M}_2(\alpha, \beta) = \left\{ a_{11}, a_{21}, a_{12}, a_{22} \;\middle|\; \begin{array}{l} a_{11} + a_{12} = \alpha_1,\; a_{21} + a_{22} = \alpha_2 \\ a_{11} + a_{21} = \beta_1,\; a_{12} + a_{22} = \beta_2 \end{array} \right\}. \tag{2.4}$$

These constraints on the exponents are imposed to preserve the homogeneity properties of (2.1); namely, they are homogenous of degree $j+m$ in the variable pair (z_{11}, z_{12}), degree $j-m$ in the variable pair (z_{21}, z_{22}), degree $j+m'$ in the variable pair (z_{11}, z_{21}), and degree $j-m'$ in the variable pair (z_{12}, z_{22}), which gives polynomials that are homogenous of total degree $2j$. We also use the space saving standard notations

$$Z^A = \prod_{i,j=1}^{n} z_{ij}^{a_{ij}},\; A! = \prod_{i,j=1}^{n} (a_{ij})!,\; \alpha! = \prod_{i=1}^{n} \alpha_i!,\; \beta! = \prod_{i=1}^{n} \beta_i!, \tag{2.5}$$

where $n = 2$ in (2.2).

The nomenclature $SU(2)$ *solid harmonics* for the polynomials defined by (2.2) is by analogy with the term $SO(3, \mathbb{R})$ solid harmonics. These latter solid harmonics are defined over all points of Cartesian 3-space $\mathbb{R}^3$ by

$$\mathcal{Y}_{l,m}(x_1, x_2, x_3) = \left[\frac{2l+1}{4\pi}(l+m)!(l-m)!\right]^{1/2} \times \sum_{k\geq 0} \frac{(-x_1 - ix_2)^{m+k}(x_1 - ix_2)^k x_3^{l-m-2k}}{2^{m+2k}(m+k)!k!(l-m-2k)!}. \tag{2.6}$$

The components (L_1, L_2, L_3) of the orbital angular momentum operator $\mathbf{L} = -i\mathbf{x}\times\boldsymbol{\nabla}$ have the standard action on the solid harmonics, which are homogeneous polynomial solutions of Laplace's equation in $\mathbb{R}^3$:

$$\begin{aligned} \mathbf{L}^2\mathcal{Y}_{lm}(x) &= l(l+1)\mathcal{Y}_{lm}(x),\; L_3\mathcal{Y}_{lm}(x) = m\mathcal{Y}_{lm}(x), \\ L_{\pm}\mathcal{Y}_{lm}(x) &= \sqrt{(l \mp m)(l \pm m + 1)}\mathcal{Y}_{l m\pm 1}(x), \\ l &\in \{0, 1, 2, \ldots\}, m = l, l-1, \ldots, -l. \end{aligned} \tag{2.7}$$

The eigenvalue relation $\mathbf{L}^2\mathcal{Y}_{lm}(x) = l(l+1)\mathcal{Y}_{lm}(x)$ is a consequence of the fact that the polynomials $\mathcal{Y}_{lm}(x)$ are homogeneous of degree l that solve Laplace's equation, since $\mathbf{L}^2$ can be written as

$$\mathbf{L}^2 = -(\mathbf{x}\cdot\mathbf{x})\boldsymbol{\nabla}^2 + (\mathbf{x}\cdot\boldsymbol{\nabla})^2 + (\mathbf{x}\cdot\boldsymbol{\nabla}), \tag{2.8}$$

which is a sum of three commuting operators, each of which is invariant under orthogonal frame transformations.

These solid harmonics in $\mathbb{R}^3$ are normalized to unity over the unit sphere, and the components L_i are Hermitian with respect to the inner product of square integral functions over the unit sphere. That only integral values of the angular momentum quantum number l can occur is a consequence of the Hermitian property and the fact that both $L_+\mathcal{Y}_{ll} = 0$ and $L_-\mathcal{Y}_{l,-l} = 0$ must be satisfied. (See Ref [5. p. 319] for more discussion of this result.)

The polynomials $D^j_{mm'}(Z), z = (z_{11}, z_{21}, z_{12}, z_{22}) \in \mathbf{C}^4$ are homogeneous of degree $2j$. The angular momentum operator $\mathbf{J}^2$, with $\mathbf{J}$ components (J_1, J_2, J_3) is given by

$$\begin{aligned} \mathbf{J}^2 &= -(\det Z)(\det \frac{\partial}{\partial Z}) + J_0(J_0+1),\; J_0 = (z\cdot\partial)/2, \\ \partial &= (\partial/\partial z_{11}, \partial/\partial z_{21}, \partial/\partial z_{12}, \partial/\partial z_{22}), \end{aligned} \tag{2.9}$$

which is a sum of two commuting operators $-(\det Z)(\det \frac{\partial}{\partial Z})$ and $J_0(J_0+1)$, each of which is invariant under $SU(2)$ transformations. The $SU(2)$ solid harmonics are homogeneous polynomials of degree $2j$ such that

$$\det\left(\frac{\partial}{\partial Z}\right) D^j_{m\,m'}(Z) = 0, \;\; \mathbf{J}^2 D^j_{m\,m'}(Z) = j(j+1) D^j_{m\,m'}(Z). \tag{2.10}$$

The components (J_1, J_2, J_3) of the angular momentum operators $\mathbf{J}$ corresponding to left transformations of Z and the components (J'_1, J'_2, J'_3) corresponding to right transformations have the standard actions on these polynomials:

$$\begin{aligned} J_{\pm} D^j_{m\,m'}(Z) &= \sqrt{j \mp m)(j \pm m + 1)}\, D^j_{m\pm1\,m'}(Z), \\ J'_{\pm} D^j_{m\,m'}(Z) &= \sqrt{j \mp m')(j \pm m' + 1)}\, D^j_{m\,m'\pm1}(Z). \end{aligned} \tag{2.11}$$

The component J_3 and J'_3 are diagonal with eigenvalues m and m'. Under either left or right $SU(2)$ transformations these polynomials give the standard unitary irreducible representations (2.1) of the group $SU(2)$.

The $SU(2)$ solid harmonics are among the most important functions in angular momentum theory. Not only do they unify the irreducible representations of $SU(2)$ in any parametrization by the appropriate definition of the indeterminates in terms of generalized coordinates, they also include the popular boson calculus realization of state vectors for quantum mechanical systems, as well as the state vectors for the symmetric rigid rotator. The role of the inner product is essential. Physical theory demands an inner product that is given in terms of integrations of wave functions over the variables of the theory, as required by the probabilistic interpretation of wave functions. It is the requirement that realizations of angular momentum operators be Hermitian with respect to the inner product for the spaces being used that assures orthogonality of functions. Thus, one can use different inner products having this property with assurance that the results for the properties of angular momentum operators will be identical, except for normalization of state vectors. For polynomials in arbitrary indeterminates it is convenient to use an inner product discussed in detail in Refs. [6,7], which we denote here by $\langle\,,\,\rangle$. With respect to this inner product, the $SU(2)$ solid harmonics satisfy the normalization rule:

$$\langle D^j_{m\,m'}, D^j_{m''\,m'''}\rangle = \delta_{m\,m''}\delta_{m',m'''}(2j)!. \tag{2.12}$$

Often, in combinatorial arguments, the inner product plays no direct role.

We next introduce natural generalizations of the $SU(2)$ solid harmonics, called $SU(n)$ solid harmonics. We will give many properties of these polynomials, which subsume for $n = 2$ properties of the $SU(2)$ solid harmonics.

3. $SU(n)$ Solid Harmonics

Generating functions codify the content of many mathematical entities in a unifying, comprehensive way. The are very popular in combinatorics, and Schwinger [8] used them extensively in his fundamental treatment of angular momentum theory. In this subsection, we present a natural generalization of the $SU(2)$ solid harmonics to a class of polynomials that are homogeneous in n^2 indeterminates. While these polynomials are of interest in their own right (see Gelfand and Graev [9]), it is their fundamental role in the addition of n kinematically independent angular momenta that motivates their introduction here. They bring an unexpected unity and coherence to angular momentum coupling and recoupling theory through their relationship to MacMahon's [10] master theorem, which was redicovered by Schwinger [8].

The natural generalization of definition (2.2) to n^2 indeterminates $Z = (z_{ij})_{1 \le i,j \le n}$ is given by

$$D^p_{\alpha\,\beta}(Z) = \sqrt{\alpha!\beta!} \sum_{A \in \mathbb{M}^p_n(\alpha,\beta)} \frac{Z^A}{A!}, \tag{3.1}$$

where α and β are sequences $\alpha = (\alpha_1, \alpha_2, \ldots, \alpha_n), \beta = (\beta_1, \beta_2, \ldots, \beta_n)$ of n nonnegative integers that sum to p, and are called compositions of p into n nonnegative parts, denoted by $\alpha \vdash p; \beta \vdash p$. For each pair (α, β) of such compositions $\alpha \vdash p$ and $\beta \vdash p$, the notation $\mathbb{M}^p_n(\alpha, \beta)$ denotes the set of all matrices A such that the entries in row i sum to α_i and those in column j to β_j, for all $i, j = 1, 2, \ldots, n$. The sequences α and β are referred to as the row sum and column sum vectors of A. The significance of the row sum vector α is that α_i is the degree of the polynomial $D^p_{\alpha\,\beta}(Z)$ in the variables $(z_{i1}, z_{i2}, \ldots, z_{in})$ in row i of Z, while the column sum vector β has the significance that β_j is the degree of the polynomial $D^p_{\alpha\,\beta}(Z)$ in the variables $(z_{1j}, z_{2j}, \ldots, z_{nj})$ in column j of Z.

The following compendium summarizes some of their important properties:

(i) Matrix of the $D^p_{\alpha\,\beta}(Z)$ polynomials: The number of compositions of the integer k into n nonnegative parts is given by $\binom{n+p-1}{p}$. The compositions in this set may be linearly ordered by the lexicographical rule $\alpha > \beta$, if the first nonzero part of $\alpha - \beta$ is positive. The polynomial $D^k_{\alpha\,\beta}(Z)$ is then the entry in row α and column β in the matrix $D^p(Z)$ of dimension

$$\dim D^p(Z) = \binom{n+p-1}{p}, \tag{3.2}$$

where, following the convention for $SU(2)$, the rows are labeled from top to bottom by the greatest to the least sequences, and the columns are labeled in the same manner as read from left to right.

(ii) Multiplication property: Chen and Louck [11] have given a purely combinatorial proof that these polynomials satisfy the following multiplication rule for arbitrary matrices X and Y, singular and nonsingular:

$$D^p(X)D^p(Y) = D^p(XY). \tag{3.3}$$

(iii) Orthogonality in the inner product $\langle\, , \rangle$:

$$\left\langle D^p_{\alpha\,\beta}, D^{p'}_{\alpha'\,\beta'} \right\rangle = \delta_{p,p'}\delta_{\alpha,\alpha'}\delta_{\beta,\beta'}p!. \tag{3.4}$$

(iv) Reduction to spinor harmonics: Specializing the matrix Z of indeterminates gives some classical results

$$D^p_{\alpha\,\beta}(0, \ldots, 0, z^{(j)}, 0, \ldots, 0) = \left(\prod_{\substack{i=1\\ i\neq j}}^n \delta_{0,\beta_i} \right) \sqrt{\frac{p!}{\alpha!}}\, z^\alpha, \tag{3.5}$$

in which all columns of Z are $\mathrm{col}(0, 0, \ldots, 0)$, except column j which is $z^{(j)} = \mathrm{col}(z_1, z_2, \ldots, z_n)$, $\alpha \vdash p$, and $z^\alpha = z_1^{\alpha_1} \cdots z_n^{\alpha_n}$. Also, we have the relation

$$D^p_{\alpha\,\beta}(\mathrm{diag}(z_1, z_2, \ldots, z_n)) = \delta_{\alpha,\beta} z^\alpha. \tag{3.6}$$

(v) Diagonal matrix: The value at the diagonal matrix is a diagonal matrix:

$$D^p(I_n) = I_{\binom{n+p-1}{p}}. \tag{3.7}$$

(vi) Transposition property: Transposition is a basic symmetry:

$$D^p(Z^T) = (D^p(Z))^T. \tag{3.8}$$

(vii) Special irreducible unitary representations of $SU(n)$:

$$D^p(U)D^p(V) = D^p(UV), \text{all } U, V \in SU(n). \tag{3.9}$$

(viii) MacMahon and Schwinger master theorems. Classical relations:

(a) Schwinger's master theorem: For any two matrices X and Y of order n, the following identities hold:

$$e^{(\partial_x:X:\partial_y)}e^{(x:Y:y)}|_{x=y=0} = \sum_{p=0}^{\infty}\sum_{\alpha,\beta\vdash p} D^p_{\alpha\,\beta}(X)D^p_{\beta\,\alpha}(Y) = \frac{1}{\det(I-XY)}, \tag{3.10}$$

$$(x:Z:y) = xZy^T = \sum_{i,j=1}^{n} z_{ij}x_iy_j. \tag{3.11}$$

(b) MacMahon's master theorem: Let X be the diagonal matrix $X = \operatorname{diag}(x_1, x_2, \ldots, x_n)$ and Y a matrix of order n. Then, the coefficient of x^α in the expansion of $\frac{1}{\det(I-XY)}$ equals the coefficient of x^α in the product y^α, where $y_i = \sum_{j=1}^{n} y_{ij}x_j$; that is,

$$\frac{1}{\det(I-XY)} = \sum_{p=0}^{\infty}\sum_{\alpha\vdash p} D^p_{\alpha\,\alpha}(Y)x^\alpha. \tag{3.12}$$

(c) Basic master theorem: Let Z be a matrix of order n, then

$$\frac{1}{\det(I-tZ)} = \sum_{p=0}^{\infty} t^p \sum_{\alpha\vdash p} D^p_{\alpha\,\alpha}(Z). \tag{3.13}$$

Schwinger's relation (3.10) follows from this relation by setting $Z = XY$ and using the multiplication property (3.3); MacMahon's relation follows from Schwinger's result by setting $X = \operatorname{diag}(x_1, x_2, \ldots, x_n)$. Of course, MacMahon's master theorem preceeding Schwinger's result by many years. The unification into the single form by using properties of the $D^k_{\alpha\,\beta}(Z)$ polynomials was observed by Louck [12]. More surprisingly, the right-hand side of relation (3.10) was already discovered for the general linear group in 1897 by Molien [13]. The properties of this relation for groups are developed extensively in Michel and Zhilinski [14].

4. Coupling of Two Angular Momenta

The coupling of two angular momenta in abstract Hilbert space is realized explicitly by $SU(2)$ solid harmonics and spinor harmonics by the following relationship:

$$\psi_{(j_1\,j_2)j\,m}(Z) = \sum_{m_1+m_2=m} C^{j_1\;j_2\;j}_{m_1\,m_2\,m} \times \psi_{j_1\,m_1}(z_{11}, z_{21})\psi_{j_2\,m_2}(z_{12}, z_{22}), \tag{4.1}$$

where $\psi_{j_1\,m_1}(z_{11}, z_{21})$ and $\psi_{j_2\,m_2}(z_{12}, z_{22})$ are obtained from the spinor harmonics defined by

$$\psi_{j\,m}(z_1, z_2) = \frac{z_1^{j+m}z_2^{j-m}}{\sqrt{(j+m)!(j-m)!}}, \tag{4.2}$$

and $\psi_{(j_1\,j_2)j\,m}(Z)$ is defined by

$$\psi_{(j_1\,j_2)j\,m}(Z) = \sqrt{\frac{2j+1}{(j_1+j_2-j)!(j_1+j_2+j+1)!}} \times (detZ)^{j_1+j_2-j} D^j_{m,j_1-j_2}(Z). \qquad (4.3)$$

Explicit knowledge of the Clebsch-Gordan coefficients (also called Wigner-Clebsch-Gordan or WCG coefficients) is not needed to prove these relationships. Their explicit form can be obtained by a purely combinatorial technique in the umbral calculus known as an *evaluation,* which goes as follows (see Roman and Rota [15]):

(i) Explicit WCG coefficients as evaluations:

(a) Rota evaluations: The evaluation at y of a divided power $x^k/k!$ of a single indeterminate x to a nonnegative integral power k is defined by

$$eval_y \frac{x^k}{k!} = \frac{(y)_k}{k!} = \frac{y(y-1)\cdots(y-k+1)}{k!} = \binom{y}{k}, \qquad (4.4)$$

where $(y)_k$ is the falling factorial. This definition is extended to products by

$$eval_{(y_1,y_2,\ldots,y_n)} \prod_{i=1}^{n} \frac{x^{k_i}}{k_i!} = \prod_{i=1}^{n} eval_{y_i} \frac{x^{k_i}}{k_i!} = \prod_{i=1}^{n} \binom{y_i}{k_i}. \qquad (4.5)$$

It is also extended by linearity to sums of such divided powers, multiplied by arbitrary numbers.

(b) A special case of Rota evaluation: The basic relation underlying relation (4.3)is

$$\frac{(\det Z)^n}{n!} \sum_{A\in\mathbb{M}_2(\alpha,\alpha')} \frac{Z^A}{A!} = \sum_{B\in\mathbb{M}_2(\beta,\beta')} \left(eval_B \frac{(\det Z)^n}{n!} \right) \frac{X^B}{B!}, \qquad (4.6)$$

in which the "line sums" β and β' of B are given in terms of the "line sums" α and α' of A by

$$\beta = (\alpha_1+n, \alpha_2+n),\ \beta' = (\alpha'_1+n, \alpha'_2+n). \qquad (4.7)$$

The evaluation operation is given by

$$eval_B \frac{(\det Z)^n}{n!} = \sum_{k_1+k_2=n} (-1)^{k_2} k_1! k_2! \binom{b_{11}}{k_1}\binom{b_{12}}{k_2}\binom{b_{21}}{k_2}\binom{b_{22}}{k_1}. \qquad (4.8)$$

This identity is a purely combinatorial, algebraic relation for arbitrary indeterminates and arbitrary row and column sum constraints on the array A as specified by $\alpha = (\alpha_1, \alpha_2)$ and $\alpha' = (\alpha'_1, \alpha'_2)$.

(c) Van der Waerden form of WCG coefficients as an evaluation: We now apply relations (4.6)-(4.8) to the case at hand, where we have $n = j_1+j_2-j, \alpha = (j+m, j-m), \alpha' = (j+j_1-j_2, j-j_1+j_2), \beta = (j_1+j_2+m, j_1+j_2-m), \beta' = (2j_1, 2j_2)$. This gives the following result:

$$C^{j_1\ j_2\ j}_{m_1\,m_2\,m} = \sqrt{\frac{(2j+1)(j+m)!(j-m)!}{(j_1+m_1)!(j_1-m_1)!(j_2+m_2)!(j_2-m_2)!}} \qquad (4.9)$$

$$\sqrt{\frac{(j_1+j_2-j)!(j_1-j_2+j)!(-j_1+j_2+j)!}{(j_1+j_2+j+1)!}} eval_A \frac{(det\,Z)^{j_1+j_2-j}}{(j_1+j_2-j)!},$$

$$eval_A \frac{(\det Z)^{j_1+j_2-j}}{(j_1+j_2-j)!} = \sum_{\substack{k_1+k_2=\\ j_1+j_2-j}} (-1)^{k_2} k_1! k_2! \binom{j_1+m_1}{k_1}\binom{j_2+m_2}{k_2} \times \binom{j_1-m_1}{k_2}\binom{j_2-m_2}{k_1}. \tag{4.10}$$

In summary, we have: *Up to multiplicative square-root factors, a WCG coefficient is the integer obtained from the evaluation at the point* $B = \begin{pmatrix} j_1+m_1 & j_2+m_2 \\ j_1-m_1 & j_2-m_2 \end{pmatrix}$ *of the divided power* $\frac{(\det Z)^{j_1+j_2-j}}{(j_1+j_2-j)!}$ *of a* 2×2 *determinant.*

5. Magic Squares and the Coupling of Two Angular Momenta

One of the most fascinating non-Lie algebraic origins of the coupling of two angular momenta is contained in Regge's [16] observation that the triangle rule on the three angular momenta (j_1, j_2, j) given by $j = j_1+j_2, j_1+j_2-1, \ldots, |j_1-j_2|$, and the sum rule on their three projections (m_1, m_2, m) given by $m_1+m_2 = m$, are realized by a 3×3 magic square with fixed line-sum J, which is given in terms of angular momentum quantum numbers by $J = j_1+j_2+j$:

$$A = \begin{pmatrix} a_{11} & a_{12} & a_{13} \\ a_{21} & a_{22} & a_{23} \\ a_{31} & a_{32} & a_{33} \end{pmatrix} = \begin{pmatrix} j_1+m_1 & j_2+m_2 & j-m \\ j_1-m_1 & j_2-m_2 & j+m \\ j_2-j_1+j & j_1-j_2+j & j_1+j_2-j \end{pmatrix}. \tag{5.1}$$

This relation gives a bijection between magic squares A of fixed line-sum J and the angular momentum quantum numbers $(j_1 m_1; j_2 m_2; j\, m)$. The abstract relationship between coupled and uncoupled state vectors is completely codified in the properties of magic squares and may be formulated as

$$\left|(a_{11}+a_{21})/2, (a_{12}+a_{22})/2; (a_{13}+a_{23})/2, (a_{23}-a_{13})/2\right\rangle \tag{5.2}$$
$$= \sum_{A_2} W_J(A) |(a_{11}+a_{21})/2, (a_{11}-a_{21})/2\rangle \otimes |(a_{12}+a_{22})/2, (a_{12}-a_{22})/2\rangle,$$

where the summation is over all matrix arrays of nonnegative integers $A_2 = \begin{pmatrix} a_{11} & a_{12} \\ a_{21} & a_{22} \end{pmatrix}$ of order 2 with row and column sums given by $\alpha_1 = J-a_{13}, \alpha_2 = J-a_{23}, \beta_1 = J-a_{31}, \beta_2 = J-a_{32}$, and the elements (a_{31}, a_{32}, a_{33}) in row 3 and the elements (a_{13}, a_{23}, a_{33}) in column 3 are held fixed. The coefficients $W_J(A)$ in relation (5.2) are not determined by the magic square structure, but all domains of definition of the entries in the ket-vectors are determined. The coefficients $W_J(A)$ themselves are WCG coefficients, as given in terms of the elements of the magic square A by

$$W_J(A) = C^{(a_{11}+a_{21})/2\ (a_{12}+a_{22})/2\ (a_{13}+a_{23})/2}_{(a_{11}-a_{21})/2\ (a_{12}-a_{22})/2\ (a_{13}-a_{23})/2}. \tag{5.3}$$

It is a quite interesting result that the number of magic squares of order 3 and line-sum J has an easy derivation as given by Stanley [17. p.92], which leads to the following formula:: Define $\Delta_J = \{\text{all triangles}(j_1, j_2, j) \,|\, j_1+j_2+j = J\}$ and $M(j_1, j_2, j) = \{(m_1, m_2) | -j_1 \le m_1 \le j_1; -j_2 \le m_2 \le j_2; -j \le m_1+m_2 \le j\}$. Then, we have the following identity:

$$\sum_{(j_1,j_2,j)\in\Delta_J} |M(j_1, j_2, j)| = \binom{J+5}{5} - \binom{J+2}{5}. \tag{5.4}$$

It is nontrivial to effect the summation on the left-hand side of this relation to obtain the right-hand side.

Thus, to each magic square A of order 3 there corresponds a unique WCG coefficient $W_J(A)$, and conversely. Thus, the full abstract structure of the addition of angular momentum is one-to-one with the structure of magic squares of order 3. These rich combinatorial footings of angular momentum theory are completed by the observation that the WCG coefficients themselves are obtained by the Regge [16] generating function for the expansion of a determinant of order 3, again there being no reference to Lie algebraic structures. *Addition of angular momentum in quantum theory can be presented in terms of magic squares and the Regge generating function for WCG coefficients without reference to Lie algebras.*

6. Kronecker Products

From a physical viewpoint, Kronecker products of representations of groups underly the theory of composite systems. Irreducible symmetry subspaces of a composite system are built from those of its individual parts by reducing the Kronecker product of the irreducible representations of the parts. The Kronecker product extends to the polynomial forms on arbitrary commuting indeterminates, so that one has the general rule for $SU(2)$ solid harmonics given by

$$\left(C^{(j_1 j_2)}\right)^T \left(D^{j_1}(Z) \otimes D^{j_2}(Z)\right) C^{(j_1 j_2)} = \sum_j \oplus (\det Z)^{j_1+j_2-j} D^j(Z), \tag{6.1}$$

in which $\oplus$ denotes the direct sum of matrices, T denotes matrix transposition, the summation is over all j that satisfy the triangle rule $j = j_1 + j_2, j_1 + j_2 - 1, \ldots, |j_1 - j_2|$, and $C^{(j_1 j_2)}$ is a matrix of WCG coefficients and $0's$ of dimension $(2j_1+1)(2j_2+1)$, appropriately defined. The validity of this result follows from the fact that the implied double WCG coefficient coupling of the solid harmonics $D^{j_1}_{m_1 m_1'}(Z)$ and $D^{j_2}_{m_2 m_2'}(Z)$ is insensitive to the detailed properties of the polynomials: It depends only on the angular momentum properties of the coupling of angular momenta. The presence of $(\det Z)^{j_1+j_2-j}$ is required by the homogeneity properties of the solid harmonics. There are three additional forms of this result, obtained by using the fact that $C^{(j_1 j_2)}$ is an orthogonal matrix, so that it can be moved in three ways to the other side of (6.1). Relation (6.1) applies to an arbitrary matrix Z, including singular ones. These relations can be specialized in many ways to obtain results for special functions (see Chen and Louck [11]) and algebras. In particular, they hold for Z an element of the groups $SU(2), U(2), GL(2,\mathbb{C})$.

7. Pfaffians and Double Pfaffians

The basic mathematical objects in the theory of recoupling coefficients are Pfaffians and double Pfaffians. Schwinger [8] observed that the calculation of $3n-j$ coefficients involves taking the square root $\sqrt{(I-AB)}$, where A and B are skew symmetric (antisymmetric) matrices of order n, but the procedure is rather obscure. The appropriate concepts for taking the square root is that of a Pfaffian and a double Pfaffian, denoted, respectively, by $Pf(A)$ and $Pf(A,B)$. The definitions require the concept of a matching of the set of integers $\{1,2,\ldots,n\}$. A matching of $\{1,2,\ldots,n\}$ is an unordered set of disjoint subsets $\{i,j\}$ containing two elements. For example, the matchings of $1,2,3$ are $\{1,2\}$, $\{1,3\}$, and $\{2,3\}$. The Pfaffian and double Pfaffian of skew symmetric matrices $A=(a_{ij})$ and $B=(b_{ij})$ of order n are defined by

$$Pf(A) = \sum_{\text{all matchings}} \varepsilon(i_1 i_2 \cdots i_n) a_{i_1,i_2} a_{i_3,i_4} \cdots a_{i_{n-1},i_n}, \tag{7.1}$$

$$\begin{aligned} Pf(A,B) = 1 + \sum_{k\geq 1} \sum_{\text{all double matchings}} & \varepsilon(i_1 i_2 \cdots i_{2k}) \varepsilon(j_1 j_2 \cdots j_{2k}) \\ & \times a_{i_1,i_2} a_{i_3,i_4} \cdots a_{i_{2k-1},i_{2k}} b_{j_1,j_2} b_{j_3,j_4} \cdots b_{j_{2k-1},j_{2k}}, \end{aligned} \tag{7.2}$$

where $\varepsilon(i_1 i_2 \cdots i_n)$ is the sign of the permutation (number of inversions), and the double matching in the double Pfaffian are the 2−subsets of matchings of a subset of $\{1, 2, \ldots, n\}$ of even length.

The relations of skew symmetric matrices A, B to Pfaffians are

$$\sqrt{\det A} = Pf(A); \; \sqrt{\det(I - AB)} = Pf(A, B). \tag{7.3}$$

It is possible to identify a skew symmetric matrix with each binary coupling scheme in the addition of n angular momenta, and the recoupling coefficients between each pair of such coupling schemes are generated by the reciprocal of $(Pf(A, B))^2$. (See the Fifth SSCPM under Ref. [4].)

8. The General D^λ−Polynomials

The D^λ−polynomials are invertible real linear transformations of the Z^A polynomials that preserve their homogeneity properties in the variables appearing in the rows and columns of Z; they have the form:

$$D\begin{pmatrix} m' \\ \lambda \\ m \end{pmatrix}(Z) = \sum_{A \in \mathbb{M}^p_{n\times n}(\alpha, \alpha')} C\begin{pmatrix} m' \\ \lambda \\ m \end{pmatrix}(A)\frac{Z^A}{A!}. \tag{8.1}$$

The symbol $\lambda = (\lambda_1, \lambda_2, \ldots, \lambda_n)$ is a partition having n parts, including parts that are zero, so that $\lambda_1 \geq \lambda_2 \geq \ldots \geq \lambda_n \geq 0$, $\binom{\lambda}{m}$ and $\binom{\lambda}{m'}$ are Gelfand-Tsetlin patterns of the same shape λ, and $\alpha = (\alpha_1, \alpha_2, \ldots, \alpha_n)$ and, similarly, α' are compositions of $p = \lambda_1 + \lambda_2 + \cdots + \lambda_n$ into n nonnegative parts. These compositions are the weights of the respective Gelfand-Tsetlin patterns: $\alpha = W\binom{\lambda}{m}$, $\alpha' = W\binom{\lambda}{m'}$. The symbol $\mathbb{M}^p_{n\times n}(\alpha, \alpha')$ denotes the set of all $n \times n$ matrix arrays of nonnegative integers having row and columns sums given by the parts of compositions α and α'. This implies that the D^λ−polynomials are homogeneous of degree

$$\alpha_i = \sum_{j=1}^{n} a_{ij} \text{ in the variables } z_i = (z_{i1}, z_{i2}, \ldots, z_{in}) \text{ in row } i \text{ of } Z,$$

$$\alpha'_j = \sum_{i=1}^{n} a_{ij} \text{ in the variables } z^j = (z_{1j}, z_{2j}, \ldots, z_{nj}) \text{ in column } j \text{ of } Z.$$

The inversion of the transformation (8.1) is given by

$$\frac{Z^A}{A!} = \sum_{\lambda \vdash p} \sum_{m, m' \in G_\lambda(\alpha, \alpha')} \frac{1}{M(\lambda)A!} C\begin{pmatrix} m' \\ \lambda \\ m \end{pmatrix}(A) D\begin{pmatrix} m' \\ \lambda \\ m \end{pmatrix}(Z),$$
$$\text{each } A \in M^p_{n\times n}(\alpha, \alpha'), \tag{8.2}$$

where $G_\lambda(\alpha, \alpha')$ is the set of double Gelfand-Tsetlin patterns of weight (α, α') and shape λ, and the quantity $M(\lambda)$ is the invariant normalizing factor defined in terms of Weyl's [18] dimension formula, Dimλ, by

$$M(\lambda) = \frac{\prod_{i=1}^{n}(\lambda_i + n - i)!}{1!2!\cdots(n-1)!\,\mathrm{Dim}\lambda},$$
$$1!2!\cdots(n-1)!\mathrm{Dim}\lambda = \prod_{1\leq i<j\leq n}(\lambda_i - \lambda_j + j - i).$$

The orthogonality and normalization of the $D^\lambda-$polynomials is expressed by

$$\left\langle D\begin{pmatrix} m' \\ \lambda \\ m \end{pmatrix}(Z),\, D\begin{pmatrix} m''' \\ \lambda' \\ m'' \end{pmatrix}(Z)\right\rangle = \delta_{m,m''}\delta_{m',m'''}\delta_{\lambda,\lambda'}M(\lambda). \tag{8.3}$$

The normalization of the $D^\lambda-$polynomials to the factor $M(\lambda)$ is made so that the $D^\lambda-$ matrices will have the property $D^\lambda(I_n) = I_{Dim\lambda}$.

There are many orthogonal transformations between the polynomials $Z^A/A!$ and a set of $D^\lambda-$polynomials that might be considered. But the polynomials we have in mind here are uniquely defined by the $C-$coefficients that occur in (8.1). These coefficients are determined by the action of a set of *fundamental shift operators* acting on a Gelfand-Tsetlin pattern. Shift operators are basic objects in combinatorial theory. They usually cannot be realized by differential operators, which is the case we have at hand. Nonetheless, the Lie algebra associated with the $D^\lambda-$polynomials can be expressed in terms of these still more basic objects, the fundamental shift operators. We cannot get into these details here. Our main point in mentioning the general $D^\lambda-$polynomials is to list some of their principal properties to show the generalization of all that we have said earlier. What we have called the $U(n)$ solid harmonics in (3.1) are the special case for a partition of one nonzero part $\lambda = (p, 0, \ldots, 0)$, where α and β are the weights of the corresponding Gelfand-Tsetlin patterns.

A short compendium of some of the more important formulas, repeated from the Fourth SSPCM in Ref. [4] for the $D^\lambda-$polynomials, which generalize the earlier list for the D^p-polynomials is the following:

(i) Multiplication property:

$$D^\lambda(X)D^\lambda(Y) = D^\lambda(XY), \text{for arbitrary } X, Y.$$

This property is fundamental, and has a combinatorial proof.

(ii) Transpositional symmetry:

$$\left(D^\lambda(Z)\right)^T = D^\lambda(Z^T).$$

(iii) Diagonal property: For Z a diagonal matrix, $Z = \text{diag}(x_1, x_2, \ldots, x_n)$, the following identity holds:

$$D^\lambda(\text{diag}(x_1, x_2, \ldots, x_n)) = \sum_{\alpha \in W_\lambda} \oplus x_1^{\alpha_1} x_2^{\alpha_2} \cdots x_n^{\alpha_n} I_{K(\lambda,\alpha)},$$

where $I_{K(\lambda,\alpha)}$ is the identity matrix of dimension $K(\lambda, \alpha)$, the Kostka number, and W_λ is the set of all weights. This result, in turn, gives $D^\lambda(I_n) = I_{\text{Dim}\lambda}$.

(iv) Reduction property: Let Z_{n-1} denote the principal submatrix of Z obtained by striking row n and column n. Then, the following relation holds between $D^\lambda-$polynomials in the n^2 variables of Z for each partition $\lambda = (\lambda_1, \ldots, \lambda_n)$ and $D^\mu-$polynomials in the $(n-1)^2$ variables in Z_{n-1} for each partition $\mu = (\mu_1, \ldots, \mu_{n-1})$:

$$D\begin{pmatrix} m' \\ \mu' \\ \lambda \\ \mu \\ m \end{pmatrix}\begin{pmatrix} Z_{n-1} & 0 \\ 0 & \text{t} \end{pmatrix} = \delta(\mu, \mu')t^{|\lambda|-|\mu|}D\begin{pmatrix} m' \\ \mu \\ m \end{pmatrix}(Z_{n-1}),$$

where $\binom{\mu}{m}$ is the Gelfand-Tsetlin pattern consisting of rows 1 through $n-1$ of the n-rowed pattern, and similarly for $\binom{\mu'}{m'}$.

(v) Kronecker product: the following identity holds:

$$C^T_{\mu\nu}(D^\mu(Z) \otimes D^\nu(Z))C_{\mu\nu} = \sum_\lambda \oplus c^\lambda_{\mu\nu} D^\lambda(Z),$$

where $C^\lambda_{\mu\nu}$ is a real orthogonal matrix of dimension

$$\mathrm{Dim} C^\lambda_{\mu\nu} = \mathrm{Dim}\mu \; \mathrm{Dim}\nu = \sum_\lambda c^\lambda_{\mu\nu} \mathrm{Dim}\ \lambda$$

that effects the reduction of the Kronecker product into a direct sum of D^λ−polynomials. We leave the summation over λ unspecified, but note that is it fully determined by the properties of the Littlewood-Richardson numbers.

(vi) Trace functions: Define $T_\lambda(Z) = \mathrm{trace} D^\lambda(Z)$. Then, these functions satisfy, in consequence of the Kronecker product relation, the identity:

$$T_\mu(Z)T_\nu(Z) = \sum_\lambda c^\lambda_{\mu\nu} T_\lambda(Z).$$

This result is a consequence of the fact that Kronecker product is brought to a direct sum by a similarity transformation.

(vii) Relation to Schur functions: The polynomials $T_\lambda(Z)$ are not symmetric functions in the variables z_{ij}, but if Z is specialized to the diagonal matrix, $Z = (\mathrm{diag}(x_1, x_2, \ldots, x_n)$, then, by the diagonal property in Item (iii), they become symmetric functions in these variables, in which case, we have the identity:

$$s_\lambda(x) = T_\lambda(\mathrm{diag}(x_1, x_2, \ldots, x_n)).$$

The Kronecker product relation in Item (vi) then gives

$$s_\mu(x)s_\nu(x) = \sum_\lambda c^\lambda_{\mu\nu} s_\lambda(x).$$

9. Concluding Remarks

The D^λ−polynomials, and there specialization to $U(n)$ and $SU(2)$ solid harmonics can be defined and all their properties deduced by the use of combinatorial methods alone, without reference to Lie algebras and groups. If the indeterminate matrices Z are specialized to be the elements of a group or algebra, then representations of that group or algebra are obtained. In the case of $U(n)$, when powers of det Z are adjoined, all irreducible unitary representations result, with similar results for the integral representations of the group $GL(n, \mathbb{C})$. It is in this sense that the representations of Lie algebras and groups have a deeper setting in combinatorial objects, thus justifying the title of our presentation.

We conclude by noting that the D^λ−polynomials can be considered to be matrix generalizations of the Schur functions. Is anyone bold enough to take on the generalization of Brian Wybourne's plethysms for Schur functions for the D^λ−polynomials?

References

[1] Wigner E P 1959 *Group Theory and Its Application to the Quantum Mechanics of Atomic Spectra* (New York: Academic Press) (Translation by Griffin J J 1931 from German edition)

[2] Wybourne W G 1970 *Symmetry Principles and Atomic Spectroscopy* (New York: Wiley)

[3] Wybourne W G 1998 *PHYSICS as a JOURNEY*, (Toruń: Nicholas Copernicus University Press)

[4] Louck J D 2001 *Combinatorial Aspects of Repesentations of the Unitary Group* (Symmetry and Structural Properties of Condensed Matter vol 6) ed T Lulek, W Florek and B Lulek (Singapore: World Scientific) p 231

Louck J D, Chen W Y C and Galbraith H W 1999 *Generating Function for* $SU(2)$ *Binary Recoupling Coefficients* (Symmetry and Structural Properties of Condensed Matter vol 5) ed T Lulek, B Lulek and A Wal (Singapore: World Scientific) p 112

Biedenharn L C, Chen W Y C, Galbraith H W and Louck J D 1997 *Combinatorial Aspects of Repesentations of the Unitary Group* (Symmetry and Structural Properties of Condensed Matter vol 4) ed T Lulek, W Florek, and B Lulek (Singapore: World Scientific) p 150

Biedenharn L C and Louck J D 1991 *Special Functions Associated with* $SU(3)$ *Wigner-Clebsch-Gordan Coefficients* (Symmetry and Structural Properties of Condensed Matter) ed W Florek, T Lulek and M Mucha (Singapore: World Scientific) p 207

[5] Biedenharn L C and Louck J D 1981 (Encyclopedia of Mathematics and Its Applications vol 8,9) ed G- C Rota (Addison-Wesley, Reading) presently by (Cambridge: Cambridge University Press)

[6] Louck J D and Galbraith H W 1972 *Rev. Mod. Phys.* **44** 540

[7] Louck J D and Biedenharn L C 1992 *Adv. Quant. Chem.* **23** 127

[8] Schwinger J 1952 *On Angular Momentum.* (U. S. Atomic Energy Commission Report NYO-3071, unpublished). Published in 1965 *Quantum Theory of Angular Momentum* ed L C Biedenharn and H van Dam (New York: Academic Press) p 229

[9] Gelfand I M and Graev M I 2000 *Dokl. Math.* **33** 336 (Transl. from 2000 *Dokl. Akad. Nauk* **372** 151)

[10] MacMahon P A 1960 *Combinatory Analysis* (New York: Chelsia Publishing Co.) (Originally published in two volumes: 1915, 1916 (Cambridge: Cambridge University Press) 25)

[11] Chen W Y C and Louck J D 1998 *Adv. Math.* **140** 207

[12] Louck J D 1996 *Adv. Appl. Math.* **17** 143

[13] Molien T 1897 *"Über die Invarianten der linearen Substitutionsgruppen"* (Sitzungber. Konig. Preuss. Akad. Wiss. **52**) p 1152

[14] Michel L and Zhilinskii B I 2001 *Physics Reports* **341** 11

[15] Roman S and Rota G -C 1978 *Adv. in Math.* **27** 96

[16] Regge T 1958 *Nuovo Cimento* **10** 544

[17] Stanley R P 1997 (Enumerative Combinatorics vol 1) (Cambridge: Cambridge University Press)

[18] Weyl H 1928 *Gruppentheorie und Quantenmechanik* (Leipzig: Hirzel), 1st ed. Translated by: Robertson H P 1931 as *The Theory of Groups and Quantum Mechanics* (London: Methuen)

On the completeness of the set of Bethe-Hulthén solutions of the linear Heisenberg system

W J Caspers[1], **M Labuz**[2] **and A Wal**[2]
[1]Enschede, The Netherlands
[2]Institute of Physics, University of Rzeszow, Poland

Abstract. In this work we formulate the standard form of the solutions of the Heisenberg chain with periodic boundary conditions and show that these solutions can be transformed into the well-known Bethe-Hulthén solutions. The standard form is found by solving the secular problem, separated according to the irreducible representations of the translation group. The relevant parameters $\exp(ik_j)$ of the Bethe-Hulthén solutions are found from a set of linear equations with coefficients derived from the standard solutions. This correspondence between standard and Bethe-Hulthén solutions realizes the completeness of the Bethe-Hulthén method.

1. Introduction

In our contribution to SSPCM 2002 [1] we already gave a historic overview of papers related to the epoch-making work of Bethe [2] and Hulthén [3]. In this work we will derive an algebraic method to determine the relevant parameters of the Bethe-Hulthén (B.H.) solutions:

$$k_j \text{ or } x_j = e^{ik_j}. \tag{1}$$

This derivation will start from a solution in the standard form, i.e. a solution found by a straightforward diagonalization of the linear Heisenberg problem. In this way we give a proof of the existence of the B.H. solutions together with a set of algebraic equations for the parameters e^{ik_j}. The solutions in standard form are given for the linear Heisenberg problem with periodic boundary conditions and an isotropic interaction between nearest neighbour interactions in the chain.

This standard solution is given in terms of an amplitude for all the spin configurations, i.e.: Ising states, for an arbitrary chosen z-axis. Ising states are the eignstates of the z components of all idnividual spins. From translational symmetry it follows that all solutions may be chosen to correspond to an irreducible representation of the translation group. All Ising states for one stationary state have the same number of inverted spins with respect to the ferromagnetic state, for which all z-components of the individual spins have the value $1/2\hbar$. So we may characterize all these Ising states for a stationary state with r deviations in a ring of N spins, by the relative positions of the inversions and the position of one inversion. The corresponding amplitude factorizes in the following way:

$$e^{ikj_1}c(d_1, d_2, ..., (d_r)) \tag{2}$$

in which the $d_j(j = 1, 2, ...r)$ are the distances between the successive deviations at positions $j_1, j_2, ...j_r$:

$$\begin{aligned} d_1 &= j_2 - j_1, d_2 = j_3 - j_2, ...d_r = N + j_1 - j_r, \\ 1 &\leqq j_1 < j_2 < j_3 < ...j_{r-1} < j_r \leqq N. \end{aligned} \tag{3}$$

The special form of the expression for the distance d_r reflects the periodic boundary conditions. The "wave number" k characterizes the irreducible representation of the translation group to which the stationary state belongs.

In identifying each stationary state in the standard form with a state in the B.H.–representation the amplitudes $c(d_1, d_2, ..., (d_r))$ will be written in terms of the x_j $(j = 1, ...r)$ of the corresponding B.H.–state. The question to be solved in this paper is whether or not this is always possible. The identification results in a set of linear equations for the elementary symmetric polynomials of the parameters x_j, which determine a unique solution for these parameters, apart from an irrelevant permutation.

The folowing section is devoted to description of the standard solution. Sections 3 and 4 contain an exposé of the Bethe-Hulthén solution for two different cases: First we give in section 3 a description of the normal case, i.e. the case in which there are no bound pairs, whereas in section 4 the solutions contain one bound pair. Solutions with more than one bound pair do not exist. These sections also give a method to determine the B.H.–parameters starting from the standard solution. In the final sections 5 and 6 we give some examples of our method.

2. Standard solutions for the Heisenberg chain.

The general form of the stationary states for given number of deviations r and a given representation of the translation group is given by:

$$\begin{aligned} \Psi_{k,r,n} &= \sum_{j=1}^{N-r+1} \sum_{d_1=1}^{N-j-r+2} \sum_{d_2=1}^{N-j-d_1-r+3} \cdots \sum_{d_{r-1}=1}^{N-j-d_1-d_2-...d_{r-2}} e^{ikj} \times \\ &\times c(d_1, d_2, d_3, ...d_{r-1}, (N - d_1 - d_2 - d_3 - ...d_{r-1})) \times \\ &\times \Phi_{j\ j+d_1\ j+d_1+d_2 ... j+d_1+d_2+...d_{r-1}} \end{aligned} \tag{4}$$

$$\begin{aligned} k &= \frac{2\pi\lambda}{N} \\ \lambda &= -\frac{N}{2}, -\frac{N}{2} + 1, ...\frac{N}{2} - 1\, (N \text{ even}) \\ \lambda &= -\frac{N-1}{2}, -\frac{N-3}{2}, ...\frac{N-1}{2}\, (N \text{ odd}) \end{aligned} \tag{5}$$

$$\begin{aligned} &\Phi_{j\ j+d_1\ j+d_1+d_2 ... j+d_1+d_2+...d_{r-1}} \\ &= S_j^- S_{j+d_1}^- S_{j+d_1+d_2}^- ... S_{j+d_1+d_2+...d_{r-1}}^- \mid + + + ... + > \end{aligned} \tag{6}$$

in which S_l^- $(l = 1, 2, 3...N)$ represents the lowering operators:

$$S_l^- = S_{l,x} - iS_{l,y} \ (l = 1, 2, 3...N). \tag{7}$$

The index n is used to distinguish between states corresponding with the same k and r.

The ferromagnetic or "pseudo vacuum" state is given by:

$$\mid + + + ... + > \tag{8}$$

$$S_{l,z} \mid +++....+ >= \frac{1}{2} \mid +++....+ > \quad (\hbar = 1) \quad (l = 1, 2, 3...N). \tag{9}$$

Periodic boundary conditions result in stationary states correponding with a given irreducible representation of the translation group:

$$\begin{aligned} & c(d_1, d_2, d_3, ...d_{r-1}, (N - d_1 - d_2 - d_3 - ...d_{r-1})) \\ = \; & e^{ikd_1} c(d_2, d_3, ...d_{r-1}, N - d_1 - d_2 - d_3 - ...d_{r-1}, (d_1)) \end{aligned} \tag{10}$$

$$d_r = N - d_1 - d_2 - d_3 - ...d_{r-1}. \tag{11}$$

The Hamiltonian of the Heisenberg chain (cf. W.J. Caspers e.a. Proc. SSPCM 2000 [5]) has the following symmetries: translational invariance, rotational invariance in spin space, time-reversal symmetry. The first symmetry implies that k is a good quantum number (already used in the representation of the stationary states in (4)). The second and third symmetry imply that we may restrict ourselves to the consideration of the stationary states with: $r = N/2$ (N even) or $r = (N-1)/2$ (N odd). Stationary states with different r may be derived from these by a rotation in spin space or time reversal, or a combination of both operations.

In the following sections we give the transformation of the standard solution into the B.H.–representation, which results in a set of equations for the **symmetrized products** of the parameters $x_n = e^{ik_n}$ in terms of the amplitudes of the standard solution:

$$c(d_1, d_2, d_3, ...d_{r-1}, (N - d_1 - d_2 - d_3 - ...d_{r-1})).$$

In section 3 this is done for the case without bound pairs and in section 4 for the case with one bound pair, covering in this way all possibilities.

The set of parameters x_n (or k_n) in the first case equals $r = [N/2]$, whereas in the second case the number of x_n to be determined equals: $r - 2 = N/2 - 2$ (N is even), one pair of deviations being bound into a pair on neighbouring positions in the chain, the pair having a resulting wave number $K_0 = -\pi$.

3. Bethe-Hulthén representation for states without a bound pair

The B.H.-solutions for given k and r has the form:

$$\begin{aligned} \Omega_{k,r,m} \; = \; & \sum_{1 \leqq j_1 < j_2 < j_3 < ... < j_r \leqq N} \sum_P e^{i(\sum_{l=1}^{r} k_{P(l)} j_l + \frac{1}{2}\sum_{l<p} \phi_{P(l)P(p)})} \times \\ & \times \Phi_{j_1 j_2 j_3 \cdots j_r} \end{aligned} \tag{12}$$

with:

$$2\cot(\frac{\phi_{lp}}{2}) = \cot(\frac{k_l}{2}) - \cot(\frac{k_p}{2}), \phi_{lp} = -\phi_{pl} \tag{13}$$

$$\sum_{l=1}^{r} k_l = k \tag{14}$$

in which formulas we have replaced the numbering of the deviations in (4) : $(j, j + d, j + d_1 + d_2, ...j + d_1 + d_2 + ... + d_{r-1})$ by: $(j_1, j_2, j_3, ...j_r)$. The symbol P stands for a permutation of the indices: $1, 2, ...r$. The individual wave numbers of the r deviations constitute the solution of a set of the B.H.-equations given in ref. 5. The question to be answered is whether or not such a B.H.-solution exists corresponding to a given standard solution. In general there are more solutions of the form (12) for given k and r , so we need an additional index m. The set of wave numbers does not contain identical $k_l = k_p$ because the state vector vanishes for this case,with the exception of $k_l = 0$, which may occur in arbitrary multiplicity. Apart from this condition there is the restriction that the wave numbers are real or appear in complex conjugated pairs.

Now we may rewrite (12) in the equivalent form:

$$\begin{aligned}\Omega_{k,r,m} &= e^{i\frac{1}{2}\sum_{l<p}\phi_{lp}} \sum_{1\leqq j_1<j_2<j_3<\ldots<j_r\leqq N} \sum_{P} \\ &\quad e^{i(\sum_{l=1}^{r} k_{P(l)}j_l+\sum_{p<q,P(p)>P(q)}\phi_{P(p)P(q)})}\Phi_{j_1j_2j_3\ldots j_r}. \end{aligned} \tag{15}$$

The alternative set of parameters:

$$e^{ik_l} = x_l, \quad e^{i\phi_{pq}} = -\frac{2x_p - x_px_q - 1}{2x_q - x_px_q - 1} \tag{16}$$

result in the following form of the state vector $\Theta_{k,r,m}$, apart from a trivial factor:

$$\begin{aligned}\Theta_{k,r,m} &= \sum_{1\leqq j_1<j_2<j_3<\ldots<j_r\leqq N} \sum_{P} \\ &\quad \prod_{l=1}^{r} x_{P(l)}^{j_l} \prod_{\substack{p<q \\ P(p)>P(q)}} \left(-\frac{2x_{P(p)} - x_{P(p)}x_{P(q)} - 1}{2x_{P(q)} - x_{P(p)}x_{P(q)} - 1}\right) \times \\ &\quad \times\Phi_{j_1j_2j_3\ldots j_r}\end{aligned} \tag{17}$$

The relation between the standard form of a solution and its B.H.-representation now follows from (4) and (17):

$$\begin{aligned} & c(d_1, d_2, d_3, \ldots d_{r-1}, (d_r)) \\ &= \sum_{P}\prod_{l=2}^{r} x_{P(l)}^{d_1+d_2+\ldots d_{l-1}} \prod_{\substack{p<q \\ P(p)>P(q)}} \left(-\frac{2x_{P(p)} - x_{P(p)}x_{P(q)} - 1}{2x_{P(q)} - x_{P(p)}x_{P(q)} - 1}\right) \\ d_r &= N - d_1 - d_2 - d_3\ldots - d_{r-1}.\end{aligned} \tag{18}$$

The coefficients $c(d_1, d_2, d_3, \ldots d_{r-1}, (d_r))$ of (15) may be rewritten:

$$c(d_1, d_2, d_3, \ldots d_{r-1}, (d_r)) = \left(\prod_{l<p} \frac{x_l - x_p}{2x_l - x_lx_p - 1}\right) \times B(d_1, d_2, d_3, \ldots d_{r-1}, (d_r)) \tag{19}$$

in which equation $B(d_1, d_2, d_3, \ldots d_{r-1}, (d_r))$ is a *symmetric polynomial* in terms of $(x_1, x_2, \ldots x_r)$, which may be expressed in terms of the elementary symmetric polynomials ESP:

$$\begin{aligned} A_1 &= -\sum_{l=1}^{r} x_l, \; A_2 = \sum_{1\leqq l<p\leqq N} x_lx_p, \; A_3 = -\sum_{1\leqq l<p<q\leqq N} x_lx_px_q, \ldots \\ A_r &= (-)^r x_1x_2\ldots x_r. \end{aligned} \tag{20}$$

The parameters x_l now obey the equation:

$$x^r + A_1x^{r-1} + A_2x^{r-2} + \ldots + A_r = 0. \tag{21}$$

The coefficients $c(d_1, d_2, d_3, ...d_{r-1}, (d_r))$ and $B(d_1, d_2, d_3, ...d_{r-1}, (d_r))$ obey identical periodic boundary conditions, which for the last read:

$$B(d_1, d_2, d_3, ...d_{r-1}, (d_r)) = e^{ikd_1} B(d_2, d_3, d_4...d_r, (d_1)). \tag{22}$$

Now it turns out that (18) gives the possibility to formulate linear equations for the ESP in terms of the amplitudes of a given solution in the standard form. It can be seen from the form of this equation that multiplication of the right member by one of the ESP results in linear combination of similar expressions for different sets of distances $(d'_1, d'_2, d'_3, ...d'_{r-1}, (N - d'_1 - d'_2 - d'_3 - ...d'_{r-1}))$. One of these r distances, however, may turn out to have the value 0. Such a set of distances could not possibly correspond to an amplitude in the standard representation, so it should be eliminated by taking a suitable linear combination of possible products of these right members and ESP. This procedure will be illustrated in section 5, in which some examples will be given.

4. Bethe-Hulthén representation for states with one bound pair

Formally a bound pair may be considered to correspond to a set of wave numbers:

$$k_1 = -\frac{\pi}{2} - i\delta, \quad k_2 = -\frac{\pi}{2} + i\delta \quad (\delta \Rightarrow \infty) \tag{23}$$

which should be combined with a regular set of finite, but generally complex wave numbers:

$$k_3, k_4, ...k_r. \tag{24}$$

This set contains only complex conjugated pairs or real k_l. There does not exist degeneracy in the spectrum of wave numbers, because this will result in a vanishing (non-physical) state vector, with the exception of zero wave numbers.

A complete set of wave numbers for this case is given by:

$$K_0 = -\pi, k_3, k_4, ...k_r \tag{25}$$

in which K_0 represents the bound pair.

It can be shown that there are no larger groups of bound deviations and that also the combination of 2 bound pairs is forbidden, because these configurations result in boundary conditions that cannot be fulfilled.

Now we may formulate an equivalent form of (15) for a stationary state with one bound pair, after we have determined the phase shift ϕ_{0l} for the interchange of the bound pair with one of the individual k_p $(p = 3, 4, ...r)$. This phase shift follows from (15) for the consecutive interchange of deviations with k_1 and k_2 of the form given in (23) and a third with wave number k_p, the resulting expression taken in the limit $\delta \Rightarrow \infty$:

$$\begin{aligned} e^{i\phi_{0l}} &= \lim_{\delta \Rightarrow \infty} \frac{2e^{i(-\frac{\pi}{2} - i\delta)} - e^{i(-\frac{\pi}{2} - i\delta)} e^{ik_l} - 1}{2e^{ik_l} - e^{i(-\frac{\pi}{2} - i\delta)} e^{ik_l} - 1} \frac{2e^{i(-\frac{\pi}{2} + i\delta)} - e^{i(-\frac{\pi}{2} + i\delta)} e^{ik_l} - 1}{2e^{ik_l} - e^{i(-\frac{\pi}{2} + i\delta)} e^{ik_l} - 1} \\ &= \frac{2 - e^{ik_l}}{-e^{ik_l}} \frac{-1}{2e^{ik_l} - 1} = \frac{2 - x_l}{x_l(2x_l - 1)} \quad (l = 3, 4, ...r). \end{aligned} \tag{26}$$

Using the value of the phase shift ϕ_{0l} defined by (26) we may formulate the expression of the wave function for a stationary state with one bound pair, omitting an irrelevant overall phase factor:

$$\begin{aligned}\Omega_{k,r,m} &= \sum_{l=1}^{r-1} \sum_{1\leqq j_1<j_2<\dots j_{l-1}<n_0<j_{l+2}-1<j_{l-1}-1<\dots j_r-1\leqq N-1} \\ & e^{-i\pi n_0}\sum_P \prod_{p<l} e^{-i\phi_{0p}} \prod_{l=3}^{r} e^{ik_{P(l)}j_l} \times \\ & \times \sum_{q<t,Pq)>P(t)}^{\prime} e^{i\phi_{P(q)P(t)}} \Phi_{j_1j_2\cdots j_{l-1}n_0j_{l+2}\cdots j_r} + \\ & + \sum_{2\leqq j_3<j_4<\dots j_{r-2}<n_0=N} \\ & \sum_P e^{-i\pi N} \prod_{p=3}^{r-2} e^{-i\phi_{0p}} \prod_{l=3}^{r} e^{ik_{P(l)}j_l} \times \\ & \times \sum_{q<t,Pq)>P(t)}^{\prime} e^{i\phi_{P(q)P(t)}} \Phi_{j_3j_4\dots j_{r-2}n_0=N}.\end{aligned} \quad (27)$$

Again we will identify the 2 representations of a stationary state in this case: On the one hand the representation given by (4) and on the other hand (27), translated in terms of the parameters $x_p (p = 3, 4, \dots r)$. The product of an ESP of the parameters and a coefficient in $\Omega_{k,r,m}$ corresponding with given distances between the deviations again results in a linear combination of coefficients, which enables us to find linear equations for the ESP. This will be illustrated by some examples in section 6.

5. Examples 1. States without a bound pair

Formula (18) will now be used to derive linear equations for the ESP with coefficients given by the amplitudes in the standard representation.

We may restrict ourselves to the cases $N = 2r$ (N even) or $N = 2r + 1$ (N odd). The case $r = 1$ will not be considered because it represents an unperturbed plane wave.

5.1. $r = 2, N = 4, 5$

There are 2 quantities in this case k and E, where E is the energy of the eigenstates of the Heisenberg Hamiltonian. They completely determine the solution, so further analysis is not necessary, if these two are known. But as a simple illustration of our analysis we write (18) for $r = 2$:

$$\begin{aligned}& c(d_1, (N - d_1)) \\ &= \sum_P x_{P(2)}^{d_1} \prod_{P(2)<P(1)} \left(-\frac{2x_{P(1)} - x_{P(1)}x_{P(2)} - 1}{2x_{P(2)} - x_{P(1)}x_{P(2)} - 1}\right) \\ &= x_2^{d_1} - x_1^{d_1} \frac{2x_2 - x_1x_2 - 1}{2x_1 - x_1x_2 - 1}, \quad d_1 = 1, 2.\end{aligned} \quad (28)$$

From this equation we derive in an easy way:

$$\begin{aligned} A_1c(1, (N - 1)) &= -x_1x_2c(0, (N)) - c(2, (N - 2)) \\ &= -A_2c(0, (N)) - c(2, (N - 2)) \\ A_1c(2, (N - 2)) &= -x_1x_2c(1, (N - 1)) - c(3, (N - 3)) \\ &= -A_2c(1, (N - 1)) - c(3, (N - 3)).\end{aligned} \quad (29)$$

The last one of these equations gives for $N = 4, 5$ respectively:

$$\begin{aligned} N &= 4 : A_1 c(2,(2)) + (e^{ik} + e^{3ik})c(1,(3)) = 0 \qquad (30) \\ N &= 5 : (A_1 + e^{3ik})c(2,(3)) + e^{ik}c(1,(4)) = 0. \qquad (31) \end{aligned}$$

These equations give the correct A_1 and A_2 for given amplitudes $c(d_1, (N - d_1))$ with the exception of the case $(N = 4, c(2,(2)) = 0)$, representing a state with a bound pair, for which we find the following values of the parameters:

$$\begin{gathered} k = -\pi \text{ or } A_2 = -1 \\ c(2,(2)) = 0 \end{gathered} \qquad (32)$$

which according to (30) should result in $c(1,(3)) = 0$, which means no solution. Formally we could construct a solution of (32) and (30) for which $A_1 = \infty$, but we prefer the description in terms of the modified B.H. scheme, analyzed in section 4.

5.2. $r = 3, N = 6, 7$

For a given solution in the standard form we know the parameter set (k, E), so we need one further equation to determine completely the B.H.-parameters in terms of the amplitudes $c(d_1, d_2, (N - d_1 - d_2))$. Formula (18) now takes the form:

$$c(d_1, d_2, (N - d_1 - d_2)) = \sum_P x_{P(2)}^{d_1} x_{P(3)}^{d_1+d_2} \prod_{\substack{1 \leqq l < p \leqq 3 \\ P(l) > P(p)}} \left(-\frac{2x_{P(l)} - x_{P(l)}x_{P(p)} - 1}{2x_{P(p)} - x_{P(l)}x_{P(p)} - 1}\right) \qquad (33)$$

from which we derive:

$$\begin{aligned} & A_1 c(d_1, d_2, (N - d_1 - d_2)) \\ = \; & -e^{ik}c(d_1 - 1, d_2, (N + 1 - d_1 - d_2)) - c(d_1 + 1, d_2 - 1, (N - d_1 - d_2)) + \\ & -c(d_1, d_2 + 1, (N - 1 - d_1 - d_2)) \qquad (34) \end{aligned}$$

$$\begin{aligned} & A_2 c(d_1, d_2, (N - d_1 - d_2)) \\ = \; & e^{ik}[c(d_1, d_2 - 1, (N + 1 - d_1 - d_2)) + c(d_1 - 1, d_2 + 1, (N - d_1 - d_2))] + \\ & +c(d_1 + 1, d_2, (N - 1 - d_1 - d_2)) \qquad (35) \end{aligned}$$

with:

$$A_1 = -(x_1 + x_2 + x_3) \quad A_2 = x_1 x_2 + x_1 x_3 + x_2 x_3. \qquad (36)$$

For $N = 7$ (34) and (35) give us 2 equations for $A_1$and A_2 taking $d_1 = d_2 = 2$. These equations contain as coefficients the amplitudes c without an entry 0. The coefficient of A_1 and A_2 in these equations: $c(2, 2, (3))$, is supposed to be unequal to 0. (N.B.: If this should be the case a modification of the procedure given for $N = 6$ will result in an alternative set of equations).

For $N = 6$ the situation is more interesting. First of all we may write down the equations for $d_1 = d_2 = 2$, making use of the periodic boundary conditions expressed by (10):

$$\begin{aligned} A_1 c(2, 2, (2)) &= -(e^{ik} + e^{3ik} + e^{5ik})c(1, 2, (3)) \\ A_2 c(2, 2, (2)) &= (e^{ik} + e^{3ik} + e^{5ik})c(1, 3, (2)). \qquad (37) \end{aligned}$$

These equations give us the ESP A_1 and A_2 in terms of the coefficients c, again with the exception of the case for which $c(2, 2, (2)) = 0$, i.e. for $k = \mp\pi/3$ and $\mp 2\pi/3$. In that case we have to make use of other values of d_1 and d_2 and the result will be a set of equations that contain coefficients c with one entry 0. These coefficients should be eliminated and the only possibility is that there appears only *one* such a coefficient in the eq. (34) and (35) and it should be the same in both equations. This may be realized with the following combination:

$$\begin{aligned} A_1 c(1,3,(2)) &= -e^{5ik} c(1,1,(4)) - e^{ik} c(0,3,(3)) \\ A_2 c(1,2,(3)) &= e^{ik} c(1,1,(4)) + e^{ik} c(0,3,(3)) \end{aligned} \tag{38}$$

which add up to:

$$A_1 c(1,3,(2)) + A_2 c(1,2,(3)) = (e^{ik} - e^{5ik}) c(1,1,(4)) \tag{39}$$

which cover all the possible cases for which $c(2,2,(2)) = 0$, again with the exception of the states with a bound pair, discussed in the section 4.

5.3. $r = 4, N = 8, 9$

Here we focus on the interesting case $N = 8$, for which we have the ESP:

$$\begin{aligned} A_1 &= -(x_1 + x_2 + x_3 + x_4) \;\; A_2 = x_1x_2 + x_1x_3 + x_1x_4 + x_2x_3 + x_2x_4 + x_3x_4 \\ A_3 &= -(x_1x_2x_3 + x_1x_2x_4 + x_1x_3x_4 + x_2x_3x_4). \end{aligned} \tag{40}$$

Apart from the quantities (k, E) we need two other relations between the $x_l (l = 1, ..4)$ to determine completely the the solution.

For this case we may derive from (18) for $r = 4$:

$$\begin{aligned} & A_1 c(d_1, d_2, d_3, (8 - d_1 - d_2 - d_3)) \\ = \; & -e^{ik} c(d_1 - 1, d_2, d_3, (9 - d_1 - d_2 - d_3)) - \\ & -c(d_1 + 1, d_2 - 1, d_3, (8 - d_1 - d_2 - d_3)) + \\ & -c(d_1, d_2 + 1, d_3 - 1, (8 - d_1 - d_2 - d_3)) + \\ & -c(d_1, d_2, d_3 + 1, (7 - d_1 - d_2 - d_3)) \end{aligned} \tag{41}$$

$$\begin{aligned} & A_2 c(d_1, d_2, d_3, (8 - d_1 - d_2 - d_3)) \\ = \; & e^{ik} [c(d_1, d_2 - 1, d_3, (9 - d_1 - d_2 - d_3)) + \\ & +c(d_1 - 1, d_2 + 1, d_3 - 1, (9 - d_1 - d_2 - d_3)) + \\ & +c(d_1 - 1, d_2, d_3 + 1, (8 - d_1 - d_2 - d_3))] + \\ & +c(d_1 + 1, d_2, d_3 - 1, (8 - d_1 - d_2 - d_3)) + \\ & +c(d_1 + 1, d_2 - 1, d_3 + 1, (7 - d_1 - d_2 - d_3)) + \\ & +c(d_1, d_2 + 1, d_3, (7 - d_1 - d_2 - d_3)) \end{aligned} \tag{42}$$

$$\begin{aligned} & A_3 c(d_1, d_2, d_3, (8 - d_1 - d_2 - d_3)) \\ = \; & -e^{ik} [c(d_1, d_2, d_3 - 1, (9 - d_1 - d_2 - d_3)) + \\ & +c(d_1, d_2 - 1, d_3 + 1, (8 - d_1 - d_2 - d_3)) + \\ & +c(d_1 - 1, d_2 + 1, d_3, (8 - d_1 - d_2 - d_3))] + \\ & -c(d_1 + 1, d_2, d_3, (7 - d_1 - d_2 - d_3)) \end{aligned} \tag{43}$$

from which it follows, again taking into account proper boundary conditions:

$$\begin{aligned}
A_1c(2,2,2,(2)) &= -(e^{ik}+e^{3ik}+e^{5ik}+e^{7ik})c(1,2,2,(3)) \\
A_2c(2,2,2,(2)) &= (e^{ik}+e^{3ik}+e^{5ik}+e^{7ik})c(1,2,3,(2))+ \\
&\quad +(e^{ik}+e^{3ik})c(1,3,1,(3)) \\
A_3c(2,2,2,(2)) &= -(e^{ik}+e^{3ik}+e^{5ik}+e^{7ik})c(1,3,2,(2))
\end{aligned} \tag{44}$$

from which we may derive the ESP, again with one possible exception, i.e.: $c(2,2,2,(2)) = 0$. For $c(2,2,2,(2)) \neq 0$ we may choose two of the equations (44) to find all $A_n (n = 1,...4)$ and consequently all $x_j (j = 1,...4)$.

For the other possibility we need two additional equations derived from (41-43). We first look for a coefficient c with one entry 0 that is generated at least two times, so that it may be eliminated. The simplest example that we could find is:

$$c(0,3,2,(3)) \tag{45}$$

that is generated in 2 ways:

$$\begin{aligned}
& A_1c(1,3,2,(2)) \\
= \;& -e^{ik}c(0,3,2,(3)) - c(1,4,1,(2) - c(1,3,3,(1)) \\
& A_3c(1,2,2,(3)) \\
= \;& -e^{ik}[c(1,2,1,(4)) + c(1,1,3,(3)) + c(0,3,2,(3))]
\end{aligned} \tag{46}$$

from which it immediately follows:

$$\begin{aligned}
& A_1c(1,3,2,(2)) - A_3c(1,2,2,(3)) \\
= \;& (e^{ik}-e^{7ik})c(1,1,3,(3)) + (e^{ik}-e^{5ik})c(1,2,1,(4)).
\end{aligned} \tag{47}$$

The next example is:

$$c(0,2,3,(3)) \tag{48}$$

which is generated again in two ways:

$$\begin{aligned}
A_1c(1,2,3,(2)) &= -e^{7ik}c(1,1,2,(4)) - (1+e^{2ik})c(1,3,2,(2)) - e^{ik}c(0,2,3,(3)) \\
A_2c(1,2,2,(3)) &= e^{ik}c(1,1,2,(4)) + (1+e^{2ik}+e^{4ik})c(1,3,2,(2))+ \\
&\quad +e^{ik}[c(0,2,3,(3)) + c(0,3,1,(4))].
\end{aligned} \tag{49}$$

But this introduces another non-vanshing amplitude with an entry 0:

$$c(0,3,1,(4)) \tag{50}$$

which in its turn also appears in:

$$A_1c(1,3,1,(3)) = -(1+e^{4ik})c(1,3,2,(2)) - (e^{ik}+e^{5ik})c(0,3,1,(4)). \tag{51}$$

Now a proper linear combination of the two equations (49) and (51) gives:

$$\begin{aligned}
& A_1[(e^{ik}+e^{5ik})c(1,2,3,(2)) + e^{ik}c(1,3,1,(3))] + A_2(e^{ik}+e^{5ik})c(1,2,2,(3)) \\
= \;& -(1-e^{2ik}+e^{4ik}-e^{6ik})c(1,1,2,(4)).
\end{aligned} \tag{52}$$

In a completely analogous way we find the equation:

$$\begin{aligned} & A_2(e^{ik}+e^{5ik})c(1,3,2,(2))+A_3[(e^{ik}+e^{5ik})c(1,2,3,(2))+e^{ik}c(1,3,1,(3))] \\ = \; & (1-e^{2ik}+e^{4ik}-e^{6ik})c(1,1,4,(2)). \end{aligned} \tag{53}$$

In the last example we consider:

$$c(0,4,2,(2)) \tag{54}$$

which is generated in the two equations:

$$\begin{aligned} & A_1c(1,1,4,(2)) \\ = \; & -e^{7ik}c(1,1,1,(5))-c(1,2,3,(2))-e^{ik}c(0,1,4,(3))-e^{2ik}c(0,4,2,(2)) \\ & A_2c(1,1,3,(3)) \\ = \; & (1+e^{2ik})c(1,2,3,(2))+ \\ & e^{ik}c(0,1,4,(3))+e^{ik}c(0,2,2,(4))+e^{2ik}c(0,3,4,(1))+e^{2ik}c(0,4,2,(2)). \end{aligned} \tag{55}$$

Now we have to look for a second way to generate:

$$c(0,2,2,(4)) \tag{56}$$

which is:

$$\begin{aligned} & A_3c(1,1,2,(4)) \\ = \; & -e^{ik}c(1,1,1,(5))-e^{2ik}c(1,2,3,(2))+ \\ & -e^{ik}c(0,2,2,(4))-e^{2ik}c(0,3,4,(1)) \end{aligned} \tag{57}$$

and it turns out that we may find a linear combination of the 2 equations of (55) together with (57) for which the amplitudes (54) and (56) as well as:

$$c(0,1,4,(3)) \text{ and } c(0,3,4,(1)) \tag{58}$$

disappear. This combination is:

$$A_1c(1,1,4,(2))+A_2c(1,1,3,(3))+A_3c(1,1,2,(4))=-(e^{ik}+e^{7ik})c(1,1,1,(5)). \tag{59}$$

N.B.1: As was stated before for the case ($N=8, r=4$) with $c(2,2,2,(2))=0$ we only need *two* additional equations to determine the complete set of ESP. We may choose two from the set (47),(52),(53),(59).

N.B. 2: Also in this case the states with a bound pair should be anlyzed according to the method of the next section.

General remark: We have not given explicit numeral examples of our analysis, but our method could be easily checked, also for it completeness, with the results we have for $N=4,6$ and 8 in other papers.

6. Examples 2. States with one bound pair

In this section we only consider 2 simple examples, i.e. *one* bound pair in combination with one or two "free" deviations, respectively corresponding with $r=3$ and $r=4$. For these cases we do not need the ESP to determine the wave numbers of these free deviations: k_3 ($r=3$) and the pair (k_3,k_4) ($r=4$).

6.1. $r = 3$

The state vector has the following form for this case, according to section 4:

$$\begin{aligned}
&\Omega_{k,3,m} \\
&= \sum_{1 \leqq n_0 < j_3 - 1 \leqq N-1} e^{-i\pi n_0} e^{ik_3 j_3} \Phi_{n_0 j_3} + \\
&+ e^{-i\phi_{03}} \sum_{2 \leqq j_3 < n_0 \leqq N} e^{-i\pi n_0} e^{ik_3 j_3} \Phi_{j_3 n_0} \qquad (60)
\end{aligned}$$

$$e^{i\phi_{03}} = \frac{2 - x_3}{x_3(2x_3 - 1)} \quad x_3 = e^{ik_3}. \qquad (61)$$

N.B.: The term $\Phi_{j_3 n_0 = N}$ represents a state with a bound pair on the sites $(N, N+1) \equiv (1, N)$.

Periodic boundary condition are now formulated in the following way. We take a term in the first sum with an index j_3 outside the prescribed range and identify this term with one in the second sum:

$$j_3 \Rightarrow j_3' + N, \; j_3' < n_0$$

and a term in the second summation with: $n_0 \Rightarrow n_0' + N, n_0' < j_3 - 1$, which may be identified with a term of the first summation:

$$n_0 \Rightarrow n_0' + N, n_0' < j_3 - 1.$$

This results in 2 boundary conditions:

$$x_3^N = \frac{x_3(2x_3 - 1)}{2 - x_3} \qquad (62)$$

and:

$$\frac{x_3(2x_3 - 1)}{2 - x_3} e^{-i\pi N} = 1 \qquad (63)$$

For even N this results in the equation:

$$x_3^N = \frac{x_3(2x_3 - 1)}{2 - x_3} = 1 \qquad (64)$$

and for odd N:

$$x_3^N = \frac{x_3(2x_3 - 1)}{2 - x_3} = -1 \qquad (65)$$

For the first case, N is even, (64) gives an acceptable solution:

$$x_3 = \mp 1 \qquad (66)$$

For odd values of N (65) gives a solution:

$$x_3 = e^{\mp i\pi/3} \quad N = 3, 9, 15, ... \qquad (67)$$

6.2. $r = 4$:

The state vector for this case according to section 4, takes the form:

$$\begin{aligned}\Omega_{k,4.m} &= \sum_{1\leqq n_0<j_3-1<j_4-1\leqq N-1} e^{-i\pi n_0}[e^{i(k_3j_3+k_4j_4)}+\\ &+e^{i\phi_{43}}e^{i(k_4j_3+k_3j_4)}]\Phi_{n_0j_3j_4}+\\ &+\sum_{1\leqq j_3<n_0<j_4-1\leqq N-1} e^{-i\pi n_0}[e^{-i\phi_{03}}e^{i(k_3j_3+k_4j_4)}+\\ &+e^{i\phi_{43}}e^{-i\phi_{04}}e^{i(k_4j_3+k_3j_4)}]\Phi_{j_3n_0j_4}+\\ &\sum_{1\leqq j_3<j_4<n_0\leqq N} e^{-i\pi n_0}e^{-i(\phi_{03}+\phi_{04})}[e^{i(k_3j_3+k_4j_4)}+\\ &+e^{i\phi_{43}}e^{i(k_4j_3+k_3j_4)}]\Phi_{j_3j_4n_0}\end{aligned} \tag{68}$$

In the first summation we now take the following index values, for j_4 outside the prescribed range:

$$j_4 \Rightarrow j_3' + N, j_3' < n_0, j_3 \Rightarrow j_4'$$

and find the following term:

$$e^{-i\pi n_0}[e^{ik_3N}e^{i\phi_{43}}e^{i(k_3j_3'+k_4j_4')} + e^{ik_4N}e^{i(k_4j_3'+k_3j_4')}]\Phi_{j_3'n_0j_4'}. \tag{69}$$

In the second summation we take:

$$j_4 \Rightarrow j_3' + N, j_3 \Rightarrow j_4', j_4' < n_0$$

which results in :

$$e^{-i\pi n_0}[e^{-i\phi_{04}}e^{ik_3N}e^{i\phi_{43}}e^{i(k_3j_3'+k_4j_4')} + e^{-i\phi_{03}}e^{ik_4N}e^{i(k_4j_3'+k_3j_4')}]\Phi_{j_3'j_4'n_0} \tag{70}$$

and finally in the third summation:

$$n_0 \Rightarrow n_0' + N, j_3 \Rightarrow j_3', j_4 \Rightarrow j_4'$$

which gives:

$$e^{-i\pi n_0}e^{-i\pi N}e^{-i(\phi_{03}+\phi_{04})}[e^{i(k_3j_3'+k_4j_4')} + e^{i\phi_{43}}e^{i(k_4j_3'+k_3j_4')}]\Phi_{n_0j_3'j_4'}. \tag{71}$$

Now we identify (69) with a term in the second summation in (68) , (70) with a term in the third summation and (71) with one in the first summation, which gives the following three equarions:

$$\begin{aligned}e^{ik_3N}e^{i\phi_{43}} &= e^{-i\phi_{03}} \quad e^{ik_4N} = e^{i\phi_{43}}e^{-i\phi_{04}}\\ e^{-i\pi N}e^{-i(\phi_{03}+\phi_{04})} &= 1\end{aligned} \tag{72}$$

For even N the translation into the variables x_3 and x_4 read:

$$\begin{aligned}x_3^N\frac{2x_4-x_3x_4-1}{2x_3-x_3x_4-1}\frac{2-x_3}{x_3(2x_3-1)} &= -1\\ x_4^N\frac{2x_3-x_3x_4-1}{2x_4-x_3x_4-1}\frac{2-x_4}{x_4(2x_4-1)} &= -1\\ \frac{2-x_3}{x_3(2x_3-1)}\frac{2-x_4}{x_4(2x_4-1)} &= 1\end{aligned} \tag{73}$$

from which it follows in an easy way:

$$(x_3x_4)^N = 1 \tag{74}$$

consistent with the periodic boundary conditions as a consequence of the fact that the total wave number should obey:

$$k = -\pi + k_3 + k_4 = 2\pi\lambda/N, \quad \lambda = -N/2, -N/2+1, ...N/2-1. \tag{75}$$

The last equation of (73) implies:

$$(1 - x_3x_4)[2(1 + x_3x_4) - (x_3 + x_4)] = 0 \tag{76}$$

which results in either one of the following possibilities:

$$\begin{aligned} x_3x_4 &= 1 \\ x_3 + x_4 &= 2(1 + x_3x_4) \end{aligned} \tag{77}$$

The wave numbers corresponding with x_3 and x_4 should be complex conjugated because of (75):

$$k_3 = \mathrm{Re(k_3)} - \mathrm{i}\delta \;\; \mathrm{k_4} = \mathrm{Re(k_3)} + \mathrm{i}\delta \;\; \delta \text{ is real} \tag{78}$$

The only acceptable solutions of (76) turn out to be:

$$\begin{aligned} x_3x_4 &= 1 \;\; k_3 = -k_4 \;\; k = -\pi \\ x_3x_4 &= -1 \;\; x_3 = -x_4 = \pm 1 \end{aligned} \tag{79}$$

The second equation results in:

$$k = 0 : K_0 = -\pi, k_3 = 0, \;\; k_4 = \pi \tag{80}$$

whereas the first one, together with the first equation of (73), results in a series of solutions given by:

$$x_3^{N-2}(2 - x_3) - (2x_3 - 1) = 0 \tag{81}$$

For the case $N = 8$ this equation takes the form:

$$x_3^7 - 2x_3^6 + 2x_3 - 1 = 0 \tag{82}$$

with solutions:

$$k_3 = (0, \;\; \mp 2.5864, \;\; \mp 1.4509, \;\; \mp i0.66662) \tag{83}$$

These results were already published in an earlier paper [4].

N.B.: In general one may treat the parts of the wave function (68) representing the individual wavelike deviations for a given position of the coupled pair, in the same way as the state vector for a solution without a bound pair, i.e. from these parts we may derive again a set of equations for the ESP of the x_l of all the deviations apart from those of the coupled pair.

References

[1] Caspers W J, Lulek T, Lulek B, Kuzma M and Wal A 2003 *Proc. of the 7th Inter. School on Theoretical Physics Symmetry and Structural Properties of Condensed Matter* eds. T. Lulek *et al* (Singapore: World Sci.) p 224

[2] Bethe H 1931 *Z. Phys.* **71** 205

[3] Hulthén L 1938 *Arkiv. Nat. Astron. Fys.* **26A** nr 1

[4] Caspers W J, Lulek B, Lulek T, Kuzma M and Wal A 2004 *Int. J. Mod. Phys.* **B18** 1277

[5] Caspers W J, Gragert P K H, Lulek T and Wal A 2001 *Proc. of the 6th Inter. School on Theoretical Physics Symmetry and Structural Properties of Condensed Matter* eds. T. Lulek *et al* (Singapore: World Sci.) p 285

Monomiality principle, Sheffer-type polynomials and the normal ordering problem

K A Penson[a], P Blasiak[a,b], G Dattoli[c], G H E Duchamp[d], A Horzela[b] and A I Solomon[a,e]

[a] Laboratoire de Physique Théorique de la Matière Condensée
Université Pierre et Marie Curie, CNRS UMR 7600
Tour 24 - 2ième ét., 4 pl. Jussieu, F 75252 Paris Cedex 05, France

[b] H. Niewodniczański Institute of Nuclear Physics, Polish Academy of Sciences
ul. Eliasza-Radzikowskiego 152, PL 31342 Kraków, Poland

[c] ENEA, Dipartimento Innovazione, Divisione Fisica Applicata
Centro Ricerche Frascati, Via E. Fermi 45, I 00044 Frascati, Rome, Italy

[d] Institut Galilée, LIPN, CNRS UMR 7030
99 Av. J.-B. Clement, F-93430 Villetaneuse, France

[e] The Open University, Physics and Astronomy Department
Milton Keynes MK7 6AA, United Kingdom

E-mail:
`penson@lptl.jussieu.fr`, `pawel.blasiak@ifj.edu.pl`, `dattoli@frascati.enea.it`, `gded@lipn-univ.paris13.fr`, `andrzej.horzela@ifj.edu.pl`, `a.i.solomon@open.ac.uk`

Abstract. We solve the boson normal ordering problem for $\left(q(a^\dagger)a + v(a^\dagger)\right)^n$ with arbitrary functions $q(x)$ and $v(x)$ and integer n, where a and $a^\dagger$ are boson annihilation and creation operators, satisfying $[a, a^\dagger] = 1$. This consequently provides the solution for the exponential $e^{\lambda(q(a^\dagger)a+v(a^\dagger))}$ generalizing the shift operator. In the course of these considerations we define and explore the monomiality principle and find its representations. We exploit the properties of Sheffer-type polynomials which constitute the inherent structure of this problem. In the end we give some examples illustrating the utility of the method and point out the relation to combinatorial structures.

1. Introduction

In this work we are concerned with one mode boson creation $a^\dagger$ and annihilation a operators satisfying the commutation relation

$$[a, a^\dagger] = 1. \tag{1}$$

We consider the normal ordering problem of a specific class of operator expressions. The normally ordered form of a general function $F(a, a^\dagger)$, denoted as $\mathcal{N}\left[F(a, a^\dagger)\right] \equiv F(a, a^\dagger)$, is defined by moving all the annihilation operators a to the right, using the commutation relation Eq.(1). Additionally we define the operation $:G(a, a^\dagger):$ which means normally order

$G(a,a^\dagger)$ *without* taking into account the commutation relations. Using the latter operation the normal ordering problem for $F(a,a^\dagger)$ is solved if we find an operator $G(a,a^\dagger)$ for which $F(a,a^\dagger) = {:}G(a,a^\dagger){:}$ is satisfied.

In 1974 J. Katriel [1] considered the normal ordering problem for powers of the number operator $N = a^\dagger a$ and pointed out its connection to combinatorics. It can be written as

$$\left(a^\dagger a\right)^n = \sum_{k=1}^{n} S(n,k)(a^\dagger)^k a^k, \tag{2}$$

where the integers $S(n,k)$ are the so called *Stirling numbers* of the second kind counting the number of ways of putting n different objects into k identical containers (none left empty).

In the coherent state representation $|z\rangle = e^{-|z|^2/2}\sum_{n=0}^{\infty}\frac{z^n}{\sqrt{n!}}|n\rangle$, where $a^\dagger a|n\rangle = n|n\rangle$, $\langle n|n'\rangle = \delta_{n,n'}$ and $a|z\rangle = z|z\rangle$ [2], we may write

$$\langle z|\left(a^\dagger a\right)^n|z\rangle = B(n,|z|^2), \tag{3}$$

where $B(n,x)$ are so called (exponential) *Bell polynomials* [3]

$$B(n,x) = \sum_{k=1}^{n} S(n,k)x^k. \tag{4}$$

Further development of this idea (see *e.g.* [4],[5]) provides the normally ordered expression for the exponential

$$e^{\lambda a^\dagger a} = \mathcal{N}\left[e^{\lambda a^\dagger a}\right] \equiv {:}\, e^{a^\dagger a(e^\lambda - 1)}\, {:}\,. \tag{5}$$

Here we shall extend these results in a very particular direction. We consider operators linear in annihilation a or creation $a^\dagger$ operators. More specifically, we consider operators which, say for linearity in a, have the form $q(a^\dagger)a + v(a^\dagger)$, where $q(x)$ and $v(x)$ are arbitrary functions. We shall find the normally ordered form of the n-th power $(q(a^\dagger)a + v(a^\dagger))^n$ and then of the exponential $e^{\lambda(q(a^\dagger)a+v(a^\dagger))}$. This is a far reaching generalization of the results of [6],[7],[8] where a special case of the operator $a^\dagger a + a^r$ was considered.

In this approach we use methods which are based on the monomiality principle [9]. First, using the methods of umbral calculus, we find a wide class of representations of the canonical commutation relation Eq.(1) in the space of polynomials. This establishes the link with Sheffer-type polynomials. Next the specific matrix elements of the above operators are derived and thereafter, with the help of coherent state representation, extended to the general form. Finally we obtain normally ordered expressions for these operators. It turns out that the exponential generating functions in the case of linear dependence on the annihilation (or creation) operator are of Sheffer-type, and that assures their convergence. In the end we give some examples with special emphasis put on their Sheffer-type origin and point out the relation to combinatorial structures.

2. Monomiality principle

Here we introduce the concept of monomiality which arises from the action of the multiplication and derivative operators on monomials. Next we provide a wide class of representations of that property in the framework of Sheffer-type polynomials. Finally we establish the correspondence to the occupation number representation.

2.1. Definition and general properties

Let us consider the Heisenberg-Weyl algebra satisfying the commutation relation

$$[P, M] = 1. \tag{6}$$

The simplest representation of Eq.(6) is by the derivative $P = D = \frac{d}{dx}$ and multiplication $M = X$ operators acting in the space of polynomials ($[D, X] = 1$). They are defined by their action on monomials

$$\begin{aligned} Xx^n &= x^{n+1}, \\ Dx^n &= nx^{n-1}. \end{aligned} \tag{7}$$

and subsequently on polynomials and formal power series.

Here we extend this framework. Suppose one wants to find the representations of Eq.(6) such that the action of M and P on certain polynomials $s_n(x)$ is analogous to the action of X and D on monomials. More specifically one searches for M and P and associated polynomials $s_n(x)$ (of degree n, $n = 0, 1, 2, ...$) which satisfy

$$\begin{aligned} Ms_n(x) = s_{n+1}(x), \\ Ps_n(x) = n\ s_{n-1}(x). \end{aligned} \tag{8}$$

The rule embodied in Eq.(8) is called the *monomiality principle*. The polynomials $s_n(x)$ are then called *quasi-monomials* with respect to operators M and P. These operators can be immediately recognized as raising and lowering operators acting on the $s_n(x)$'s.

The definition Eq.(8) implies some general properties besides fulfilling commutation relation of Eq.(6). First the operators M and P obviously satisfy Eq.(6). Further consequence of Eq.(8) is the eigenproperty of MP

$$MPs_n(x) = ns_n(x). \tag{9}$$

The polynomials $s_n(x)$ may be obtained through the action of M^n on s_0

$$s_n(x) = M^n s_0 \tag{10}$$

and consequently the exponential generating function of $s_n(x)$'s is

$$G(\lambda, x) \equiv \sum_{n=0}^{\infty} s_n(x) \frac{\lambda^n}{n!} = e^{\lambda M} s_0. \tag{11}$$

Also, if we write the quasimonomial $s_n(x)$ explicitly as

$$s_n(x) = \sum_{k=0}^{n} s_{n,k}\ x^k, \tag{12}$$

then

$$s_n(x) = \left(\sum_{k=0}^{n} s_{n,k}\ X^k \right) 1. \tag{13}$$

Several types of such polynomial sequences were studied recently using this monomiality principle [10],[11],[12],[13],[14].

2.2. Monomiality principle representations: Sheffer-type polynomials

Here we show that if $s_n(x)$ are of *Sheffer-type* then it is possible to give explicit representations of M and P. Conversely, if $M = M(X, D)$ and $P = P(D)$ then $s_n(x)$ of Eq.(8) are necessarily of Sheffer-type.

Properties of Sheffer-type polynomials are naturally handled within the so called *umbral calculus* [15],[16],[17]. They are usually defined through their exponential generating function

$$G(\lambda, x) = \sum_{n=0}^{\infty} s_n(x)\frac{\lambda^n}{n!} = A(\lambda)e^{xB(\lambda)} \tag{14}$$

with some (formal) power series $A(\lambda) = \sum_{n=0}^{\infty} a_n \frac{\lambda^n}{n!}$ and $B(\lambda) = \sum_{n=0}^{\infty} b_n \frac{\lambda^n}{n!}$ such that $b_0 = 0$, $b_1 \neq 0$ and $a_0 \neq 0$.

Here we focus on their ladder structure aspect and recall the following equivalent definition. Suppose we have a polynomial sequence $s_n(x)$, $n = 0, 1, 2, \ldots$ ($s_n(x)$ being a polynomial of degree n). It is called of a Sheffer A-type zero [18],[19] (which we shall call Sheffer-type) if there exists a function $f(x)$ such that

$$f(D)s_n(x) = ns_{n-1}(x). \tag{15}$$

Operator $f(D)$ plays the role of the lowering operator. This characterization is not unique, *i.e.* there are many Sheffer-type sequences $s_n(x)$ satisfying Eq.(15) for given $f(x)$. We can further classify them by postulating the existence of the associated raising operator. A general theorem [15],[20] states that a polynomial sequence $s_n(x)$ satisfying the monomiality principle Eq.(8) with an operator P given as a function of the derivative operator only $P = P(D)$ is *uniquely* determined by two (formal) power series $f(\lambda) = \sum_{n=0}^{\infty} f_n \frac{\lambda^n}{n!}$ and $g(\lambda) = \sum_{n=0}^{\infty} g_n \frac{\lambda^n}{n!}$ such that $f_0 = 0$, $f_1 \neq 0$ and $g_0 \neq 0$. The exponential generating function of $s_n(x)$ is then equal to

$$G(\lambda, x) = \sum_{n=0}^{\infty} s_n(x)\frac{\lambda^n}{n!} = \frac{1}{g(f^{-1}(\lambda))}\, e^{xf^{-1}(\lambda)}, \tag{16}$$

and their associated raising and lowering operators of Eq.(8) are given by

$$\begin{aligned} P &= f(D), \\ M &= \left[X - \tfrac{g'(D)}{g(D)}\right] \tfrac{1}{f'(D)}\ . \end{aligned} \tag{17}$$

Observe that X enters M only linearly and the order of X and D in $M(X, D)$ matters.

By direct calculation one may check that any pair M, P from Eq.(17) automatically satisfies Eq.(6). The detailed proof can be found in [15],[20].

Here are some examples we have obtained of representations of the monomiality principle Eq.(8) and their associated Sheffer-type polynomials:

a) $M(X, D) = 2X - D, \qquad P(D) = \frac{1}{2}D,$
   $s_n(x) = H_n(x)$ - Hermite polynomials;
   $G(\lambda, x) = e^{2\lambda x - \lambda^2}.$

b) $M(X, D) = -XD^2 + (2X - 1)D - X + 1, \qquad P(D) = -\sum_{n=1}^{\infty} D^n,$
   $s_n(x) = n!L_n(x)$ - where $L_n(x)$ are Laguerre polynomials;
   $G(\lambda, x) = \frac{1}{1-\lambda}e^{x\frac{\lambda}{\lambda-1}}.$

c) $M(X, D) = X\frac{1}{1-D}, \qquad P(D) = -\frac{1}{2}D^2 + D,$
   $s_n(x) = P_n(x)$ - Bessel polynomials [21];
   $G(\lambda, x) = e^{x(1-\sqrt{1-2\lambda})}.$

d) $M(X,D) = X(1+D), \qquad P(D) = \ln(1+D),$
$s_n(x) = B_n(x)$ - (exponential) Bell polynomials;
$G(\lambda, x) = e^{x(e^\lambda - 1)}$.

e) $M(X,D) = Xe^{-D}, \qquad P(D) = e^D - 1,$
$s_n(x) = x^{\underline{n}}$ - the lower factorial polynomials [22];
$G(\lambda, x) = e^{x\ln(1+\lambda)}$.

f) $M(X,D) = (X - \tan(D))\cos^2(D), \qquad P(D) = \tan(D),$
$s_n(x) = R_n(x)$ - Hahn polynomials [23];
$G(\lambda, x) = \frac{1}{\sqrt{1+\lambda^2}} e^{x \arctan(\lambda)}$.

g) $M(X,D) = X\frac{1+W_L(D)}{W_L(D)}D, \qquad P(D) = W_L(D),$
where $W_L(x)$ is the Lambert W function [24];
$s_n(x) = I_n(x)$ - the idempotent polynomials [3];
$G(\lambda, x) = e^{x\lambda e^\lambda}$.

2.3. Monomiality vs Fock space representations

We have already called operators M and P satisfying Eq.(8) the rising and lowering operators. Indeed, their action rises and lowers the index n of the quasimonomial $s_n(x)$. This resembles the property of creation $a^\dagger$ and annihilation a operators in the Fock space given by

$$\begin{aligned} a|n\rangle &= \sqrt{n}\,|n-1\rangle, \\ a^\dagger|n\rangle &= \sqrt{n+1}\,|n+1\rangle. \end{aligned} \tag{18}$$

These relations are almost the same as Eq.(8). There is a difference in coefficients. To make them analogous it is convenient to redefine the number states $|n\rangle$ as

$$\widetilde{|n\rangle} = \sqrt{n!}\,|n\rangle. \tag{19}$$

(Note that $\widetilde{|0\rangle} \equiv |0\rangle$).
Then the creation and annihilation operators act as

$$\begin{aligned} a\widetilde{|n\rangle} &= n\,\widetilde{|n-1\rangle}, \\ a^\dagger\widetilde{|n\rangle} &= \widetilde{|n+1\rangle}. \end{aligned} \tag{20}$$

Now this exactly mirrors the relation of Eq.(8). So, we make the correspondence

$$\begin{aligned} P &\longleftrightarrow a \\ M &\longleftrightarrow a^\dagger \\ s_n(x) &\longleftrightarrow \widetilde{|n\rangle}\,, \qquad n = 0,1,2,\ldots\ . \end{aligned} \tag{21}$$

We note that this identification is purely algebraic, *i.e.* we are concerned here only with the commutation relation Eqs.(1) or (6). We neither impose the scalar product in the space of polynomials nor consider the conjugacy properties of the operators. These properties are irrelevant for our proceeding discussion. We note only that they may be rigorously introduced, see *e.g.* [15].

3. Normal ordering via monomiality

In this section we shall exploit the correspondence of Section 2.3 to obtain the normally ordered expression of powers and exponential of the operators $a^\dagger q(a) + v(a)$ and (by the conjugacy

property) $q(a^\dagger)a + v(a^\dagger)$. To this end we shall apply the results of Section 2.2 to calculate some specific coherent state matrix elements of operators in question and then, through the exponential mapping property, we shall extend them to general matrix elements. In conclusion we shall also comment on other forms of linear dependence on a or $a^\dagger$.
We use the correspondence of Section 2.3 cast in the simplest form for $M = X$, $P = D$ and $s_n(x) = x^n$, *i.e.*

$$\begin{array}{ccl} D & \longleftrightarrow & a, \\ X & \longleftrightarrow & a^\dagger, \\ x^n & \longleftrightarrow & \widetilde{|n\rangle}\,, \qquad n = 0, 1, 2, \ldots\,. \end{array} \tag{22}$$

Now, recall the representation Eq.(17) of operators M and P in terms of X and D. Applying the correspondence of Eq.(22) it takes the form

$$\begin{array}{l} P(a) = f(a), \\ M(a, a^\dagger) = \left[a^\dagger - \frac{g'(a)}{g(a)}\right]\frac{1}{f'(a)}\,. \end{array} \tag{23}$$

From Eqs.(10),(12) and (13) we get

$$\left[M(a, a^\dagger)\right]^n |0\rangle = \sum_{k=0}^{n} s_{n,k}\,(a^\dagger)^k |0\rangle. \tag{24}$$

In the coherent state representation it yields

$$\langle z|\left[M(a, a^\dagger)\right]^n|0\rangle = s_n(z^*)\langle z|0\rangle. \tag{25}$$

Exponentiating $M(a, a^\dagger)$ and using Eq.(16) we obtain

$$\langle z|e^{\lambda M(a,a^\dagger)}|0\rangle = \frac{1}{g(f^{-1}(\lambda))}\, e^{z^* f^{-1}(\lambda)}\langle z|0\rangle. \tag{26}$$

Observe that in Eq.(25) we obtain Sheffer-type polynomials (modulus coherent states overlapping factor $\langle z|0\rangle$). This connection will be explored in detail in Section 4.2.
The result of Eq.(26) can be further extended to the general matrix element $\langle z|e^{\lambda M(a,a^\dagger)}|z'\rangle$. To this end recall the property $|z\rangle = e^{-\frac{|z|^2}{2}} e^{za^\dagger}|0\rangle$ and write

$$\begin{aligned} \langle z|e^{\lambda M(a,a^\dagger)}|z'\rangle &= e^{-\frac{1}{2}|z'|^2}\langle z|e^{\lambda M(a,a^\dagger)}e^{z'a^\dagger}|0\rangle \\ &= e^{-\frac{1}{2}|z'|^2}\langle z|e^{z'a^\dagger}e^{-z'a^\dagger}e^{\lambda M(a,a^\dagger)}e^{z'a^\dagger}|0\rangle \\ &= e^{z^*z'-\frac{1}{2}|z'|^2}\langle z|e^{-z'a^\dagger}e^{\lambda M(a,a^\dagger)}e^{z'a^\dagger}|0\rangle. \end{aligned}$$

Next, using the exponential mapping formula $e^{-xa^\dagger}F(a, a^\dagger)e^{xa^\dagger} = F(a + x, a^\dagger)$ (see [2],[25]) we arrive at

$$\begin{aligned} \langle z|e^{\lambda M(a,a^\dagger)}|z'\rangle &= e^{z^*z'-\frac{1}{2}|z'|^2}\langle z|e^{\lambda M(a+z',a^\dagger)}|0\rangle \\ &= e^{z^*z'-\frac{1}{2}|z'|^2}\langle z|e^{\lambda\left(a^\dagger - \frac{g'(a+z')}{g(a+z')}\right)\frac{1}{f'(a+z')}}|0\rangle. \end{aligned}$$

Now we are almost ready to apply Eq.(26) to evaluate the matrix element on the r.h.s. of the above equation. Before doing so we have to appropriately redefine functions $f(x)$ and $g(x)$ in the following way (z' - fixed)

$$\begin{array}{lcl} f(x) & \rightarrow & \tilde{f}(x) = f(x + z') - f(z'), \\ g(x) & \rightarrow & \tilde{g}(x) = g(x + z')/g(z'). \end{array}$$

Then $\tilde{f}(0)=0$, $\tilde{f}'(0)\neq 0$ and $\tilde{g}(0)=1$ as required by the Sheffer property for $\tilde{f}(x)$ and $\tilde{g}(x)$. Observe that these conditions are not fulfilled by $f(x+z')$ and $g(x+z')$. This step imposes (analytical) constraints on z', *i.e.* it is valid whenever $\tilde{f}'(z')\neq 0$ (although, we note that for formal power series approach this does not present any difficulty). Now we can write

$$\begin{aligned}
\langle z|e^{\lambda\left(a^\dagger-\frac{g'(a+z')}{g(a+z')}\right)\frac{1}{f'(a+z')}}|0\rangle &= \langle z|e^{\lambda\left(a^\dagger-\frac{\tilde{g}'(a)}{\tilde{g}(a)}\right)\frac{1}{\tilde{f}'(a)}}|0\rangle \\
&= \frac{1}{\tilde{g}(\tilde{f}^{-1}(\lambda))}\, e^{z^*\tilde{f}^{-1}(\lambda)}\langle z|0\rangle.
\end{aligned}$$

By going back to the initial functions $f(x)$ and $g(x)$ this readily gives the final result

$$\langle z|e^{\lambda M(a,a^\dagger)}|z'\rangle = \frac{g(z')}{g(f^{-1}(\lambda+f(z')))}e^{z^*[f^{-1}(\lambda+f(z'))-z']}\langle z|z'\rangle, \tag{27}$$

where $\langle z|z'\rangle = e^{z^*z'-\frac{1}{2}|z'|^2-\frac{1}{2}|z|^2}$ is the coherent states overlapping factor.
To obtain the normally ordered form of $e^{\lambda M(a,a^\dagger)}$ we apply the crucial property of the coherent state representation $\langle z|F(a,a^\dagger)|z'\rangle = \langle z|z'\rangle\ G(z^*,z') \Longrightarrow F(a,a^\dagger) = :G(a^\dagger,a):$ (see [2]). Then Eq.(27) provides the central result

$$e^{\lambda M(a,a^\dagger)} = :e^{a^\dagger[f^{-1}(\lambda+f(a))-a]}\frac{g(a)}{g(f^{-1}(\lambda+f(a)))}: . \tag{28}$$

Let us point out again that $a^\dagger$ appears linearly in $M(a,a^\dagger)$, see Eq.(23).
For simplicity we put

$$\begin{aligned}
q(x) &= \frac{1}{f'(x)}, \\
v(x) &= \frac{g'(x)}{g(x)}\frac{1}{f'(x)},
\end{aligned}$$

and define

$$\begin{aligned}
T(\lambda,x) &= f^{-1}(\lambda+f(x)), \\
G(\lambda,x) &= \frac{g(x)}{g(T(\lambda,x))}.
\end{aligned}$$

This allows us to rewrite the main normal ordering formula of Eq.(28) as

$$e^{\lambda\left(a^\dagger q(a)+v(a)\right)} = :e^{a^\dagger[T(\lambda,a)-a]}\ G(\lambda,a): \tag{29}$$

where the functions $T(\lambda,x)$ and $G(\lambda,x)$ fulfill the following differential equations

$$\frac{\partial T(\lambda,x)}{\partial\lambda} = q(T(\lambda,x))\ , \qquad T(0,x)=x\ , \tag{30}$$

$$\frac{\partial G(\lambda,x)}{\partial\lambda} = v(T(\lambda,x))\cdot G(\lambda,x)\ , \qquad G(0,x)=1\ . \tag{31}$$

Eq.(29) in the coherent state representation takes the form

$$\langle z'|e^{\lambda\left(a^\dagger q(a)+v(a)\right)}|z\rangle = \langle z'|z\rangle\ e^{z'^*[T(\lambda,z)-z]}G(\lambda,z). \tag{32}$$

We conclude by making a comment on other possible forms of linear dependence on a or $a^\dagger$. By hermitian conjugation of Eq.(29) we obtain the expression for the normal form of $e^{\lambda\left(q(a^\dagger)a+v(a)\right)}$. This amounts to the formula

$$e^{\lambda\left(q(a^\dagger)a+v(a)\right)} = \,:G(\lambda, a^\dagger)e^{[T(\lambda,a^\dagger)-a^\dagger]a}: \tag{33}$$

with the same differential equations Eqs.(30) and (31) for functions $T(\lambda, x)$ and $G(\lambda, x)$. In the coherent state representation it yields

$$\langle z'|e^{\lambda\left(q(a^\dagger)a+v(a)\right)}|z\rangle = \langle z'|z\rangle\; G(\lambda, z'^*)e^{[T(\lambda,z'^*)-z'^*]z} \tag{34}$$

We also note that all other operators linearly dependent on a or $a^\dagger$ may be written in just considered forms using $[a, F(a, a^\dagger)] = \frac{\partial}{\partial a^\dagger}\,F(a, a^\dagger)$ which yields $aq(a^\dagger) + v(a^\dagger) = q(a^\dagger)a + q'(a^\dagger) + v(a^\dagger)$ and $q(a)a^\dagger + v(a) = a^\dagger q(a) + q'(a) + v(a)$.
Observe that from analytical point of view certain limitations on the domains of z, z' and λ should be put in some specific cases (locally around zero all the formulas hold true). Also we point out the fact that functions $q(x)$ and $v(x)$ (or equivalently $f(x)$ and $g(x)$) may be taken as the formal power series.
In the end we note that the reverse process, *i.e.* derivation of the normally ordered form from the substitution theorem, is also possible, see [26].

4. Sheffer-type polynomials and normal ordering: Examples

We now proceed to examples putting special emphasis on their Sheffer-type origin.

4.1. Examples

We start with enumerating some examples of the evaluation of the coherent state matrix elements of Eqs.(25) and (32). We choose the $M(a, a^\dagger)$'s as in the list a) - g) in Section 2.2:

a) $\langle z|(-a+2a^\dagger)^n|0\rangle = H_n(z^*)\langle z|0\rangle$, Hermite polynomials;
$\langle z|e^{\lambda(-a+2a^\dagger)}|z'\rangle = e^{\lambda(2z^*-z')-\lambda^2}\langle z|z'\rangle$.

b) $\langle z|\left[-a^\dagger a^2+(2a^\dagger-1)a-a^\dagger+1\right]^n|0\rangle = n!L_n(z^*)\langle z|0\rangle$, Laguerre polynomials;
$\langle z|e^{\lambda[-a^\dagger a^2+(2a^\dagger-1)a-a^\dagger+1]}|z'\rangle = \frac{z'^2-\lambda z'+1}{(1-z')(1-\lambda(z'-1))}e^{z^*\lambda\frac{(1-z')^2}{(\lambda(1-z')-1)}}\langle z|z'\rangle$.

c) $\langle z|\left(a^\dagger\frac{1}{1-a}\right)^n|0\rangle = P_n(z^*)\langle z|0\rangle$, Bessel polynomials;
$\langle z|e^{\lambda\left(a^\dagger\frac{1}{1-a}\right)}|z'\rangle = e^{z^*[1-\sqrt{1-2(\lambda+z'-\frac{1}{2}z'^2)}-z']}\langle z|z'\rangle$.

d) $\langle z|(a^\dagger a+a^\dagger)^n|0\rangle = B_n(z^*)\langle z|0\rangle$, Bell polynomials;
$\langle z|e^{\lambda(a^\dagger a+a^\dagger)}|z'\rangle = e^{z^*(z'+1)(e^\lambda-1)}\langle z|z'\rangle$.

e) $\langle z|(a^\dagger e^{-a})^n|0\rangle = (z^*)^{\underline{n}}\langle z|0\rangle$, the lower factorial polynomials;
$\langle z|e^{\lambda(a^\dagger e^{-a})}|z'\rangle = e^{z^*[\ln(e^{z'}+\lambda)-z']}\langle z|z'\rangle$.

f) $\langle z|\left[(a^\dagger-\tan(a))\cos^2(a)\right]^n|0\rangle = R_n(z^*)\langle z|0\rangle$, Hahn polynomials;
$\langle z|e^{\lambda(a^\dagger-\tan(a))\cos^2(a)}|z'\rangle = \frac{\cos[\arctan(\lambda+\tan(z'))]}{\cos(z')}e^{z^*[\arctan(\lambda+\tan(z'))-z']}\langle z|z'\rangle$.

g) $\langle z|\left[a^\dagger\frac{1+W_L(a)}{W_L(a)}a\right]^n|0\rangle = I_n(z^*)\langle z|0\rangle$, the idempotent polynomials;
$\langle z|e^{\lambda a^\dagger\frac{1+W_L(a)}{W_L(a)}a}|z'\rangle = e^{z^*[\lambda e^{\lambda+W_L(z')}+z'(e^\lambda-1)]}\langle z|z'\rangle$.

Note that for $z'=0$ we obtain the exponential generating functions of appropriate polynomials multiplied by the coherent states overlapping factor $\langle z|0\rangle$, see Eq.(27).
These examples show how the Sheffer-type polynomials and their exponential generating functions arise in the coherent state representation. This generic structure is the consequence of Eqs.(25) and (27) or in general Eqs.(32) or (34) and it will be investigated in more detail now. Afterwards we shall provide more examples of combinatorial origin.

4.2. Sheffer polynomials and normal ordering
First recall the definition of the family of Sheffer-type polynomials $s_n(z)$ defined through the exponential generating function (see Eq.(14)) as

$$G(\lambda,z)=\sum_{n=0}^{\infty}s_n(z)\frac{\lambda^n}{n!}=A(\lambda)\;e^{zB(\lambda)} \tag{35}$$

where functions $A(\lambda)$ and $B(\lambda)$ satisfy: $A(0)\neq 0$ and $B(0)=0$, $B'(0)\neq 0$.
Returning to normal ordering, recall that the coherent state expectation value of Eq.(33) is given by Eq.(34). When one fixes z' and takes λ and z as indeterminates, then the r.h.s. of Eq.(34) may be read off as an exponential generating function of Sheffer-type polynomials defined by Eq.(35). The correspondence is given by

$$A(\lambda)=g(\lambda,z'^*), \tag{36}$$
$$B(\lambda)=\left[T(\lambda,z'^*)-z'^*\right]. \tag{37}$$

This allows us to make the statement that the coherent state expectation value $\langle z'|...|z\rangle$ of the operator $\exp\left[\lambda(q(a^\dagger)a+v(a^\dagger))\right]$ for any fixed z' yields (up to the overlapping factor $\langle z'|z\rangle$) the exponential generating function of a certain sequence of Sheffer-type polynomials in the variable z given by Eqs.(36) and (37). The above construction establishes the connection between the coherent state representation of the operator $\exp\left[\lambda(q(a^\dagger)a+v(a^\dagger))\right]$ and a family of Sheffer-type polynomials $s_n^{(q,v)}(z)$ related to q and v through

$$\langle z'|e^{\lambda[q(a^\dagger)a+v(a^\dagger)]}|z\rangle=\langle z'|z\rangle\left(1+\sum_{n=1}^{\infty}s_n^{(q,v)}(z)\frac{\lambda^n}{n!}\right), \tag{38}$$

where explicitly (again for z' fixed):

$$s_n^{(q,v)}(z)=\langle z'|z\rangle^{-1}\langle z'|\left[q(a^\dagger)a+v(a^\dagger)\right]^n|z\rangle. \tag{39}$$

We observe that Eq.(39) is an extension of the seminal formula of Katriel [1],[27] where $v(x)=0$ and $q(x)=x$. The Sheffer-type polynomials are in this case Bell polynomials expressible through the Stirling numbers of the second kind Eq.(4).
Having established relations leading from the normal ordering problem to Sheffer-type polynomials we may consider the reverse approach. Indeed, it turns out that for any Sheffer-type sequence generated by $A(\lambda)$ and $B(\lambda)$ one can find functions $q(x)$ and $v(x)$ such that the coherent state expectation value $\langle z'|\exp\left[\lambda(q(a^\dagger)a+v(a^\dagger))\right]|z\rangle$ results in a corresponding exponential generating function of Eq.(35) in indeterminates z and λ (up to the overlapping factor $\langle z'|z\rangle$ and z' fixed). Appropriate formulas can be derived from Eqs.(36) and (37) by substitution into Eqs.(30) and (31):

$$q(x)\;=\;B'(B^{-1}(x-z'^*)), \tag{40}$$
$$v(x)\;=\;\frac{A'(B^{-1}(x-z'^*))}{A(B^{-1}(x-z'^*))}. \tag{41}$$

One can check that this choice of $q(x)$ and $v(x)$, if inserted into Eqs. (30) and (31), results in

$$T(\lambda, x) = B(\lambda + B^{-1}(x - z'^*)) + z'^*, \tag{42}$$

$$g(\lambda, x) = \frac{A(\lambda + B^{-1}(x - z'^*))}{A(B^{-1}(x - z'^*))}, \tag{43}$$

which reproduce

$$\langle z'|e^{\lambda[q(a^\dagger)a+v(a^\dagger)]}|z\rangle = \langle z'|z\rangle A(\lambda) e^{zB(\lambda)}. \tag{44}$$

The result summarized in Eqs.(36) and (37) and in their 'dual' forms Eqs.(40)-(43) provide us with a considerable flexibility in conceiving and analyzing a large number of examples.

4.3. Combinatorial examples

In this section we will work out examples illustrating how the exponential generating function $G(\lambda) = \sum_{n=0}^{\infty} a_n \frac{\lambda^n}{n!}$ of certain combinatorial sequences $(a_n)_{n=0}^{\infty}$ appear naturally in the context of boson normal ordering. To this end we shall assume specific forms of $q(x)$ and $v(x)$ thus specifying the operator that we exponentiate. We then give solutions to Eqs.(30) and (31) and subsequently through Eqs.(36) and (37) we write the exponential generating function of combinatorial sequences whose interpretation will be given.

a) Choose $q(x) = x^r$, $r > 1$ (integer), $v(x) = 0$ (which implies $g(\lambda, x) = 1$). Then $T(\lambda, x) = x\left[1 - \lambda(r-1)x^{r-1}\right]^{\frac{1}{1-r}}$. This gives

$$\mathcal{N}\left[e^{\lambda(a^\dagger)^r a}\right] \equiv : \exp\left[\left(\frac{a^\dagger}{(1-\lambda(r-1)(a^\dagger)^{r-1})^{\frac{1}{r-1}}} - 1\right) a\right] :$$

as the normally ordered form. Now we take $z' = 1$ in Eqs.(36) and (37) and from Eq.(44) we obtain

$$\langle 1|z\rangle^{-1}\langle 1|e^{\lambda(a^\dagger)^r a}|z\rangle = \exp\left[z\left(\frac{1}{(1-\lambda(r-1))^{\frac{1}{r-1}}} - 1\right)\right],$$

which for $z = 1$ generates the following sequences:

$$\begin{array}{ll} r=2: & a_n = 1, 1, 3, 13, 73, 501, 4051, ... \\ r=3: & a_n = 1, 1, 4, 25, 211, 2236, 28471, ... \end{array} \quad , \text{ etc.}$$

These sequences enumerate r-ary forests [28],[29],[30].

b) For $q(x) = x\ln(ex)$ and $v(x) = 0$ (implying $g(\lambda, x) = 1$) we have $T(\lambda, x) = e^{e^\lambda - 1}x^{e^\lambda}$. This corresponds to

$$\mathcal{N}\left[e^{\lambda a^\dagger \ln(ea^\dagger) a}\right] \equiv : \exp\left[\left(e^{e^\lambda - 1}(a^\dagger)^{e^\lambda} - 1\right) a\right] : ,$$

whose coherent state matrix element with $z' = 1$ is equal to

$$\langle 1|z\rangle^{-1}\langle 1|e^{\lambda a^\dagger \ln(ea^\dagger) a}|z\rangle = \exp\left[z\left(e^{e^\lambda - 1} - 1\right)\right],$$

which for $z = 1$ generates $a_n = 1, 1, 3, 12, 60, 385, 2471, ...$ enumerating partitions of partitions [29], [28], [30].

The following two examples refer to the reverse procedure, see Eqs.(40)-(43). We choose first a Sheffer-type exponential generating function and deduce $q(x)$ and $v(x)$ associated with it.

c) $A(\lambda) = \frac{1}{1-\lambda}$, $B(\lambda) = \lambda$, see Eq.(35). This exponential generating function for $z = 1$ counts the number of arrangements $a_n = n!\sum_{k=0}^{n}\frac{1}{k!} = 1, 2, 5, 65, 326, 1957, ...$ of the set of n elements [3]. The solutions of Eqs.(40) and (41) are: $q(x) = 1$ and $v(x) = \frac{1}{2-x}$. In terms of bosons it corresponds to

$$\mathcal{N}\left[e^{\lambda\left(a+\frac{1}{2-a^\dagger}\right)}\right] \equiv :\frac{2-a^\dagger}{2-a^\dagger-\lambda}e^{\lambda a}: = \frac{2-a^\dagger}{2-a^\dagger-\lambda}e^{\lambda a}.$$

d) For $A(\lambda) = 1$ and $B(\lambda) = 1 - \sqrt{1-2\lambda}$ one gets the exponential generating function of the Bessel polynomials [21]. For $z = 1$ they enumerate special paths on a lattice [31]. The corresponding sequence is $a_n = 1, 1, 7, 37, 266, 2431, ...$. The solutions of Eqs.(40) and (41) are: $q(x) = \frac{1}{2-x}$ and $v(x) = 0$. It corresponds to

$$\mathcal{N}\left[e^{\lambda\frac{1}{2-a^\dagger}a}\right] \equiv :e^{\left(1-\sqrt{(2-a^\dagger)-2\lambda}\right)a}:$$

in the boson formalism.

These examples show that any combinatorial structure which can be described by a Sheffer-type exponential generating function can be cast in boson language. This gives rise to a large number of formulas of the above type which put them in a quantum mechanical setting.

5. Conclusions

We have solved the boson normal ordering problem for the powers and exponentials of the operators linear either in the creation or in the annihilation operator, *i.e.* $(q(a^\dagger)a + v(a^\dagger))^n$ and $e^{\lambda(q(a^\dagger)a+v(a^\dagger))}$ where $q(x)$ and $v(x)$ are arbitrary functions. This was done by the use of umbral calculus methods [15] in finding representations of the monomiality principle (*i.e.* representations of the Heisenberg-Weyl algebra in the space of polynomials) and application of the coherent state methodology. Moreover, we have established one-to-one connection between this class of normal ordering problems and the family of Sheffer-type polynomials and provided a wealth of combinatorial examples.

References

[1] Katriel J 1974 *Lett. Nuovo Cimento* **10** 565

[2] Klauder J R and Skagerstam B-S 1985 *Coherent States. Application in Physics and Mathematical Physics* (Singapore: World Scientific)

[3] Comtet L 1974 *Advanced Combinatorics* (Dordrecht: Reidel)

[4] Katriel J 2002 *J. Opt. B: Quantum Semiclass. Opt.* **4** S200

[5] Blasiak P 2005 *Concepts of Physics* **2** (PhD Thesis, Institut of Nuclear Physics in Kraków and University of Paris VI, arXiv:quant-ph/0507206)

[6] Mikhailov V V 1983 *J. Phys. A : Math. Gen.* **16** 3817

[7] Mikhailov V V 1985 *J. Phys. A : Math. Gen.* **18** 231

[8] Katriel J 1983 *J. Phys. A : Math. Gen.* **16** 4171

[9] Blasiak P and Dattoli G and Horzela A and Penson K A 2005 arXiv:quant-ph/0504009

[10] Dattoli G and Ottaviani P L and Torre A and Vasquez L 1997 *Riv. Nuovo Cim.* **20** 1

[11] Dattoli G Torre A and Mazzacurati G 1997 *Nuov. Cim. B* **112** 133

[12] Dattoli G 1999 *Hermite-Bessel and Laguerre-Bessel functions: a by-product of the monomiality principle* (Advanced Special Functions and Applications) ed D Cocolicchio, G Dattoli and H M Srivastava (Rome: Aracne Rditrice)

[13] Dattoli G and Srivastava H M and Cesarano C 2001 *Appl. Math. Comput.* **124** 117

[14] Cesarano C 2000 Monomiality principle and Legendre polynomials (Advanced Special Functions and Integration Methods) ed G Dattoli, H M Srivastava and C Cesarano (Rome: Aracne Editrice)

[15] Roman S 1984 *The Umbral Calculus* (Orlando: Academic Press)

[16] Roman S and Rota G -C 1978 *Adv. Math.* **27** 95

[17] Di Bucchianico A 1998 *Introduction to Umbral Calculus* (http://www.win.tue.nl/~sandro/)

[18] Sheffer I M 1939 *Duke Math. Journal* **5** 590

[19] Rainville E D 1965 *Special Functions* (New York: The Macmillan Company)

[20] Ben Cheikh Y 2003 *Appl. Math. Comput.* **141** 63

[21] Grosswald E 1978 *Bessel Polynomials* (Berlin: Springer)

[22] Turbiner A and Smirnov Y 1995 *Mod. Phys. Lett. A* **10** 1795

[23] Bender C M 1991 *Solution of operator equations of motion*(Rigorous Results in Quantum Dynamics) ed J Dittrich and P Exner (Singapore: World Scientific) pp 99-112

[24] Corless R M and Gonnet G H and Hare D E G and Jeffrey D J and Knuth D E 1996 *Adv. Comp. Math.* **5** 329

[25] Louisell W H 1964 *Radiation and Noise in Quantum Electronics* (New York: McGrow-Hill Co.)

[26] Blasiak P and Horzela A and Penson K A and Duchamp G H E and Solomon A I 2005 *Phys. Lett. A* **338** 108

[27] Katriel J 2000 *Phys. Lett. A* **237** 159

[28] Sloane N J A 2005 *Encyclopedia of Integer Sequences* (http://www.research.att.com/~njas/sequences)

[29] Stanley R P 1999 Enumerative Combinatorics (Cambridge: University Press)

[30] Flajolet P and Sedgewick R 2005 *Analytic Combinatorics* (http://algo.inria.fr/flajolet/Publications/books.html)

[31] Pittman J 1999 *A lattice path model for the Bessel polynomials* (U.C. Berkeley, USA) nr 551

Finite-Size scaling at Yang-Lee singularities of 2D discrete spin models

Tomasz Wydro[(1)] **and John F McCabe**[(2)]

[(1)] Laboratoire de Physique Moléculaire et des Collisions Université de Metz 1 bvd Arago, 57078 Metz, France

[(2)] 412 Morris Ave., # 34, Summit, NJ 07901, USA

E-mail: wydro@sciences.univ-metz.fr

E-mail: jfmccabe2@earthlink.net

Abstract. We numerically study the Hamiltonian limits of the two dimensional (2D) Ising and 3-state Potts models in complex magnetic fields. From the Phenomenological Renormalization Group, we find the critical field values associated with the Yang-Lee singularity (YLS) in these models. We also determine the low-lying part of the excitation spectrum at the YLS. Finally, we compare the resulting patterns of energy levels to predictions for conformal non-unitary minimal models.

1. Introduction

In their seminal papers [1, 2], Yang and Lee related the analytical properties of the partition function to phase properties of statistical systems. Yang and Lee showed that grand canonical partition functions of lattice gases have zeros at complex fugacities z. Yang and Lee also showed that these zeros become dense at edges of a gap about the positive z axis and that the gap disappears at the transition temperature. The edges of this gap have become known as the Yang–Lee edge singularities (YLSs).

While Yang and Lee showed that the zeros of the partition function have a role in ordinary phase transitions, it has also become clear that YLSs themselves are critical points [3]. Thus, YLSs became other critical points of interest in studies of statistical systems. Indeed, YLSs appear in many magnetic systems for complex values of the magnetic field. Furthermore, it is now clear that the effects of YLSs are directly accessible via experiments [4].

The critical nature of YLSs has raised considerable interest. The interest is even larger in 2D where there is a classification of many critical points. A large class of 2D statistical models have critical points that are described by the conformal field theories (CFTs) of the ADE classification[8], i.e., the minimal CFTs [6, 7].

Cardy identified the YLS of the 2D Ising model with the (A_4, A_1) minimal CFT [5]. Indeed, some properties of this YLS have been studied numerically [9, 10]. Nevertheless, other properties of this YLS had not been studied prior to our work. Also, the YLS of the next simplest 2D spin model, i.e., the 3-state Potts model, had not been studied at all prior to our work.

In this paper, we review our numerical studies of the YLSs of the 2D Ising and 3-state Potts models. In particular, we discuss numerical determinations of the low-lying excitation spectrum at theses YLSs and compare our numerical determinations with predictions from CFT.

This paper is organized as follows. In section 2, we introduce the quantum spin chains for the 2D Ising and 3-state Potts models. Via universality, the criticality of a 2D model is equivalent to that of its quantum spin chain, which is more convenient for numerical studies. In the section 3, we provide a list of candidate CFTs for the YLS of the 2D 3-state Potts model based on the ADE classification.[7, 8, 6] In section 4, we review finite-size scaling [11], and provide predictions for low-energy excitation spectra from CFT. In section 5, we comment on numerical methods herein. In section 6, we report our measurements of the low-lying excitation spectrum at the YLS of the 2D Ising model. In section 7, we report our measurements of the low-lying excitation spectrum and the exponent ν at the YLS of the 2D 3-state Potts model and determine the associated CFT. In section 8, we summarize our results.

2. Quantum spin chain for the Ising and 3-state Potts models

We will consider the Hamiltonian limit of the transfer matrix for $2D$ discrete spin models on a square lattice [12, 13].

For the 2D Ising model, the associated N-site, quantum spin chain has the Hamiltonian given by [14]

$$H_{Ising} = -\sum_{n=1}^{N}\{t\sigma_z(n)\sigma_z(n+1) + B\sigma_z(n) + \sigma_x(n)\}\ . \tag{1}$$

Here, $\sigma_x(n)$ and $\sigma_z(n)$ are 2x2 Pauli spin matrices at site n.

For the 2D 3-state Potts model, the associated N-site, quantum spin chain has the Hamiltonian given by:[14, 13]

$$H_{\text{Potts chain}} = -\sum_{n=1}^{N}\{t[W(n)^{\dagger}W(n+1) + H.C.] + L(n) + BD(n)\}\ . \tag{2}$$

Here, t is a positive coupling associated with same row ferromagnetic spin-spin interactions, and B is an external magnetic field. $W(n)$, $L(n)$, and $D(n)$ are 3x3 matrices associated with intra-row spin-spin interactions, single inter-row spin flips, and the external magnetic field interaction, respectively. These matrices are given by:

$$W = \begin{pmatrix} \omega & 0 & 0 \\ 0 & \omega^2 & 0 \\ 0 & 0 & 1 \end{pmatrix},\ L = \begin{pmatrix} 0 & 1 & 1 \\ 1 & 0 & 1 \\ 1 & 1 & 0 \end{pmatrix} \text{ and } D = \begin{pmatrix} 1 & 0 & 0 \\ 0 & 0 & 0 \\ 0 & 0 & 0 \end{pmatrix} \tag{3}$$

Here, w is the 3rd root of unity, i.e., $w = \exp(i2\pi/3)$.

Quantum chains have sparse Hamiltonian matrices and can be studied with powerful numerical matrix techniques, e.g., the Lanczos algorithm [15]. Both the quantum spin chain and its associated $2D$ spin model have the same critical points via universality.

3. Candidate Conformal Field Theories

The YLS of the 2D Ising model is described by a conformal minimal model [5, 10]. This suggests that the YLS of the 2D 3-state Potts model is also described by one of the conformal minimal models.

Each conformal minimal model is classified by a central charge and a primary field content. The central charges of the conformal minimal models satisfy [6, 7, 8]

$$c = 1 - \frac{6(p-p')^2}{pp'}\ , \tag{4}$$

where p and p' are coprime integers that are greater than 1. The primary fields are denoted by pairs of conformal dimensions $(h_{r,s}, \bar{h}_{r',s'})$, given by:

$$h_{r,s} = h_{p'-r,p-s} = \frac{(rp - sp')^2 - (p-p')^2}{4pp'} \tag{5a}$$

$$\bar{h}_{r',s'} = \bar{h}_{p'-r',p-s'} = \frac{(r'p - s'p')^2 - (p-p')^2}{4pp'} \,. \tag{5b}$$

Here, s, r, s', and r' are integers satisfying: $1 \leq r,\ r' \leq p' - 1$ and $1 \leq s, s' \leq p - 1$. When $|p - p'|>1$, the associated conformal minimal model is non-unitary.

The YLSs are described by non-unitary CFTs, because they occur at complex values of magnetic fields. These singularities are associated with non-Hermitian transfer matrices. For this reason, we assumed that the Yang-Lee edge singularity of the 2D 3-state Potts model would be described by a conformal minimal model in which $|p - p'|>1$.

Since we expected that the YLS of the 2D 3-state Potts model would be described by a relatively simple CFT, we searched for candidate CFTs among conformal minimal models with small values of p and p'.

The conformal minimal models are described by the ADE classification. The ADE classification lists the sets of $(h_{r,s}, \bar{h}_{r',s'})$ of the conformal minimal models as being in an $(A_{p'-1}, A_{p-1})$ series; two $(D_{p'/2+1}, A_{p-1})$ series; and a (E_n, A_{p-1}) series [8]. For each series, the ADE classification provides the primary field content. The simplest models of the $(A_{p'-1}, A_{p-1})$ and $(D_{p'/2+1}, A_{p-1})$ series are listed in Table1.

Table 1. Simplest Conformal Minimal Models.

$(\mathbf{p}', \mathbf{p})$	**Central Charge**	**Unitary**	**Comments**
		$(\mathrm{A}_{\mathrm{p}'-1}, \mathrm{A}_{\mathrm{p}-1})$ series	
$(3,2)$	0	Yes	Trivial - not candidate
$(4,3)$	1/2	Yes	Standard critical point of Ising
$(5,2)$	-22/5	No	$(\mathbf{A_4}, \mathbf{A_1})$, Ising Yang-Lee singularity - **candidate**
$(5,3)$	-3/5	No	$(\mathbf{A_4}, \mathbf{A_2})$ - **candidate**
$(5,4)$	7/10	Yes	(A_4, A_3) Standard critical point of tricritical Ising
		1st $(\mathrm{D}_{1+\mathrm{p}'/2}, \mathrm{A}_{\mathrm{p}-1})$ series, p'/2 odd	
$(6,5)$	4/5	Yes	Standard critical point of 3-state Potts
$(10,3)$	-44/5	No	$(\mathbf{D_6}, \mathbf{A_2})$ - **candidate**
$(10,7)$	8/35	No	$(\mathbf{D_6}, \mathbf{A_6})$ - **candidate**
$(10,9)$	14/15	Yes	Unitary - not a candidate
		2nd $(\mathrm{D}_{1+\mathrm{p}'/2}, \mathrm{A}_{\mathrm{p}-1})$ series, p'/2 even	
$(8,3)$	-21/4	No	$(\mathbf{D_5}, \mathbf{A_2})$ - **candidate**
$(8,5)$	-7/20	No	$(\mathbf{D_5}, \mathbf{A_4})$ - **candidate**
$(8,7)$	25/28	Yes	unitary - not candidate
		$(\mathrm{E}_n, \mathrm{A}_{\mathrm{p}-1})$ series	
		Many primary fields - not promising candidates	

In Table 1, non-unitary models are candidates for the YLS of the 2D 3-state Potts model [13]. Among these models, we did not expect the complex (E_n, A_p) models to describe the YLS of the simple 2D 3-state Potts model. Thus, we took as candidates the (A_4, A_1), (A_4, A_2), (D_6, A_2), (D_6, A_6) (D_5, A_2) and (D_5, A_4) conformal minimal models.

4. Finite-size Scaling at the YLS

The phenomenological renormalization group (PRG) and finite-size scaling enabled our numerical study of YLSs. In particular, we studied scaling in the length, N, of the quantum spin chain. In such scaling, the correlation length varies linearly with the length, N. The PRG fixes imposes such scaling via the energy gap and the equation [19, 10]:

$$(N-1)m(B_{YL}(N), N-1) = Nm(B_{YL}(N), N) \; . \tag{6}$$

The energy gap, $m(B,N)$, is defined by $m(B,N) = [E_1(B,N) - E_0(B,N)]$, where $E_0(B,N)$ and $E_1(B,N)$ are energies of the ground state and first excited state of the quantum spin chain. The PRG equation relates energy gaps of quantum spin chains of lengths N and $N-1$ at a critical field, $B_{YL}(N)$. Thus, the PRG equation fixes the critical value of the magnetic field, $B_{YL}(N)$, for each chain length, N. The complex solutions for $B_{YL}(N)$ scale towards to YLS.

At these critical field values, CFT predicts how excitation energies of the quantum spin chain scale. For a state "i" of the quantum spin chain, CFT predicts [20] that $E_i(N) - E_0(N)$, will scale as:

$$E_i(N) - E_0(N) = \zeta 2\pi \frac{h_i + \bar{h}_i - (h + \bar{h})}{N} \; . \tag{7}$$

Above, h_i and $\bar{h}_i$ are the dimensions of primary fields, and h and $\bar{h}$ are the dimensions of the primary field of lowest dimension. The constant ζ is a non-universal and depends on a normalization.

For each candidate CFT, the low-lying excitation spectrum is from the associated partition function. For each low-lying state "i", Table 2 lists the ratio of the excitation energy of state "i" over the excitation energy of the lowest excited state "1". That is, excitation energies are normalized with respect to the lowest excitation. The excitation energies of Table 2 are directly

Table 2. Low-lying excitation spectra of Candidate CFTs

Name of CFT	Lowest Excitation Energies					
(A_4, A_1)	0	1	2.5	5.0	6.0	7.5
(A_4, A_2)	0	1	5.0	10.0	15.0	16.0
(D_5, A_2)	0	1	1.6	3.2	3.4	4.2
(D_5, A_4)	0	1	1.875	6.0	10.0	11.875
(D_6, A_2)	0	1	2.0	2.5	3.5	5.0
(D_6, A_6)	0	1	3.0	10.0	15.0	17.5

measurable via finite-size scaling at the $B_{YL}(N)$'s, of Eq. (6).

The critical field values, $B_{YL}(N)$, approach a limit, $B_{YL}(\infty)$, as the length, N, of the quantum spin chain becomes large. At $B_{YL}(N)$, derivatives of the energy gap, $m(B,N)$, provide estimates, $\nu(N)$, of the exponent ν that defines the scaling behavior of the spin-spin correlation ξ as [19, 10]:

$$1/\nu(N) = \frac{log[m'(B,N)/m'(B,N-1)]|_{B=B_{YL}(N)}}{log[N/(N-1)]} + 1. \tag{8}$$

Here, $m'(B,N) = \dfrac{dm(B,N)}{dB}$. The estimates, $\nu(N)$, converge to ν as $N \to \infty$.

5. Numerical methods

The measurements of the critical magnetic fields, $B_{YL}(N)$, were obtained by numerically solving PRG Eq.(6) for quantum spin chains of different lengths. For the numerical solutions, the Lanczos algorithm determined the energies of the quantum Ising and Potts chains [15]. Then, the critical field values B_{YL}'s were used to find the low-lying excitation spectra of Ising and 3-state Potts quantum spin chains of various lengths.

While Lanczos can provide the low-lying excitation spectra, it is inconvenient for determining the complete spectrum. In particular, an evaluation of the complete low-lying excitation spectrum via the Lanczos algorithm would require finding a first Krylov space and then, finding other Krylov spaces orthogonal to the first space. To avoid such numerical complications, we used a commercial package of Maple$_{TM}$ 9.5 to generate spectra based on previously determined values of the critical magnetic field. The measured excitation energies have been normalized by a division by the lowest excitation energy as already described to remove dependence on the non-universal constant ζ of Eq. (7).

6. Ising YLS

For the Ising quantum spin chain, we numerically evaluated the critical field, $B_{YL}(N)$, for t = 0.1 and various values of N and then, numerically evaluated the complete low-lying excitation spectra as shown in Table 3 [17].

Table 3. The table shows the low-lying excitation energies of states A - G for Ising quantum spin chains with 6 to 12 sites. Degeneracies of the states are as shown in the left column.

State/[Degeneracy]	$N = 6$	$N = 7$	$N = 8$	$N = 9$	$N = 10$	$N = 11$	$N = 12$
A/[2]	2.68432	2.64386	2.61415	2.59207	2.57540	2.56260	2.55253
B/[1]	4.18193	4.27896	4.36713	4.44474	4.51197	4.56977	4.61912
C/[2]	4.51738	4.63236	4.70368	4.75182	4.78652	4.81281	4.83329
D/[2]	5.85889	5.89208	5.91240	5.92644	5.93703	5.94544	5.95210
E/[2]	–	5.68559	6.03104	6.27270	6.45018	6.58573	6.69223
F/[2]	–	6.24252	6.35798	6.46344	6.55966	6.64694	6.72535

We used a Bulirsch and Stoer (BST) algorithm[16] to extrapolate energies of state types A – F as $1/N \to 0$. The BST analysis indicated that the normalized energies of states of types A, B, C, D, E, and F scale to 2.4995(5), 5.005(1), 5.003(3), 5.99(1), 7.54(8), and 7.60(7), respectively.

Table 4. Low-lying excitation spectra of (A_4, A_1) minimal CFT

CFT	(A_4, A_1)					
Normalized Energies	0	1	2.5	5.0	6.0	7.5
Degeneracy	1	1	2	3	2	4

Table 3 shows the state sets A, B & C, D, and E & F, which have different normalized excitation energies as $1/N \to 0$ and their respective degeneracies of 2, 3, 2, and 4. Our numerical PRG measurements of excitation energies and associated degeneracies at the YLS of the Ising quantum spin chain are in excellent agreement with the normalized low-lying excitation spectrum of the (A_4, A_1) CFT as shown in Table 4.

7. Potts YLS

For the three-state Potts quantum spin chain, we again numerically evaluated the critical field, $B_{YL}(N)$, for t = 0.1 and various values of N and then, numerically evaluated the complete low-lying excitation spectra [18]. The low-lying excitation spectra is shown in Table 5.

From the spectra of Table 5, we determined how excitation energies of low-lying eigenstates scale with the chain length N. This determination requires identifying corresponding eigenstates in chains of different length, N. Our identification of eigenstate correspondences was based on two rules. The first rule was that the energies of corresponding eigenstates vary smoothly and monotonically with N. The second rule was that the correspondences of eigenstates for different N should properly account for eigenstate degeneracies. In particular, old eigenstates do not disappear as N increases. That is, degeneracies remain constant or increase with N. These rules identifying corresponding low-lying energy eigenstates for different chain length, N.

Table 5. Low-lying excitation energies of 3-state Potts quantum chain. Here, "g" is the degeneracy and "s" labels a state of an associated CFT.

$N=4$	g	s	$N=5$	g	s	$N=6$	g	s	$N=7$	g	s	$N=8$	g	s
2.26083	2	A	2.32641	2	A	2.36741	2	A	2.39452	2	A	2.41354	2	A
3.06697	1	B	3.56700	2	B	3.88032	2	B	4.09253	2	B	4.24476	2	B
4.56147	1	C	4.68089	1	C	4.38630	1	F	4.81263	1	C	4.84954	1	C
5.03441	1	D	5.47927	2	D	4.75981	1	C	4.94654	2	F	5.35160	2	F
5.94491	1		6.30921	2	E	5.66128	2	D	5.76468	2	D	5.72334	1	
5.94494	2		7.20262	2		6.59225	2	E	6.78325	2	E	5.82923	2	D

Applying the above rules enables separating the low-lying eigenstates of Table 5 into sets A – F. Different ones of the sets A – F first appeared at different chain lengths, N, as expected. For example, the set D first appeared in chains of length 4 whereas the set E first appeared in the chain of length 5.

The states of sets A – E are plotted as a function of $1/N$ in Figures 1 – 4. [1]

CFT describes the excitation spectra near criticality or as $1/N \to 0$. Thus, we extrapolated the excitation energies of the eigenstates in Figures 1 – 4 to $1/N = 0$ to determine the low-lying excitation energies at the Yang-Lee edge singularity of the 2D 3-state Potts model. Referring to Figure 1, the doubly degenerate A eigenstates have energies that extrapolate towards about 2.5 as $1/N \to 0$. Referring to Figure 2, both the doubly degenerate B eigenstates and single

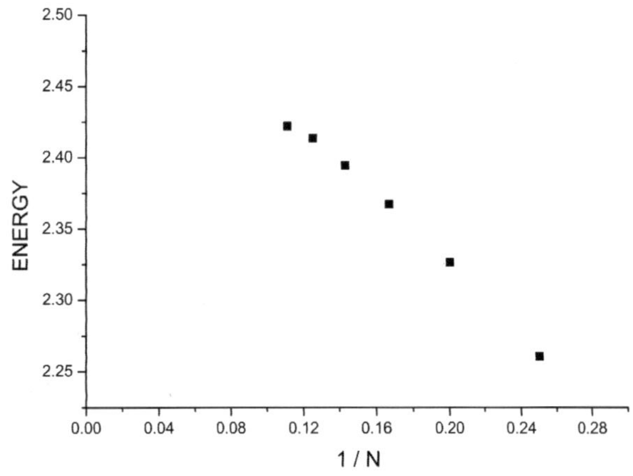

Figure 1. Measured energies of type A states as a function of number of sites, N.

C eigenstate have energies that extrapolate to about 5.1 as $1/N \to 0$. Referring to Figure 3, the doubly degenerate D eigenstates have energies that extrapolate to about 6.2 as $1/N \to 0$. Finally, referring to Figure 4, the E eigenstates have energies that extrapolate to about 7.8 as $1/N \to 0$. Thus, Figures 1 – 4 provide an observed low-lying excitation spectra of {1, 2.5, 5.1, 6.2, 7.8}.[2] These results are in very good agreement with the values {1, 2.5, 5, 6, 7.5} as predicted for the excitation spectrum of the (A_4, A_1) minimal CFT of Table 4. Also, Figures 1 – 4 show

[1] Figures 1 and 2 also plot some energies of eigenstates in 3-state Potts quantum chains of length 9 and 10. These values were obtained via the Lanczos method from which we did not attempt to obtain a complete spectra.

[2] The lowest excited eigenstate, which automatically has a normalized energy of 1 in our normalization, is not shown in Figures 1 – 4. This state was however, observed to be singly degenerate as expected.

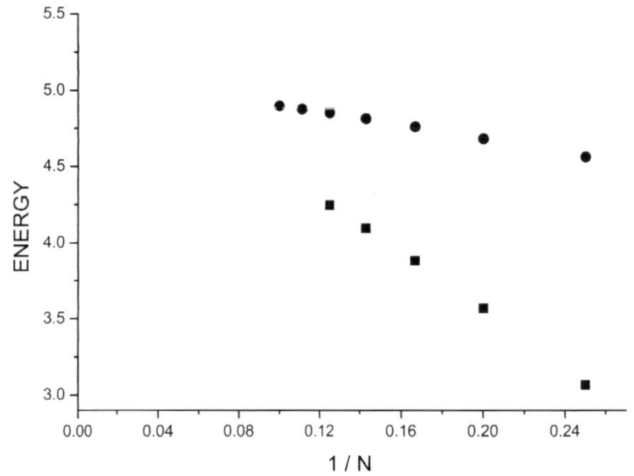

Figure 2. Measured energies of type B states (squares) and type C states (circles) as a function of number of sites, N.

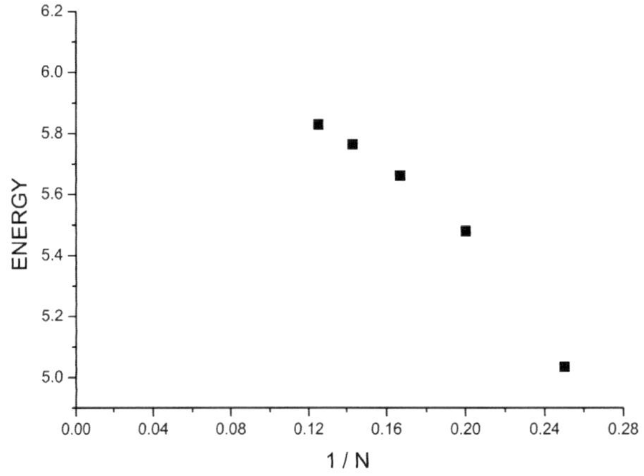

Figure 3. Measured energies of type D states as a function of number of sites, N.

that degeneracies of these same excitation states are (1, 2, 3, 2, ?),[3] which also agrees with the degeneracies of the (A_4, A_1) minimal CFT.

Thus, both degeneracies and energies of the low-lying excitation spectrum at the Yang-Lee edge singularity of the 3-state Potts model agree with those of the (A_4, A_1) minimal CFT. No other candidate minimal CFTs of Table 2 has a low-lying excitation spectra that agrees with the finite-size scaling spectrum that we measured at the YLS of the 2D 3-state Potts model.

Finally, we also found the exponent, ν, at the YLS of 3-state Potts model by solving Eq. (8). For $N = 5$, 6, and 7, our estimates of $\nu(N)$ are 0.4105, 0.4123, and 0.4134, respectively. These estimates vary due to finite-size corrections to Eq. (8). An extrapolation of the measured

[3] The observed degeneracy of the states of energy 7.5 is indicated as unknown, because our measurements were not to high enough chain lengths and energies to determine this degeneracy.

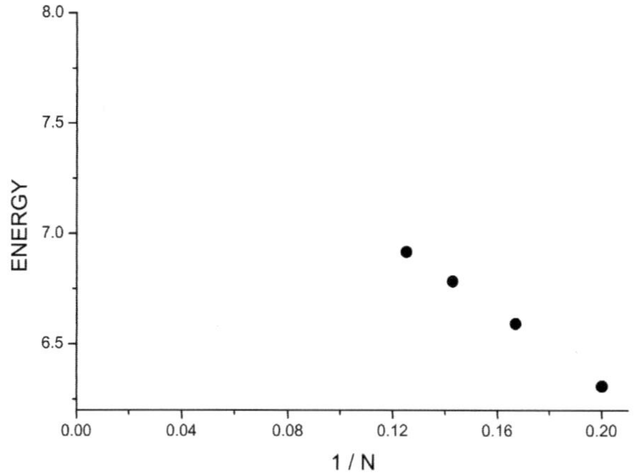

Figure 4. Measured energies of type E states as a function of number of sites, N.

$\nu(N)$'s to $1/N = 0$ provides the best estimate of the measured value for ν. We used a BST algorithm to perform the extrapolation. The BST algorithm predicted that the $\nu(N)$'s converge to a value in the interval [0.417, 0.418] as $1/N \to 0$. This result also agrees very well with the ν of 5/16, i.e., 0.4167, in the (A_4, A_1) minimal CFT.

8. Conclusions

Our finite-size scaling measurements of the low-lying excitation spectra and the exponent ν show that the YLS of the $2D$ 3-state Potts and Ising models are described by the (A_4, A_1) conformal minimal model. In particular, both 2D discrete spin models have Yang-Lee singularities in the same universality class.

References

[1] Yang C N and Lee T D 1952 *Phys. Rev.* **87** 404
[2] Yang C N and Lee T D 1952 *Phys. Rev.* **87** 410
[3] Fisher M E 1978 *Phys. Rev. Lett.* **40** 1610
[4] Ch. Binek 1998 *Phys. Rev. Lett.* **81** 5644; Ch. Binek, W. Kleemann, and H. A. Katori 2001 *J. Phys.* **B13** L811
[5] Cardy J L 1985 Phys. *Rev. Lett.* **54** 1354
[6] Belavin AA, Polyakov AM, and Zamolodchikov AB 1984 *Nucl. Phys.* **B 241** 333
[7] Friedan D, Qiu Z and Shenker S 1984 Phys. Rev. Lett. **52** 1575; 1986 *Comm. Math. Phys.* **107** 535; 1985 *Phys. Lett.* **B 151** 37
[8] Cappelli A, Itzykson C, Zuber J-B 1987 *Nucl. Phys.* **B 280** 445; 1987 *Comm. Math. Phys.* **113** 1; Kato A 1987 *Mod. Phys. Lett.* **A 2** 585; for a review see e.g., Itzykson C and Drouffe J-M 1989 *Statistical Field Theory* (Cambridge University) chapter IX, table II
[9] Uzelac K and Jullien R 1981 J. Phys. **A 14** L151
[10] Itzykson C, Saleur H, and Zuber J-B 1986 *Europhys. Lett.* **2** 91
[11] Fisher M E and Barber M N 1972 *Phys. Rev. Lett.***28** 1516; for a brief review see, e.g., Henkel M 1999 *Conformal Invariance and Critical Phenomena*(Springer-Verlag) chapter 3
[12] Fradkin E and Susskind L 1978 *Phys. Rev.* **D 17** 2637; Suzuki M 1971 *Prog. Theor. Phys.* **46** 1337; for a review see e.g., Henkel M 1999 *Conformal Invariance and Critical Phenomena* (Springer-Verlag) chapters 8-10
[13] McCabe J and Wydro T 2003 in *Proc. of the 7th Int. School on Theoretical Physics : Symmetry and Structural Properties of Condensed Matter* (World Scientific) p 9
[14] von Gehlen G, Rittenberg V, and Vescan T 1987 *J. Phys.* **A 20** 2577

[15] For a review see e.g.: Parlett P N, *The Symmetric Eigenvalue Problem* (Prentice-Hall 1998) 261-321; Jennings A, *Matrix Computation for Engineers and Scientists* (Wiley 1977) 316-319; Kerner W 1989 *J. Comp. Phys.* **85** 1
[16] Burlisch M N and Stoer J 1964 *Numer. Math.* **6** 413
[17] J. McCabe and T. Wydro 2005 *Preprint* cond-mat/0510201 (submitted to *Int. Journ. of Mod. Physics* **B**)
[18] T. Wydro and J. McCabe 2005 *Int. Journ. of Mod. Physics B* **B 19** 3021
[19] Nightingale M P 1976 *Physica* **83A** 561; Derrida B and de Seze L 1982 *J. Physique* **43** 475; see, e.g., Nightingale M P 1990 *Finite Size Scaling and Numerical Simulation of Statistical Systems* (World Scientific) chapter VII
[20] Cardy J L 1984 *J. Phys.* **A 17** L385

Feynman graphs and related Hopf algebras

G H E Duchamp[1], P Blasiak[2], A Horzela[2], K A Penson[3] and A I Solomon[3,4]

[1] Institut Galilée, LIPN, CNRS UMR 7030
99 Av. J.-B. Clement, F-93430 Villetaneuse, France

[2] H. Niewodniczański Institute of Nuclear Physics, Polish Academy of Sciences
ul. Eliasza-Radzikowskiego 152, PL 31342 Kraków, Poland

[3] Laboratoire de Physique Théorique de la Matière Condensée
Université Pierre et Marie Curie, CNRS UMR 7600
Tour 24 - 2ième ét., 4 pl. Jussieu, F 75252 Paris Cedex 05, France

[4] The Open University, Physics and Astronomy Department
Milton Keynes MK7 6AA, United Kingdom

E-mail: ghed@lipn-univ.paris13.fr, pawel.blasiak@ifj.edu.pl, andrzej.horzela@ifj.edu.pl, penson@lptl.jussieu.fr, a.i.solomon@open.ac.uk

Abstract. In a recent series of communications we have shown that the reordering problem of bosons leads to certain combinatorial structures. These structures may be associated with a certain graphical description. In this paper, we show that there is a Hopf Algebra structure associated with this problem which is, in a certain sense, unique.

1. Introduction

In a relatively recent paper Bender, Brody and Meister [3] introduced a special Field Theory described by a product formula (a kind of Hadamard product for two exponential generating functions - EGF) in the purpose of proving that any sequence of numbers could be described by a suitable set of rules applied to some type of Feynman graphs. Inspired by this idea, we have worked out combinatorial consequences of the product and exponential formulas in a recent series of papers [12, 13, 14, 15, 16, 17, 18].

Here, we consider two aspects of the product formula for formal power series applied to combinatorial field theories. Firstly, we remark that the case when the functions involved in the product formula have a constant term (equal to one) is of special interest as often these functions give rise to substitutional groups. The groups arising from the normal ordering problem of boson strings are naturally associated with explicit vector fields, or their conjugates, in the case when there is only one annihilation operator [14, 17]. We also consider one-parameter groups of operators when several annihilators are present. Secondly, we discuss the Feynman-type graph representation resulting from the product formula. We show that there is a correspondence between the packed integer matrices of the theory of noncommutative symmetric functions and the labelled version of these Feynman-type graphs.

We thus obtain a new Hopf algebra structure over the space of matrix quasi-symmetric functions that is a natural cocommutative Hopf algebra structure on the space of diagrams themselves which originates from the formal doubling of variables in the product formula.

2. Single and double exponentials

2.1. One parameter groups and the connected graph theorem

2.1.1. Substitutions The Weyl algebra W is the $\mathbb{C}$-associative algebra (with unit) defined by two generators a and a^+ and the unique relation $[a, a^+] = 1$. This algebra is of Gelfand-Kirillov dimension 2 and has a basis consisting of the following family $\left\{(a^+)^k a^l\right\}_{k,l\geq 0}$.
It is known that it is impossible to represent faithfully a, a^+ by bounded operators in a Banach space, but one often uses the representation $a \mapsto \frac{\mathrm{d}}{\mathrm{dx}}$; $a^+ \mapsto x$ as operators acting "on the line" or, better said, on the space of polynomials $\mathbb{C}[x]$. Through this representation (faithful and coined under the name "Bargmann-Fock"), one sees that we can define a grading on W by the weight function $w(a) = -1$; $w(a^+) = 1$.
A homogeneous operator (under this grading) $\Omega \in W$ is then of the form

$$\Omega = \sum_{k,l;\ k-l=e} c(k,l)(a^+)^k a^l \tag{1}$$

According to whether the excess e is positive or negative, the normal ordering of Ω^n reads

$$\mathcal{N}(\Omega^n) = (a^+)^{ne}\left(\sum_{k=0}^{\infty} S_\Omega(n,k)(a^+)^k a^k\right) \text{ or } \left(\sum_{\mathrm{k}=0}^{\infty} \mathrm{S}_\Omega(\mathrm{n},\mathrm{k})(\mathrm{a}^+)^{\mathrm{k}}\mathrm{a}^{\mathrm{k}}\right)(\mathrm{a})^{\mathrm{n}|\mathrm{e}|} \tag{2}$$

We get combinatorial quantities with two indices i.e. an infinite $\mathbb{N}\times\mathbb{N}$ matrix $\{S_\Omega(n,k)\}_{n,k\geq 0}$ which we will call the *generalized Stirling matrix of* Ω. In fact, it is easily checked that, if the coefficients $c(k,l)$ of Ω are non-negative integers, so are the entries $(S_\Omega(n,k))$ of this matrix.

Let us give some examples of these generalized Stirling matrices.
For $\Omega = a^+a$, one gets the usual matrix of the Stirling numbers of the second kind

$$\begin{bmatrix} 1 & 0 & 0 & 0 & 0 & 0 & 0 & \cdots \\ 0 & 1 & 0 & 0 & 0 & 0 & 0 & \cdots \\ 0 & 1 & 1 & 0 & 0 & 0 & 0 & \cdots \\ 0 & 1 & 3 & 1 & 0 & 0 & 0 & \cdots \\ 0 & 1 & 7 & 6 & 1 & 0 & 0 & \cdots \\ 0 & 1 & 15 & 25 & 10 & 1 & 0 & \cdots \\ 0 & 1 & 31 & 90 & 65 & 15 & 1 & \cdots \\ \vdots & \vdots & \vdots & \vdots & \vdots & \vdots & \vdots & \ddots \end{bmatrix} \tag{3}$$

For $\Omega = a^+aa^+ + a^+$, we obtain

$$\begin{bmatrix} 1 & 0 & 0 & 0 & 0 & 0 & 0 & \cdots \\ 2 & 1 & 0 & 0 & 0 & 0 & 0 & \cdots \\ 6 & 6 & 1 & 0 & 0 & 0 & 0 & \cdots \\ 24 & 36 & 12 & 1 & 0 & 0 & 0 & \cdots \\ 120 & 240 & 120 & 20 & 1 & 0 & 0 & \cdots \\ 720 & 1800 & 1200 & 300 & 30 & 1 & 0 & \cdots \\ 5040 & 15120 & 12600 & 4200 & 630 & 42 & 1 & \cdots \\ \vdots & \vdots & \vdots & \vdots & \vdots & \vdots & \vdots & \ddots \end{bmatrix} \tag{4}$$

and for $w = a^+ aaa^+ a^+$

$$\begin{bmatrix} 1 & 0 & 0 & 0 & 0 & 0 & 0 & 0 & 0 & \cdots \\ 2 & 4 & 1 & 0 & 0 & 0 & 0 & 0 & 0 & \cdots \\ 12 & 60 & 54 & 14 & 1 & 0 & 0 & 0 & 0 & \cdots \\ 144 & 1296 & 2232 & 1296 & 306 & 30 & 1 & 0 & 0 & \cdots \\ 2880 & 40320 & 109440 & 105120 & 45000 & 9504 & 1016 & 52 & 1 & \cdots \\ \vdots & \vdots & \vdots & \vdots & \vdots & \vdots & \vdots & \vdots & \vdots & \ddots \end{bmatrix} \tag{5}$$

Let $\Omega = \sum_{k,l \geq 0} c(k,l)(a^+)^k a^l$ (finite supported sum) be a general term of W in normal form and let us call a *dominant term*, the sum of monomials with maximum length $k+l$. It is not difficult to prove that, if Ω is homogeneous, the dominant term consists of a single monomial $c(k_0, l_0)(a^+)^{k_0} a^{l_0}$. Thus, the dominant term of Ω^n must be $c(k_0, l_0)^n (a^+)^{nk_0} a^{nl_0}$. Then, for example in the case when $e = k_0 - l_0 \geq 0$, in the generalized Stirling matrix of Ω, the rightmost non-zero coefficient of the line n has address $(n, n.l_0)$ and bears the coefficient $c(k_0, l_0)^n$. All these matrices are row-finite and triangular iff $l_0 = 1$ (which means that no monomial possesses more than one a).

Remark 2.1 *i) There is a beautiful combinatorial expression of the normal form of w^n in case w is a string in a and a^+. The normal form of w is*

$$\mathcal{N}(w) = \sum_{k \geq 0} r(B,k)(a^+)^{r-k} a^{s-k} \tag{6}$$

where $r(B,k)$ is the kth rook number of a certain board B constructed after w (see [13, 25], and $r = |w|_{a^+}$; $s = |w|_a$ are the number of occurences of a^+ and a in w.
ii) To each matrix $M \in \mathbb{C}^{\mathbb{N}\times\mathbb{N}}$ of this kind and more generally "row finite" matrices (which means that, for each n, the family $(M(n,k))_{k\in\mathbb{N}}$ is finite supported), one can associate a transformation of EGFs (see [14, 17]*) $f \mapsto \hat{f}$ such that, if $f = \sum_{n\geq 0} a_n \frac{z^n}{n!}$ then $\hat{f} = \sum_{n \geq 0} b_n \frac{z^n}{n!}$ (with $b_n = \sum_{k\geq 0} M(n,k) a_k$).*
iii) It can be shown that, if no monomial of Ω possesses more than one a, the action of the transformation induced by Ω (through the Bargmann-Fock representation) can be expressed in terms of vector fields or their conjugates, thus the one-parameter group $e^{\lambda\Omega}$ acts by substitutions and products [14, 17].

2.1.2. Combinatorial matrices and one-parameter groups One can also draw generalized Stirling matrices from another source, namely from the combinatorial graph theory.
Let C be a class of graphs such that

$$\Gamma \in C \text{ iff every connected component of } \Gamma \text{ is in } C \tag{7}$$

For these classes of graphs, one has the exponential formula [9, 23, 21] saying roughly that

$$\text{EGF(all graphs)} = e^{\text{EGF(Connected Graphs)}} \tag{8}$$

This implies, in particular, that the matrix

$$M(n,k) = \text{ number of graphs with } n \text{ vertices and having } k \text{ connected components} \tag{9}$$

is the matrix of a substitution (see [14, 17]). One can prove, using a Zariski-like argument (a polynomial vanishing for every integer vanishes everywhere), that, if M is such a matrix (with identity diagonal) then, all its powers (positive, negative and fractional) are substitution

matrices and form a one-parameter group of substitutions, thus coming from a vector field on the line which can be computed.
But no nice combinatorial principle seems to emerge.
For example, beginning with the Stirling substitution $z \mapsto e^z - 1$, we know that there is a unique one-parameter group of substitutions $s_\lambda(z)$ such that, for λ integer, one has the value ($s_2(z) \leftrightarrow$ partition of partitions)

$$s_2(z) = e^{e^z-1} - 1 \; ; \; s_3(z) = e^{e^{e^z-1}-1} - 1 \; ; \; s_{-1}(z) = ln(1+z) \tag{10}$$

but we have no nice description of this group nor of the vector field generating it.

2.2. A product formula

The Hadamard product of two sequences $(a_n)_{n\geq 0}$; $(b_n)_{n\geq 0}$ is given by the pointwise product $(a_n b_n)_{n\geq 0}$. We can at once transfer this law on EGFs by

$$\Big(\sum_{n\geq 0} a_n \frac{x^n}{n!}\Big) \odot_{exp} \Big(\sum_{n\geq 0} b_n \frac{x^n}{n!}\Big) := \sum_{n\geq 0} a_n b_n \frac{x^n}{n!} \tag{11}$$

In the following, we will omit the subscript (in $\odot_{exp}$) as this will be the only kind of Hadamard product under consideration.

But, it is not difficult to check that the family

$$\left(\frac{(y\frac{\partial}{\partial x})^n}{n!} \frac{x^m}{m!}\right)_{n,m\in\mathbb{N}} \tag{12}$$

is summable in $\mathbb{C}[[x,y]]$ (the space of formal power series in x and y) as we have

$$\frac{(y\frac{\partial}{\partial x})^n}{n!} \frac{x^m}{m!} = \begin{cases} 0 & \text{if } n > m \\ \frac{y^n x^{m-n}}{n!(m-n)!} & \text{otherwise} \end{cases} \tag{13}$$

and therefore, for $F(x) = \sum_{n\geq 0} a_n \frac{x^n}{n!}$ and $G(x) = \sum_{n\geq 0} b_n \frac{x^n}{n!}$ one gets the *product formula*

$$(F \odot G)(x) := F(x\frac{\partial}{\partial y})G(y)|_{y=0} = \sum_{n\geq 0} a_n b_n \frac{x^n}{n!} \tag{14}$$

With this product, the set of series forms a commutative associative algebra with unit, which is actually the product algebra $\mathbb{C}^{\mathbb{N}}$.

2.3. The double exponential formula

The case $F(0) = G(0) = 1$ will be of special interest in our study. Every series with constant term 1 can be represented by an exponential $exp(\sum_{n\geq 1} L_n \frac{x^n}{n!})$ which can be expanded using Bell polynomials and Faà di Bruno coefficients. Let us now recall some facts about these combinatorial notions.
We still consider the alphabet $\mathbb{L} = \{L_1, L_2, \cdots\} = \{L_i\}_{i\geq 1}$, then the complete Bell polynomials [7] are defined by

$$exp(\sum_{m\geq 1} L_m \frac{x^m}{m!}) = \sum_{n\geq 0} Y_n(\mathbb{L}) \frac{x^n}{n!} \tag{15}$$

We will denote alternatively $Y_n(L_1, \cdots L_n)$ for $Y_n(\mathbb{L})$ as this polynomial is independent from the subalphabet $(L_m)_{m>n}$. We know [7] that

$$Y_n(\mathbb{L}) = Y_n(L_1, \cdots L_n) = \sum_{||\alpha||=n} ((\alpha))\mathbb{L}^\alpha = \sum_{||\alpha||=n} ((\alpha)) L_1^{\alpha_1} L_2^{\alpha_2} \cdots L_n^{\alpha_n} \tag{16}$$

where $\alpha = (\alpha_1, \alpha_2, \cdots \alpha_n)$ is an integral vector, $||\alpha|| := \sum_{j=1}^{m} j\alpha_j$, $\mathbb{L}^\alpha = L_1^{\alpha_1} L_2^{\alpha_2} \cdots L_n^{\alpha_n}$ is the multiindex standard notation and

$$((\alpha)) = \frac{||\alpha||!}{(1!)^{\alpha_1}(2!)^{\alpha_2}\cdots(n!)^{\alpha_n}(\alpha_1)!\cdots(\alpha_n)!} \tag{17}$$

is the Faà di Bruno coefficient [8, 20] which will be interpreted, in the next section, as enumerating structures called *set partitions.*
Combining (15) and (16) one gets

$$exp\left(\sum_{m\geq 1} L_m \frac{(y\frac{\partial}{\partial x})^m}{m!}\right) exp\left(\sum_{n\geq 1} V_n \frac{x^n}{n!}\right)\Big|_{x=0} = \sum_{k\geq 0} \frac{y^k}{k!}\left(\sum_{||\alpha||=||\beta||=k} ((\alpha))((\beta))\mathbb{L}^\alpha \mathbb{V}^\beta\right) \tag{18}$$

Formula (18) will be called in the sequel the *double exponential formula.*

2.4. Monomial expansion of the double exponential formula

In this paragraph, we will use unordered and ordered set partitions. By an unordered partition P of the set X we mean a finite subset $P \subset (\mathfrak{P}(X) - \{\emptyset\})$ ($\mathfrak{P}(X)$ is the set of all subsets of X [7]) such that

$$\bigcup_{Y\in P} Y = X \text{ and } (Y_1, Y_2 \in P,\ Y_1 \neq Y_2 \Longrightarrow Y_1 \cap Y_2 = \emptyset) \tag{19}$$

this explains why *without any convention* the classical Stirling number of the second kind $S(0,0)$ equals 1. The elements of P are called *blocks.*
Following Comtet ([8] p 39), we will say that a partition P is of type $\alpha = (\alpha_1, \alpha_2, \cdots, \alpha_m)$ iff there is no j-block for $j > m$ and α_j j-block(s) for each $j \leq m$. This implies in particular that the set X is of cardinality $||\alpha|| := \sum_{j=1}^{m} j\alpha_j$.

Here one can see easily that the number of blocks of a partition of type α is $|\alpha| = \sum_{j=1}^{m} \alpha_j$.
An ordered partition of type α of the set X is just a partition in which the blocks are labelled from 1 to $|\alpha|$.
In other words, one could say that an ordered partition is a *list* of subsets and an unordered partition is a *set* of subsets.
To every ordered partition $P = (B_1, B_2, \cdots, B_{|\alpha|})$ corresponds an unordered one $\Phi_p(P) = \{B_1, B_2, \cdots, B_{|\alpha|}\}$ where Φ_p is the "forgetful" function which forgets the order. Now to a pair $(P^{(1)}, P^{(2)})$ of ordered partitions of the same set (call it X)

$$P^{(1)} = (B_1^{(1)}, B_2^{(1)}, \cdots, B_{k_1}^{(1)}) \qquad\qquad P^{(2)} = (B_1^{(2)}, B_2^{(2)}, \cdots, B_{k_2}^{(2)}) \tag{20}$$

one can associate the intersection matrix $IM_o(P^{(1)}, P^{(2)})$ such that the entry of address (i,j) is the number of elements of the intersection of the block i of the first partition and the block j of the second. For example with partitions of $X = \{1,2,3,4,5,6\}$, and specifying

$$P^{(1)} = (\{1,2,5\},\{3,4,6\}) \qquad\qquad P^{(2)} = (\{1,2\},\{3,4\},\{5,6\}),$$

one gets

	$\{1,2\}$	$\{3,4\}$	$\{5,6\}$
$\{1,2,5\}$	2	0	1
$\{3,4,6\}$	0	2	1

and hence the matrix

$$\begin{pmatrix} 2 & 0 & 1 \\ 0 & 2 & 1 \end{pmatrix} \tag{21}$$

Formally, $IM_o(P^{(1)}, P^{(2)})$ is the matrix of size $k_1 \times k_2$ such that

$$IM_o(P^{(1)}, P^{(2)})[i,j] = \mathrm{card}(\mathrm{B}_\mathrm{i}^{(1)} \cap \mathrm{B}_\mathrm{j}^{(2)}) \tag{22}$$

The matrices obtained in such a way form the set of packed matrices defined in [19] as, indeed, one sees that every packed matrix can be obtained through the matching procedure illustrated above.

If we consider now a pair of unordered partitions $(Q^{(1)}, Q^{(2)})$, we cannot associate to them a single matrix but rather a class of matrices obtained from the preimages of $(Q^{(1)}, Q^{(2)})$ under $\Phi_p \times \Phi_p$. In a compact formulation, the set of matrices so obtained is

$$\left\{ IM_0(P^{(1)}, P^{(2)}) \right\}_{\Phi_p(P^{(1)})=Q^{(1)};\ \Phi_p(P^{(2)})=Q^{(2)}} \tag{23}$$

For example, with

$$(Q^{(1)}, Q^{(2)}) = (\{\{1,2,5\},\{3,4,6\}\}, \{\{1,2\},\{3,4\},\{5,6\}\}) \tag{24}$$

one gets the 12 preimages $(P_i^{(1)}, P_j^{(2)})$, where $P_i^{(1)}$ are among the two preimages of $Q^{(1)}$ and $P_j^{(2)}$ are among the 6 preimages of $Q^{(2)}$. Explicitely

$$P_1^{(1)} = (\{1,2,5\},\{3,4,6\}) \qquad P_2^{(1)} = (\{3,4,6\},\{1,2,5\})$$

are the preimages of $Q^{(1)}$ and

$$\begin{array}{ll} P_1^{(2)} = (\{1,2\},\{3,4\},\{5,6\}) & P_2^{(2)} = (\{1,2\},\{5,6\},\{3,4\}) \\ P_3^{(2)} = (\{3,4\},\{1,2\},\{5,6\}) & P_4^{(2)} = (\{3,4\},\{5,6\},\{1,2\}) \\ P_5^{(2)} = (\{5,6\},\{1,2\},\{3,4\}) & P_6^{(2)} = (\{5,6\},\{3,4\},\{1,2\}) \end{array}$$

are the preimages of $Q^{(2)}$.

The set of matrices so obtained reads

$$\begin{aligned} IM_u(M) \;=\; & \left\{ \begin{pmatrix} 2 & 0 & 1 \\ 0 & 2 & 1 \end{pmatrix}, \begin{pmatrix} 2 & 1 & 0 \\ 0 & 1 & 2 \end{pmatrix}, \begin{pmatrix} 1 & 2 & 0 \\ 1 & 0 & 2 \end{pmatrix}, \right. \\ & \left. \begin{pmatrix} 0 & 2 & 1 \\ 2 & 0 & 1 \end{pmatrix}, \begin{pmatrix} 0 & 1 & 2 \\ 2 & 1 & 0 \end{pmatrix}, \begin{pmatrix} 1 & 0 & 2 \\ 1 & 2 & 0 \end{pmatrix} \right\}. \end{aligned} \tag{25}$$

This is the orbit of one of them under permutation of lines and columns.

The correspondence which, to a pair of unordered partitions, associates a class of matrices (under permutations of lines and columns) will be denoted IM_u.

Thus, one gets a commutative diagram of mappings

$$\begin{array}{ccc} \text{Pairs of ordered partitions} & \xrightarrow{\Phi_p \times \Phi_p} & \text{Pairs of unordered partitions} \\ {\scriptstyle IM_o}\downarrow & & \downarrow{\scriptstyle IM_u} \\ \text{Packed matrices} & \xrightarrow{Class} & \text{Classes of packed matrices} \\ {\scriptstyle Dg_o}\downarrow & & \downarrow{\scriptstyle Dg_u} \\ \text{Labelled diagrams} & \xrightarrow{\Phi_d} & \text{Diagrams} \end{array} \tag{26}$$

The scheme presented above shows how to associate to a pair of ordered (resp. unordered) set partitions, a packed matrix (resp. a class of packed matrices). The packed matrices can be alternatively represented by *labelled diagrams* which are bipartite multigraphs built from two sets of vertices being a column of white spots (WS) and column of black spots (BS) as shown below.

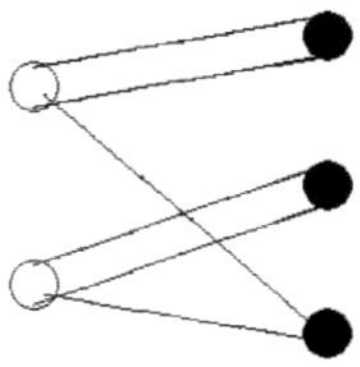

Labelled diagram of the matrix $\begin{pmatrix} 2 & 0 & 1 \\ 0 & 2 & 1 \end{pmatrix}$

Let us explain how to associate to a (drawn) diagram a packed matrix. The white (resp. black) spots are labelled from 1 to r (resp. 1 to c) from top to bottom and the number of lines from the i-th white spot to the j-th black spot is exactly the entry a_{ij} of the matrix. Conversely, a packed matrix of dimension $r \times c$ being given, one draws r white spots (resp. c black spots) and (with the labelling as above) join the i-th white spot to the j-th black by a_{ij} lines. This gives exactly the one-to-one correspondence between (drawn) diagrams and packed matrices.

In the sequel, we set $Diag_u := Dg_u \circ IM_u$ and $Diag_o := Dg_o \circ IM_o$ for the mappings which associate diagrams to pairs of partitions. Now, the multiplicity of a diagram $\mathcal{D}$ is the number of pairs $(P^{(1)}, P^{(2)})$ of unordered partitions such that $Dg_u(IM_u(P^{(1)}, P^{(2)})) = \mathcal{D}$.
Let us call bitype of a diagram $\mathcal{D}$ the pair $(\alpha(P^{(1)}), \alpha(P^{(2)}))$ where $Dg_u(IM_u(P^{(1)}, P^{(2)})) = \mathcal{D}$ (remark that it does not depend on the choosen premiage inside the formula) and we will refer it as the bitype $(\alpha(\mathcal{D}), \beta(\mathcal{D})))$ of $\mathcal{D}$. In a similar way $\alpha(\mathcal{D})$ (resp. $\beta(\mathcal{D})$) will be called the left (resp. the right) type of $\mathcal{D}$ and $|\mathcal{D}|$ $(= |\alpha(\mathcal{D})| = |\beta(\mathcal{D})|)$ will denote the number of edges of $\mathcal{D}$. The product formula now reads

$$exp\Big(\sum_{m\geq 1} \frac{L_m}{m!}\left(y\frac{\partial}{\partial x}\right)^m\Big)exp\Big(\sum_{n\geq 1}\frac{V_n}{n!}x^n\Big)\Bigg|_{x=0} =$$
$$\sum_{n\geq 0}\frac{y^n}{n!}\Big(\sum_{\substack{\mathcal{D}\text{diagram}\\ |\mathcal{D}|=n}} mult(\mathcal{D})\mathbb{L}^{\alpha(\mathcal{D})}\mathbb{V}^{\beta(\mathcal{D})}\Big) = \sum_{\mathcal{D}\text{ diagram}}\frac{mult(\mathcal{D})}{|\mathcal{D}|!}m(\mathcal{D}, \mathbb{L}, \mathbb{V}, y) \quad (27)$$

with

$$m(\mathcal{D}, \mathbb{L}, \mathbb{V}, y) := \mathbb{L}^{\alpha(\mathcal{D})}\mathbb{V}^{\beta(\mathcal{D})}y^{|\mathcal{D}|}. \quad (28)$$

3. Diagrammatic expansion of the double exponential formula
The main interest of the expansion (27) is that we can impose (at least) two types of rules on the diagrams

- on the diagrams themselves (selection rules) : on the outgoing degrees, ingoing degrees, total or partial weights (the graph is supposed oriented from white to black spots)
- on the set of diagrams (composition and decomposition rules): product and coproduct on the space of diagrams.

We have already such a structure on the space of monomials (i.e. the polynomials).
The (usual) product of polynomials is well known and amounts to the addition of the multidegrees. The (usual) coproduct is given by the substitution of a "doubled" variable to each variable [4, 20]. For example, with $P = x^2y^3$, we first form $(x_1 + x_2)^2(y_1 + y_2)^3$, expand and then separate (on the left) the "1" labelled variables and (on the right) the "2" labelled. As

$$\begin{aligned} P = {} & x_1^2y_1^3 + 3x_1^2y_1^2y_2 + 3x_1^2y_1y_2^2 + x_1^2y_2^3 + 2x_1y_1^3x_2 + 6x_1y_1^2x_2y_2 + \\ & 6x_1y_1x_2y_2^2 + 2x_1x_2y_2^3 + y_1^3x_2^2 + 3y_1^2x_2^2y_2 + 3y_1x_2^2y_2^2 + x_2^2y_2^3 \end{aligned} \tag{29}$$

one gets, with Δ the coproduct operator,

$$\begin{aligned} \Delta(P) \;=\; & x^2y^3 \otimes 1 + 3x^2y^2 \otimes y + 3x^2y \otimes y^2 + x^2 \otimes y^3 + \\ & 2xy^3 \otimes x + 6xy^2 \otimes xy + 6xy \otimes xy^2 + 2x \otimes xy^3 + \\ & y^3 \otimes x^2 + 3y^2 \otimes x^2y + 3y \otimes x^2y^2 + 1 \otimes x^2y^3. \end{aligned} \tag{30}$$

The space of polynomials with product and coproduct (and other items like neutrals, co-neutrals and antipode, which will be made more precise in the next paragraph) is endowed with the structure of a Hopf algebra.
The last consideration suggests the following question:

Is it possible to structure the (spaces of) diagrams into a Hopf algebra ?

Is it possible that this structure be compatible, in some sense, with the mapping $(\mathcal{D}, \mathbb{L}, \mathbb{V}, y) \to m(\mathcal{D}, \mathbb{L}, \mathbb{V}, y)$ *?*

Answer is yes. To establish it, we have to proceed in three steps.

- First Step : Define the space(s)
- Second Step : Define a product
- Third Step : Define a coproduct

3.1. Algebra structure

FIRST STEP. — Let $Diag_{\mathbb{C}}$ (resp. $LDiag_{\mathbb{C}}$) be the $\mathbb{C}$-vector space freely generated by the diagrams (resp. labelled diagrams) i.e.

$$Diag_{\mathbb{C}} := \bigoplus_{d \text{ diagram}} \mathbb{C} \qquad\qquad LDiag_{\mathbb{C}} := \bigoplus_{d \text{ labelleddiagram}} \mathbb{C}d \tag{31}$$

at this stage, we have a linear mapping (linear arrow) $LDiag_{\mathbb{C}} \to Diag_{\mathbb{C}}$ provided by the linear extension of Φ_d and an arrow (linear, by construction) $m(., \mathbb{L}, \mathbb{V}, z) : Diag_{\mathbb{C}} \to \mathbb{C}[\mathbb{L} \cup \mathbb{V} \cup \{z\}]$ provided by the linear extension of $m(., \mathbb{L}, \mathbb{V}, z)$.

SECOND STEP. — We remark that, if

$$d_1 \star d_2 = \boxed{\begin{array}{c} d_1 \\ d_2 \end{array}} \tag{32}$$

denotes the superposition of the diagrams, then

$$m(d_1 \star d_2, \mathbb{L}, \mathbb{V}, z) = m(d_1, \mathbb{L}, \mathbb{V}, z)m(d_2, \mathbb{L}, \mathbb{V}, z). \tag{33}$$

The law (32) makes sense as well for labelled and unlabelled diagrams. In the first case, it amounts to computing the blockdiagonal product of packed matrices. Indeed, for M_1, M_2 being packed matrices, one has

$$Dg_o\left(\begin{pmatrix} M_1 & 0 \\ 0 & M_2 \end{pmatrix}\right) = Dg_o(M_1) \star Dg_o(M_2). \tag{34}$$

This product yields the product of monomials in the following way. From $\mathcal{D}$ a diagram and all the other parameters fixed, with the setting of (28), we get a polynomial.

The product (32) is associative with unit (the empty diagram), it is compatible with the arrow Φ_d and so defines the product on *Diag* which, in turn is compatible with the product of monomials.

$$\begin{array}{ccccc} \text{Labelled diagrams}^2 & \xrightarrow{\Phi_d \times \Phi_d} & \text{Diagrams}^2 & \xrightarrow{m(?,\mathbb{L},\mathbb{V},z)\times m(?,\mathbb{L},\mathbb{V},z)} & \text{Monomials}^2 \\ \text{product}\big\downarrow & & \text{product}\big\downarrow & & \text{product}\big\downarrow \\ \text{Labelled diagrams} & \xrightarrow{\Phi_d} & \text{Diagrams} & \xrightarrow{m(?,\mathbb{L},\mathbb{V},z)} & \text{Monomials} \end{array} \tag{35}$$

Remark 3.1 *One sees easily that the set of labelled diagrams (resp. diagrams) is closed under products and therefore is a monoid. The spaces $LDiag_{\mathbb{C}}$ and $Diag_{\mathbb{C}}$ are thus algebras of these monoids [1, 2, 4].*

3.2. Admissible coproducts

For the coproduct on *LDiag*, we have several possibilities:

(i) split with respect to the white spots (two ways : by intervals and by subsets)

(ii) split with respect to the black spots (two ways : by intervals and by subsets)

(iii) split with respect to the edges

The discussion goes as follows:

a) (iii) does not give a nice identity with the monomials (when applying $d \mapsto m(d, ?, ?, ?)$) nor do (ii) and (iii) by intervals.

b) (i) and (ii) are essentially the same (because of the WS $\leftrightarrow$ BS symmetry).

In fact (i) and (ii) by subsets give a good representation and, moreover, they are appropriate for several physics models.

In the next section, we develop the possibility (i) and (ii) by subsets.

4. Hopf algebra structures associated with Δ_{WS} and Δ_{BS}

4.1. The philosophy of bi- and Hopf algebras thru representation theory

Let $\mathcal{A}$ be a k-algebra (k is a field as $\mathbb{R}$ or $\mathbb{C}$). In this paragraph, we consider associative algebras with unit (AAU). A representation of $\mathcal{A}$ is here a pair (V, ρ_V) where V is a k-vector space and $\rho_V : \mathcal{A} \to End_k(V)$ a morphism of k-algebras (AAU).

One can make operations with representations as direct sums and quotient of a representation by a sub-representation (a sub-representation is a subspace which is closed under the action of $\mathcal{A}$). In general, one does not know how to endow the tensor product (of two representations) and the dual (of a representation) with the structure of representation.

It is however classical in two cases: groups and Lie algebras.

If G is a group, a representation of G is a pair (V, ρ_V) where V is a k-vector space and $\rho_V : G \to Aut_k(V)$ a morphism of groups. If G is a Lie algebra, a representation of G is a pair (V, ρ_V) where V is a k-vector space and $\rho_V : G \to End_k(V)$ a morphism of Lie algebras

(i.e. $\rho_V([u,v]) = \rho_V(u)\rho_V(v) - \rho_V(v)\rho_V(u)$). These two cases enter the scheme of (AAU) as a representation of a group can be extended uniquely as a representation of its algebra kG and a representation of a Lie algebra as a representation of $\mathcal{U}_k(G)$, its envelopping algebra. These two constructions (kG and $\mathcal{U}_k(G)$) are (AAU).
For the sake of readibility let us denote in all cases $\rho_V(g)(u)$ by $g.u$ ($g \in G$ and $u \in V$).
If G is a group and V, W two representations, we construct a representation of G on $V \otimes W$ by

$$g.(u \otimes v) = g.u \otimes g.v \tag{36}$$

If G is a Lie algebra and V, W two representations, we construct a representation of G on $V \otimes W$ by

$$g.(u \otimes v) = g.u \otimes v + u \otimes g.v \tag{37}$$

This can be rephrased in saying that the action of g in the first case (group) is $g \otimes g$ and in the second (Lie algebra) $g \otimes 1 + 1 \otimes g$ (1 is here for the appropriate identity mapping). In the two cases, it amounts to give a linear mapping $\Delta : \mathcal{A} \to \mathcal{A} \otimes \mathcal{A}$ which will be called a coproduct.
One can show [6] that, if we want that this new operation enjoy "nice" properties (associativity of the tensor product etc...), one has to suppose that this coproduct is a morphism of (AAU) ($\mathcal{A} \otimes \mathcal{A}$ has received the structure of - non twisted - tensor product of algebras), that it is coassociative with a counit [6]. Let us make these requirements more precise.
The first says that for all $x, y \in \mathcal{A}$ one has $\Delta(xy) = \Delta(x)\Delta(y)$, the second that the two compositions

$$\mathcal{A} \xrightarrow{\Delta} \mathcal{A} \otimes \mathcal{A} \xrightarrow{\Delta \otimes 1_{\mathcal{A}}} \mathcal{A} \otimes \mathcal{A} \otimes \mathcal{A} \text{ and } \mathcal{A} \xrightarrow{\Delta} \mathcal{A} \otimes \mathcal{A} \xrightarrow{1_{\mathcal{A}} \otimes \Delta} \mathcal{A} \otimes \mathcal{A} \otimes \mathcal{A} \tag{38}$$

are equal, the third says that there is a mapping (linear form) $\varepsilon : \mathcal{A} \to k$ such that the compositions

$$\mathcal{A} \xrightarrow{\Delta} \mathcal{A} \otimes \mathcal{A} \xrightarrow{\Delta \otimes \varepsilon} \mathcal{A} \otimes k \xrightarrow{\text{nat}} \mathcal{A} \text{ and } \mathcal{A} \xrightarrow{\Delta} \mathcal{A} \otimes \mathcal{A} \xrightarrow{\varepsilon \otimes \Delta} k \otimes \mathcal{A} \xrightarrow{\text{nat}} \mathcal{A} \tag{39}$$

(where nat is for the natural mappings) are equal to the identity $1_{\mathcal{A}}$.
An algebra (AAU) together with a coproduct Δ and a counity ε which fulfills the three requirements above is called a **bialgebra**.

If, moreover one wants to have a nice dualization of the representations (i.e. nice structures for the duals $V^* = Hom(V, k)$), it should exist an element of $Hom(\mathcal{A}, \mathcal{A})$ such that the compositions

$$\mathcal{A} \xrightarrow{\Delta} \mathcal{A} \otimes \mathcal{A} \xrightarrow{\alpha \otimes 1_{\mathcal{A}}} \mathcal{A} \otimes \mathcal{A} \xrightarrow{\mu} \mathcal{A} \text{ and } \mathcal{A} \xrightarrow{\Delta} \mathcal{A} \otimes \mathcal{A} \xrightarrow{1_{\mathcal{A}} \otimes \alpha} \mathcal{A} \otimes \mathcal{A} \xrightarrow{\mu} \mathcal{A} \tag{40}$$

are equal to $e_A\varepsilon$ (where e_A denotes the unit of A). When a bialgebra possesses such an element (unique), it is called the **antipode** and the bialgebra a **Hopf algebra**. For more details and connections to physics, one can consult [6].

One can prove that the bialgebras constructed below possess an antipode and then are Hopf algebras.

4.2. Bialgebra structures on LDiag and Diag
The space spanned by the packed matrices has already received a structure of Hopf algebra, the algebra **MQSym** [19]. We briefly review the structure of this Hopf algebra.
We describe in details Δ_{WS} as the other coproduct is actually got by the same process but applied on the columns instead of the lines. Let M be a packed matrix of dimensions $k_1 \times k_2$ for every subset $X \in [1..k_1]$ we consider the matrix $\pi_X(M) := pack(M[X, [1..k_2]])$, the restriction

to the lines of X and then packed (with this restriction to the lines, we only need to perform a horizontal packing). Thus, the coproduct Δ_{WS} reads

$$\Delta_{WS}(M) = \sum_{X+Y=[1..k_1]} \pi_X(M) \otimes \pi_Y(M) \tag{41}$$

To avoid confusion we will call the supporting space $\mathcal{H}_{WS}$ (= **MQSym**). We keep the (total) grading of **MQSym** by the total weight (i.e. the sum of the coefficients) of the matrices. The packed matrices are a linear basis of $\mathcal{H}_{WS}$ = **MQSym**, thus every element expresses uniquely

$$x = \sum_{M \text{ packed}} \lambda_M(x) M \tag{42}$$

The coproduct above is cocommutative and with counit $\lambda_\emptyset$ where $\emptyset$ is the void matrix corresponding to the void diagram. This particular matrix will be denoted $1_{\mathcal{H}_{WS}}$.
For example, with the packed matrix above one has

$$\begin{aligned}\Delta_{WS}(\begin{pmatrix}2&0\\0&2\\1&1\end{pmatrix}) = & \begin{pmatrix}2&0\\0&2\\1&1\end{pmatrix} \otimes 1_{\mathcal{H}_{WS}} + (2) \otimes \begin{pmatrix}0&2\\1&1\end{pmatrix} + (2) \otimes \begin{pmatrix}2&0\\1&1\end{pmatrix} + \\ & (1\ \ 1) \otimes \begin{pmatrix}2&0\\0&2\end{pmatrix} + \begin{pmatrix}2&0\\0&2\end{pmatrix} \otimes (1\ \ 1) + \\ & \begin{pmatrix}2&0\\1&1\end{pmatrix} \otimes (2) + \begin{pmatrix}0&2\\1&1\end{pmatrix} \otimes (2) + 1_{\mathcal{H}_{WS}} \otimes \begin{pmatrix}2&0\\0&2\\1&1\end{pmatrix}\end{aligned}$$

This coproduct is compatible with the usual coproduct on the monomials for the constant alphabet $\mathbb{V} = \mathbf{1}_{\mathbb{N}}$ defined by $V_n = 1$ for all $n \geq 0$. Then, using Sweedler's notation, for this particular $\mathbb{V}$, if $\Delta_{WS}(d) = \sum d_{(1)} \otimes d_{(2)}$, one has

$$m(d, \mathbb{L}' + \mathbb{L}'', \mathbf{1}_{\mathbb{N}}, z) = \sum m(d_{(1)}, \mathbb{L}', \mathbf{1}_{\mathbb{N}}, z) m(d_{(2)}, \mathbb{L}'', \mathbf{1}_{\mathbb{N}}, z) \tag{43}$$

Thus, one sees easily that, with this structure (product with unit, coproduct and the counit), $LDiag_{\mathbb{C}}$ is a bialgebra graded in finite dimensions and then a Hopf algebra.
The arrow $LDiag_{\mathbb{C}} \to Diag_{\mathbb{C}}$ endows $Diag_{\mathbb{C}}$ with a structure of Hopf algebra.

5. Conclusion

The structure of the Hopf algebras $LDiag_{\mathbb{C}}, Diag_{\mathbb{C}}$, by a theorem of Cartier, Milnor and Moore [5, 22], is that of envelopping algebras of their primitive elements ($Diag_{\mathbb{C}}$, being commutative, is thus an algebra of polynomials).
Moreover, it appears that the structure described above is the starting point for a series of connections with mathematical and physical Hopf algebras. The coproduct Δ_{WS} is the cristallisation ($q = 1$) of a one-parameter deformation of coproducts (all coassociative) on $LDiag_{\mathbb{C}} \simeq$ **MQSym**, the other end ($q = 0$) being an infinitesimal coproduct isomorphic to $\Delta_{\mathbf{MQSym}}$. Recently, **FQSym** (a subalgebra of **MQSym**) has been established by Foissy [10] as a case in a family of Hopf algebras of decorated planar trees which is strongly related to other Hopf algebras like Connes-Kreimer's and Connes-Moscovici's [10, 11].

Acknowledgments

We would like to express our gratitude to Jean-Louis Loday, Jean-Bernard Zuber, Jean-Yves Thibon and Florent Hivert for stimulating interactions on the subject presented in this paper.

References

[1] Berstel J and Reutenauer C 1988 *Rational Series and their Languages EATCS Monographs on Theoretical Computer Science* (Springer)

[2] Bertelle C, Duchamp G H E and Khatatneh K Tables, Memorized Semirings and Applications *Preprint* arXiv: cs.MA/0502081

[3] Bender C M, Brody D C and Meister B K 1999 Quantum field theory of partitions *J. Math. Phys.* Vol **40**

[4] Bourbaki N 1989 *Elements of Mathematics: Algebra Chapters 1-3* (Springer-Verlag)

[5] Cartier P 1955-56 *Séminaire "Sopus Lie", 2ème année* (Faculté des Sciences de Paris)

[6] Chari V and Pressley A 1994 *A Guide to Quantum Groups* (Cambridge)

[7] Comtet L 1970 *Analyse Combinatoire* PUF Tome 1

[8] Comtet L 1970 *Analyse Combinatoire* PUF, Tome 2

[9] Flajolet P http://algo.inria.fr/flajolet/

[10] Foissy L *Bidendriform bialgebras, trees, and free quasi-symmetric functions* (LM-Reims) *Preprint* arXiv: math.RA/0505207

[11] Foissy L 2002 *Les Algèbres de Hopf des Arbres Enracinés Decorés* PhD Memoir (Reims University)

[12] Solomon A I, Blasiak P, Duchamp G, Horzela A and Penson K A 2004 Combinatorial Physics, Normal Order and Model Feynman Graphs *Proc. of the Symp. 'Symmetries in Science XIII'* Bregenz, Austria, eds B. Gruber, G. Marmo and N. Yoshinaga (Kluwer Academic Publishers) p 527 *Preprint* arXiv: quant-ph/0310174

[13] Solomon A I, Duchamp G, Blasiak P, Horzela A and Penson K A 2004 Normal Order: Combinatorial Graphs Quantum Theory and Symmetries *Proc. of the 3rd Int. Symp.* eds P.C. Argyres, T.J. Hodges, F. Mansouri, J.J. Scanio, P. Suranyi, and L.C.R. Wijewardhana (World Scientific Publishing) p 398 *Preprint* arXiv: quant-ph/0402082

[14] Duchamp G, Solomon A I, Penson K A, Horzela A and Blasiak P 2004 One-parameter groups and combinatorial physics *Proc. of the Symp. Third International Workshop on Contemporary Problems in Mathematical Physics (COPROMAPH3)* Porto-Novo, Benin, eds J. Govaerts, M. N. Hounkonnou and A. Z. Msezane (World Scientific Publishing) p 436 *Preprint* arXiv: quant-ph/04011262

[15] Solomon A I, Duchamp G, Blasiak P, Horzela P and Penson K A 2004 Partition functions and graphs: A combinatorial approach *Proc. of the XI Int. Conf. on Symmetry Methods in Physics (SYMPHYS-11)* Prague, Czech Republic, eds C. Burdik, O. Navratil and S. Posta (Dubna: JINR Publishers) *Preprint* arXiv: quant-ph/0409082

[16] Horzela A, Blasiak P, Duchamp P, Penson K A and Solomon A I 2004 A product formula and combinatorial field theory *Proc. of the XI Int. Conf. on Symmetry Methods in Physics (SYMPHYS-11)* Prague, Czech Republic eds C. Burdik, O. Navratil and S. Posta (Dubna: JINR Publishers) *Preprint* arXiv: quant-ph/0409152

[17] Blasiak P, Horzela A, Penson K A, Duchamp G H E, Solomon A I 2005 Boson normal ordering via substitutions and Sheffer-Type Polynomials *Phys. Lett.* A **338** 108

[18] Blasiak P, Penson K A, Solomon A I, Horzela A and Duchamp G H E 2005 Some useful formula for bosonic operators *Jour. Math. Phys.* **46** 052110

[19] Duchamp G, Hivert F and Thibon J Y 2002 Non commutative functions VI: Free quasi-symmetric functions and related algebras *International Journal of Algebra and Computation* Vol **12** No 5

[20] Joni S A and Rota G C 1979 Colgebras and Bialgebras in Combinatorics *Studies in Applied Mathematics* **61** 93-139

[21] Knuth D 1981 *The Art of Computer Programming* Tome I (Addison-Wesley)

[22] Milnor J and Moore J 1965 On the structure of Hopf Algebras *Ann. of Math.* **81** 211-264

[23] Stanley R P 2004 *Enumerative Combinatorics* vol. **2** (Cambridge Univ. Press)

[24] Weinzierl S 2004 Hopf algebra structures in particle physics *Eur. Phys. J.* C **33** *Preprint* arXiv: hep-th/0310124

[25] Varvak A Rook numbers and the normal ordering problem *Preprint* arXiv: math. CO/0402376

Paths, Virasoro characters and fermionic expressions

Trevor A Welsh

School of Mathematics, University of Southampton, Southampton SO17 1BJ, UK

E-mail: `taw@maths.soton.ac.uk`

Abstract. Baxter's corner transfer matrix method enables the one-point functions of certain 2D statistical models to be expressed as generating functions of weighted paths. In the case of the Forrester-Baxter models, the generating functions are Virasoro characters associated with the conformal minimal models $M(p,p')$. By combinatorially manipulating the paths, we obtain a quasiparticle description of the ensemble of paths. This directly leads to fermionic expressions for the Virasoro characters, and thence to interesting q-series identities.

1. Background

The models of Andrews, Baxter and Forrester [4], and of Forrester and Baxter [15] are 2-dimensional statistical models in which each grid site can be in one of a finite number of states. As specified in [15, eqn. (1.2.1)], interactions between the sites take place locally, only around each fundamental grid square. The interactions have been chosen so as to satisfy the star-triangle relation (also known as the Yang-Baxter relation). As a consequence, the models can be *solved*. In particular, the probability that a site takes a specific value may be found at temperatures close to the critical point. This probability is expressed in terms of the so-called one-point function and is most conveniently calculated using Baxter's corner transfer matrix method [5]. In the limit of an infinite size grid, the corner transfer matrices can be diagonalised, resulting in the one-point function being expressed as the generating function of certain infinite-length weighted paths [15]. In regime III, these one-point functions turn out to be the Virasoro characters $\chi^{p,p'}_{r,s}$ that arise in the minimal models of conformal field theory. In this talk, we combinatorially manipulate the paths to obtain expressions for the characters $\chi^{p,p'}_{r,s}$ that are of fermionic form. Equating these expressions with other well-known expressions for $\chi^{p,p'}_{r,s}$ results in q-series identities, the simplest of which are the celebrated Rogers-Ramanujan identities.

2. Paths

A path h of length L is an integer sequence $(h_0, h_1, h_2, \ldots, h_L)$ for which $|h_i - h_{i-1}| = 1$ for $1 \le i \le L$. By linking the points $(0, h_0)$, $(1, h_1), \ldots, (N, h_N)$ on the i-j plane we obtain a graph that conveniently depicts the path.

The Forrester-Baxter models are parametrised by integers p, p', a, b, c, L for which $1 \le p < p'$ with p and p' coprime, $1 \le a, b, c < p'$ with $c = b \pm 1$, $L \ge 0$ and $L \equiv (a-b) \bmod 2$. For these, we consider the set $\mathcal{P}^{p,p'}_{a,b,c}(L)$ of length L paths h for which $h_0 = a$, $h_L = b$ and $1 \le h_i < p'$ for $0 \le i \le L$.[1]

[1] The values of p and c do not feature here, but they are used later to specify how the paths are weighted.

For $1 \leq k \leq p'-1$, we refer to the region of the i-j plane between $j=k$ and $j=k+1$ as the kth band. Thus, when drawn on the i-j plane, the paths from $\mathcal{P}^{p,p'}_{a,b,c}(L)$ lie in the $p'-2$ bands between $j=1$ and $j=p'-1$. For $1 \leq r < p$, we shade the $\lfloor rp'/p \rfloor$th band and refer to it as the rth odd band. Bands that are not odd bands are referred to as even bands. The band structure for the case $(p,p')=(3,8)$, along with a typical path, is shown in Fig. 1.

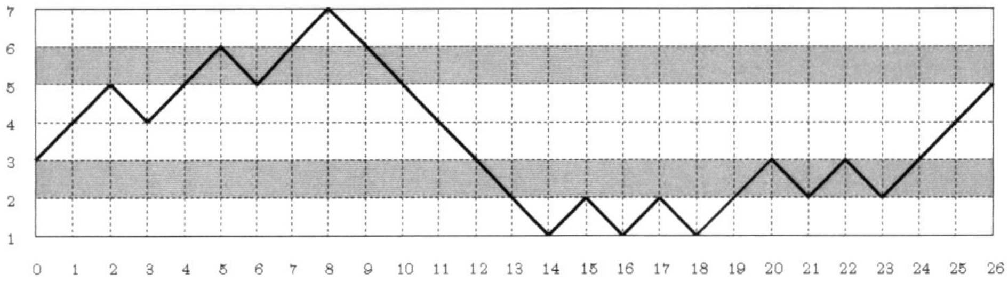

Figure 1. Typical path with shaded bands.

Below in Section 4, we will define a weight function on the set $\mathcal{P}^{p,p'}_{a,b,c}(L)$ of paths. However, we first examine one of the simplest cases in detail. This special case leads to one of the most famous results in mathematics.

3. A special case

In this section, we will consider the case where $p=2$, $p'=5$, $a=b=2$, $c=3$ and $L \in 2\mathbb{Z}_{\geq 0}$. Our approach is a recasting of that described in [7]. The band structure appropriate to this case, and a particular path are shown in Fig. 2.

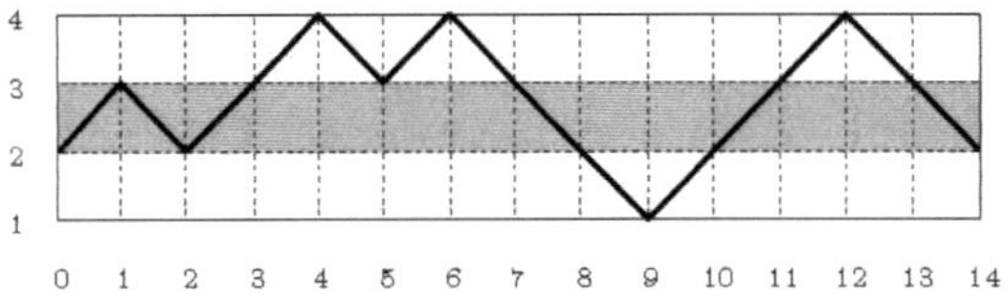

Figure 2. Typical element of $\mathcal{P}^{2,5}_{2,2,3}(14)$.

For this section, we define the weight $wt(h)$ of a path $h \in \mathcal{P}^{2,5}_{2,2,3}(L)$ by:

$$wt(h) = \sum_{\substack{i=1 \\ h_i \in \{1,4\}}}^{L} i. \tag{1}$$

Then the weight of the path depicted in Fig. 2 is $4+6+9+12=31$.

For $h \in \mathcal{P}^{2,5}_{2,2,3}(L)$ we say that there is a particle at position $i>0$ if and only if $h_i \in \{1,4\}$. Thus $wt(h)$ is the sum of the positions of the particles in h.

The generating function for $\mathcal{P}^{2,5}_{2,2,3}(L)$ is then defined by:

$$\chi^{2,5}_{2,2,3}(L) = \sum_{h \in \mathcal{P}^{2,5}_{2,2,3}(L)} q^{wt(h)}. \tag{2}$$

Now, for $n \geq 0$, consider the subset $\mathcal{P}^{2,5}_{2,2,3}(L)_n$ of $\mathcal{P}^{2,5}_{2,2,3}(L)$ that comprises those paths having precisely n particles. It is easy to see that $\mathcal{P}^{2,5}_{2,2,3}(L)_n$ has a unique element of minimal weight. It is the path $h^{\min}$ having particles at positions $1, 3, 5, \ldots, 2n-1$. Thus $h^{\min}_i = 2$ when i is even,

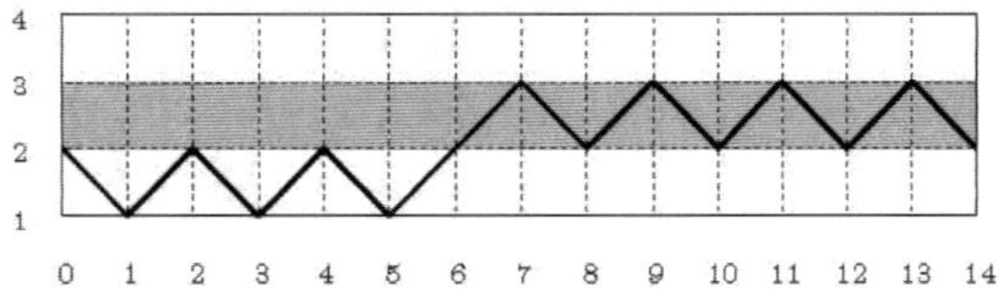

Figure 3. $h^{\min}$ in the case $n = 3$

$h^{\min}_i = 1$ for i odd and $i < 2n$, and $h^{\min}_i = 3$ for i odd and $i > 2n$. For $n = 3$, the path $h^{\min}$ is shown in Fig. 3. Using (1), we obtain

$$wt(h^{\min}) = 1 + 3 + 5 + \cdots + (2n-1) = n^2. \tag{3}$$

Other elements of $\mathcal{P}^{2,5}_{2,2,3}(L)_n$ can now be obtained from a sequence of local deformations of the path $h^{\min}$, with these deformations corresponding to a particle moving one position to its right. For instance, on moving the rightmost particle of the path given in Fig. 3 by a sequence of three steps, we obtain the paths given in Fig. 4. Note that it is possible to move this rightmost

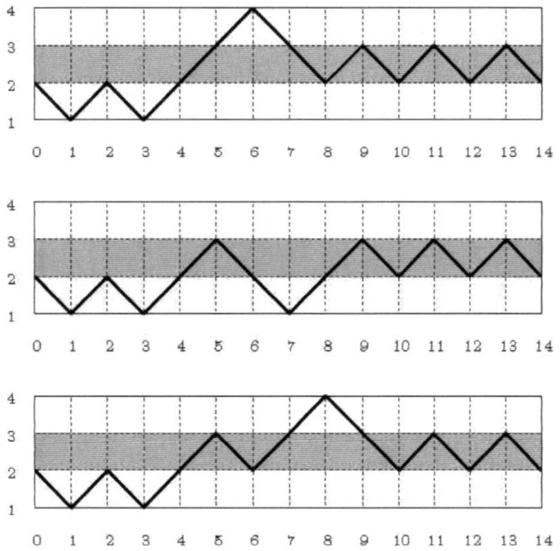

Figure 4. Three paths obtained from $h^{\min}$ by moving rightmost particle

particle at most $L - 2n$ steps and that with each step, the weight of the path increases by exactly one.

Having moved the rightmost particle, the next rightmost can be moved. For example, on moving the second particle in the final path of Fig. 4 by two steps, we obtain the two paths given in Fig. 5. Note that this second particle can move by at most the number of steps that the first particle has moved. This process continues with each of the particles moving at most as far as the previous. This process is akin to the movement of beads on the rungs of an abacus, as depicted in Fig. 6. Here the filled dots represent the particles and the unfilled dots represent

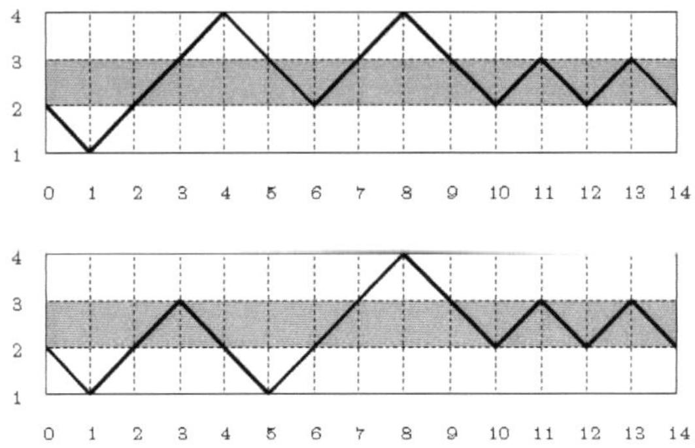

Figure 5. Paths obtained from $h^{\min}$ by moving two rightmost particles

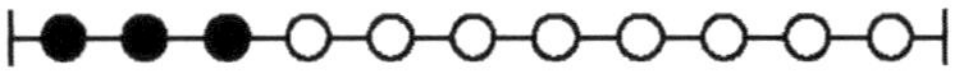

Figure 6. Particles on an abacus

vacancies. We see that there are $\binom{L-n}{n}$ configurations of the particles and vacancies. Moreover, since each movement of a particle by one step to the right increases the weight by one, we obtain:[2]

$$\sum_{h\in\mathcal{P}^{2,5}_{2,2,3}(L)_n} q^{wt(h)} = q^{n^2}\begin{bmatrix} L-n \\ n \end{bmatrix}_q, \tag{4}$$

where, as usual, the Gaussian polynomial $\begin{bmatrix} A \\ B \end{bmatrix}_q$ is defined to be:

$$\begin{bmatrix} A \\ B \end{bmatrix}_q = \begin{cases} \displaystyle\prod_{i=1}^{B} \frac{1-q^{A-B+i}}{1-q^i} & \text{if } 0\le B\le A; \\ 0 & \text{otherwise.} \end{cases} \tag{5}$$

By summing (4) over all values of $n \ge 0$, (2) yields:

$$\chi^{2,5}_{2,2,3}(L) = \sum_{n=0}^{\infty} q^{n^2}\begin{bmatrix} L-n \\ n \end{bmatrix}_q. \tag{6}$$

The expression for $\chi^{2,5}_{2,2,3}(L)$ that has just been obtained is known as a *fermionic* expression. This is due to the interpretation in terms of particles which cannot occupy identical positions. However, there exist other expressions for $\chi^{2,5}_{2,2,3}(L)$. For example, it may be shown that:[3]

$$\chi^{2,5}_{2,2,3}(L) = \sum_{k\in\mathbb{Z}} (-1)^k q^{\frac{1}{2}k(5k+1)} \begin{bmatrix} L \\ \lfloor \frac{1}{2}(L-5k)\rfloor \end{bmatrix}_q. \tag{7}$$

[2] This relies on the fact that the generating function for weighted partitions in an $M\times N$ box is given by $\begin{bmatrix} M+N \\ M \end{bmatrix}_q$. This fact is proved in [3, Theorem 3.1].

[3] First note that the path picture implies that $\chi^{2,5}_{2,2,3}(L) = q^{L-1}\chi^{2,5}_{2,2,3}(L-2) + \chi^{2,5}_{2,3,2}(L-1)$ for $L \in 2\mathbb{Z}_{>0}$ and $\chi^{2,5}_{2,2,3}(0) = 1$; and that $\chi^{2,5}_{2,3,2}(L) = q^{L-1}\chi^{2,5}_{2,3,2}(L-2) + \chi^{2,5}_{2,2,3}(L-1)$ for $L \in 2\mathbb{Z}_{>0}+1$ and $\chi^{2,5}_{2,3,2}(1) = 1$. These recurrence relations are then uniquely solved with *both* $\chi^{2,5}_{2,2,3}(L)$ and $\chi^{2,5}_{2,3,2}(L)$ given by the right side of (7).

Comparing (6) and (7) yields the identity (see [21] and [3, §9, Ex. 4]):

$$\sum_{n=0}^{\infty} q^{n^2} \begin{bmatrix} L-n \\ n \end{bmatrix}_q = \sum_{k\in\mathbb{Z}} (-1)^k q^{\frac{1}{2}k(5k+1)} \begin{bmatrix} L \\ \lfloor \frac{1}{2}(L-5k) \rfloor \end{bmatrix}_q, \tag{8}$$

which, although we have only derived it here for even L, holds for all $L \geq 0$. Taking the limit $L \to \infty$ then yields:

$$\sum_{n=0}^{\infty} \frac{q^{n^2}}{(q)_n} = \frac{1}{(q)_\infty} \sum_{k\in\mathbb{Z}} (-1)^k q^{\frac{1}{2}k(5k+1)}, \tag{9}$$

where $(q)_0 = 1$, $(q)_n = \prod_{i=0}^{n}(1-q^i)$ and $(q)_\infty = \prod_{i=0}^{\infty}(1-q^i)$. Note that the right side of this expression is a sum over terms of alternating sign. This is characteristic of a *bosonic* expression. (9) is then an example of a bosonic-fermionic identity.

By applying Jacobi's triple product identity [16, eqn. (II.28)] to the right side of (9), we obtain the first of the celebrated Rogers-Ramanujan identities:

$$\sum_{n=0}^{\infty} \frac{q^{n^2}}{(q)_n} = \prod_{n=1}^{\infty} \frac{1}{(1-q^{5n-4})(1-q^{5n-1})}. \tag{10}$$

As we will see below, the q-series given by either side of (9) or (10) is the Virasoro character $\chi_{1,2}^{2,5}$. In the next section, we report on efforts to extend the above results to other Virasoro characters $\chi_{r,s}^{p,p'}$.

4. The general case

For the case of general p, p', a, b, c, the weight function for the paths is best specified in terms of an (x, y)-coordinate system which is inclined at 45° to the original (i, j)-coordinate system and whose origin is at the path's initial point at $(i = 0, j = a)$. Specifically, $x = \frac{1}{2}(i - j + a)$ and $y = \frac{1}{2}(i + j - a)$. Note that at each step in the path, either x or y is incremented and the other is constant. In this system, the path depicted in Fig. 1 has its first few vertices at $(0, 1)$, $(0, 2)$, $(1, 2)$, $(1, 3)$, $(1, 4)$, $(2, 4)$, $(2, 5)$, $(2, 6)$, …

Now, for $1 \leq i \leq L$, we define the weight $v_i = v(h_{i-1}, h_i, h_{i+1})$ of the ith vertex according to its shape, the shading between h_i and h_{i+1}, and its (x, y)-coordinate, as specified in Table 1 (the nature of the vertex at $i = L$ is determined upon setting $h_{L+1} = c$). In this table, the bands

Vertex	v_i	Vertex	v_i	Vertex	v_i	Vertex	v_i
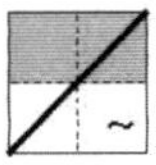	x	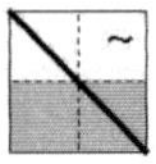	y	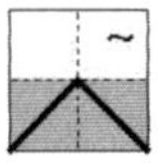	0		0
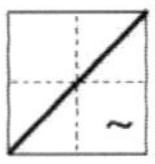	0	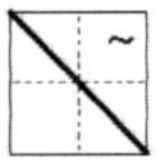	0	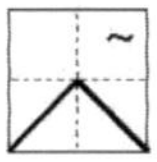	x		y

Table 1. Vertex weights.

highlighted with the $\sim$ symbol can be either even or odd. We shall refer to those four types of vertices for which, in general, the weight is non-zero, as *scoring* vertices. The other four types

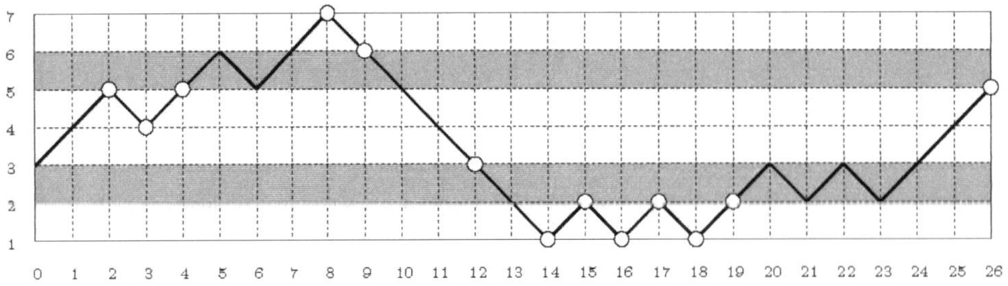

Figure 7. Typical path with scoring vertices circled.

of vertices will be termed *non-scoring*. In Fig. 7, the path of Fig. 1 has been reproduced with each of its scoring vertices circled.

We now define:

$$wt(h) = \sum_{i=1}^{L} v_i. \tag{11}$$

To illustrate this procedure, consider again the path $h \in \mathcal{P}^{3,8}_{3,5,6}(26)$ depicted in Figs. 1 and 7. Using Table 1, the scoring vertices at $i = 2, 3, 4, 8, 9, 12, 14, 15, 16, 17, 18, 19$ and 26 lead to: $wt(h) = 0 + 2 + 1 + 2 + 6 + 6 + 6 + 8 + 7 + 9 + 8 + 10 + 12 = 77$.

The generating function $\chi^{p,p'}_{a,b,c}(L)$ for the set of paths $\mathcal{P}^{p,p'}_{a,b,c}(L)$ is defined to be:

$$\chi^{p,p'}_{a,b,c}(L;q) = \sum_{h\in\mathcal{P}^{p,p'}_{a,b,c}(L)} q^{wt(h)}. \tag{12}$$

Often, we drop the base q from the notation so that $\chi^{p,p'}_{a,b,c}(L) = \chi^{p,p'}_{a,b,c}(L;q)$. The same will be done for other functions without comment.

We also define the set $\mathcal{P}^{p,p'}_{a,b,c}(L,m)$ to comprise those elements of $\mathcal{P}^{p,p'}_{a,b,c}(L)$ that have precisely m non-scoring vertices. The corresponding generating function is $\chi^{p,p'}_{a,b,c}(L,m;q) = \sum_{h\in\mathcal{P}^{p,p'}_{a,b,c}(L,m)} q^{wt(h)}$.

5. Vertex words and path manipulations

The vertex word $w(h)$ of a path h is the word in the letters S and N corresponding to the sequence of scoring and non-scoring vertices in h. The lengths of $w(h)$ and h are thus equal. For example, for the path $h \in \mathcal{P}^{3,8}_{3,5,6}(26)$ depicted in Figs. 1 and 7, we have

$$w(h) = NSSSNNNSSNNSNSSSSSSNNNNNNS. \tag{13}$$

By working from right to left, it is easy to see that the map from a particular $\mathcal{P}^{p,p'}_{a,b,c}(L)$ to the set of vertex words is injective. In view of this, we can manipulate paths by specifying changes to $w(h)$.

For the remainder of this section, we restrict our path manipulations to the cases for which $p' > 2p$, so that there are no neighbouring odd bands.

We now associate particles with pairs of two or more scoring vertices in a path: where there occur $d \geq 2$ consecutive scoring vertices, we associate $\lfloor d/2 \rfloor$ particles (when d is odd, the ambiguity over which pairs are the actual particles is seen to be irrelevant once the movement of the particles are considered). The movement of a particle by one step is defined to be the change corresponding to the local change $SSN \to NSS$ in $w(h)$, or vice-versa. It is easily checked that the corresponding altered path is also an element of $\mathcal{P}^{p,p'}_{a,b,c}(L)$. In addition, for the change $SSN \to NSS$, the weight of the path increases by precisely 1.[4]

As in Section 3, we then again have the analogy with beads moving on an abacus. Among all the paths obtained from h by consecutive particle movements, there is a unique path $h^{\min} \in \mathcal{P}^{p,p'}_{a,b,c}(L)$ of minimal weight obtained by moving all particles as far to the left as possible. The generating function for this set of paths is then $q^{wt(h^{\min})} \begin{bmatrix} n+m \\ n \end{bmatrix}_q$, where n is the total number of particles, and m is the number of non-scoring vertices in h (or $h^{\min}$).

In the case of the path $h \in \mathcal{P}^{3,8}_{3,5,6}(26)$ depicted in Fig. 7, $n = 5$, $m = 13$ and

$$w(h^{\min}) = SSSSSSSSSSNSNNNNNSNNNNNNNS. \tag{14}$$

Correspondingly, $h^{\min} \in \mathcal{P}^{3,8}_{3,5,6}(26)$ here is given in Fig. 8.

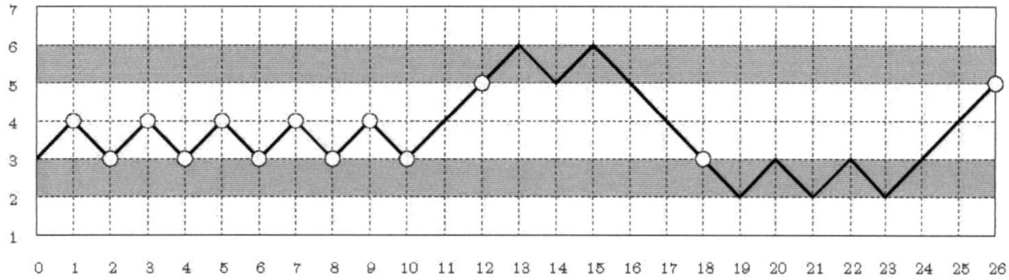

Figure 8. Typical $h^{\min}$.

Note that $h^{\min}$ begins with either $2n$ (or $2n+1$) consecutive scoring vertices. We now define the path $h^{\text{cut}} \in \mathcal{P}^{p,p'}_{a,b,c}(L-2n, m)$ by removing the first $2n$ vertices of $h^{\min}$. For the case of $h^{\min}$ depicted in Fig. 8, the resulting h^{cut} is shown in Fig. 9. Lemma A.1 shows that:

$$wt(h^{\min}) = wt(h^{\text{cut}}) + n(L-n-m). \tag{15}$$

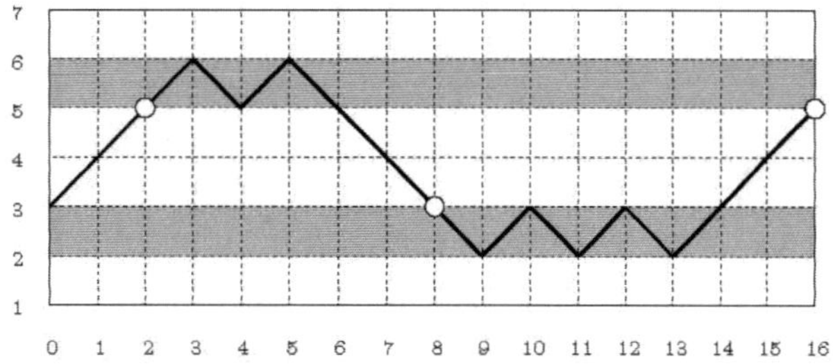

Figure 9. Typical h^{cut}.

Our construction above has ensured that $w(h^{\text{cut}})$ contains no consecutive pair SS. Our next aim is to construct, from such a path $h^{\text{cut}} \in \mathcal{P}^{p,p'}_{a,b,c}(L', m)$, a path $h^{\text{sqz}} \in \mathcal{P}^{p,p'-p}_{a',b',c'}(m, 2m-L')$ for

[4] Although not required here, an explicit description of the moves in terms of changes to the path may be found in [11, §2.3] or [23, §3.3].

certain a', b' and c' defined below, by shriking the features of the path. This is done by specifying that $w(h^{\rm sqz})$ is obtained from $w(h^{\rm cut})$ by removing the letter N immediately preceding each letter S (restrictions below will ensure that $w(h^{\rm cut})$ does not begin with S). This is possible because, in passing from $h^{\rm cut}$ to $h^{\rm sqz}$, the reduction in the distance between scoring vertices is compatible with the reduced distance between odd bands in the underlying grid: nearest neighbour odd bands become separated by one fewer even bands. The map from the path $h^{\rm cut} \in \mathcal{P}^{3,8}_{3,5,6}(16,13)$ of Fig. 9 to the path $h^{\rm sqz} \in \mathcal{P}^{3,5}_{2,3,4}(13,10)$ given in Fig. 10 provides an example of this construction.

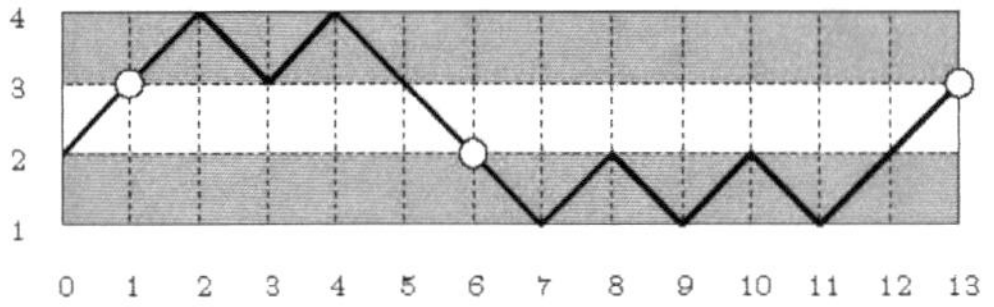

Figure 10. Typical $h^{\rm sqz}$.

However, in some case, complications arise at the leftmost and rightmost extremities of the path. Here, we avoid these by imposing restrictions on the values of a, b, c.[5]

To describe the restrictions, and to specify a', b' and c', define:

$$r^{p,p'}_{b,c} = \left\lfloor \frac{pc}{p'} \right\rfloor + \frac{b-c+1}{2}, \quad \rho^{p,p'}_j = \left\lfloor \frac{(j+1)p}{p'} \right\rfloor, \quad \text{and} \quad \beta^{p,p'}_{a,b,c} = r^{p,p'}_{b,c} - \rho^{p,p'}_a. \tag{16}$$

First, we set $a' = a - \rho^{p,p'}_a$ and $b' = b - r^{p,p'}_{b,c}$. For $2 \le j \le p'-2$, we designate j as *intereven*, *interfacial* or *interodd* according to whether $\lfloor (j+1)p/p' \rfloor - \lfloor (j-1)p/p' \rfloor$ is 0, 1 or 2 respectively. These cases correspond to j being between two even bands, between an even and an odd band, and between two odd bands respectively. If j is interfacial, then $\rho^{p,p'}_j$ gives the particular odd band that j borders. The upper and lower bands of the underlying grid are even or odd depending on whether $p' > 2p$ or $p' < 2p$. Correspondingly, in these two cases, we designate $j \in \{1, p'-1\}$ as *extraeven* or *extraodd*. Note that if $p' > 2p$ and $1 \le j < p'$ then j is either intereven, interfacial or extraeven.

For a, we restrict to the cases where a and a' are both extraeven, where a and a' are both interfacial, or where a is interfacial and a' is extraodd. It may be seen that these restrictions ensure that $w(h^{\rm cut})$ does not start with the letter S. For b and c, we restrict to the cases where b and b' are of the same nature (intereven, interfacial or extraeven), or where b is interfacial and either b' is interodd or extraodd. In the case where b and b' are of the same nature, we set $c' = c - r^{p,p'}_{b,c}$. In the other cases, we set $c' = b' + 1$ if $pb/p' < r$, and $c' = b - 1$ otherwise (this ensures that if b and c straddle the rth odd band, then so do b' and c').

In these restricted cases, it is easily checked that each symbol S_x (resp. S_y) in $w^*(h^{\rm cut})$ is immediately preceded by a N_x (resp. N_y). Application of Lemma A.2 then shows that:

$$wt(h^{\rm cut}) = wt(h^{\rm sqz}) + \frac{1}{4}((L'-m)^2 - \beta^2), \tag{17}$$

where $\beta = \beta^{p,p'}_{a,b,c}$. With $L' = L - 2n$, Combining (15) and (17) yields $wt(h^{\rm min}) = wt(h^{\rm sqz}) + \frac{1}{4}((L-m)^2 - \beta^2)$. The bijective nature of the three maps constructed above (taking into account the value of n and the sequence of moves required to obtain h from $h^{\rm min}$) now leads to:

[5] Some of these restrictions can be lifted after extending the constructions described here. These extensions and the ensuing complications are tackled to increasing degrees in [13, 14, 23].

Theorem 5.1 *Let $p' > 2p$. If a, b, c, a', b' and c' are restricted as above then:*

$$\chi^{p,p'}_{a,b,c}(L,m) = \sum_{n=0}^{\infty} q^{\frac{1}{4}((L-m)^2-\beta^2)} \chi^{p,p'-p}_{a',b',c'}(m, 2m+2n-L) \begin{bmatrix} m+n \\ m \end{bmatrix}_q . \tag{18}$$

We refer to this transformation between generating functions as a $\mathcal{B}$-transform.

6. The $\mathcal{D}$-transform

The $\mathcal{D}$-transform maps a path $h \in \mathcal{P}^{p,p'}_{a,b,c}(L,m)$ to a path $h^{\mathrm{dul}} \in \mathcal{P}^{p'-p,p'}_{a,b,c}(L, L-m)$. This map is defined by specifying that $w(h^{\mathrm{dul}})$ is obtained from $w(h)$ by exchanging each S for an N and vice-versa. Note that, since each band has changed parity, the shapes of h and h^{dul} are identical. For example, the $\mathcal{D}$-transform maps the path $h^{\mathrm{sqz}} \in \mathcal{P}^{3,5}_{2,3,4}(13)$ given in Fig. 10 to the path $h^{\mathrm{dul}} \in \mathcal{P}^{2,5}_{2,3,4}(13)$ given in Fig. 11.

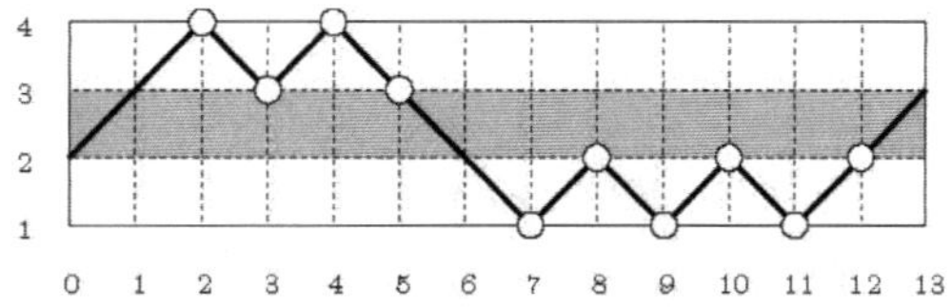

Figure 11. Typical h^{dul}.

Lemma A.3 shows that $wt(h^{\mathrm{dul}}) = \frac{1}{4}(L^2 - (a-b)^2) - wt(h)$. This leads to:

Theorem 6.1

$$\chi^{p,p'}_{a,b,c}(L,m;q) = q^{\frac{1}{4}(L^2-(a-b)^2)} \chi^{p'-p,p'}_{a,b,c}(L, L-m; q^{-1}). \tag{19}$$

Repeated use of the $\mathcal{B}$- and $\mathcal{D}$-transforms enables us to write a particular $\chi^{p,p'}_{a,b,c}(L,m)$ in terms of simpler expressions. In fact, if we define r_0 and s_0 to be the smallest non-negative integers such that $|p'r_0 - ps_0| = 1$, then we can express $\chi^{p,p'}_{s_0,s_0,s_0+1}(L,m)$ in terms of the trivial $\chi^{1,3}_{1,1,2}(L_*, m_*)$.[6] This latter generating function is easily seen to be given by:

$$\chi^{1,3}_{1,1,2}(L_*, m_*) = q^{\frac{1}{4}L_*^2} \delta_{m_*,0}. \tag{20}$$

The sequence of $\mathcal{B}$- and $\mathcal{D}$-transforms that is used depends on the continued fraction of p'/p.

7. Continued fractions

For p and p' coprime integers with $1 \le p < p'$, we say that $[c_0, c_1, c_2, \ldots, c_n]$ is the *continued fraction* for p'/p if:

$$\frac{p'}{p} = c_0 + \cfrac{1}{c_1 + \cfrac{1}{c_2 + \cfrac{1}{\ddots \; \cfrac{}{c_{n-1} + \cfrac{1}{c_n}}}}}$$

[6] A minor extension of our definitions enables us to use the even simpler $\chi^{1,2}_{1,1,2}(L_*, m_*) = \delta_{L_*,0}\delta_{m_*,0}$.

with $c_i \geq 1$ for $0 \leq i < n$, and $c_n \geq 2$.

We then define $t = c_0 + c_1 + \cdots + c_n - 2$ and also define:

$$t_k = -1 + \sum_{i=0}^{k-1} c_i, \tag{21}$$

for $0 \leq k \leq n+1$. Then $t_{n+1} = t+1$ and $t_n \leq t-1$.

We now use these values to define a $t \times t$ matrix $\boldsymbol{C}$, a $(t - t_1 - 1) \times (t - t_1 - 1)$ matrix $\overline{\boldsymbol{C}}$, and a $t_1 \times t_1$ matrix $\boldsymbol{B}$. First define C_{ji} for $0 \leq i, j \leq t$ by, when the indices are in this range,

$$\begin{array}{llll} C_{j,j-1} = -1, & C_{j,j} = 1, & C_{j,j+1} = 1, & \text{if } j = t_k, \quad k = 1, 2, \ldots, n; \\ C_{j,j-1} = -1, & C_{j,j} = 2, & C_{j,j+1} = -1, & 0 \leq j < t \text{ otherwise}, \end{array} \tag{22}$$

and set $C_{ji} = 0$ for $|i - j| > 1$. Also define $B_{ji} = \min\{i, j\}$ for $1 \leq i, j \leq t_1$. Then define the matrices:

$$\boldsymbol{C} = \{C_{ji}\}_{\substack{0 \leq j < t \\ 0 \leq i < t}}, \qquad \overline{\boldsymbol{C}} = \{C_{ji}\}_{\substack{t_1+1 \leq j < t \\ t_1+1 \leq i < t}}, \qquad \boldsymbol{B} = \{B_{ji}\}_{\substack{1 \leq j \leq t_1 \\ 1 \leq i \leq t_1}}.$$

Note that $\boldsymbol{C}$ and $\overline{\boldsymbol{C}}$ are both tri-diagonal. The matrices $\boldsymbol{C}$, $\overline{\boldsymbol{C}}$ and $\boldsymbol{B}^{-1}$ may be viewed as minor generalisations of Cartan matrices of type A.

We now obtain $\chi^{p,p'}_{s_0,s_0,s_0+1}(L, m)$ by applying the following sequence of $\mathcal{B}$- and $\mathcal{D}$-transforms to the expression for $\chi^{1,3}_{1,1,2}(L_*, m_*)$ given in (20):

$$\mathcal{B}^{c_0-1}\mathcal{D}\mathcal{B}^{c_1}\mathcal{D}\mathcal{B}^{c_2}\mathcal{D}\cdots\mathcal{D}\mathcal{B}^{c_{n-1}}\mathcal{D}\mathcal{B}^{c_n-2}. \tag{23}$$

Having thus obtained $\chi^{p,p'}_{s_0,s_0,s_0+1}(L, m)$, a final sum of all *even* m yields the following result (the proof is detailed in [11]):

Theorem 7.1 *Let $1 \leq p < p'$ with p and p' coprime. If L is even then:*

$$\chi^{p,p'}_{s_0,s_0,s_0+1}(L) = \sum_{\boldsymbol{m} \in (2\mathbb{Z}_{\geq 0})^{t-1}} q^{\frac{1}{4}\hat{\boldsymbol{m}}^T \boldsymbol{C} \hat{\boldsymbol{m}} - \frac{1}{4}L^2} \prod_{j=1}^{t-1} \begin{bmatrix} m_j - \frac{1}{2}(\boldsymbol{C}\hat{\boldsymbol{m}})_j \\ m_j \end{bmatrix}_q, \tag{24}$$

where the sum is over all vectors $\boldsymbol{m} = (m_1, m_2, \ldots, m_{t-1})^T$, each of whose integer components is even, and where we set $\hat{\boldsymbol{m}} = (L, m_1, m_2, \ldots, m_{t-1})^T$. Here, the components of $\boldsymbol{C}\hat{\boldsymbol{m}}$ are labelled by $0, \ldots, t-1$.

8. Minimal model Virasoro characters

For $1 < p < p'$ with p and p' coprime, and $1 \leq r < p$ and $1 \leq s < p'$, the Virasoro character $\chi^{p,p'}_{r,s}$ is given by the following *bosonic* expression [9, 18, 10]:

$$\chi^{p,p'}_{r,s} = \frac{1}{(q)_\infty} \sum_{\lambda=-\infty}^{\infty} \left(q^{\lambda^2 pp' + \lambda(p'r - ps)} - q^{(\lambda p + r)(\lambda p' + s)}\right). \tag{25}$$

The significance of our generating functions $\chi^{p,p'}_{a,b,c}(L)$ is that, if we set $r = r^{p,p'}_{b,c}$ (using (16)) then for $1 \leq r < p$,

$$\lim_{L \to \infty} \chi^{p,p'}_{a,b,c}(L) = \chi^{p,p'}_{r,a}, \tag{26}$$

assuming that $|q| < 1$.[7]

[7] As detailed in [11, 23], an explicit (*bosonic*) generating function for $\chi^{p,p'}_{a,b,c}(L)$ can be established, from which (26) readily follows.

In view of (26), we can now derive fermionic expressions for the characters $\chi^{p,p'}_{r_0,s_0}$ by taking the $L\to\infty$ limit of the expressions for $\chi^{p,p'}_{s_0,s_0,s_0+1}(L)$ given in Theorem 7.1. In the two cases $p=2$ and $p>2$, this yields:

Theorem 8.1 *If $k\geq 2$ then:*

$$\chi^{2,2k+1}_{1,k} = \sum_{\boldsymbol{n}\in\mathbb{Z}^{k-1}_{\geq 0}} \frac{q^{\boldsymbol{n}^T B\boldsymbol{n}}}{(q)_{n_1}(q)_{n_2}\cdots(q)_{n_{k-1}}}, \tag{27}$$

where $\boldsymbol{n}=(n_1,n_2,\ldots,n_{k-1})^T$.

Theorem 8.2 *If $2<p<p'$ with p and p' coprime then:*

$$\chi^{p,p'}_{r_0,s_0} = \sum_{\substack{\boldsymbol{n}\in\mathbb{Z}^{t_1}_{\geq 0}\\ \overline{\boldsymbol{m}}\in(2\mathbb{Z}_{\geq 0})^{t-t_1-1}}} q^{\tilde{\boldsymbol{n}}^T B\tilde{\boldsymbol{n}}+\frac{1}{4}\overline{\boldsymbol{m}}^T\overline{\boldsymbol{C}}\overline{\boldsymbol{m}}} \frac{1}{(q)_{m_{t_1+1}}} \prod_{j=1}^{t_1}\frac{1}{(q)_{n_j}} \prod_{j=t_1+2}^{t-1} \begin{bmatrix} m_j-\frac{1}{2}(\overline{\boldsymbol{C}}\overline{\boldsymbol{m}})_j \\ m_j\end{bmatrix}_q, \tag{28}$$

where $\boldsymbol{n}=(n_1,n_2,\ldots,n_{t_1})^T$ and $\overline{\boldsymbol{m}}=(m_{t_1+1},m_{t_1+2},\ldots,m_{t-1})^T$, using which we define $\tilde{\boldsymbol{n}}=(n_1,n_2,\ldots,n_{t_1}+\frac{1}{2}m_{t_1+1})^T$. Here, the components of $\overline{\boldsymbol{C}}\overline{\boldsymbol{m}}$ are labelled by $t_1+1,\ldots,t-1$.

By equating the expressions given in Theorems 8.1 and 8.2 with the corresponding instances of (25) we obtain q-series identities of bosonic-fermionic type.

In addition, for certain cases, a product form is also available for $\chi^{p,p'}_{r,s}$ (see [23, §1.1] for more information).

For example, application of Jacobi's triple product identity [16, eqn. (II.28)] to the expression for $\chi^{2,2k+1}_{1,k}$ given by (25) leads, via Theorem 8.1, to the following identities of Andrews and Gordon [1, 17]:

$$\sum_{n_1,n_2,\ldots,n_{k-1}\in\mathbb{Z}_{\geq 0}} \frac{q^{N_1^2+N_2^2+\cdots+N_{k-1}^2}}{(q)_{n_1}\cdots(q)_{n_{k-2}}(q)_{n_{k-1}}} = \prod_{\substack{n=1\\ n\not\equiv 0,\pm k \bmod (2k+1)}}^{\infty} \frac{1}{1-q^n}, \tag{29}$$

where $k\geq 2$ and $N_i=n_i+n_{i+1}+\cdots+n_{k-1}$ for $1\leq i<k$. The Rogers-Ramanujan identity (10) is the case $k=2$ here.

In addition, application of Watson's quintuple product identity [16, ex. 5.6] to the expressions for $\chi^{3,3k+1}_{1,k}$ and $\chi^{3,3k+2}_{1,k+1}$ given by (25) leads, via Theorem 8.2, to the following two sequences of identities:

$$\sum_{n_1,n_2,\ldots,n_k\in\mathbb{Z}_{\geq 0}} \frac{q^{N_1^2+N_2^2+\cdots+N_{k-1}^2+2N_k^2}}{(q)_{n_1}(q)_{n_2}\cdots(q)_{n_{k-1}}(q)_{2n_k}} = \prod_{\substack{n=1\\ n\not\equiv 0,\pm k\,(\bmod 6k+2)\\ n\not\equiv\pm(4k+2)\,(\bmod 12k+4)}}^{\infty} \frac{1}{1-q^n} \tag{30}$$

(c.f. [12, eqn. (A.4)] and [8, eqns. (2.1) & (2.5)]), and

$$\sum_{n_1,n_2,\ldots,n_k\in\mathbb{Z}_{\geq 0}} \frac{q^{N_1^2+N_2^2+\cdots+N_{k-1}^2+N_k^2}}{(q)_{n_1}(q)_{n_2}\cdots(q)_{n_{k-1}}(q)_{2n_k}} = \prod_{\substack{n=1\\ n\not\equiv 0,\pm(k+1)\,(\bmod 6k+4)\\ n\not\equiv\pm(4k+2)\,(\bmod 12k+8)}}^{\infty} \frac{1}{1-q^n} \tag{31}$$

(see [2]), where in both cases $k\geq 1$, and $N_i=n_i+n_{i+1}+\cdots+n_k$ for $1\leq i\leq k$.

9. Further results

In this talk, we have derived fermionic expressions for the characters $\chi^{p,p'}_{r_0,s_0}$ where r_0 and s_0 are dependent on p and p'. Over the past 15 years, fermionic expressions for a gradually widening range of $\chi^{p,p'}_{r,s}$ have been obtained (see [23, §1.1] for references). Recently, in [23], fermionic expressions for all Virasoro characters $\chi^{p,p'}_{r,s}$ have been obtained by extending the techniques presented in this talk. In general, these fermionic expressions are sums over terms similar to those appearing on the right side of (28). Intriguingly, as exhibited in [6, 14], only one such term occurs if and only if s (or $p'-s$) and r are respectively a *Takahashi length* and a *truncated Takahashi length* [22] associated with the continued fraction of p'/p.

Acknowledgments

I thank Omar Foda, Keith Lee and Yaroslava Pugai for collaboration on [11], upon which part of this talk is loosely based. It is also partly based on [23].

Appendix A

Knowledge of $w(h)$ alone is not sufficient to determine the weight $wt(h)$ of a path h. Instead, we can use the augmented vertex word $w^*(h)$ obtained from $w(h)$ by appending x or y to each letter in $w(h)$ depending on the direction of the path immediately preceding the corresponding vertex in h. For example, for the path $h \in \mathcal{P}^{3,8}_{3,5,6}(26)$ depicted in Fig. 7, we have:

$$w^*(h) = N_yS_yS_xS_yN_yN_xN_yS_yS_xN_xN_xS_xN_xS_xS_yS_xS_yS_xS_yN_yN_xN_yN_xN_yN_yS_y. \tag{A.1}$$

The definition of the weight given in (11) then implies that:

$$wt(h) = \#\{N_x \cdots S_y\} + \#\{S_x \cdots S_y\} + \#\{N_y \cdots S_x\} + \#\{S_y \cdots S_x\}, \tag{A.2}$$

where the symbol $\#\{A \cdots B\}$ denotes the number of pairs A, B of subscripted letters in $w^*(h)$ with A to the left of B.

Lemma A.1 *If the leftmost length $2n$ subword of $w^*(h^{\min})$ is $S_xS_yS_xS_y \cdots S_xS_y$ or $S_yS_xS_yS_x \cdots S_yS_x$, and $w^*(h^{\text{cut}})$ is obtained from $w^*(h)$ by removing this subword, then*

$$wt(h^{\min}) = wt(h^{\text{cut}}) + n(L - n - m), \tag{A.3}$$

where L is the length of h and m the number of its non-scoring vertices.

Proof: Since the two cases are very similar, we only consider the case where $w^*(h^{\min})$ begins with S_x. Consider the change in the right side of (A.2) in passing from $w^*(h^{\min})$ to $w^*(h^{\text{cut}})$. From S_x and S_y amongst the first $2n$ subscripted letters, $\#\{S_x \cdots S_y\}$ loses $1+2+\cdots+n = n(n+1)/2$ contributions, and $\#\{S_y \cdots S_x\}$ loses $0+1+\cdots+(n-1) = n(n-1)/2$ contributions. Apart from the first $2n$ subscripted letters, $w^*(h^{\min})$ contains $L-2n-m$ symbols S_x or S_y. The contribution from each of these symbols S_x to $\#\{S_y \cdots S_x\}$ decreases by n; and likewise the contribution from each of these symbols S_y to $\#\{S_x \cdots S_y\}$ decreases by n. Eqn. (A.3) follows. □

In the following two proofs, we let $\#S_x$ be the number of symbols S_x in $w^*(h)$ etc.,

Lemma A.2 *Let $h \in \mathcal{P}^{p,p'}_{a,b,c}(L,m)$ be such that there are no consecutive pairs SS in $w(h)$. If $h^{\text{sqz}} \in \mathcal{P}^{p,p'-p}_{a',b',c'}(m, 2m-L)$ is such that $w^*(h^{\text{sqz}})$ is formed from $w^*(h)$ by removing a symbol N_x immediately preceding each symbol S_x, and removing a symbol N_y immediately preceding each symbol S_y, then:*

$$wt(h^{\text{sqz}}) = wt(h) - \frac{1}{4}((L-m)^2 - \beta^2), \tag{A.4}$$

where $\beta = \beta^{p,p'}_{a,b,c}$.

Proof: In passing from h to h^{sqz}, (A.2) shows that the weight is reduced by the total number of pairs S_x and S_y in $w^*(h)$. Therefore:

$$wt(h) - wt(h^{\mathrm{sqz}}) = \#S_x \#S_y = \frac{1}{4}((\#S_x + \#S_y)^2 - (\#S_x - \#S_y)^2).$$

Since $\#S_x + \#S_y = L - m$, the lemma is proved if it can be shown that $\#S_y - \#S_x = \beta$. We do this using induction on path length.

Consider a path $h^\dagger \in \mathcal{P}^{p,p'}_{a,b^\dagger,c^\dagger}(L)$ and let $c^\ddagger = c^\dagger \pm 1$. Define $h^\ddagger \in \mathcal{P}^{p,p'}_{a,c^\dagger,c^\ddagger}(L+1)$ by $h^\ddagger_i = h^\dagger_i$ for $0 \le i \le L$ and $h^\ddagger_{L+1} = c^\dagger$. Let $\#S^\dagger_x$ and $\#S^\dagger_y$ be the number of symbols S_x and S_y in $w(h^\dagger)$, and define $\#S^\ddagger_x$ and $\#S^\ddagger_y$ correspondingly. With $\Delta\sigma = (\#S^\ddagger_y - \#S^\ddagger_x) - (\#S^\dagger_y - \#S^\dagger_x)$ and $\Delta r = r^{p,p'}_{b^\ddagger,c^\ddagger} - r^{p,p'}_{b^\dagger,c^\dagger} = \lfloor c^\ddagger p/p' \rfloor - \lfloor c^\dagger p/p' \rfloor + (2c^\dagger - b^\dagger - c^\ddagger)/2$, we claim that $\Delta\sigma = \Delta r$. To show this, we consider the nature of the $(L+1)$th vertex of $h^\ddagger$. The eight possible cases are those considered in Table 1. In the top left case, we have $\Delta\sigma = 1$. Since $c^\dagger = b^\dagger + 1$ and $c^\ddagger = b^\dagger + 2$, we also have $\Delta r = 1$ establishing the claim in this case. For the bottom left case, we obtain $\Delta\sigma = 0 = \Delta r$ as required. In the third case on the top row, $\Delta\sigma = 0$, and since $c^\dagger = b^\dagger + 1$ and $c^\ddagger = b^\dagger$, also $\Delta r = 0$. For the third case on the bottom row, we obtain $\Delta\sigma = 1 = \Delta r$ as required. The other four cases follow similarly. Since $\beta^{p,p'}_{a,b^\ddagger,c^\ddagger} - \beta^{p,p'}_{a,b^\dagger,c^\dagger} = \Delta r$, the induction step is complete, and the lemma is proved. □

Lemma A.3 *For $h \in \mathcal{P}^{p,p'}_{a,b,c}(L)$, let h^{dul} be such that $w^*(h^{\mathrm{dul}})$ is obtained from $w^*(h)$ by replacing each S_x, S_y, N_x and N_y by N_x, N_y, S_x and S_y respectively. Then:*

$$wt(h^{\mathrm{dul}}) = \frac{1}{4}(L^2 - (a-b)^2) - wt(h). \tag{A.5}$$

Proof: With $wt(h)$ given by (A.2), $wt(h^{\mathrm{dul}})$ is given by the same expression with the symbols S and N interchanged. Thus each pair of symbols in $w^*(h)$ that have different subscripts contributes to exactly one of $wt(h)$ or $wt(h^{\mathrm{dul}})$. Thus $wt(h) + wt(h^{\mathrm{dul}})$ is the number of such pairs, which is given by:

$$(\#S_x + \#N_x)(\#S_y + \#N_y) = \frac{1}{4}\{(\#S_x + \#N_x + \#S_y + \#N_y)^2 - (\#S_x + \#N_x - \#S_y - \#N_y)^2\}.$$

Since $L = \#S_x + \#N_x + \#S_y + \#N_y$ and $a - b = \#S_x + \#N_x - \#S_y - \#N_y$, this immediately yields (A.5). □

References

[1] Andrews G E 1974 An analytic generalization of the Rogers-Ramanujan identities for odd moduli *Proc. Nat. Acad. Sci. USA* **71** 4082–5

[2] Andrews G E 1984 Multiple series Rogers-Ramanujan type identities *Pac. J. Math.* **114** 267–283

[3] Andrews G E 1976 *The Theory of Partitions* (Reading MA: Addison-Wesley)

[4] Andrews G E, Baxter R J and Forrester P J 1984 Eight-vertex SOS model and generalized Rogers-Ramanujan-type identities *J. Stat. Phys.* **35** 193–266.

[5] Baxter R J 1982 *Exactly Solved Models in Statistical Mechanics* (London: Academic Press)

[6] Berkovich A, McCoy B M and Schilling A 1998 Rogers-Schur-Ramanujan type identities for the $M(p,p')$ minimal models of conformal field theory *Commun. Math. Phys.* **191** 325–395

[7] Bressoud D M 1989 Lattice paths and the Rogers-Ramanujan identities *Proceedings of the international Ramanujan centenary conference 1987 Madras* ed K Alladi Lecture Notes in Mathematics **1395**, (New York: Springer)

[8] Christe P 1991 Factorized characters and form factors of descendant operators in perturbed conformal systems *Int. J. Mod. Phys. A* **6** 5271–86

[9] Feigin B L and Fuchs D B 1983 Verma modules over the Virasoro algebra *Funct. Anal. Appl.* **17** 241–2
[10] Felder G 1989 BRST approach to minimal models *Nucl. Phys.* **B317** 215–236
[11] Foda O, Lee K S M, Pugai Y and Welsh T A 2000 Path generating transforms *Contemp. Math.* **254** 157–186
[12] Foda O and Quano Y-H 1997 Virasoro character identities from the Andrews-Bailey construction *Int. J. Mod. Phys. A* **12** 1651–75
[13] Foda O and Welsh T A 1999 Melzer's identities revisited *Contemp. Math.* **248** 207–234
[14] Foda O and Welsh T A 2000 On the combinatorics of Forrester-Baxter models *Physical Combinatorics* 1999 Kyoto ed M. Kashiwara and T. Miwa *Prog. Math.* **191** (Boston: Birkhäuser) pp 49–103
[15] Forrester P J and Baxter R J 1985 Further exact solutions of the eight-vertex SOS model and generalizations of the Rogers-Ramanujan identities *J. Stat. Phys.* **38** 435–472
[16] Gasper G and Rahman M 1990 *Basic Hypergeometric Series* (Cambridge: Cambridge University Press)
[17] Gordon B 1961 A combinatorial generalization of the Rogers-Ramanujan identities *Amer. J. Math.* **83** 393–9
[18] Rocha-Caridi A 1985 Vacuum vector representations of the Virasoro algebra *Vertex Operators in Mathematics and Physics* ed J Lepowsky *et al* (Berlin: Springer) pp 451–473
[19] Rogers L J 1894 Second memoir on the expansion of certain infinite products *Proc. London Math. Soc.* **25** 318–343
[20] Rogers L J and Ramanujan S 1919 Proof of certain identities in combinatory analysis *Proc. Cambridge Philos. Soc.* **19** 211–6
[21] Schur I 1917 Ein Beitrag zur additiven Zahlentheorie und zur Theorie der Kettenbrüche *S.-B. Preuss. Akad. Wiss. Phys.-Math. Kl.* 302–321 (Reprinted 1973 *Gesammelte Abhandlungen* vol 2 (Springer: Berlin) pp 117–136)
[22] Takahashi M and Suzuki M 1972 One-dimensional anisotropic Heisenberg model at finite temperatures *Prog. Theor. Phys.* **48** 2187–2209
[23] Welsh T A 2005 Fermionic expressions for minimal model Virasoro characters *Mem. Amer. Math. Soc.* **175** (no. 827)

Effects of symmetry lowering of spin systems and comparison with experiment.

S Thanos and A K Theophilou
Democritos NCSR, Institute of Materials Science, 15310, Athens, Greece

Abstract. The Heisenberg Hamiltonian is a highly symmetric Hamiltonian where group theory is very effective for the qualitative as well as the quantitative study of its eigenstates. For this matter, a system of nested algebras related to the Irreps of the symmetry group of the Hamiltonian was developed. Recently, experimental facts on magnetic materials showed that interactions lowering the symmetry, like the Dzialoshinski term, are necessary for their interpretation. A perturbation theory is developed for such Hamiltonians and applied to experimental data with very good agreement between theory and experiment.

1. Introduction

The Heisenberg Hamiltonian

$$H = \sum_{ij} J_{ij} \mathbf{S}_i \cdot \mathbf{S}_j \tag{1}$$

in spite of its simplicity, has been a successful model Hamiltonian for studying the magnetic properties of solids, molecules and clusters. This is a highly symmetric Hamiltonian as it is invariant under a permutation group reflecting the geometric symmetries of the physical system under consideration and a rotation group in the spin space. The geometric symmetries of the physical system under consideration can be represented by a permutation group of the spin indices. In practice it is more convenient to think in terms of the group of rotations and translation instead of the permutation group. Thus, one can replace the index i by the position $\mathbf{R}_i$ and define the translations T_{R_i} as follows

$$T_{R_i} \mathbf{S}_{R_j} T_{R_i}^{-1} = \mathbf{S}_{R_i + R_j} \tag{2}$$

while the rotation operator is defined accordingly,

$$\mathbf{R}_\omega \mathbf{S}_{R_j} \mathbf{R}_\omega^{-1} = \mathbf{S}_{(R_\omega R_j)}. \tag{3}$$

The rotations $\widehat{R}_\theta$ in the spin space are defined as

$$\widehat{R}_\theta \mathbf{S}_{R_j} R_\theta^{-1} = \mathbf{S}'_{R_j} \tag{4}$$

or explicitly

$$\begin{bmatrix} S_j^{\prime x} \\ S_j^{\prime y} \\ S_j^{\prime z} \end{bmatrix} = \widehat{R}_\theta \begin{bmatrix} S_j^x \\ S_j^y \\ S_j \end{bmatrix} \tag{5}$$

where $\widehat{R}_\theta$ is the rotation matrix in the three dimensional Euclidean space.

Beyond these symmetries, [illegible] also exist [1, 2]. In most studies particular emphasis is given on the Bethe ansatz. The physical picture of this type of solution is that it considers the holes (spin down lattice sites) as particles possessing a linear momentum. However, the existence of such type of solutions is not implied by any symmetry group of the Hamiltonian. The success of the Bethe ansatz is that it starts from a good physical picture, related to the fact that the number of holes is preserved under the action of the Hamiltonian H, as the total angular momentum $S^z = \sum S_j^z$ commutes with H. Beyond this, no advantage is made of the symmetry group of the Hamiltonian. The Bethe ansatz solutions have been investigated extensively in a series of papers by [1, 2, 3, 4]. In this paper we shall follow a different approach.

In a series of papers[5, 6, 7, 8, 9] we made full exploitation of the symmetry. Further, we classified the algebras of the spins in a sequence of algebras $\mathcal{A}_i$ with the inclusion relation, i.e. $\mathcal{A}_i \subset \mathcal{A}_{i+1}$. The advantage of this classification of algebras is that one can write part of the Heisenberg Hamiltonian H in terms of the Casimir operators of these algebras. The next step is to construct eigenstates of the Casimir operators which also belong to the Irreps of the permutation group of the indices. In this way, a part of the Hamiltonian H_0, is a linear combination of these operators and is automatically diagonal. The next step is to find the nonvanishing matrix elements of $H_a = H - H_0$ and diagonalize it. For this purpose it is much more convenient to write the operators of the various algebras as Irreducible tensor operators of G.

Thus the operators

$$\mathbf{b}_k = \sum e^{ik \cdot R_i} \mathbf{S}_{R_i} \tag{6}$$

transform according to the $\mathbf{k}$ Irrep of the group of the lattice translations. By taking into account the commutation relations

$$[\mathbf{S}^x_{R_i}, \mathbf{S}^y_{R_j}] = i\delta_{ij}\mathbf{S}^z_{R_j} \tag{7}$$

we find

$$[b^x_{\mathbf{k}}, b^y_{\mathbf{k}'}] = ib^z_{\mathbf{k}+\mathbf{k}'} \tag{8}$$

for the case of the space groups. Then it follows that the operators $\mathbf{b_0}$ form the $\mathcal{A}_0$ algebra since

$$[b^x_{\mathbf{0}}, b^y_{\mathbf{0}}] = ib^z_0 \tag{9}$$

Similarly when $\mathbf{K}$ is a reciprocal lattice vector, it follows that

$$\mathbf{b}_{\mathbf{K}+\mathbf{k}} = \mathbf{b}_{\mathbf{k}} \tag{10}$$

and therefore

$$[b^x_{\mathbf{K/2}}, b^y_{\mathbf{K/2}}] = ib^z_0 \tag{11}$$

since the label $\mathbf{K}$ is equivalent to 0 as far as the Irrep of the translation group is concerned. The new operators have the following commutation relations with those of the $\mathcal{A}_0$ algebra

$$[b^x_0, b^y_{\mathbf{K/2}}] = ib^z_{\mathbf{K}/2}. \tag{12}$$

Notice that $\mathbf{b}_{-\mathbf{K/2}} = \mathbf{b}_{K/2}$ since $\mathbf{b}_{-\mathbf{K}+\mathbf{K/2}} = \mathbf{b}_{\mathbf{K}/2}$.

In this way we get the $\mathcal{A}_1$ algebra containing $\mathcal{A}_0$ when our system is the one-dimensional lattice with periodic boundary conditions to be referred to as the linear ring in the following. For the multidimensional lattice where the reciprocal lattice vectors are $\mathbf{K}_1$,$\mathbf{K}_2$, $\mathbf{K}_3$, the vectors of the Brillouin zone are the $\frac{1}{2}\mathbf{K}_i$ and their linear combinations together with 0 label the $\mathcal{A}_1$ algebra.

The Casimir operators are $C_0 = \mathbf{b_0} \cdot \mathbf{b_0}$ for $\mathcal{A}_0$ and $C_1 = \mathbf{b_0} \cdot \mathbf{b_0} + \mathbf{b_{K/2}} \cdot \mathbf{b_{K/2}}$ and in this way the eigenvalues of $\mathbf{b_0} \cdot \mathbf{b_0}$ and $\mathbf{b_{K/2}} \cdot \mathbf{b_{K/2}}$ can be directly determined from the values of the Casimir operators.

In general. the operators $B_n = C_n - C_{n-1}$ form a commuting set. For this reason the eigenvalues and eigenstates of the operator

$$H_0 = \sum \Lambda_n B_n \tag{13}$$

with Λ_n real numbers can be found and they can be used as a complete orthonormal set by means of which one can find the exact eigenstates by diagonalizing the matrix D resulting from the matrix elements of $H - H_0$. In this way one can assign a physical meaning to the eigenstates considering the predominant eigenstate coefficient in the expansion of the exact solution. More details in the implementation of the hierarchy of algebras method can be found in Refs [5, 6]

2. Results and comparison with experiments.

One of the crystals extensively studied experimentally is the (N2H5)CuCl3 compound. We shall use here the data given by Hagiwara and coworkers[10]. For this crystal we adopted the model of 16-spin-1/2 ring with nearest neighbor (nn) and next nearest neighbor (nnn) interactions[11]. This model can simulate with excellent accuracy the magnetic behavior of this system. We have approached this problem with two different methods. First, we calculated the susceptibility of the 16-spin ring from the exact eigenstates with the appropriate coupling constants j_1 and j_2. Second, the same physical quantity was calculated for a 32-spin-1/2 ring, in which the interaction of the nearest neighbors was treated as a perturbation, because $j_1 < j_2$ according to Hagiwara [10]. In the latter case, we used the Landau thermodynamic perturbation theory for the free energy and the Gibbs average for the thermodynamic quantities, i.e. the thermal average of a physical quantity A is given by

$$< A >_T = \frac{Tr[Ae^{-\beta H}]}{Tr[e^{-\beta H}]} \tag{14}$$

where $\beta = 1/kT$. Our theoretical results about the magnetic susceptibility as a function of temperature are presented in figure 1. The relation $j_1 = 0.2j_2$ gives the best fitting of the theory to the experimental results. Both models give practically the same results for the magnetic susceptibility and are in good agreement with those derived experimentally[10]. In the low temperature limit the 32-spin model gives a better fit to the experimental results, as expected.

We come now to the NENP crystal, Ni(C2H6N2)NO2(ClO4), which is an one dimensional antiferromagnet with S=1, with only nn interactions. This crystal was studied experimentally by Nojiri and coworkers [12]. By adopting an 8-spin-1 rings model and applying the methodology of hierarchy of algebras we determined the exact eigenstates and eigenvalues of this system which are $3^8 = 6561$. We used these energy eigenstates to determine the thermal magnetization vs. magnetic field by applying the Gibbs thermal average. In figure 2 we present our theoretical results for NENP and compare them with the experimental ones from [12], using $j_1 = 45.5^o$K and g=2.23. The theoretically derived critical magnetic field is 122 Tesla in complete agreement to its experimental value. The coincidence between experimental and theoretical results is remarkable although a model of only 8 spins was used.

Another application of our method was to square lattice compounds and in particular to $K_2V_2O_5$. The above system presents a weak ferromagnetic behavior at low temperatures, when the external magnetic field is perpendicular to the z-axis, i.e. parallel to the xy plane of the square lattice [13]. This ferromagnetism was attributed to an additional interaction H_{DM}, introduced by Dzialoshinski and Moriya (DM) [14] which has the form

$$H_{DM} = j_{DM} \sum_{nn} (\mathbf{S}_{R_i} \times \mathbf{S}_{R_j})^z. \tag{15}$$

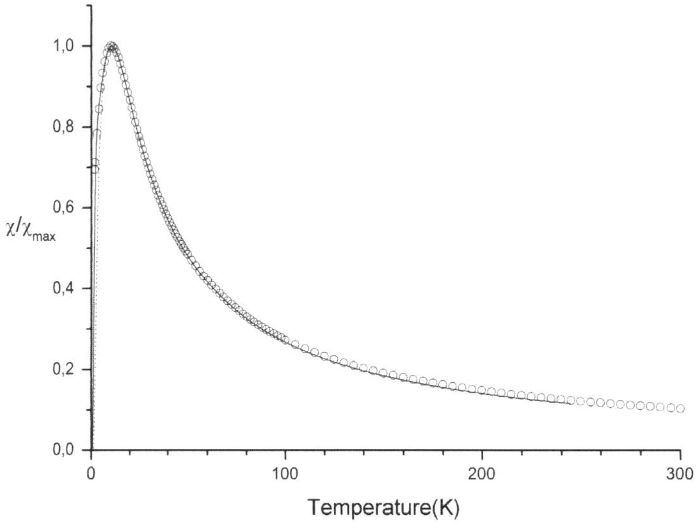

Figure 1. Magnetic susceptibility vs. temperature of $(N_2H_5)CuCl_3$ (circles) divided by its maximum value [10]. The solid line is the magnetic susceptibility of a 32-spin-1/2 ring and the dashed line that of a 16-spin-1/2, obtained theoretically by including nnn interaction and $j_1 = 0.2j_2$.

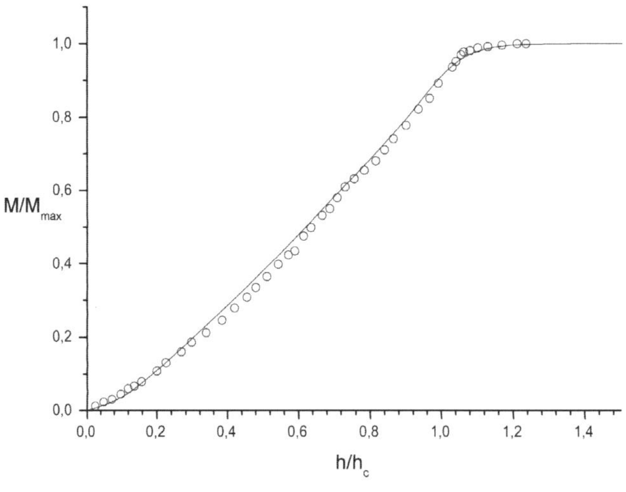

Figure 2. Magnetization of NENP at T=10°K vs. magnetic field. The magnetic field is divided by its critical value and the magnetization by its maximum value. The circles are the experimental values [12] and the solid line the theoretical results of an 8-spin-1 ring, without nnn-interaction.

In fact this is the smallest symmetry breaking interaction. Thus, the Hamiltonian considered is the following

$$H = \sum_{ij} J_{ij}\mathbf{S}_i \cdot \mathbf{S}_j + j_{DM} \sum_{nn} (\mathbf{S}_{R_i} \times \mathbf{S}_{R_j})^z - gh \sum \mathbf{S}^z_{R_i} \qquad (16)$$

where the first term is the nearest neighbor Heisenberg Hamiltonian, the second term the DM interaction between nearest neighbors and the third term is due to the external magnetic field. It is evident that the above Hamiltonian has lower symmetry than the typical Heisenberg one, due to the DM term, which is invariant only under rotations about the z-axis, while the Heisenberg one is invariant under any rotations. One can show that this lowering of symmetry causes the weak ferromagnetic effect at low temperatures when the external magnetic field is normal to the z-axis..

We dealt with the DM term of the Hamiltonian supposing that every spin at a site of the lattice interacts with the thermal average of the others. We treated this interaction in second order perturbation theory and the susceptibility of the crystal was calculated self-consistently.

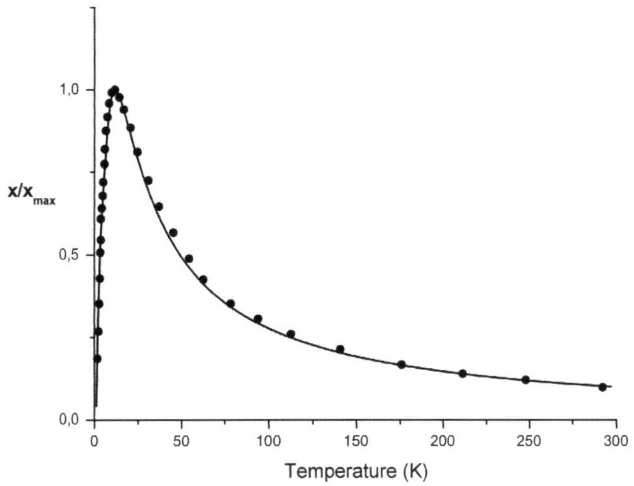

Figure 3. Experimental magnetic susceptibility vs. temperature of $K_2V_2O_5$ divided by its maximum value (circles), when the external magnetic field (H=100 Oe) is in the z axis of the crystal [13]. The solid line is the theoretical magnetic susceptibility, including the DM interaction.

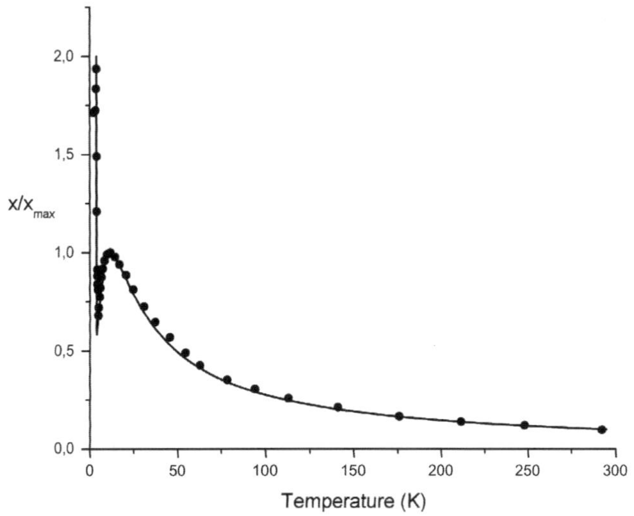

Figure 4. Theoretical (solid curve) and experimental (circles) magnetic susceptibility of $K_2V_2O_5$, when the external magnetic field (H=100 Oe) is in the xy plane of the crystal [13].

In figure 3 we present the theoretically calculated magnetic susceptibility together with the experimental one, for a magnetic field of 100 Oe in the z-direction of the crystal. No ferromagnetism appears as expected. One can see from this figure that, our theoretical results for the square lattice are in remarkable agreement with the experimental results presently available. For the z-direction of the external magnetic field the DM interaction does not change the qualitative characteristics of the crystal. This is easily concluded from group theoretical arguments. The coupling constant j with the best fit of the theoretical to the experimental results of Ref [13] is equal to 11.2°K, while the value adopted by these authors is of 12.6°K.

In figure 4 we present the same physical quantities, when the external field of 100 Oe is in the xy-plane [13]. For the value of j_{DM}= 0.12j, our theory gives a change of phase from the ferromagnetic state to the antiferromagnetic one at 4°K, in complete agreement with the experiment.

References

[1] Caspers W J, Kuzma M, Lulek B and Lulek T 1998 *Physica* A **252** 477

[2] Lulek B 1991 *Acta. Phys. Pol.* **22**, 371
Lulek B 1992 *J. Phys. Condensed Matter* **4** 8734

[3] Lulek T, Kuzma M and Olchawa R 1995 *Proc. Int. Conf. Symmetry and Structural Properties of Condensed Matter* ed T Lulek, W Florek *et al* (Singapore: World Scientific) p 433

[4] Tielen G J, Lulek T, Traa M R M J, Kuzma M and Caspers W J 1997 *Physica* A **246** 199

[5] Theophilou A K and Thanos S 1994 *Physica* B **202** 41

[6] Thanos S 1994 *Physica* B **202** 47 (1994)
Thanos S 1993 *Physica* B **192** 333

[7] Moustanis P N and Thanos S 1994 *Physica* B **202** 75
Moustanis P N and Thanos S 1999 *Physica* A **271** 418

[8] Moustanis P N, Theophilou A and Thanos S 1995 *Proc. Int. Conf. Symmetry and Structural Properties of Condensed Matter* ed T Lulek, W Florek *et al* (Singapore: World Scientific) p 97
Moustanis P N, Theophilou A and Thanos S 1995 *Proc. Int. Conf. Symmetry and Structural Properties of Condensed Matter* ed T Lulek, W Florek *et al* (Singapore: World Scientific) p 107

[9] Thanos S and Moustanis P N 1997 *Proc. Int. Conf. Symmetry and Structural Properties of Condensed Matter* ed T Lulek, W Florek *et al* (Singapore: World Scientific) p 180

[10] Hagiwara M, Narumi Y, Kindo K, Maechina N, Okunishi K, Sakai T and Takahachi M 2001 *Physica* B **83-86** 294

[11] Thanos S 2003 *Physica* A **327** 491

[12] Nojiri H, Shinamoto Y, Miura N and Ajiro Y 1995 *J. Phys.: Condens. Matter* **7** 5881

[13] Lumsden M D, Sales B C, Mandrus D, Nagler S E and Thomson J R 2001 *Phys. Rev. Lett.* **86** 159

[14] Moriya T 1960 *Phys. Rev.* **120** 91

Exact calculations for the ground state properties of the frustrated 2-dimensional Ising model

J A Blackman[1] and J Poulter[2]

[1] Department of Physics, University of Reading, Whiteknights, Reading RG6 6AF, UK
[2] Department of Mathematics, Mahidol University, Rama 6 Road, Bangkok 10400, Thailand

E-mail: [1]`j.a.blackman@reading.ac.uk`, [2]`scjpt@mahidol.ac.th`

Abstract. We have developed a technique that enables the exact calculation of the ground state properties of the 2-dimensional $\pm J$ Ising model. This paper gives an overview of the methodology and describes the latest results on the correlation function.

1. Introduction

The 2-dimensional Ising model is one of the most widely studied systems in statistical physics largely because it is one of the few non-trivial models that admit an exact solution. The original method of Onsager [1] based on Lie algebras was superseded by a combinatorial approach [2, 3] in which the partition function is expressed in terms of a determinant (or Pfaffian). One can alternatively express the formalism in terms of a Grassmann algebra [4, 5], in which the system takes the form of a non-interacting field theory in anticommuting variables (or an interacting field theory in three dimensions). Although the early formulation considered specifically the perfect Ising lattice, problems of disordered systems can also be addressed using the technique.

Perhaps the most studied disordered system is the so-called spin-glass problem [6] in which the key element is frustration. This is characterized by conflicts in the spin configurations needed to minimize the energy. Frustration can arise from the geometry of the lattice or from a quenched random distribution of positive (ferromagnetic) and negative (antiferromagnetic) bonds. One of the canonical models of frustrated systems is the 2-dimensional $\pm J$ Ising model.

We showed [7] a number of years ago how certain features of the 2D-dimensional $\pm J$ Ising model are particularly transparent when expressed within the determinantal formalism [3], and developed a technique for the exact calculation of the ground state energy and entropy of this system and applied it to the square [8, 9] and triangular [10] lattices. Recently we have extended these ideas to evaluate the ground state correlation functions [11]. This paper reviews the methodology and discusses the latest results.

2. Background

2.1. The model

The Hamiltonian for the system is written as

$$H = -\sum_{<ij>} J_{ij}\sigma_i\sigma_j, \tag{1}$$

where i and j label sites and the σ variables take the usual Ising model values of ± 1. The exchange interactions J_{ij} couple nearest-neighbour only and are quenched random variables of fixed magnitude but random sign ($\pm J$). The bonds are negative with a probability $p \in [0, 0.5]$ and we focus here on the square lattice. The model with $p = 0.5$ on the square lattice has attracted the most interest and it is generally accepted that spin glass behaviour occurs at zero temperature [12, 13, 14, 15, 16, 17], persisting down to a critical probability p_c of around 0.11 [9, 18, 19, 20, 21, 22, 23, 24, 25, 26, 27, 28].

The ground state properties we wish to address are the energy, entropy, and the correlation between spins separated by distance R which is expected to decay algebraically with a characteristic exponent η according to

$$[< \sigma_0 \sigma_R >^2]_{av} \sim R^{-\eta}. \qquad (2)$$

The subscript 'av' indicates a statistical average over configurations of bonds. The advantage of our approach is that it gives direct access to the ground state without the need to extrapolate from finite temperatures as is necessary, for example, in Monte Carlo based methods.

2.2. Formalism

The partition function for the nearest neighbour Ising model on an N-site square lattice at temperature T can be written generally as

$$Z = [\prod_{<ij>} \cosh(J_{ij}/kT)][Tr \prod_{<ij>} (1 + t_{ij}\sigma_i\sigma_j)], \qquad (3)$$

where the products are over the $2N$ pairs of nearest neighbour sites $< ij >$ and $t_{ij} = \tanh(J_{ij}/kT)$. The trace is over the 2^N spin configurations. The second factor in Eq. (3) can be written diagrammatically as a sum over closed polygons and this can be expressed in closed form as a determinant [2, 3]

$$Z = 2^N [\prod_{<ij>} \cosh(J_{ij}/kT)](\det D)^{1/2}. \qquad (4)$$

D is a $4N \times 4N$ skew-symmetric matrix. Its form can be visualised if we decorate each lattice site with four nodes (as shown in figure 1). D contains N blocks, Ω, down the diagonal; the elements of each block connect nodes associated with a single lattice site.

$$\Omega = \begin{pmatrix} 0 & -i & i & i \\ i & 0 & -i & i \\ -i & i & 0 & -i \\ -i & -i & i & 0 \end{pmatrix} \qquad (5)$$

D also contains elements that link nodes 1(3) of one site with 2(4) of neighbouring sites along the $x(y)$ Cartesian directions. For pairs of sites $< ij >$ these elements are $\mp it_{ij}$ where the minus and plus signs correspond respectively to the positive and negative Cartesian directions.

In the development that follows it is useful to consider the eigenvalues of D. They occur in pairs, $\pm\epsilon$, which are real since D is Hermitian (we can alternatively define D as real skew symmetric and perform a 2×2 block diagonalisation). For the perfect system (J_{ij} independent of $< ij >$), the eigenvalues fall into two bands whose upper and lower bounds are shown as a function of t in figure 2. The critical temperature is at $t = \sqrt{2} - 1$ where there is a zero eigenvalue.

One can also calculate the correlation function for spins at any pair of sites 0 and R. It is given by

$$< \sigma_0 \sigma_R >= \left[\frac{\det C}{\det D}\right]^{1/2} \prod_{path} t_{ij}, \tag{6}$$

where the product is over the bonds along an arbitrary path between the two sites, and C is obtained from D by replacing t_{ij} by $1/t_{ij}$ for all elements corresponding to bonds along that path. The reason for this is easy to see; to calculate the correlation function we require the quantity

$$Z' = [\prod_{<ij>} \cosh(J_{ij}/kT)][Tr \prod_{<ij>} \sigma_0 \sigma_R (1 + t_{ij}\sigma_i\sigma_j)], \tag{7}$$

Since $\sigma_i^2 = 1$, we can introduce a term $(\sigma_a^2 \sigma_b^2 \cdots)$ for all spins along a path joining σ_0 and σ_R. Then grouping these spins with the terms in parenthesis corresponding to links along the path, we can write $\sigma_i\sigma_j(1 + t_{ij}\sigma_i\sigma_j) = t_{ij}(1 + t_{ij}^{-1}\sigma_i\sigma_j)$, and the result immediately follows.

In the frustrated system we average over many configurations of quenched disorder and prefer to define correlations in the form given in Eq. (2), and so we will base our discussion on the expression

$$< \sigma_0 \sigma_R >^2 = \frac{\det C}{\det D} \prod_{path} t_{ij}^2. \tag{8}$$

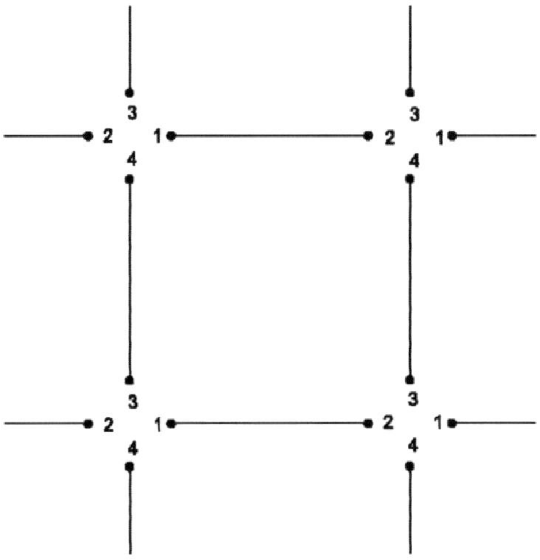

Figure 1. The labelling of nodes associated with a lattice site.

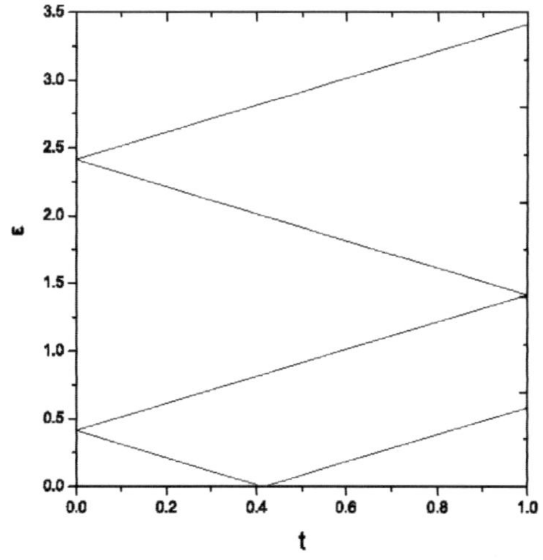

Figure 2. Spectral bounds (positive eigenvalues) for the perfect Ising system.

2.3. Simple defects

We consider the introduction of a few negative bonds into an otherwise perfect lattice. Defect eigenvalues appear at $\epsilon = 0$ in the $T \to 0$ $(t \to 1)$ limit. At small temperatures each eigenvalue can be written

$$\epsilon = \frac{1}{2} X \exp(-2rJ/kT), \tag{9}$$

where r is an integer and X is a real number. The number of defect eigenvalues is exactly equal to the number of frustrated plaquettes (elementary squares on the lattice containing an odd number of negative bonds).

Two examples will illustrate the behaviour. There are two frustrated plaquettes in figure 3 and two defect eigenvalues ($\pm\epsilon$); ϵ is as in Eq. (9) with $r = 1$ and $X = 1$. In figure 4 there are four frustrated plaquettes and so two pairs of defect eigenvalues of the form of Eq. (9); one has $r = 1$, $X = 1$ and the other $r = 3$, $X = 5$.

The free energy of the perfect lattice is given by $F = -2NJ$ (there are $2N$ bonds). With negative defects present there are unsatisfied bonds and the free energy can be expressed as $F = -2NJ + \Delta F$. It is straightforward to show that

$$\Delta F = 2J \sum_d r_d, \tag{10}$$

where the sum is over all defect eigenvalues. It can also be shown [8] that the ground state degeneracy, M, or entropy, S, can be written in terms of the factors, X, in Eq. (9)

$$M = \prod_d X_d, \tag{11}$$

$$S = k \sum_d \ln X_d. \tag{12}$$

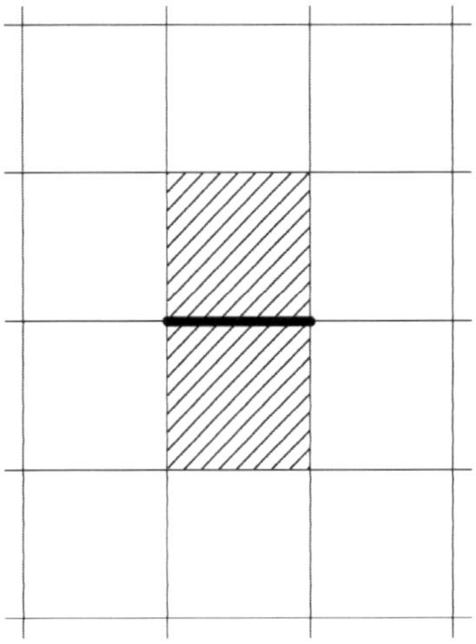

Figure 3. A single negative bond (denoted by a heavy line) in a lattice of positive bonds. Shading indicates frustrated plaquettes.

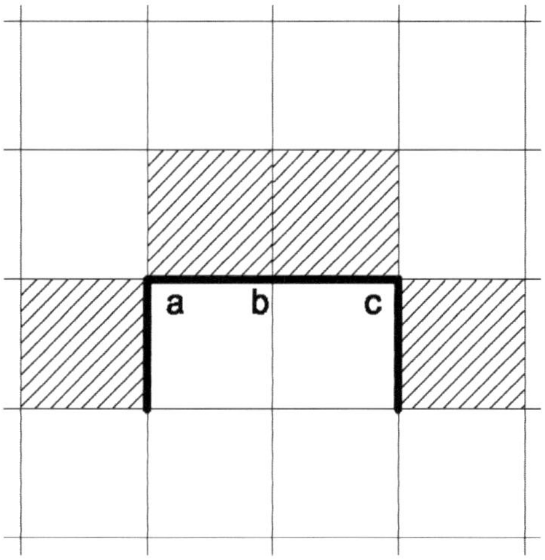

Figure 4. A more complex defect configuration with four negative bonds. Notation is as in figure 3.

Referring back to figure 3, we see that $r = 1$ and $X = 1$ imply that $\Delta F = 2J$ (there is one 'wrong' bond) and $M = 1$ (there is only one spin configuration that achieves the minimum energy). In figure 4, with ($r = 1$, $X = 1$) and ($r = 3$, $X = 5$), we get $\Delta F = 2J \times (1 + 3)$ (there are four 'wrong' bonds), and $M = 1 \times 5$. In this case flipping spins on sites a, b, c, or both a and c produces new configurations while leaving the ground state energy unchanged. In the next section we shall see how this rather simple structure can be developed to deal with an arbitrary concentration, p, of negative bonds.

3. The $\pm J$ system (energy and entropy)

There are several features of the procedure described above that allow us to develop a method that can deal with a system with an arbitrary concentration of negative bonds. The first is the form of the eigenvalues, Eq. (9), which suggests $\exp(-2J/kT)$ as a small parameter in the problem, and the use of perturbation theory to lift the degeneracy of the defect eigenvalues thus determining r (which will be the order of perturbation theory at which the degeneracy is lifted), and the associated prefactor, X. It should be noted that, although we are using perturbation theory, it is an *exact* solution for the ground state properties that emerges; once the perturbation expansion has been taken to a sufficiently high level to lift the degeneracy of all $T = 0$ eigenvalues, we have a complete solution.

The second feature, which makes it feasible to perform calculations on rather large lattices, is the fact that the zero temperature eigenstates are completely localised on the frustrated plaquettes. Let us make this statement a little more specific. We introduced a node basis as displayed in figure 1, and denote a basis state associated with a lattice site (x, y) as $\mid x, y; k\rangle$, where $k = 1 \cdots 4$. It is convenient to work in a bond basis. We form bond basis states as linear combinations of the node states associated with a particular bond. For the bonds connecting sites (x, y) and $(x + 1, y)$ along the x-axis, the new basis states are $[\mid x, y; 1\rangle \pm \mid x + 1, y; 2\rangle]$. We can associate one of these with the plaquette above the bond and one with the plaquette below. A similar procedure is adopted for the y direction with the linear combination $[\mid x, y; 3\rangle \pm \mid x, y + 1; 4\rangle]$.

We can associate four bond basis states with each plaquette and, assuming a judicious choice of assigning bond states to plaquettes is made, we find that the zero eigenvalue defect states (denoted by $\mid d\rangle$) are completely localised on their respective frustrated plaquette.

The perturbation that we have to deal with in considering small finite temperatures is also highly local. Denoting the $T = 0$ value of D by D_0 (by definition, $D_0 \mid d\rangle = 0$), we can write D as

$$D = D_0 + \delta D_1, \tag{13}$$

where $\delta = 1 - \tanh(J/kT)$. Eq. (13) is exact, but at low temperatures it will also be convenient to use δ as a small parameter in the perturbation expansion. The matrix elements of D_1 are zero except between the two basis states associated with a particular bond. If there are n frustrated plaquettes on an N site lattice, there are n defect states and $4N - n$ non-zero eigenstates which we refer to as continuum states: $D_0 \mid c\rangle = \epsilon_c \mid c\rangle$. Note all states occur as positive and negative pairs; we need only discuss the 2N positive ones.

3.1. Degenerate-state perturbation theory

The perturbation formalism used to lift the degeneracy of the defect states and determine r_d and X_d is developed in this section. The theory is based on Eq. (13). We write the ground-state defect eigenstates $\mid d\rangle$ more specifically as $\mid r, i\rangle$, where r is the order of perturbation theory at which degeneracy is lifted and i labels eigenstates of that order.

The eigenvalue equation is formally written

$$D \mid \Psi\rangle = \lambda \mid \Psi\rangle, \tag{14}$$

and λ and $\mid \Psi\rangle$ can be expanded in powers of δ

$$\lambda = \delta^n \epsilon_n^j + O(\delta^{n+1}) \tag{15}$$

$$\mid \Psi\rangle = \mid \Psi_0\rangle + \delta \mid \Psi_1\rangle + \cdots \tag{16}$$

where the eigenvalue is the jth defect eigenvalue of order n. $\mid \Psi_0\rangle = \mid n, j\rangle$ is the leading term for the corresponding eigenvector, that is $D_0 \mid n, j\rangle = 0$.

For $1 \leq m \leq n$, equating powers of δ, we obtain

$$D_0 \mid \Psi_m\rangle + D_1 \mid \Psi_{m-1}\rangle = \delta_{mn}\epsilon_n^j \mid n, j\rangle \tag{17}$$

By definition, $D_0 \mid r, i\rangle = 0$, so that

$$\langle r, i \mid D_1 \mid \Psi_{m-1}\rangle = \delta_{mn}\delta_{rn}\delta_{ij}\epsilon_n^j \tag{18}$$

Then, for $m = n = r = 1$ we now have

$$\langle 1, i \mid D_1 \mid 1, j\rangle = \delta_{ij}\epsilon_1^j \tag{19}$$

and the first order states are determined by diagonalization of D_1 in the defect basis set, that is the localized vectors associated with the frustrated plaquettes. This determines all the $r = 1$ states of Eq. (9).

For the zero eigenstates whose degeneracy is not lifted at first order, we proceed order by order until all degeneracy is lifted. At second order we need to consider intermediate states, the continuum states, $\mid c\rangle$. We introduce a Green's function $G_c(E)$ defined as

$$G_c(E) = \sum_c \mid c\rangle (E - \epsilon_c)^{-1} \langle c \mid, \tag{20}$$

where E is an energy parameter and ϵ_c are the eigenvalues of the continuum states.

Since we are concerned with the lifting of the degeneracy of the zero eigenvalue states, we require specifically $G_c(E = 0)$, which henceforward will be designated simply as G_c. By detailed consideration of the matrix elements we can show that G_c takes a simple form. It can be expressed as

$$G_c = (1 - P)g_c(1 - P), \tag{21}$$

where P is a projection operator,

$$P = \sum_d \mid d\rangle\langle d \mid, \tag{22}$$

and g_c is a quantity we term the continuum propagator. Detailed expressions for g_c within the bond basis are available in ref. [8].

Now, acting with a continuum state on equation (17), we have

$$\epsilon_c\langle c \mid \Psi_m\rangle = -\langle c \mid D_1 \mid \Psi_{m-1}\rangle \tag{23}$$

Then using closure

$$\sum_{r,i} \mid r, i\rangle\langle r, i \mid + \sum_c \mid c\rangle\langle c \mid = 1, \tag{24}$$

and the definition of the continuum Green's function in equation (20), we obtain

$$\mid \Psi_m\rangle = G_c D_1 \mid \Psi_{m-1}\rangle + \sum_{r,i} Z_r^i \mid r, i\rangle, \tag{25}$$

where Z_r^i is equal to $\langle r, i \mid \Psi_m\rangle$, but for our purposes these are just coefficients to be determined. Then, using equation (18) and (21), we can write in terms of a new set of coefficients Z_r^i,

$$| \Psi_m\rangle = g_c D_1 | \Psi_{m-1}\rangle + \sum_{r,i} Z_r^i | r, i\rangle. \tag{26}$$

Now, acting on Eq. (26) with $\langle r, i | D_1$, and using Eq. (18), we obtain

$$\delta_{m,n-1}\delta_{rn}\delta_{ij}\epsilon_n^j = \langle r, i | D_2 | \Psi_{m-1}\rangle + \delta_{r1} Z_1^i \epsilon_1^i, \tag{27}$$

where

$$D_2 = D_1 g_c D_1. \tag{28}$$

With $r = n = 2$ and $m = 1$, we obtain

$$\epsilon_2^j \delta_{ij} = \langle 2, i | D_2 | 2, j\rangle, \tag{29}$$

which completes the analysis as far as second order ($r = 2$).

For orders higher than 2, we insert $r = 1$ into Eq. (27) which gives

$$Z_1^i = -\langle 1, i | D_2 | \Psi_{m-1}\rangle / \epsilon_1^i, \tag{30}$$

and hence from Eq. (26)

$$| \Psi_m\rangle = (1 + G_1 D_1) g_c D_1 | \Psi_{m-1}\rangle + \sum_{r>1,i} Z_r^i | r, i\rangle. \tag{31}$$

G_1 is a first-order Green's function, which is a special case of the general definition

$$G_r = -\sum_i | r, i\rangle (1/\epsilon_r^i) \langle r, i | \,. \tag{32}$$

Acting on Eq. (31) with $\langle r, i | D_2$ and using Eq. (27) gives

$$\delta_{m,n-2}\delta_{rn}\delta_{ij}\epsilon_n^j = \langle r, i | D_3 | \Psi_{m-1}\rangle + \delta_{r2} Z_2^i \epsilon_2^i, \tag{33}$$

where

$$D_3 = D_2(1 + G_1 D_1) g_c D_1 \tag{34}$$

With $r = n = 3$ and $m = 1$, we obtain

$$\epsilon_3^j \delta_{ij} = \langle 3, i | D_3 | 3, j\rangle, \tag{35}$$

which completes the analysis as far as third order ($r = 3$).

The analysis to arbitrary order follows through in a similar fashion with the rth order eigenvalue being given by

$$\epsilon_r^j \delta_{ij} = \langle r, i | D_r | r, j\rangle, \tag{36}$$

where

$$D_r = D_{r-1}(1 + G_{r-2} D_{r-2}) \cdots (1 + G_1 D_1) g_c D_1 \tag{37}$$

3.2. *Results*

We have performed numerical calculations [8, 9] based on the above theory to obtain the full set of r and X values for large finite lattices of size $L \times L$. The largest value of r (order of perturbation theory necessary to complete the calculation) varies between configurations of disorder but is seldom larger than 12. Averages are performed over many configurations of disorder. Calculations were performed with $L = 64, 128$ and 256 and the energy and entropy extrapolated to an infinite lattice using the forms

$$F(L) = F + \alpha L^{-1}, \tag{38}$$

$$S(L) = S + \beta L^{-1}, \tag{39}$$

where $F(L)$ and $S(L)$ are the values for an $L \times L$ lattice and F and L are the $L \to \infty$ limits. The results obtained for a selection of values of p (the concentration of negative bonds) is shown in Table 1. Between 300 and 1000 samples are used in each configurational average. The perfect Ising ferromagnet corresponds to $p = 0$ and has values of F and S of -2 and 0 respectively.

Table 1. Ground state energy F (in units of J) and entropy S (in units of k) per spin for selected values of p.

p (%)	F	S
5	-1.8024±0.0005	0.0103±0.0002
8	-1.6924±0.0005	0.0252±0.0003
10	-1.6274±0.0003	0.0363±0.0004
11	-1.5994±0.0003	0.0405±0.0004
12	-1.5750±0.0003	0.0440±0.0004
15	-1.5182±0.0003	0.0528±0.0005
20	-1.4590±0.0005	0.0620±0.0003
50	-1.4021±0.0002	0.0709±0.0004

3.3. *Critical value of p*

The perfect Ising ferromagnet has $p = 0$ while $p = 50\%$ for the canonical $\pm J$ spin-glass. In between there is a critical concentration, p_c, that marks the transition from pure Ising to spin-glass behaviour. There is no singular behaviour in either F or S to flag this transition so we take a different approach.

As noted earlier the eigenvectors associated with the zero eigenvalue states are localised on the respective frustrated plaquettes. Lifting the degeneracy as described produces eigenvectors that are linear combinations of these basis states. As frustration is increased by introducing more negative bonds these eigenvectors become more extended. This suggests that p_c might be marked by a transition from localised to extended states indicating the demarcation of a regime in which frustration is a local perturbation to one in which frustration dominates the whole lattice.

We denote the linear size of an eigenvector by l and the number of states on an $L \times L$ lattice whose size is larger than l by $N(l)$. The procedure for obtaining l numerically is described elsewhere [8, 9]. We find that, if $p > p_c$, $N(l)$ fits rather well to a power law

$$N(l) \sim L^2 l^{-\rho}. \tag{40}$$

The number of defect eigenvectors whose size is comparable to that of the lattice is $\sim L^{(2-\rho)}$. This approaches zero or infinity as $L \to \infty$ according to whether $\rho > \rho_c$ or $\rho < \rho_c$, where $\rho_c = 2$. We thus take $\rho < \rho_c$ as the criterion for the presence of extended states.

The results for a number of values of p are shown in figures 5 and 6. The fall-off in the data at higher values of l is due to the finite size of the samples. We find good straight-line behaviour over the concentration range $11.5\% \leq p \leq 50\%$ with a value of the exponent of $\rho = 1.66 \pm 0.02$ over the entire range. At lower concentrations ($p \leq 10.5\%$), as shown in figure 6, a straight-line fit to the data is no longer feasible, and clearly there is a faster fall-off in $N(l)$ with increasing l than a power law indicating that the effects of frustration are extremely localised. This is strong confirmation that $p_c = 11.0 \pm 0.5\%$.

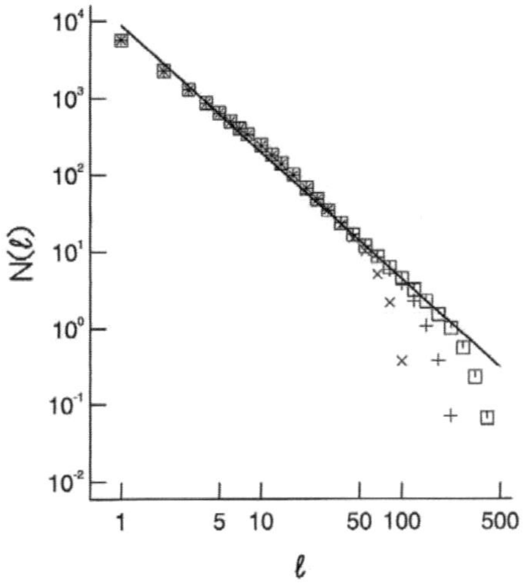

Figure 5. Distribution $N(l)$ (normalised to a 256×256 sample size) as a function of l for $p = 50\%$ for samples with $L = 64(\times), 128(+), 256(\square)$. A straight line fit to the $L = 256$ data points is also shown.

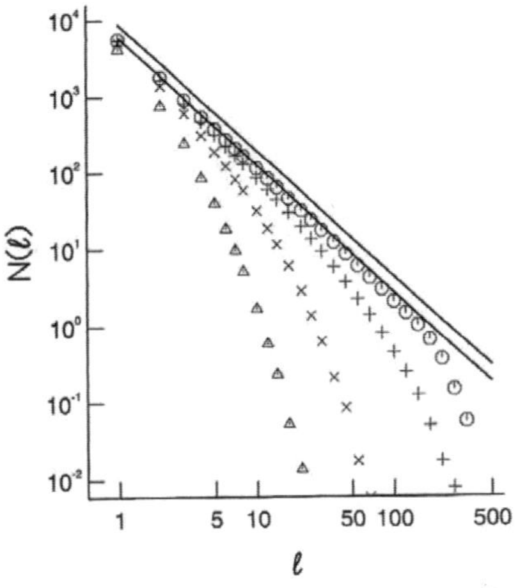

Figure 6. Distribution $N(l)$ for $L = 256$ samples as a function of l for $p = 11\%(\circ), p = 10\%(+), p = 8\%(\times), p = 5\%(\triangle)$. Straight line fits to $p = 11.5\%$ data and $p = 50\%$ data (from figure 5) are shown.

4. The $\pm J$ system (correlation function)

We now turn to recent developments [11] of our approach to address spin correlations. To our knowledge the only direct computations of spin correlation functions in the ground state are those of Ozeki [29]. This work used a numerical transfer matrix method with long thin samples of circumference L wrapped around a cylinder of length $9L$ with open ends. The maximum possible circumference for this study was limited to $L = 12$, due to the transfer matrix computational requirements scaling exponentially with L.

All other attempts to study spin correlations in the ground state have been indirect, involving extrapolation from finite temperature. Monte Carlo techniques have been applied to obtain results for quite low temperatures, for example $0.86J$ [30] and $0.63J$ [31]. However, the reliability of the extrapolations to zero temperature is hard to assess.

The transfer matrix method has also been used to obtain spin correlations functions at finite temperature [23, 24, 32], although the sample shape restrictions remain severe with cylindrical circumferences limited to $L \leq 20$. A related approach where the computational requirements scale as L^3, rather than exponentially, is the network model [26]. This has been successful in studying systems without severe shape restrictions, but only at finite temperature. The zero temperature limit remains inaccessible.

4.1. Development of the formalism

We need to consider the matrix C that appears in Eq. (8) for the correlation function. At zero temperature $C = D_0$ since $t = 1$ at $T = 0$. At finite temperature we again consider an expansion in powers of δ. For path bonds only we expand t^{-1} while for other bonds it is t as before. The expansion is given by

$$C = D_0 + \delta(D_1 + 2V) + (\delta^2 + \delta^3 + \cdots)V, \tag{41}$$

where $V = -D_1$ across a path bond. All other elements of V are zero.

Unlike D in equation (13) which is written exactly in terms of a perturbation linear in δ, C is expressed as a series in powers of δ, at least as far as the path bonds are concerned. At first sight, it appears that lifting the ground state eigenvalue degeneracy while keeping track of the powers of δ in a perturbation expansion is going to be an intractable task. However there exists a remarkable property of the continuum propagator, g_c, that was introduced in Eq. (21) that does allow us to efficiently achieve this goal. It can be written as the sum of two terms

$$g_c = g_{c1} + g_{c2}. \tag{42}$$

Like many of the other quantities the occur in the formalism, g_c is spatially very localised. g_{c1} is 4×4 block diagonal in the four bond basis states associated within a plaquette and g_{c2} is 2×2 block diagonal in the two bond basis states associated with a bond. We can now state an important theorem concerning the determination of $\det C$.

Theorem. *In the ground state limit, the solution for* $\det C$ *using*

$$\begin{aligned} C &= D_0 + \delta(D_1 + 2V) + (\delta^2 + \delta^3 + \cdots)V \\ g_c &= g_{c1} + g_{c2} \end{aligned} \tag{43}$$

is exactly the same as the solution using

$$\begin{aligned} C &= D_0 + \delta(D_1 + 2V) \\ g_c &= g_{c1} \end{aligned} \tag{44}$$

Basically, the effects of g_{c2} exactly cancel the awkward nonlinear terms in δ. We are able to simply rerun the perturbation theory for $D_1 + 2V$ instead of D_1, noting that this only involves sign changes across path bonds (since $V = -D_1$). The proof of this theorem arises within an extension of the procedure described in Section 3.1 and the full details are given in ref. [11].

In ref. [8] it was proven that $\det D$ is proportional to the ground state degeneracy (an equivalent statement is Eq. (12)). In analogy, we can consider $\det C$ as proportional to an effective degeneracy for the reciprocal defect system. In fact we can usually write

$$< \sigma_0 \sigma_R >^2 = \exp(2(S_C(R) - S_D)) \tag{45}$$

where S_D is the actual system entropy and S_C is the analogue for the reciprocal defect system. This result is valid if the effective energy of the reciprocal defect system is equal to actual energy. Otherwise the correlation function is zero. We are unaware of previous discussions of this result elsewhere in the literature.

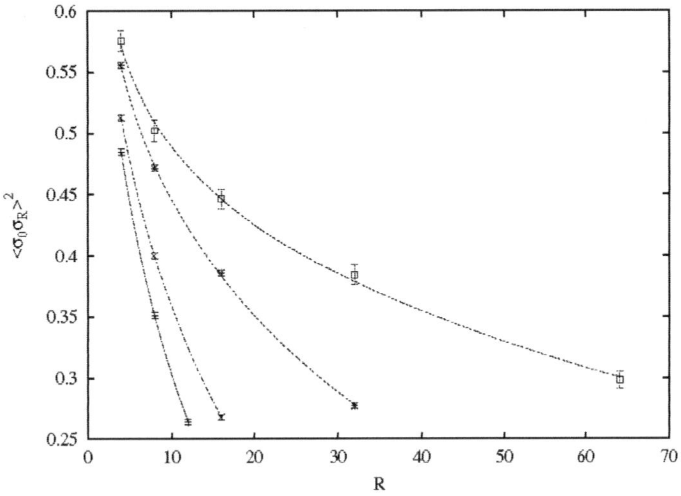

Figure 7. The spin correlation function on $9L \times L$ lattices with $L = 12(+), 16(\times), 32(*)$ and $64(\square)$. The curves are a best fit to Eq. (46).

4.2. Results

We have calculated correlation functions with paths of length R, in units of the lattice spacing, parallel to the axes of long thin cylinders of dimensions $9L \times L$ with $L = 12$, 16, 32 and 64. The results are displayed in figure 7. The error bars indicate statisitical uncertainties equal to two standard deviations, meaning that we have 95.4 per cent confidence intervals. The configurational averages are over 10^5 random samples, except for $L = 64$ where only 10^4 samples were obtained. The uncertainties are approximately independent of either L or R, being only strongly dependent on the number of samples. With 10^5 samples, they are all in the range 0.0021 to 0.0026. Consistently, the corresponding interval with 10^4 samples is 0.0073 to 0.0086, about a factor of $\sqrt{10}$ larger. The curves are fits of the data to the expected scaling form

$$[< \sigma_0 \sigma_R >^2]_{av} \sim R^{-\eta} \exp(-R/\xi), \tag{46}$$

where the exponential factor is included to deal with the effects of finite sample size. The χ^2 nonlinear fits were done according to the Levenberg-Marquardt method [33] with statistical uncertainties fully incorporated.

It is found that both the coefficient A and the exponent η are approximately independent of L. The values of η are 0.12(6), 0.14(3), 0.15(1) and 0.14(2). In contrast, the correlation length ξ does not vary like L as would be expected from finite size scaling theory. In fact we find that it varies much like $L^{\frac{3}{2}}$. Indeed, the respective values of $\xi/L^{\frac{3}{2}}$ are 0.41(6), 0.41(4), 0.40(3) and 0.47(10). It is also found that the finite size corrections to the entropy for the $9L \times L$ system scale like $L^{-\frac{3}{2}}$ (scaling as Eq. (39) applies with the $L \times L$ lattice). It is clear that the correlation functions remain intimately associated with these corrections to the entropy as is highlighted in equation (45). It is interesting to note that Lukic et al [34] also observe finite size corrections to the ground state entropy scaling like $L^{-\frac{3}{2}}$, although the system geometry is different.

Figure 8 shows the same data plotted against $L^{-\frac{3}{2}}$ for $R = 4$, 8, 16 and 32 and provides good confirmation of the behaviour of the correlation length. The χ^2 fits are to the form $\alpha(R)\exp(-\beta(R)L^{-\frac{3}{2}})$ and were done initially with linearized data before polishing with nonlinear fits which gave the uncertainties for the fitting parameters $\alpha(R)$ and $\beta(R)$. A

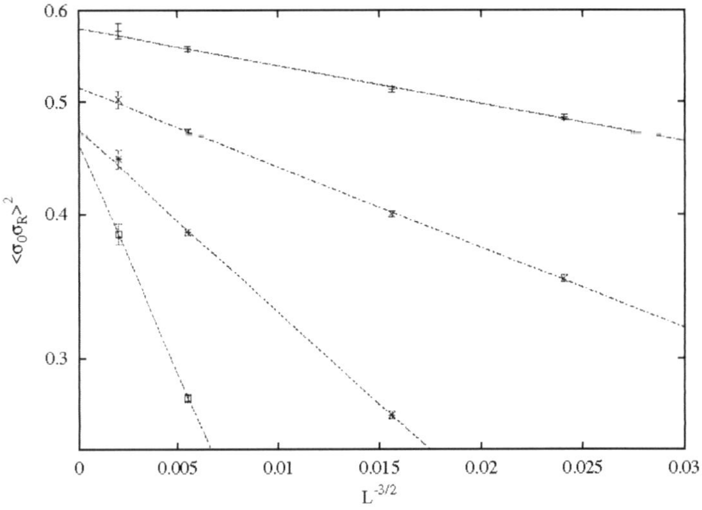

Figure 8. As figure 7 with the data plotted against $L^{3/2}$ for $R = 4(+), 8(\times), 16(*)$ and $32(\square)$. The lines are a best fit to $\alpha(R)\exp(-\beta(R)L^{-3/2})$.

subsequent fit of $\alpha(R)$ to a power function $R^{-\eta}$ gave $\eta = 0.14(1)$ where the error estimate is that accumulated from both stages.

Our best estimate of the exponent η is 0.14 ± 0.01. A wide range of values for η have been published in the literature over a number of years. Most of these indicate a value around 0.2, somewhat higher than ours, although in most cases extrapolated Monte Carlo data was used rather than working directly on the ground state. It is also interesting to add that a recent calculation [17] (which involved extrapolated Monte Carlo data) reported $\eta = 0.138 \pm 0.005$ in apparent agreement with our result.

5. Conclusions

The formalism at the level expressed in Eqs. (4) and (6) has, of course, very general validity. The development we have described from that starting point is very specific to the the $\pm J$ model, and it is one of the very few approaches available that enables one to access directly the ground state without the uncertainties of an extrapolation from finite temperatures. This is particularly important for the calculation of the spin correlations. Monitoring the behaviour of the defect states also gives one a very direct insight into the underlying physics of the problem because of the way these states reflect very closely the spatial range over which the frustration exerts an influence.

We have also used the method to study the $\pm J$ model on a triangular lattice [10]. Unlike the square lattice for which the concentration region $0.5 \leq p \leq 1$ is identical to the $0 \leq p \leq 0.5$ range via a gauge transformation, the whole range of p is distinct for the triangular lattice. At $p = 0$ we have a pure ferromagnet, while $p = 1$ results in an antiferromagnet that is fully frustrated. We find that there are two critical concentrations between which we have a spin-glass regime that has much in common to the system described in this paper; one can, for example, study the extension of the defect states in the intermediate regime and extract an exponent ρ that is essentially identical in value to that discussed here, implying a common universality class. One can also identify the higher critical concentration as marking a percolation threshold for frustrated plaquettes.

References

[1] Onsager L 1944 *Phys. Rev.* **65** 117
[2] Kac M and Ward J C 1952 *Phys. Rev.* **88** 1332
[3] Green H S and Hurst C A 1964 *Order-Disorder Phenomena* (London, Interscience)
[4] Samuel S 1980 *J. Math. Phys.* **21** 2806
[5] Itzykson C 1982 *Nucl. Phys.* B **210** [FS6] 448
[6] Edwards S F and Anderson P W 1975 *J. Phys. F: Met. Phys.* **5** 965
[7] Blackman J A 1982 *Phys. Rev.* B **26** 4987
[8] Blackman J A and Poulter J 1991 *Phys. Rev.* B **44** 4374
[9] Blackman J A, Gonçalves J R and Poulter J 1998 *Phys. Rev.* E **58** 1502
[10] Poulter J and Blackman J A 2001 *J. Phys. A: Maths. Gen.* **34** 7527
[11] Poulter J and Blackman J A 2005 *Phys. Rev.* B **72** 104424
[12] Kawashima N and Rieger H 1997 *Europhys. Lett.* **39** 85
[13] Kitatani H and Sinada A 2000 *J. Phys. A: Math. Gen.* **33** 3545
[14] Hartmann A K and Young A P 2001 *Phys. Rev.* B **64** 180404(R)
[15] Houdayer J 2001 *Eur. Phys. J.* B **22** 479
[16] Sungthong R and Poulter J 2003 *J. Phys. A: Math. Gen.* **36** 6347
[17] Katzgraber H G and Lee L W 2005 *Phys. Rev.* B **71** 134404
[18] Kirkpatrick S 1977 *Phys. Rev.* B **16** 4630
[19] Morgenstern I and Binder K 1980 *Phys. Rev.* B **22** 288
[20] Ozeki Y and Nishimori H 1987 *J. Phys. Soc. Japan* **56** 3265
[21] Kitatani H and Oguchi T 1990 *J. Phys. Soc. Japan* **59** 3823
[22] Ueno Y and Ozeki Y 1991 *J. Stat. Phys.* **64** 227
[23] Aarao Reis F D A, de Queiroz S L A and dos Santos R R 1999 *Phys. Rev.* B **60** 6740
[24] Honecker A, Picco M and Pujol P 2001 *Phys. Rev. Lett.* **87** 047201
[25] Nobre F D 2001 *Phys. Rev.* E **64** 046108
[26] Merz F and Chalker J T 2002 *Phys. Rev.* B **65** 054425
[27] Nishimori H and Nemoto K 2002 *J. Phys. Soc. Japan* **71** 1198
[28] Amoruso C and Hartmann A K 2004 *Phys. Rev.* B **70** 134425
[29] Ozeki Y 1990 *J. Phys. Soc. Japan* **59** 3531
[30] McMillan W L 1983 *Phys. Rev.* B **28** 5216
[31] Wang J-S and Swendsen R H 1988 *Phys. Rev.* B **37** 7745; 1988 *Phys. Rev.* B **38** 4840; 2004 cond-mat/0407273
[32] de Queiroz S L A and Stinchcombe R B 2003 *Phys. Rev.* B **68** 144414
[33] Press W H, Teukolsky S A, Vettering W T and Flannery B P 1992 *Numerical Recipes in Fortran* (Cambridge University Press).
[34] Lukic J, Galluccio A, Marinari E, Martin O C and Rinaldi G 2004 *Phys. Rev. Lett.* **92** 117202

Stationary or static space-times and Young tableaux

Bernd Fiedler

Eichelbaumstr. 13, D-04249 Leipzig, Germany. URL: http://www.fiemath.de/

E-mail: bfiedler@fiemath.de

Abstract. Algebraic curvature tensors possess generators which can be formed from symmetric or alternating tensors S, A or tensors θ with an irreducible (2 1)-symmetry. In differential geometry examples of curvature formulas are known which contain generators on the basis of S or A realized by differentiable tensor fields in a natural way. We show that certain curvature formulas for stationary or static space-times contain such differentiable realizations of generators based on θ. The tensor θ is connected with the timelike Killing vector field of the space-time. θ lies in a special symmetry class from the infinite family of irreducible (2 1)-symmetry classes. We determine characteristics of this class. In particular, this class allows a maximal reduction of the length of the curvature formulas. We use a projection formalism by Vladimirov, Young symmetrizers and Littlewood-Richardson products. Computer calculations were carried out by means of the packages Ricci and PERMS.

1. Introduction

The subject of the present paper is the search for examples of Riemann tensors which contain terms with a structure of certain generators of algebraic curvature tensors.

Let V be a finite-dimensional $\mathbb{K}$-vector space, where $\mathbb{K}$ is the field of real or complex numbers. We denote by $\mathcal{T}_r V$ the $\mathbb{K}$-vector space of covariant tensors of order r over V.

Definition 1.1 *The $\mathbb{K}$-vector space $\mathcal{A}(V) \subset \mathcal{T}_4 V$ of all algebraic curvature tensors is the set of all tensors $\mathfrak{R} \in \mathcal{T}_4 V$ which satisfy for all $w, x, y, z \in V$*

$$\mathfrak{R}(w,x,y,z) \;=\; -\mathfrak{R}(w,x,z,y) \;=\; \mathfrak{R}(y,z,w,x) \quad (1)$$

$$\mathfrak{R}(w,x,y,z) + \mathfrak{R}(w,y,z,x) + \mathfrak{R}(w,z,x,y) \;=\; 0\,. \quad (2)$$

For algebraic curvature tensors several types of generators are known. For instance, algebraic curvature tensors can be generated by the following tensors:

$$\gamma(S)_{\kappa\lambda\mu\nu} \;:=\; S_{\kappa\nu}S_{\lambda\mu} - S_{\kappa\mu}S_{\lambda\nu} \qquad , \; S \in \mathcal{S}^2(V)\,, \quad (3)$$

$$\alpha(A)_{\kappa\lambda\mu\nu} \;:=\; 2A_{\kappa\lambda}A_{\mu\nu} + A_{\kappa\mu}A_{\lambda\nu} - A_{\kappa\nu}A_{\lambda\mu} \qquad , \; A \in \Lambda^2(V)\,, \quad (4)$$

where $\mathcal{S}^p(V)$, $\Lambda^p(V)$ denotes the spaces of totally symmetric/alternating p-forms over V. P. Gilkey [1, pp.41-44, P.236] and B. Fiedler [2] gave different proofs for

Theorem 1.2 $\mathcal{A}(V) = \mathrm{Span}_{S\in\mathcal{S}^2(V)}\{\gamma(S)\} = \mathrm{Span}_{A\in\Lambda^2(V)}\{\alpha(A)\}$.

But $\mathcal{A}(V)$ possesses also generators on the basis of products $U \otimes w$ or $w \otimes U$, $U \in \mathcal{T}_3 V$, $w \in \mathcal{T}_1 V$, where U has a so-called irreducible (2 1)-symmetry. B. Fiedler proved in [3]

Theorem 1.3 [1] *Let $\mathfrak{r} \subset \mathbb{K}[\mathcal{S}_3]$ be a minimal right ideal belonging to the partition $(2\,1) \vdash 3$ and let $\mathcal{T}_{\mathfrak{r}}$ be the symmetry class of tensors $U \in \mathcal{T}_3 V$ that is defined by $\mathfrak{r}$. Further, let y_t be the Young symmetrizer of the Young tableau*

$$t \;:=\; \begin{array}{|c|c|} \hline 1 & 3 \\ \hline 2 & 4 \\ \hline \end{array}\,. \tag{5}$$

Then the following statements are equivalent

(i) $\mathcal{A}(V) = \mathrm{Span}_{U \in \mathcal{T}_{\mathfrak{r}}, w \in \mathcal{T}_1 V}\{y_t^*(U \otimes w)\} = \mathrm{Span}_{U \in \mathcal{T}_{\mathfrak{r}}, w \in \mathcal{T}_1 V}\{y_t^*(w \otimes U)\}$.

(ii) $\mathfrak{r}$ *is different from the right ideal* $\mathfrak{r}_0 := f_0 \cdot \mathbb{K}[\mathcal{S}_3]$ *which is generated by the idempotent* $f_0 := \frac{1}{2}\{\mathrm{id} - (1\,3)\} - \frac{1}{6}\sum_{p \in \mathcal{S}_3} \mathrm{sign}(p)\, p$.

In differential geometry or general relativity theory (GR), many examples of Riemann tensors are known, in which expressions of the type (3) or (4) occur (see e.g. [4, Thm.2.1],[5, Thm.2],[6, Sec.3]). In the present paper we want to show that there exist also curvature formulas in which generators from Theorem 1.3 appear realized by differentiable tensor fields in a natural way. We found such generators in curvature formulas of stationary and static space-times of GR.

Let (M, g) be a 4-dimensional pseudo-Riemannian manifold of class C^∞ whose fundamental tensor g has signature $(+---)$. Let ∇ be the Levi-Civita connection of g.

Definition 1.4 *(i) (M, g) is called a stationary space-time if there exists a timelike Killing field ξ on M characterized by the conditions*

$$\xi_{\mu;\nu} + \xi_{\nu;\mu} \;=\; 0 \quad , \quad g_{\mu\nu}\xi^\mu\xi^\nu \;>\; 0\,. \tag{6}$$

(ii) A stationary (M, g) is called a static space-time if the Killing field ξ is hypersurface-orthogonal, i.e. ξ satisfies the additional condition $\xi_{[\mu;\nu}\xi_{\lambda]} = 0$.

If (M, g) is stationary then one can construct local coordinates t, x^1, x^2, x^3 around every point $p \in M$ such that $g_{\mu\nu} = g_{\mu\nu}(x^1, x^2, x^3)$ and $\xi = \partial_t$. If (M, g) is static, then we can choose these local coordinates t, x^1, x^2, x^3 in such a way that

$$\begin{array}{rcl} ds^2 \;=\; g_{\mu\nu}dx^\mu dx^\nu & = & f(x^1, x^2, x^3)^2 dt^2 - h_{ab}(x^1, x^2, x^3)dx^a dx^b \\ \mu, \nu = 0, 1, 2, 3 & & a, b = 1, 2, 3\,. \end{array} \tag{7}$$

Here $d\sigma^2 := h_{ab}dx^a dx^b$ is a positive definite, 3-dimensional metric.

Now we formulate the main results of our paper.

Proposition 1.5 *Let $\tau^\mu := (\xi_\alpha\xi^\alpha)^{-1/2}\xi^\mu$ be the timelike unit vector field which is proportional to the above Killing field ξ^μ. Further, let $Z_{\kappa\lambda\mu\nu}$ be the (covariant) spatial projection (35) of the covariant Riemann tensor $R_{\kappa\lambda\mu\nu}$, let $P_{\kappa\lambda\mu\nu}$ be the (covariant) 3-dimensional curvature tensor from (34) and let F_μ be the field (26). Then the following formula holds:*

$$\begin{aligned} -Z_{\kappa\lambda\mu\nu} \;=\; P_{\kappa\lambda\mu\nu} \;-\; & \tfrac{3}{4}F_\lambda F_\nu \tau_\kappa \tau_\mu + \tfrac{3}{4}F_\kappa F_\nu \tau_\lambda \tau_\mu + \tfrac{3}{4}F_\lambda F_\mu \tau_\kappa \tau_\nu - \tfrac{3}{4}F_\kappa F_\mu \tau_\lambda \tau_\nu - \\ & F_\nu \tau_\mu \tau_{[\kappa;\lambda]} + F_\mu \tau_\nu \tau_{[\kappa;\lambda]} - \tfrac{1}{2}F_\nu \tau_\lambda \tau_{[\kappa;\mu]} + \tfrac{1}{2}F_\lambda \tau_\nu \tau_{[\kappa;\mu]} + \\ & \tfrac{1}{2}F_\mu \tau_\lambda \tau_{[\kappa;\nu]} - \tfrac{1}{2}F_\lambda \tau_\mu \tau_{[\kappa;\nu]} + \tfrac{1}{2}F_\nu \tau_\kappa \tau_{[\lambda;\mu]} - \tfrac{1}{2}F_\kappa \tau_\nu \tau_{[\lambda;\mu]} - \\ & \tfrac{1}{2}F_\mu \tau_\kappa \tau_{[\lambda;\nu]} + \tfrac{1}{2}F_\kappa \tau_\mu \tau_{[\lambda;\nu]} - F_\lambda \tau_\kappa \tau_{[\mu;\nu]} + F_\kappa \tau_\lambda \tau_{[\mu;\nu]} - \\ & 2\tau_{[\kappa;\lambda]}\tau_{[\mu;\nu]} + \tau_{[\kappa;\nu]}\tau_{[\lambda;\mu]} - \tau_{[\kappa;\mu]}\tau_{[\lambda;\nu]} \end{aligned} \tag{8}$$

[1] See section 4 for details about symmetry classes of tensors, Young tableaux and Young symmetrizers.

Theorem 1.6 *Let M_p be the tangent space of M at a point $p \in M$.*

(i) At every $p \in M$ the tensor $(\tau_\lambda \tau_{[\mu;\nu]})|_p$ belongs to a symmetry class $\mathcal{T}_\mathfrak{r} \subseteq \mathcal{T}_3 M_p$ whose defining right ideal $\mathfrak{r} \subset \mathbb{K}[\mathcal{S}_3]$ has a decomposition $\mathfrak{r} = \mathfrak{r}_1 \oplus \mathfrak{r}_2$ into 2 minimal right ideals $\mathfrak{r}_i$ which is described by the Littlewood-Richardson product

$$[1^2][1] \sim [2\,1] + [1^3]\,. \tag{9}$$

(ii) At every $p \in M$ the tensor $(F_\kappa \tau_\lambda \tau_{[\mu;\nu]})|_p$ lies in a symmetry class $\mathcal{T}_\mathfrak{r} \subseteq \mathcal{T}_4 M_p$ whose defining right ideal $\mathfrak{r} \subset \mathbb{K}[\mathcal{S}_4]$ has a decomposition $\mathfrak{r} = \mathfrak{r}_1 \oplus \ldots \oplus \mathfrak{r}_5$ into 5 minimal right ideals $\mathfrak{r}_i$ which is described by the Littlewood-Richardson products

$$[2\,1][1] \sim [3\,1] + [2^2] + [2\,1^2] \quad , \quad [1^3][1] \sim [2\,1^2] + [1^4]\,. \tag{10}$$

At most the product $[2\,1][1]$ yields a contribution to the symmetry class $\mathcal{A}(V)$ of algebraic curvature tensors which belongs to $[2^2]$.

Lemma 1.7 *For the above fields ξ^μ and τ^μ the conditions $\xi_{[\lambda}\xi_{\mu;\nu]} = 0$ and $\tau_{[\lambda}\tau_{\mu;\nu]} = 0$ are equivalent.*

Theorem 1.8 *A stationary space-time is static iff $\tau_{[\lambda}\tau_{\mu;\nu]} = 0$ or, equivalently, iff the part $[1^3]$ of (9) vanishes. In a static space-time the tensor $\tau_\lambda \tau_{[\mu;\nu]}$ belongs to an irreducible $(2\,1)$-symmetry class.*

Because of (9) the tensor $\tau_\lambda \tau_{[\mu;\nu]}$ possesses a decomposition

$$\tau_\lambda \tau_{[\mu;\nu]} = \theta_{\lambda\mu\nu} + \tau_{[\lambda}\tau_{\mu;\nu]}\,, \tag{11}$$

where $\theta_{\lambda\mu\nu} := \tau_\lambda \tau_{[\mu;\nu]} - \tau_{[\lambda}\tau_{\mu;\nu]}$ is the unique part of $\tau_\lambda \tau_{[\mu;\nu]}$ which has an irreducible $(2\,1)$-symmetry. (10) leads to the remarkable consequence that $\tau_{[\lambda}\tau_{\mu;\nu]}$ does not yield a contribution to (8) even in the case of a stationary space-time.

Theorem 1.9 *The substitution (11) transforms (8) into*

$$\begin{aligned}
-Z_{\kappa\lambda\mu\nu} = P_{\kappa\lambda\mu\nu} \; - \; & \frac{3}{4}F_\lambda F_\nu \tau_\kappa \tau_\mu + \frac{3}{4}F_\kappa F_\nu \tau_\lambda \tau_\mu + \frac{3}{4}F_\lambda F_\mu \tau_\kappa \tau_\nu - \frac{3}{4}F_\kappa F_\mu \tau_\lambda \tau_\nu - \\
& F_\nu \theta_{\mu\kappa\lambda} + F_\mu \theta_{\nu\kappa\lambda} - \frac{1}{2}F_\nu \theta_{\lambda\kappa\mu} + \frac{1}{2}F_\lambda \theta_{\nu\kappa\mu} + \\
& \frac{1}{2}F_\mu \theta_{\lambda\kappa\nu} - \frac{1}{2}F_\lambda \theta_{\mu\kappa\nu} + \frac{1}{2}F_\nu \theta_{\kappa\lambda\mu} - \frac{1}{2}F_\kappa \theta_{\nu\lambda\mu} - \\
& \frac{1}{2}F_\mu \theta_{\kappa\lambda\nu} + \frac{1}{2}F_\kappa \theta_{\mu\lambda\nu} - F_\lambda \theta_{\kappa\mu\nu} + F_\kappa \theta_{\lambda\mu\nu} - \\
& 2\tau_{[\kappa;\lambda]}\tau_{[\mu;\nu]} + \tau_{[\kappa;\nu]}\tau_{[\lambda;\mu]} - \tau_{[\kappa;\mu]}\tau_{[\lambda;\nu]}
\end{aligned} \tag{12}$$

In (12) the tensor $\tau_{[\lambda}\tau_{\mu;\nu]}$ does not appear.

Theorem 1.10 *Let $\theta_{\lambda\mu\nu} := \tau_\lambda \tau_{[\mu;\nu]} - \tau_{[\lambda}\tau_{\mu;\nu]}$ be the tensor field from (11).*

(i) The symmetry properties of θ are described by the relations

$$\begin{aligned}
0 &= \theta_{\lambda\mu\nu} + \theta_{\lambda\nu\mu} \\
0 &= -\theta_{\lambda\mu\nu} + \theta_{\nu\mu\lambda} + \theta_{\mu\lambda\nu} \\
0 &= \theta_{\lambda\mu\nu} - \theta_{\nu\mu\lambda} + \theta_{\mu\nu\lambda} \\
0 &= \theta_{\nu\mu\lambda} + \theta_{\nu\lambda\mu}
\end{aligned} \tag{13}$$

(*ii*) *Applying* (13) *we can reduce the* 20 *summands of* (12) *to the* 12 *summands*

$$\begin{aligned} -Z_{\kappa\lambda\mu\nu} = P_{\kappa\lambda\mu\nu} & - \frac{3}{4}F_\lambda F_\nu \tau_\kappa \tau_\mu + \frac{3}{4}F_\kappa F_\nu \tau_\lambda \tau_\mu + \frac{3}{4}F_\lambda F_\mu \tau_\kappa \tau_\nu - \frac{3}{4}F_\kappa F_\mu \tau_\lambda \tau_\nu \\ & + \tau_{[\kappa;\nu]}\tau_{[\lambda;\mu]} - \tau_{[\kappa;\mu]}\tau_{[\lambda;\nu]} - 2\tau_{[\kappa;\lambda]}\tau_{[\mu;\nu]} \\ & \underbrace{-\frac{3}{2}F_\lambda \theta_{\kappa\mu\nu} + \frac{3}{2}F_\kappa \theta_{\lambda\mu\nu} + \frac{3}{2}F_\nu \theta_{\mu\lambda\kappa} - \frac{3}{2}F_\mu \theta_{\nu\lambda\kappa}}_{=\frac{1}{2}y_t^*(\theta \otimes F)_{\kappa\lambda\mu\nu} \text{ according to Theorem 1.3}} \quad . \end{aligned} \tag{14}$$

2. A projection formalism

In investigations of stationary or static space-times one can use a projection formalism which is described for instance in [7, pp. 180] or [8, pp.49]. We apply the formalism from the book [8, pp.49] by Yu. S. Vladimirov in our paper.

The formalism of Vladimirov starts with the assumption that a timelike unit vector field τ^μ is given on M which describes the 4-speed of a "*continuum of observers*". If we define

$$h_{\mu\nu} := \tau_\mu \tau_\nu - g_{\mu\nu} \quad , \quad h^\mu_\nu := g^{\mu\alpha} h_{\alpha\nu} \quad , \quad h^{\mu\nu} := g^{\mu\alpha} g^{\nu\beta} h_{\alpha\beta} \, , \tag{15}$$

then we obtain the following decompositions of the metric tensors:

$$g_{\mu\nu} := \tau_\mu \tau_\nu - h_{\mu\nu} \quad , \quad g^{\mu\nu} := \tau^\mu \tau^\nu - h^{\mu\nu} \, . \tag{16}$$

A simple consequence of $\tau^\mu \tau_\mu = g_{\mu\nu}\tau^\mu\tau^\nu = 1$ is

$$h_{\mu\nu}\tau^\nu = 0 \quad , \quad h^\mu_\nu \tau^\nu = 0 \quad , \quad h^{\mu\nu}\tau_\nu = 0 \, . \tag{17}$$

If $B_{\mu_1 \ldots \mu_r}{}^{\nu_1 \ldots \nu_s}$ are the coordinates of a r-times *covariant* and s-times *contravariant* tensor B then we can define the *time component* of B

$$\tilde{B} := B_{\mu_1 \ldots \mu_r}{}^{\nu_1 \ldots \nu_s} \tau^{\mu_1} \ldots \tau^{\mu_r} \tau_{\nu_1} \ldots \tau_{\mu_s} \tag{18}$$

and the *spatial projection* of B

$$\tilde{B}_{\alpha_1 \ldots \alpha_r}{}^{\beta_1 \ldots \beta_s} := (-1)^{r+s} h^{\mu_1}_{\alpha_1} \ldots h^{\mu_r}_{\alpha_r} h^{\beta_1}_{\nu_1} \ldots h^{\beta_s}_{\nu_s} B_{\mu_1 \ldots \mu_r}{}^{\nu_1 \ldots \nu_s} \, . \tag{19}$$

Furthermore we can form mixed projections, for instance

$$\tilde{B}_{\alpha_1 \ldots \alpha_{r-1}}{}^{\beta_1 \ldots \beta_s} := (-1)^{r+s-1} h^{\mu_1}_{\alpha_1} \ldots h^{\mu_{r-1}}_{\alpha_{r-1}} \tau^{\mu_r} h^{\beta_1}_{\nu_1} \ldots h^{\beta_s}_{\nu_s} B_{\mu_1 \ldots \mu_r}{}^{\nu_1 \ldots \nu_s} \, . \tag{20}$$

Because of (17) the projections (19) and (20) fulfill

$$\tau^{\alpha_i} \tilde{B}_{\ldots \alpha_i \ldots}{}^{\ldots\ldots} = 0 \quad , \quad \tau_{\beta_i} \tilde{B}_{\ldots\ldots}{}^{\ldots \beta_i \ldots} = 0 \, . \tag{21}$$

If we use the time component $d\tilde{\tau} := \tau_\mu dx^\mu$ and the spatial projection $d\tilde{x}^\nu := -h^\nu_\mu dx^\mu$ of dx^μ and set $dl^2 := h_{\mu\nu} d\tilde{x}^\mu d\tilde{x}^\nu$ then we obtain

$$ds^2 = (\tau_\mu \tau_\nu - h_{\mu\nu}) \, dx^\mu dx^\nu = d\tilde{\tau}^2 - h_{\mu\nu} d\tilde{x}^\mu d\tilde{x}^\nu = d\tilde{\tau}^2 - dl^2 \, . \tag{22}$$

Finally, (15), (17) and $4 = g_{\mu\nu}g^{\mu\nu} = 1 + h_{\mu\nu}h^{\mu\nu}$ lead to

$$h^\mu_\alpha h_{\mu\beta} = -h_{\alpha\beta} \quad , \quad h_{\mu\alpha} h^{\alpha\nu} = -h^\nu_\mu \quad , \quad h_{\mu\nu} h^{\mu\nu} = 3 \, . \tag{23}$$

Now we consider the decomposition

$$\tau_{\mu;\nu} = \frac{1}{2}(\tau_{\mu,\nu} - \tau_{\nu,\mu}) + \frac{1}{2}(\tau_{\mu;\nu} + \tau_{\nu;\mu}) \tag{24}$$

of the covariant derivative $\tau_{\mu;\nu}$. If we covariantly differentiate $1 = \tau_\mu \tau^\mu$ we obtain immediately

$$\tau_{\mu;\nu}\tau^\mu = 0 \quad , \quad (\tau_{\mu,\nu} - \tau_{\nu,\mu}) \, \tau^\mu \tau^\nu = 0 \quad , \quad (\tau_{\mu;\nu} + \tau_{\nu;\mu}) \, \tau^\mu \tau^\nu = 0 \, . \tag{25}$$

Definition 2.1 *The following quantities $F_\alpha, A_{\alpha\beta}, D_{\alpha\beta}$ are usefull in formulas for the Riemann tensor:*

$$F_\alpha \;:=\; -\left(\tau_{\mu,\nu}-\tau_{\nu,\mu}\right)\tau^\nu h^\mu_\alpha \;=\; -\left(\tau_{\mu;\nu}+\tau_{\nu;\mu}\right)\tau^\mu h^\nu_\alpha \;=\; \left(\tau_{\alpha,\nu}-\tau_{\nu,\alpha}\right)\tau^\nu\,, \tag{26}$$

$$A_{\alpha\beta} \;:=\; \frac{1}{2}\left(\tau_{\mu,\nu}-\tau_{\nu,\mu}\right)h^\mu_\alpha h^\nu_\beta\;,\quad D_{\alpha\beta}\;:=\;-\frac{1}{2}\left(\tau_{\mu;\nu}+\tau_{\nu;\mu}\right)h^\mu_\alpha h^\nu_\beta\,. \tag{27}$$

Obviously, D is a symmetric tensor, $D_{\alpha\beta}=D_{\beta\alpha}$, whereas A is skew-symmetric, $A_{\alpha\beta}=-A_{\beta\alpha}$.

Lemma 2.2 *The tensor $A_{\alpha\beta}$ satisfies (see [8, p.51])*

$$A_{\alpha\beta} \;=\; \frac{1}{2}\left(\tau_{\alpha,\beta}-\tau_{\beta,\alpha}\right)+\frac{1}{2}\left(\tau_\alpha F_\beta-\tau_\beta F_\alpha\right)\,. \tag{28}$$

PROOF. Taking into account (15) and (27) we can write

$$\begin{aligned} A_{\alpha\beta} \;=\; &\frac{1}{2}\left(\tau_{\mu,\nu}-\tau_{\nu,\mu}\right)\tau^\mu\tau^\nu\tau_\alpha\tau_\beta-\frac{1}{2}\left(\tau_{\mu,\nu}-\tau_{\nu,\mu}\right)\tau^\mu\tau_\alpha\delta^\nu_\beta \\ &-\frac{1}{2}\left(\tau_{\mu,\nu}-\tau_{\nu,\mu}\right)\tau^\nu\tau_\beta\delta^\mu_\alpha+\frac{1}{2}\left(\tau_{\mu,\nu}-\tau_{\nu,\mu}\right)\delta^\mu_\alpha\delta^\nu_\beta\,, \end{aligned} \tag{29}$$

where δ^ν_μ denotes the Kronecker symbol. The first summand of (29) vanishes because of (25). But then (26) leads to (28). □

We use the following definition of the Christoffel symbols and the Riemann tensor of ∇:

Definition 2.3 *Let ∇ be the Levi-Civita connection of g. Then we define the Christoffel symbols $\Gamma^\lambda_{\mu\nu}$ and the Riemann tensor $R_{\mu\nu\alpha}{}^\lambda$ of ∇ by*

$$\Gamma^\lambda_{\mu\nu} \;:=\; \frac{1}{2}g^{\lambda\gamma}\left(\partial_\mu g_{\nu\gamma}+\partial_\nu g_{\mu\gamma}-\partial_\gamma g_{\mu\nu}\right) \tag{30}$$

$$R_{\mu\nu\alpha}{}^\lambda \;:=\; \partial_\mu\Gamma^\lambda_{\nu\alpha}-\partial_\nu\Gamma^\lambda_{\mu\alpha}+\Gamma^\epsilon_{\nu\alpha}\Gamma^\lambda_{\mu\epsilon}-\Gamma^\epsilon_{\mu\alpha}\Gamma^\lambda_{\nu\epsilon}\,. \tag{31}$$

The book [8] by Vladimirov uses (30), too, but defines the Riemannian curvature tensor by

$$\breve{R}^\lambda{}_{\alpha\nu\mu} \;:=\; \partial_\nu\Gamma^\lambda_{\alpha\mu}-\partial_\mu\Gamma^\lambda_{\alpha\nu}+\Gamma^\epsilon_{\alpha\mu}\Gamma^\lambda_{\epsilon\nu}-\Gamma^\epsilon_{\alpha\nu}\Gamma^\lambda_{\epsilon\mu}\,. \tag{32}$$

The transformation between (31) and (32) reads

$$R_{\mu\nu\alpha}{}^\lambda \;=\; -\breve{R}^\lambda{}_{\alpha\nu\mu}\,. \tag{33}$$

Lemma 2.4 *The Christoffel symbols $\Gamma^\lambda_{\mu\nu}$ can be expressed by $\tau^\mu, h_{\mu\nu}, F_\alpha, A_{\alpha\beta}$ and $D_{\alpha\beta}$ in the following way*

$$\Gamma^\mu_{\alpha\beta} \;=\; \left(L^\mu_{\alpha\beta}+h^\mu_{\alpha,\beta}\right)-\tau^\mu\left(A_{\alpha\beta}-D_{\alpha\beta}+F_\alpha\tau_\beta\right)+F^\mu\tau_\alpha\tau_\beta+\left(\tau_\alpha A^\mu{}_\beta+\tau_\beta A^\mu{}_\alpha\right)-\tau_\alpha\tau^\mu{}_{,\beta}\,,$$

where $L^\mu_{\alpha\beta}\;:=\;\frac{1}{2}h^{\mu\epsilon}\left(\partial_\beta h_{\alpha\epsilon}+\partial_\alpha h_{\beta\epsilon}-\partial_\epsilon h_{\alpha\beta}\right)$ denotes the "Christoffel symbols" of the "3-dimensional metric" h. (See [8, p.53].)

Definition 2.5 *We denote by P the tensor*

$$P_{\alpha\beta\rho}{}^\lambda \;:=\; -h^\mu_\alpha h^\nu_\beta h^\sigma_\rho h^\lambda_\epsilon\left(\partial_\mu\tilde{L}^\epsilon_{\sigma\nu}-\partial_\nu\tilde{L}^\epsilon_{\sigma\mu}+\tilde{L}^\epsilon_{\kappa\mu}\tilde{L}^\kappa_{\sigma\nu}-\tilde{L}^\epsilon_{\kappa\nu}\tilde{L}^\kappa_{\sigma\mu}\right)\;,\quad \tilde{L}^\epsilon_{\sigma\nu}\;:=\;L^\epsilon_{\sigma\nu}+h^\epsilon_{\sigma,\nu}\,, \tag{34}$$

which can be considered the curvature tensor assigned to h. (See [8, p.55]).

The right-hand side of (34) is equal to the right-hand side of formula (3.30) in [8, p.55]. We adapted only the left-hand side of (34) to our definition (31) of the curvature tensor by means of (33). Note that $\tilde{L}^{\epsilon}_{\sigma\nu}$ is not symmetric with respect to σ, ν in general.

The Riemann tensor R possesses three spatial projections.

Definition 2.6 *We denote by $Z_{\mu\nu\kappa}{}^{\lambda}$, $Y_{\mu\nu\kappa}$, $X_{\nu\kappa}$ the following three spatial projections of the Riemann tensor R:*

$$Z_{\mu\nu\kappa}{}^{\lambda} := h^{\alpha}_{\mu} h^{\beta}_{\nu} h^{\gamma}_{\kappa} h^{\lambda}_{\delta} R_{\alpha\beta\gamma}{}^{\delta} \ , \ Y_{\mu\nu\kappa} := h^{\alpha}_{\mu} h^{\beta}_{\nu} h^{\gamma}_{\kappa} \tau_{\delta} R_{\alpha\beta\gamma}{}^{\delta} \ , \ X_{\nu\kappa} := -h^{\beta}_{\nu} h^{\gamma}_{\kappa} \tau^{\alpha} \tau_{\delta} R_{\alpha\beta\gamma}{}^{\delta} . \quad (35)$$

In the present paper we consider only $Z_{\mu\nu\kappa}{}^{\lambda}$.

Proposition 2.7 *The spatial projection Z of the Riemann tensor R satisfies*

$$-Z_{\epsilon\gamma\kappa}{}^{\lambda} = P_{\epsilon\gamma\kappa}{}^{\lambda} + 2\, A^{\lambda}{}_{\kappa} A_{\epsilon\gamma} + (D^{\lambda}_{\epsilon} + A_{\epsilon}{}^{\lambda})\,(D_{\gamma\kappa} + A_{\gamma\kappa}) - (D^{\lambda}_{\gamma} + A_{\gamma}{}^{\lambda})\,(D_{\epsilon\kappa} + A_{\epsilon\kappa}) \quad (36)$$

$$-Z_{\epsilon\gamma\kappa\lambda} = P_{\epsilon\gamma\kappa\lambda} + 2\, A_{\lambda\kappa} A_{\epsilon\gamma} + (D_{\epsilon\lambda} + A_{\epsilon\lambda})\,(D_{\gamma\kappa} + A_{\gamma\kappa}) - (D_{\gamma\lambda} + A_{\gamma\lambda})\,(D_{\epsilon\kappa} + A_{\epsilon\kappa}) \quad (37)$$

where $P_{\epsilon\gamma\kappa\lambda} = g_{\lambda\sigma} P_{\epsilon\gamma\kappa}{}^{\sigma} = -h_{\lambda\sigma} P_{\epsilon\gamma\kappa}{}^{\sigma}$.

Proof. Relation (36) is equal to the relation (3.39) in [8, p.56], in which a transformation (33) of the left-hand side was carried out. From (36) we obtain (37) by lowering of λ by means of $g_{\lambda\sigma}$. The relation $P_{\epsilon\gamma\kappa\lambda} = -h_{\lambda\sigma} P_{\epsilon\gamma\kappa}{}^{\sigma}$ is a consequence of (16) and (17). □

3. Stationary and static space-times

Now we apply the projection formalism of Section 2 to a stationary space-time. Let ξ^{μ} be the timelike Killing field of such a space-time and τ^{μ} be the timelike unit vector field which is proportional to ξ^{μ}, i.e. we have $\xi^{\mu} = \phi\, \tau^{\mu}$.

3.1. Proof of Lemma 1.7

From $\xi^{\mu} = \phi\, \tau^{\mu}$ we obtain

$$\xi_{\mu;\nu} = (\partial_{\nu}\phi)\tau_{\mu} + \phi \tau_{\mu;\nu} \, . \quad (38)$$

and $\xi_{[\lambda}\xi_{\mu;\nu]} = \phi\tau_{[\lambda}\tau_{\mu}(\partial_{\nu]}\phi) + \phi^2 \tau_{[\lambda}\tau_{\mu;\nu]} = \phi^2 \tau_{[\lambda}\tau_{\mu;\nu]}$, since $\tau_{[\lambda}\tau_{\mu]} = 0$. Consequently, $\xi_{[\lambda}\xi_{\mu;\nu]} = 0$ and $\tau_{[\lambda}\tau_{\mu;\nu]} = 0$ are equivalent. □

3.2. Proof of Proposition 1.5

First we show

Lemma 3.1 *If τ^{μ} is proportional to a Killing field ξ^{μ}, i.e. $\xi^{\mu} = \phi\, \tau^{\mu}$ is fulfilled, then $D_{\mu\nu} = 0$.*

Proof. The Killing equation $\xi_{(\mu;\nu)} = 0$ and (38) lead to $0 = \tau_{(\mu}\partial_{\nu)}\phi + \phi\tau_{(\mu;\nu)}$. But then (27) and (17) yield $D_{\alpha\beta} = -\tau_{(\mu;\nu)} h^{\mu}_{\alpha} h^{\nu}_{\beta} = \phi^{-1} \tau_{(\mu}\partial_{\nu)}\phi\, h^{\mu}_{\alpha} h^{\nu}_{\beta} = 0$. □

Now Proposition 1.5 can be proved in the following way. Because of $D_{\mu\nu} = 0$ we can transform (37) into

$$-Z_{\epsilon\gamma\kappa\lambda} = P_{\epsilon\gamma\kappa\lambda} + 2\, A_{\lambda\kappa} A_{\epsilon\gamma} + A_{\epsilon\lambda} A_{\gamma\kappa} - A_{\gamma\lambda} A_{\epsilon\kappa} \, . \quad (39)$$

If we substitute $A_{\alpha\beta}$ by means of (28) in (39) and use the notation $\tau_{[\mu;\nu]} = \frac{1}{2}(\tau_{\mu,\nu} - \tau_{\nu,\mu})$, then we obtain (8). We determined the long formula (8) by means of the Mathematica package Ricci [9]. The Mathematica notebook of this calculation can be downloaded from [10]. □

4. Symmetry classes of tensors

We denote by $\mathbb{K}[\mathcal{S}_r]$ the group ring of the symmetric group $\mathcal{S}_r$.

Definition 4.1 *If $T \in \mathcal{T}_r V$ and $a = \sum_{p\in\mathcal{S}_r} a(p)\, p \in \mathbb{K}[\mathcal{S}_r]$, then we denote by aT the r-times covariant tensor*

$$(aT)(v_1,\ldots,v_r) := \sum_{p\in\mathcal{S}_r} a(p)T(v_{p(1)},\ldots,v_{p(r)}) \quad , \quad (aT)_{i_1\ldots i_r} = \sum_{p\in\mathcal{S}_r} a(p)T_{i_{p(1)}\ldots i_{p(r)}}\,. \tag{40}$$

Because of (40), the group ring elements $a \in \mathbb{K}[\mathcal{S}_r]$ are called symmetry operators for the tensors $T \in \mathcal{T}_r V$. Further we denote by $* : \mathbb{K}[\mathcal{S}_r] \to \mathbb{K}[\mathcal{S}_r]$ the operator

$$* : a = \sum_{p\in\mathcal{S}_r} a(p)p \;\mapsto\; a^* := \sum_{p\in\mathcal{S}_r} a(p)p^{-1}\,. \tag{41}$$

Definition 4.2 *Let $\mathfrak{r} \subseteq \mathbb{K}[\mathcal{S}_r]$ be a right ideal of $\mathbb{K}[\mathcal{S}_r]$. Then the tensor set*

$$\mathcal{T}_\mathfrak{r} := \{aT \mid a \in \mathfrak{r}\,,\; T \in \mathcal{T}_r V\} \tag{42}$$

is called the symmetry class of r-times covariant tensors defined by $\mathfrak{r}$. $\mathcal{T}_\mathfrak{r}$ is called irreducible iff $\mathfrak{r}$ is minimal.

Proposition 4.3 [2] *Let $e \in \mathbb{K}[\mathcal{S}_r]$ be a generating idempotent of a right ideal $\mathfrak{r} \subseteq \mathbb{K}[\mathcal{S}_r]$. Then a tensor $T \in \mathcal{T}_r V$ lies in the symmetry class $\mathcal{T}_\mathfrak{r}$ of $\mathfrak{r}$ iff $eT = T$.*

Important special symmetry operators are Young symmetrizers, which are defined by means of Young tableaux. A *Young tableau* t of $r \in \mathbb{N}$ is an arrangement of r boxes such that

(i) the numbers λ_i of boxes in the rows $i = 1,\ldots,l$ form a decreasing sequence $\lambda_1 \geq \lambda_2 \geq \ldots \geq \lambda_l > 0$ with $\lambda_1 + \ldots + \lambda_l = r$,

(ii) the boxes are filled with the numbers $1, 2, \ldots, r$ in any order.

For instance, the following graphic shows a Young tableau of $r = 15$.

$\lambda_1 = 5$	11	2	5	4	12
$\lambda_2 = 4$	9	6	13	3	
$\lambda_3 = 4$	8	15	1	7	
$\lambda_4 = 2$	10	14			

$= t\,.$

Obviously, the unfilled arrangement of boxes, the *Young frame*, is characterized by a partition $\lambda = (\lambda_1,\ldots,\lambda_l) \vdash r$ of r.

If a Young tableau t of a partition $\lambda \vdash r$ is given, then the *Young symmetrizer* y_t of t is defined by[3]

$$y_t := \sum_{p\in\mathcal{H}_t}\sum_{q\in\mathcal{V}_t} \mathrm{sign}(q)\, p \circ q \tag{43}$$

where $\mathcal{H}_t$, $\mathcal{V}_t$ are the groups of the *horizontal* or *vertical permutations* of t which only permute numbers within rows or columns of t, respectively. The Young symmetrizers of $\mathbb{K}[\mathcal{S}_r]$ are essentially idempotent and define decompositions

$$\mathbb{K}[\mathcal{S}_r] = \bigoplus_{\lambda\vdash r}\bigoplus_{t\in\mathcal{ST}_\lambda} \mathbb{K}[\mathcal{S}_r]\cdot y_t \quad , \quad \mathbb{K}[\mathcal{S}_r] = \bigoplus_{\lambda\vdash r}\bigoplus_{t\in\mathcal{ST}_\lambda} y_t \cdot \mathbb{K}[\mathcal{S}_r] \tag{44}$$

[2] See [11, pp.127], [12, Lemma III.2.2], [13].

[3] We use the convention $(p \circ q)(i) := p(q(i))$ for the product of two permutations p, q.

of $\mathbb{K}[\mathcal{S}_r]$ into minimal left or right ideals. In (44), the symbol $\mathcal{ST}_\lambda$ denotes the set of all standard tableaux of the partition λ. *Standard tableaux* are Young tableaux in which the entries of every row and every column form an increasing number sequence.[4]

The inner sums of (44) are minimal two-sided ideals

$$\mathfrak{a}_\lambda \;:=\; \bigoplus_{t\in\mathcal{ST}_\lambda} \mathbb{K}[\mathcal{S}_r]\cdot y_t \;=\; \bigoplus_{t\in\mathcal{ST}_\lambda} y_t\cdot\mathbb{K}[\mathcal{S}_r] \tag{45}$$

of $\mathbb{K}[\mathcal{S}_r]$. The set of all Young symmetrizers y_t which lie in $\mathfrak{a}_\lambda$ is equal to the set of all y_t whose tableau t has the frame $\lambda\vdash r$. Furthermore two minimal left ideals $\mathfrak{l}_1,\mathfrak{l}_2\subseteq\mathbb{K}[\mathcal{S}_r]$ or two minimal right ideals $\mathfrak{r}_1,\mathfrak{r}_2\subseteq\mathbb{K}[\mathcal{S}_r]$ are *equivalent* iff they lie in the same ideal $\mathfrak{a}_\lambda$. Now we say that a symmetry class $\mathcal{T}_\mathfrak{r}$ *belongs to* $\lambda\vdash r$ iff $\mathfrak{r}\subseteq\mathfrak{a}_\lambda$.

S.A. Fulling, R.C. King, B.G.Wybourne and C.J. Cummins showed in [14]

Theorem 4.4 (Fulling, King, Wybourne, Cummins)[5]
Let y_t be the Young symmetrizer of the standard tableau (5). Then a tensor $T\in\mathcal{T}_4V$ lies in $\mathcal{A}(V)$ if and only if $\frac{1}{12}y_t^*T = T$.

The group ring element $\frac{1}{12}y_t^*$ is a primitive idempotent. Since $y_t\in\mathfrak{a}_{(2^2)}$, the minimal ideals $\mathfrak{l}:=\mathbb{K}[\mathcal{S}_4]\cdot y_t$, $\mathfrak{r}=\mathfrak{l}^*=y_t^*\cdot\mathbb{K}[\mathcal{S}_4]$ satisfy $\mathfrak{l},\mathfrak{r}\subset\mathfrak{a}_{(2^2)}$, i.e. the symmetry class $\mathcal{A}(V)$ of algebraic curvature tensors belongs to the partition $\lambda=(2^2)\vdash 4$.

The following proposition guarantees that we can use Littlewood-Richardson products to determine information about symmetry classes which contain product tensors such as $\tau_\lambda\tau_{[\mu;\nu]}$ and $F_\kappa\tau_\lambda\tau_{[\mu;\nu]}$.

Proposition 4.5[6] *Let $\mathfrak{r}_i\subseteq\mathbb{K}[\mathcal{S}_{r_i}]$ $(i=1,\ldots,m)$ be right ideals and $T^{(i)}\in\mathcal{T}_{\mathfrak{r}_i^*}\subseteq\mathcal{T}_{r_i}V$ be r_i-times covariant tensors from the symmetry classes characterized by the $\mathfrak{r}_i$. Consider the product*

$$T \;:=\; T^{(1)}\otimes\ldots\otimes T^{(m)} \;\in\; \mathcal{T}_rV \quad,\quad r:=r_1+\ldots+r_m\,. \tag{46}$$

For every i we define an embedding

$$\iota_i:\mathcal{S}_{r_i}\to\mathcal{S}_r \quad,\quad (\iota_i s)(k):=\begin{cases}\Delta_i+s(k-\Delta_i) & \text{if } r_{i-1}<k\le r_i\\ k & \text{else}\end{cases} \tag{47}$$

where $\Delta_i:=r_0+\ldots+r_{i-1}$ and $r_0:=0$. Then all product tensors (46) belong to the symmetry class $\mathcal{T}_\mathfrak{r}$ of the right ideal

$$\mathfrak{r} \;:=\; \left(\mathbb{K}[\mathcal{S}_r]\cdot\mathcal{L}\{\tilde{\mathfrak{l}}_1\cdot\ldots\cdot\tilde{\mathfrak{l}}_m\}\right)^* \;=\; \left(\mathbb{K}[\mathcal{S}_r]\cdot(\tilde{\mathfrak{l}}_1\otimes\ldots\otimes\tilde{\mathfrak{l}}_m)\right)^* \tag{48}$$

where $\tilde{\mathfrak{l}}_i:=\iota_i(\mathfrak{l}_i)$ are the embeddings of the left ideals $\mathfrak{l}_i=\mathfrak{r}_i^*$ into $\mathbb{K}[\mathcal{S}_r]$ induced by the ι_i. If $\dim V\ge r$, then the right ideal $\mathfrak{r}$ does not contain a proper right subideal $\tilde{\mathfrak{r}}\subset\mathfrak{r}$ such that all tensors (46) lie also in the symmetry class $\mathcal{T}_{\tilde{\mathfrak{r}}}$ of $\tilde{\mathfrak{r}}$.

Let $\breve{\omega}_G: G\to GL(\mathbb{K}[G])$ denote the *regular representation* of a finite group G defined by $\breve{\omega}_g(f):=g\cdot f$, $g\in G$, $f\in\mathbb{K}[G]$. If we use the left ideals $\mathfrak{l}_i=\mathfrak{r}_i^*$, $\mathfrak{l}=\mathfrak{r}^*$ to define subrepresentations $\alpha_i:=\breve{\omega}_{\mathcal{S}_{r_i}}|_{\mathfrak{l}_i}$, $\beta:=\breve{\omega}_{\mathcal{S}_r}|_{\mathfrak{l}}$, then the representation β is equivalent to the *Littlewood-Richardson product* of the α_i (see B. Fiedler [12, Sec.III.3.2]):

$$\beta \;\sim\; \alpha_1\alpha_2\ldots\alpha_m \;:=\; \alpha_1\#\ldots\#\alpha_m\uparrow\mathcal{S}_r \tag{49}$$

('#' denotes the outer tensor product of the above representations.) This result corresponds to statements of S.A. Fulling et al. [14]. Relation (49) allows us to determine information about the structure of the right ideal (48) by means of the *Littlewood-Richardson rule*[7].

[4] About Young symmetrizers and Young tableaux see for instance the references given in [3, Footnote 6].

[5] Theorem 4.4 is a special case of a more general theorem about symmetry classes of $R_{ijkl\,;\,(s_1\ldots s_u)}$ in [14].

[6] See B. Fiedler [12, Sec.III.3.2] and B. Fiedler [15].

[7] About the Littlewood-Richardson rule see for instance the references in [3, Footnote 13].

5. On symmetry classes for $\tau_\lambda\tau_{[\mu;\nu]}$ and $F_\kappa\tau_\lambda\tau_{[\mu;\nu]}$

5.1. Proof of Theorem 1.6

Because of Proposition 4.5 we can determine information about a symmetry class containing $\tau_\lambda\tau_{[\mu;\nu]}$ from the corresponding Littlewood-Richardson product. Since the symmetries of τ_λ and $\tau_{[\mu;\nu]}$ are generated by the Young symmetrizers of the tableaux $\boxed{1}$, $\begin{array}{|c|}\hline 1\\\hline 2\\\hline\end{array}$ we have to calculate the Littlewood-Richardson product $[1^2][1]$. The use of the Littlewood-Richardson rule yields the graphic

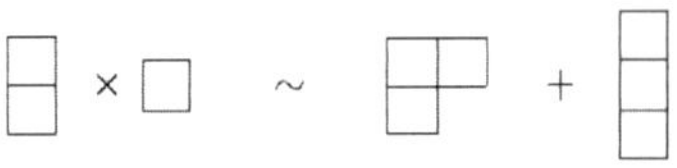

which can be translated into (9). The relation (9) means that the right ideal $\mathfrak{r}$ of the symmetry class of $\tau_\lambda\tau_{[\mu;\nu]}$ possesses a decomposition $\mathfrak{r} = \mathfrak{r}_1 \oplus \mathfrak{r}_2$ into 2 minimal right ideals which belong to the partitions $(2\,1), (1^3) \vdash 3$.

Now we see that information about the right ideal $\mathfrak{r}$ of a symmetry class for $F_\kappa\tau_\lambda\tau_{[\mu;\nu]}$ can be gained from the Littewood-Richardson products $[2\,1][1]$ and $[1^3][1]$. For these products the Littlewood-Richardson rule yields (10) since it leads to the graphics

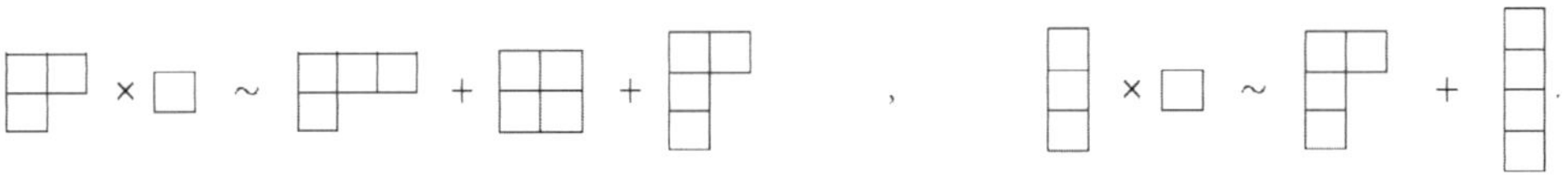

Theorem 4.4 tells us that the symmetry class $\mathcal{A}(V)$ is defined by the right ideal $\mathfrak{r} = y_t^* \cdot \mathbb{K}[\mathcal{S}_4]$ generated by the Young symmetrizer of the tableau (5). This right ideal $\mathfrak{r}$ is minimal and belongs to $(2^2) \vdash 4$ because $y_t \in \mathfrak{a}_{(2^2)}$. But only $[2\,1][1]$ possesses a part $[2^2]$ which belongs to a minimal right ideal of (2^2). This proves the last assertion of Theorem 1.6. □

5.2. Proof of Theorem 1.8

One and only one symmetry class $\mathcal{T}_{\tilde{\mathfrak{t}}}$ of $\mathcal{T}_3V$ belongs to the partition $(1^3) \vdash 3$. It is defined by the right ideal $\tilde{\mathfrak{t}} := y_{\tilde{t}} \cdot \mathbb{K}[\mathcal{S}_3]$ where $y_{\tilde{t}}$ is the Young symmetrizer of the Young tableau

$$\tilde{t} \;=\; \begin{array}{|c|}\hline 1\\\hline 2\\\hline 3\\\hline\end{array}\,.$$

$\frac{1}{6}y_{\tilde{t}}$ yields the alternation of a tensor from $\mathcal{T}_3V$, i.e. it holds $\frac{1}{6}(y_{\tilde{t}}T)_{\lambda\mu\nu} = \tau_{[\lambda}\tau_{\mu;\nu]}$ if we abbreviate $T_{\lambda\mu\nu} := \tau_\lambda\tau_{[\mu;\nu]}$. Thus, $\tau_{[\lambda}\tau_{\mu;\nu]} = 0$ means that the part of $\tau_\lambda\tau_{[\mu;\nu]}$ which lies in $\mathcal{T}_{\tilde{\mathfrak{t}}}$ vanishes. □

5.3. Proof of Theorem 1.9

Because of (10) the Littlewood-Richardson product $[1^3][1]$ has no part that belongs to the partition $(2^2) \vdash 4$. This leads us to the assumption that the terms $\tau_{[\lambda}\tau_{\mu;\nu]}$ will possibly fall out of (8) if we carry out the substitution (11) in (8). We verified this by a Mathematica calculation using the tensor package Ricci [9]. In this calculation we showed that the lines 3, 4, 5 of (8) vanish if we replace all expressions $\tau_\lambda\tau_{[\mu;\nu]}$ by the coordinates of an arbitrary alternating tensor $a_{\lambda\mu\nu}$ of order 3. A record of this calculation can be found in [10, curvterms.nb]. □

5.4. Proof of Theorem 1.10

The proof uses results from [16, 3]. In [16] we showed by means of *discrete Fourier transforms* for symmetric groups that every minimal right ideal $\mathfrak{r} \subset \mathfrak{a}_{(2\,1)} \subset \mathbb{K}[\mathcal{S}_3]$ is generated by exactly one element of the following set of (primitive) idempotents[8]

$$\begin{aligned}
\zeta_\nu &:= \frac{1}{3}\{[1,2,3]+\nu[1,3,2]+(1-\nu)[2,1,3] \\
&\quad -\nu[2,3,1]+(-1+\nu)[3,1,2]-[3,2,1]\} \ , \quad \nu \in \mathbb{K} \\
\eta &:= \frac{1}{3}\{[1,2,3]-[2,1,3]-[2,3,1]+[3,2,1]\} \, .
\end{aligned}$$

On the other hand $\theta_{\lambda\mu\nu} = \tau_\lambda \tau_{[\mu;\nu]} - \tau_{[\lambda}\tau_{\mu;\nu]}$ is generated by the idempotent symmetry operator $\rho := \frac{1}{2}\{[1,2,3]-[1,3,2]\} - \frac{1}{6}\sum_{p\in\mathcal{S}_3} \mathrm{sign}(p)\, p$ from $\tau_\lambda\tau_{\mu;\nu}$. We showed by means of the Mathematica package PERMS [17] in the notebook [10, curvterms.nb] that $\eta\cdot\rho \neq \rho$, $\rho\cdot\eta \neq \eta$ and

$$\zeta_\nu \cdot \rho = \rho \ \text{ and } \ \rho\cdot\zeta_\nu = \zeta_\nu \quad \Leftrightarrow \quad \nu = -1 \, .$$

Consequently, the tensor $\theta_{\lambda\mu\nu}$ belongs to the symmetry class $\mathcal{T}_\mathfrak{r}$ which is defined by the right ideal $\mathfrak{r} := \zeta_{-1}\cdot\mathbb{K}[\mathcal{S}_3]$ with generating idempotent ζ_{-1}.

This right ideal $\mathfrak{r}$ is different from the right ideal $\mathfrak{r}_0$ in Theorem 1.3 since a result of [3] says that $\mathfrak{r}_0 = \zeta_{1/2}\cdot\mathbb{K}[\mathcal{S}_3]$. In the notebook [10, part16a.nb] for the paper [3] we proved that all tensors of the symmetry class of ζ_{-1} satisfy the identities (13). Using (13) we transformed (12) into (14) by means of Ricci [9] in [10, curvterms.nb]. Finally, we see by a comparison with formulas in [3, Sec.4.3] that the last line of (14) is equal to $\frac{1}{2}y_t^*(\theta\otimes F)_{\kappa\lambda\mu\nu}$, where t is the tableau (5). □

Remark : In [3] we showed that the symmetry operator ζ_{-1} produces tensors $U_{\lambda\mu\nu}$ of a (2 1)-symmetry class which admits the *index commutation symmetry* $U_{\lambda\nu\mu} = -U_{\lambda\mu\nu}$. Obviously the above tensor $\theta_{\lambda\mu\nu}$ possesses this index commutation symmetry. A further result of [3] says that then the coordinates of algebraic curvature tensors $y_t^*(w\otimes U)$ (t given by (5)) can be reduced to a sum with a minimal length of 4 summands. This effect can be observed in the above reduction of the number of terms $F_\kappa\theta_{\lambda\mu\nu}$ in formula (12), too.

Acknowledgements. I would like to thank Prof. P. B. Gilkey for important and helpful discussions and for valuable suggestions for future investigations.

References

[1] Gilkey P B 2001 *Geometric Properties of Natural Operators Defined by the Riemann Curvature Tensor* (Singapore: World Scientific)

[2] Fiedler B 2002 Determination of the structure of algebraic curvature tensors by means of Young symmetrizers *Séminaire Lotharingien de Combinatoire* **B48d** 20 pages `http://www.mat.univie.ac.at/~slc` (*Preprint* math.CO/0212278).

[3] Fiedler B 2004 Generators of algebraic curvature tensors based on a (2 1)-symmetry *Preprint* math.DG/0411056

[4] Gilkey P B 2000 Geometric properties of the curvature operator *Proc. Conf. Geometry and Topology of Submanifolds (Beijing)* vol 10 ed Chen W H et al (Singapore: World Scientific) pp 62–70

[5] Díaz-Ramos J C and García-Río E 2004 A note on the structure of algebraic curvature tensors *Linear Algebra Appl.* **382** 271

[6] Díaz-Ramos J, Fiedler B, García-Río E and Gilkey P B 2004 The structure of algebraic covariant derivative curvature tensors *Int. J. Geom. Meth. Mod. Phys.* **1**(6) 711–20 (*Preprint* math.DG/0404060)

[7] Kramer D, Stephani H, MacCallum M and Herlt E 1980 *Exact Solutions of Einstein's Field Equations* (Berlin: Deutscher Verlag der Wissenschaften)

[8] Владимиров Ю С 1982 *Системы отсчёта в теории гравитации* (Москва: Энергоисдат)

[9] Lee J M, Lear D, Roth J, Coskey J and Nave L 1998 *Ricci. A Mathematica package for doing tensor calculations in differential geometry. User's Manual. Version 1.32 ©1992–1998* (Seattle WA: Department of Mathematics Box 354350 University of Washington Seattle WA 98195-4350) `http://www.math.washington.edu/~lee/Ricci/`

[8] In [16, 3] the idempotents ζ_ν were denoted by ξ_ν.

[10] Fiedler B Examples of calculations by means of PERMS (Mathematica notebooks) *Internet page* `http://www.fiemath.de/pnbks.htm`

[11] Boerner H 1955 *Darstellungen von Gruppen* (Die Grundlehren der mathematischen Wissenschaften in Einzeldarstellungen vol 74) (Berlin: Springer-Verlag)

[12] Fiedler B 1999 *An Algorithm for the Decomposition of Ideals of Semi-Simple Rings and its Application to Symbolic Tensor Calculations by Computer (Habilitationsschrift)* (Leipzig: Fakultät für Mathematik und Informatik Universität Leipzig Germany)

[13] Fiedler B 2001 Ideal decompositions and computation of tensor normal forms *Séminaire Lotharingien de Combinatoire* **B45g** 16 pages `http://www.mat.univie.ac.at/~slc` (*Preprint* math.CO/0211156)

[14] Fulling S A, King R C, Wybourne B G and Cummins C J 1992 Normal forms for tensor polynomials: I. The Riemann tensor *Class. Quantum Grav.* **9** 1151–97

[15] Fiedler B 2001 Characterization of tensor symmetries by group ring subspaces and computation of normal forms of tensor coordinates *Proc. Euroconf. ALCOMA Algebraic Combinatorics and Applications (Gößweinstein Germany 1999)* ed Laue R et al (Berlin: Springer-Verlag) pp 118–33

[16] Fiedler B 2003 Generators of algebraic covariant derivative curvature tensors and Young symmetrizers *Preprint* math.CO/0310020

[17] Fiedler B 2005 *PERMS 2.4 (4.8.2005) ©1994–2005* (Leipzig: Mathematisches Institut Universität Leipzig Germany)

Multiple CSLs for the body centered cubic lattice

Peter Zeiner

Faculty of Mathematics, Bielefeld University, Universitätsstraße 25, 33615 Bielefeld, Germany and

Institute of Theoretical Physics & CMS, TU Wien, Wiedner Hauptstraße 8–10, 1040 Wien, Austria

E-mail: pzeiner@math.uni-bielefeld.de

Abstract. Ordinary Coincidence Site Lattices (CSLs) are defined as the intersection of a lattice Γ with a rotated copy $R\Gamma$ of itself. They are useful for classifying grain boundaries and have been studied extensively since the mid sixties. Recently the interests turned to so-called multiple CSLs, i.e. intersections of n rotated copies of a given lattice Γ, in particular in connection with lattice quantizers. Here we consider multiple CSLs for the 3-dimensional body centered cubic lattice, in particular we discuss the spectrum of coincidence indices and their multiplicity, in particular we show that the latter is a multiplicative function and give an explicit expression of it for some special cases.

1. Introduction

Ordinary coincidence site lattices (CSLs) have been studied intensively since the 1960s (see e.g. [1, 2] and references therein), because they are an important tool to characterize and analyze the structure of grain boundaries in crystals. Hence there is a vast literature on 3–dimensional CSLs, in particular on cubic lattices [3, 4, 5, 6, 7, 8]. CSLs in higher dimensions have been studied as well (see [7, 9] and references therein) with possible applications to quasiperiodic structures. In addition, the concept of CSLs has been generalized for modules, again to cover the needs of quasiperiodic structures [10, 7].

We want to discuss another generalization of CSLs here. Loosely speaking, CSLs are the intersection of two mutually rotated lattices. It is thus natural to consider the intersection of n mutually rotated lattices. This question has recently been raised in connection with quantizing procedures [11, 12], where it seems useful to represent a complex lattice as the intersection of simpler lattices. In the meantime this question has been answered in detail for several 2–dimensional lattices [13], so the next step is to consider a 3–dimensional example. We choose a cubic lattice here, namely the body–centered cubic lattice, since the cubic case is one of the best studied 3–dimensional cases. It might seem more natural to discuss the primitive cubic case first, but it turns out that the body centered case can be treated more elegantly. Moreover most results hold for all three cubic lattices. Details for the other cubic lattices shall be published elsewhere [14].

2. Ordinary CSLs

We recall some definitions first [7]. Let $\Gamma \subseteq \mathbb{R}^n$ be an n-dimensional lattice and $R \in SO(n)$ a rotation (we restrict our considerations here to proper rotations for simplicity). Then R is called a *coincidence rotation* if $\Gamma(R) := \Gamma \cap R\Gamma$ is a lattice of finite index in Γ. The corresponding lattice $\Gamma(R) = \Gamma \cap R\Gamma$ is called a *coincidence site lattice*. The *coincidence index* $\Sigma(R)$ is defined as the index of $\Gamma(R)$ in Γ. Note in passing that the set of coincidence rotations forms a group under matrix multiplication.

In the following let Γ be a body centered cubic lattice. Then $R \in SO(3)$ is a coincidence rotation if and only if R is a matrix with rational entries (see [3, 4, 5, 6, 7, 8]). They can be parameterized by integral quaternions $\boldsymbol{q} = (\kappa, \lambda, \mu, \nu)$, i.e., by quaternions with integral coefficients $\kappa, \lambda, \mu, \nu$ in the following way [15, 16, 17, 7, 8]:

$$R(\boldsymbol{q}) = \frac{1}{|\boldsymbol{q}|^2} \begin{pmatrix} \kappa^2 + \lambda^2 - \mu^2 - \nu^2 & -2\kappa\nu + 2\lambda\mu & 2\kappa\mu + 2\lambda\nu \\ 2\kappa\nu + 2\lambda\mu & \kappa^2 - \lambda^2 + \mu^2 - \nu^2 & -2\kappa\lambda + 2\mu\nu \\ -2\kappa\mu + 2\lambda\nu & 2\kappa\lambda + 2\mu\nu & \kappa^2 - \lambda^2 - \mu^2 + \nu^2 \end{pmatrix}, \tag{1}$$

where $|\boldsymbol{q}|^2 = \kappa^2 + \lambda^2 + \mu^2 + \nu^2$ is called the norm of $\boldsymbol{q}$. Note that we will call a quaternion an integer quaternion if it is an integral quaternion $(\kappa, \lambda, \mu, \nu)$ or the sum of an integral quaternion with the quaternion $(1, 1, 1, 1)/2$. We call an integral quaternion $\boldsymbol{r} = (\kappa, \lambda, \mu, \nu)$ primitive if the greatest common divisor of $\kappa, \lambda, \mu, \nu$ equals 1. If not stated otherwise every (integral) quaternion will be assumed to be a primitive quaternion. Furthermore let $\boldsymbol{H} \subset \boldsymbol{Q}$ denote the ring of integer quaternions and the ring of real quaternions, respectively.

One can show that the coincidence index is given by $\Sigma(R(\boldsymbol{q})) = |\boldsymbol{q}|^2/2^\ell$, where ℓ is the maximal power such that 2^ℓ divides $|\boldsymbol{q}|^2$ (see e.g. [4, 6, 7]), i.e. $\Sigma(R)$ is always odd. On the other hand $\Sigma(R)$ runs over all positive odd integers if R runs over all coincidence rotations, i.e. the spectrum of coincidence rotations is the set of all positive odd integers. Let O denote the cubic symmetry group. Then $\Gamma(RQ) = \Gamma(R)$ if and only if $Q \in O$, i.e. R and RQ generate the same CSL, which motivates to call R and RQ *strongly equivalent*. In general one calls R and R' *equivalent* if there exist Q, Q' such that $R = QR'Q'$. For any $R(\boldsymbol{q})$ we can find a strongly equivalent $R' = R(\boldsymbol{q}')$ such that $|\boldsymbol{q}'|^2$ is odd, i.e. $\Sigma(R') = |\boldsymbol{q}'|^2$. Hence we will assume in the following that $|\boldsymbol{q}|^2$ is odd.

If we define the projection $P : \boldsymbol{Q} \to \mathbb{R}^3$ by $P(q_0, q_1, q_2, q_3) = (q_1, q_2, q_3)$ then $\Gamma = P\boldsymbol{H}$, i.e. the body centered cubic lattice is obtained by a projection of $\boldsymbol{H}$ onto $\mathbb{R}^3$. Moreover Lemma 5.1 of [8] tells us that $\Gamma(R(\boldsymbol{q})) = P(\boldsymbol{qH})$, where $\boldsymbol{qH}$ is a left ideal of $\boldsymbol{H}$. In fact this establishes a one to one correspondence of CSLs and left ideals of $\boldsymbol{H}$ (see [14]), which is a key in the discussion of the body centered CSLs. In particular, finding the number $f(\Sigma)$ of different CSLs of given index Σ is equivalent to counting the corresponding left ideals of $\boldsymbol{H}$. One can show that $f(\Sigma)$ is a multiplicative function, i.e. $f(mn) = f(m)f(n)$ if m and n are coprime, and in particular we have $f(1) = 1$, $f(2) = 0$ and $f(p^r) = (p+1)p^{r-1}$ for all odd primes p [7, 8].

3. Multiple CSLs

Now we turn to multiple CSLs, which we define as follows [13, 14]:

Definition 3.1 *Let Γ be an n-dimensional lattice and R_i, $i = 1, \ldots m$ coincidence rotations of Γ. Then the lattice*

$$\Gamma(R_1, \ldots, R_m) := \Gamma \cap R_1\Gamma \cap \ldots \cap R_m\Gamma = \Gamma(R_1) \cap \ldots \cap \Gamma(R_m) \tag{2}$$

is called a multiple CSL (MCSL). Its index in Γ is denoted by $\Sigma(R_1, \ldots, R_m)$.

Note that $\Sigma(R_1, \ldots, R_m)$ is finite since $\Gamma(R_1, \ldots, R_m)$ is a finite intersection of mutually commensurate lattices [7]. In particular, it follows from the second homomorphism theorem that

$$\Sigma(R_1, R_2) = \frac{\Sigma(R_1)\Sigma(R_2)}{\Sigma_+(R_1, R_2)}, \tag{3}$$

where $\Sigma_+(R_1, R_2)$ is the index of the direct sum $\Gamma_+(R_1, R_2) = \Gamma(R_1) + \Gamma(R_2)$ in Γ. In general one shows

$$\Sigma(R_1, \ldots, R_m) = \frac{\Sigma(R_1, \ldots, R_{m-1})\Sigma(R_m)}{\Sigma_+(R_1, \ldots, R_{m-1}; R_m)}, \tag{4}$$

where $\Sigma_+(R_1, \ldots, R_{m-1}; R_m)$ is the index of $\Gamma_+(R_1, \ldots, R_{m-1}; R_m) = \Gamma(R_1, \ldots, R_{m-1}) + \Gamma(R_m)$ in Γ. In particular, $\Sigma(R_1, \ldots, R_m)$ divides $\Sigma(R_1) \cdot \ldots \cdot \Sigma(R_m)$. In case of the 3–dimensional cubic lattices this implies immediately that the spectrum is the same for multiple and ordinary CSLs, i.e. $\Sigma(R_1, \ldots, R_m)$ runs over all odd positive integers, too. However, new lattices emerge and the multiplicity of a given index will increase. Note that the spectrum is preserved also for the square lattice and the 4–dimensional hypercubic lattices.

Having determined the spectrum we can attack the second main problem, the number $f_m(\Sigma)$ of different MCSLs. To this end we have to determine all possible MCSLs. We first note that $\Gamma_+(R_1, R_2) = P(\boldsymbol{q}_1\boldsymbol{H} + \boldsymbol{q}_2\boldsymbol{H}) = P(\boldsymbol{q}\boldsymbol{H})$, where $\boldsymbol{q}$ is the greatest left common divisor (glcd) of $\boldsymbol{q}_1$ and $\boldsymbol{q}_2$. Hence (recall that we may assume that $|\boldsymbol{q}_i|^2$ is odd)

$$\Sigma(R_1, R_2) = \frac{|\boldsymbol{q}_1|^2|\boldsymbol{q}_2|^2}{|\boldsymbol{q}|^2} \quad \text{with } \boldsymbol{q} = \text{glcd}(\boldsymbol{q}_1, \boldsymbol{q}_2). \tag{5}$$

In case that $|\boldsymbol{q}_1|^2$ and $|\boldsymbol{q}_2|^2$ are relatively prime this reduces to $\Sigma(R_1, R_2) = |\boldsymbol{q}_1|^2|\boldsymbol{q}_2|^2$ which suggests the following lemma [14]:

Lemma 3.1 *Let $|\boldsymbol{q}_1|^2$ and $|\boldsymbol{q}_2|^2$ be relatively prime. Then there exists a quaternion $\boldsymbol{q}$, such that $\Gamma(R_1, R_2) = \Gamma(R(\boldsymbol{q}))$.*

The proof makes use of the fact there exists a right least common multiple $\boldsymbol{q}$ of $\boldsymbol{q}_1$ and $\boldsymbol{q}_2$. Note that in this case the MCSL is equal to an ordinary CSL. This theorem can be immediately generalized for arbitrary m. Conversely we have [14]

Lemma 3.2 *Let $p_1^{\alpha_1} \cdots p_\ell^{\alpha_\ell}$ be the prime decomposition of $|\boldsymbol{q}|^2$. Then there exist quaternions $\boldsymbol{q}_i$ such that $|\boldsymbol{q}_i|^2 = p_i^{\alpha_i}$ and $\Gamma(R(\boldsymbol{q})) = \Gamma(R(\boldsymbol{q}_1)) \cap \ldots \cap \Gamma(R(\boldsymbol{q}_\ell))$.*

This decomposition is unique. More generally we have [14]

Lemma 3.3 *For any $\Gamma(R_1, \ldots, R_m)$ with $\Sigma(R_1, \ldots, R_m) = p_1^{\alpha_1} \cdots p_\ell^{\alpha_\ell}$ (all p_i distinct) there exists a unique decomposition $\Gamma(R_1, \ldots, R_m) = \Gamma_1 \cap \ldots \cap \Gamma_\ell$ such that Γ_i is a MCSL with index $\Sigma_i = p_i^{\alpha_i}$.*

Thus the analysis of MCSLs can be reduced to the study of the MCSLs with prime power index. Another consequence is the multiplicity of f_m:

Theorem 3.4 *Let $f_m(\Sigma)$ be the number of different MCSLs for a given index Σ. Then $f_m(nn') = f_m(n)f_m(n')$ if n and n' are relatively prime.*

Since the analysis of general MCSLs with prime power index is rather cumbersome we confine our discussion to the intersection of two CSLs. The general discussion will be published elsewhere [14]. We may confine our discussion to the case that neither $\boldsymbol{q}_1$ nor $\boldsymbol{q}_2$ is a right multiple of the other, i.e. there does not exist an integer quaternion $\boldsymbol{r}$ such that $\boldsymbol{q}_1 = \boldsymbol{q}_2\boldsymbol{r}$ or vice versa, since otherwise $\Gamma(R_1, R_2)$ reduces to $\Gamma(R_2)$ or $\Gamma(R_1)$, respectively. We first mention a representation of $\Gamma(R_1, R_2)$. Here $\bar{\boldsymbol{q}} = (\kappa, -\lambda, -\mu, -\nu)$ denotes the conjugate of $\boldsymbol{q} = (\kappa, \lambda, \mu, \nu)$.

Lemma 3.5 *Let $|\boldsymbol{q}_i|^2 = p^{\alpha_i}$, $i = 1, 2$, p prime, none of the $\boldsymbol{q}_i$ a right multiple of the other one. Choose $\boldsymbol{r}$ such that $\boldsymbol{q}_1\boldsymbol{r}\bar{\boldsymbol{q}}_2$ is a primitive quaternion and let $\boldsymbol{q}$ be the least right common multiple of $\boldsymbol{q}_1$ and $\boldsymbol{q}_2$. Then $\Gamma(R_1, R_2) = \Gamma(R(\boldsymbol{q}_1)) \cap \Gamma(R(\boldsymbol{q}_2)) = P(\boldsymbol{q}\boldsymbol{H} + \boldsymbol{q}_1\boldsymbol{r}\bar{\boldsymbol{q}}_2\mathbb{Z})$.*

Such an $\boldsymbol{r}$ is by no means unique. Its existence follows from the uniqueness of the left (or right) prime power decomposition of integer quaternions. Alternatively we may decompose $\Gamma(R_1, R_2)$ as follows:

Lemma 3.6 *Under the conditions of the previous lemma, we have* $\Gamma(R_1, R_2) = P(\boldsymbol{q}\boldsymbol{H} + \boldsymbol{q}_1\boldsymbol{H}\bar{\boldsymbol{q}}_2) = P(\boldsymbol{q}\boldsymbol{H} + \boldsymbol{q}_2\boldsymbol{H}\bar{\boldsymbol{q}}_1)$.

Note that $\boldsymbol{q}\boldsymbol{H} + \boldsymbol{q}_2\boldsymbol{H}\bar{\boldsymbol{q}}_1$ is no ideal and hence $\Gamma(R_1, R_2)$ is neither an ordinary CSL nor a multiple of an ordinary CSL. Note further that $P(\boldsymbol{q}\boldsymbol{H} + \boldsymbol{q}_1\boldsymbol{H}\bar{\boldsymbol{q}}_2)/P(\boldsymbol{q}\boldsymbol{H})$ is a cyclic group of order $\frac{|q|^2}{\max(|q_1|^2, |q_1|^2)}$ and that $P(\boldsymbol{q}\boldsymbol{H})$ is a multiple of an ordinary CSL ($\boldsymbol{q}$ is not primitive here!). The next lemma tells us under which conditions different pairs of CSLs give rise to different MCSLs:

Lemma 3.7 *Let* $\boldsymbol{q}_i$ *be primitive quaternions with* $|\boldsymbol{q}_i|^2 = p^{\alpha_i}$, *where* p *is a prime and* $\alpha_1 \geq \alpha_2 \geq \alpha_4, \alpha_3 \geq \alpha_4$. *Let* $\boldsymbol{q}_{ij}$ *with* $|\boldsymbol{q}_{ij}|^2 = p^{\alpha_{ij}}$ *be the greatest left common divisor of* $\boldsymbol{q}_i$ *and* $\boldsymbol{q}_j$. *Then* $\Gamma(R_1) \cap \Gamma(R_2) = \Gamma(R_3) \cap \Gamma(R_4)$ *if and only if (in case of* $\alpha_1 = \alpha_2$ *possibly after interchanging* $\boldsymbol{q}_1$ *and* $\boldsymbol{q}_2$*)* $\alpha_1 = \alpha_3, \alpha_2 - \alpha_{12} = \alpha_4 - \alpha_{34}, \alpha_1 - \alpha_{13} \leq \min(\alpha_4 - \alpha_{34}, \alpha_{34})$ *and* $\alpha_2 - \alpha_{24} \leq \min(\alpha_4 - \alpha_{34}, \alpha_{34})$ *are satisfied.*

Thus we can calculate the number $f_2(\Sigma)$ of different MCSLs which are intersections of at most two ordinary CSLs:

Theorem 3.8 *Let* p *be an odd prime number. Then*

$$\begin{aligned} f_2(p^r) =& (r/2 + 1/2)\,(p+1)p^{r-1} + (r/2 - 1)p^{r-2} + (r/2 - [r/2])p^{r-4} \\ &+ \frac{p^{r-1} - p^{r-2[r/3]-1}}{p^2 - 1} + \frac{p^{4[r/3]-r+2} - p^{4[r/2]-r-2}}{2(p^2-1)}, \end{aligned} \tag{6}$$

where $[x]$ *is Gauss' symbol denoting the largest integer* n *such that* $n \leq x$.

Note that $f_2(p^r) = f_m(p^r)$ for $r = 1, 2$. Thus we know $f_m(\Sigma)$ for all Σ that are free from third powers. The more complex general case will be presented elsewhere [14].

Acknowledgments

The author is very grateful to Michael Baake for interesting and stimulating discussions on the present subject. Financial support by the Austrian Academy of Sciences (APART-program) and the EU Research Training Network "Quantum Probability with Applications to Physics, Information Theory and Biology" is gratefully acknowledged.

References

[1] Bollmann W 1970 *Crystal Defects and Crystalline Interfaces* (Berlin: Springer)

[2] Bollmann W 1982 *Crystal Lattices, Interfaces, Matrices* published by the author, Geneva.

[3] Grimmer H 1974 Disorientations and coincidence rotations for cubic lattices *Acta Cryst.*, A **30** 685–8

[4] Grimmer H, Bollmann W, and Warrington D H 1974 Coincidence-site lattices and complete pattern-shift lattices in cubic crystals *Acta Cryst.* A **30** 197–207

[5] Bleris G L and Delavignette P 1981 A new formulation for the generation of coincidence site lattices (CSL's) in the cubic system *Acta Cryst.* A **37** 779–86

[6] Grimmer H 1984 The generating function for coincidence site lattices in the cubic system *Acta Cryst.* A **40** 108–12.

[7] Baake M 1997 Solution of the coincidence problem in dimensions $d \leq 4$. In *The Mathematics of Long-Range Aperiodic Order* ed. R V Moody (Dordrecht: Kluwer) pp 9–44

[8] Zeiner P Symmetries of coincidence site lattices of cubic lattices *Z. Kristallogr.* **220** 915–25

[9] Zeiner P Coincidences of hypercubic lattices in 4 dimensions *Z. Kristallogr.*, in press

[10] Pleasants P A B, Baake M, and Roth J 1996 Planar coincidences for n–fold symmetry *J. Math. Phys.* **37** 1029–58.

[11] Diggavi S N, Sloane N J A, and Vaishampayan V A 2002 Asymmetric multiple description lattice vector quantizers *IEEE Transactions Information Theory* **48** 174–91
[12] Sloane N J A and Beferull-Lozano B Quantizing using lattice intersections arXiv: math.CO/0207147
[13] Baake M and Grimm U 2005 Multiple planar coincidences with N–fold symmetry preprint
[14] Zeiner P Multiple CSLs for cubic lattices, in preparation
[15] Koecher M and Remmert R 1991 Hamilton's Quaternions In *Numbers* ed. Ebbinghaus H-D et al pp. 155–81 (Springer)
[16] Hurwitz A 1919 *Vorlesungen über die Zahlentheorie der Quaternionen* (Berlin: Springer)
[17] du Val P 1964 *Homographies, Quaternions and Rotations* (Oxford: Clarendon Press)

On the operation of composition of distributions

A Kamiński[1] **and S Sorek**[2]

[1]Institute of Mathematics, University of Rzeszow, Rejtana 16A, 35-310 Rzeszow, Poland
[2]Institute of Mathematics, University of Rzeszow, Rejtana 16A, 35-310 Rzeszow, Poland

E-mail: [1]`akaminsk@univ.rzeszow.pl`

E-mail: [2]`ssorek@univ.rzeszow.pl`

Abstract. The proofs of the results of P. Antosik [3] on the distributional composition of distributions (in the sense of Mikusiński's theory of irregular operations), which contained essential gaps, are completed due to some measure theory techniques, and the results are generalized. The obtained theorems can be applied to prove some formulas, which may be interesting to physicists, concerning the substitution of measures (in particular, the Dirac delta distribution) to continuous functions.

1. Introduction

The theory of distributions, born as a mathematical theory in the middle of 20th century, fulfilled expectations and requirements of many mathematicians and physicists for whom the frames of the classical theory of functions were too narrow. The mathematicians, inspired by various problems in the classical analysis and differential equations, wanted to extend the possibility of differentiation of functions and give a sense certain divergent integrals. The physicists needed the mathematical justification of some objects like impuls, point mass, dipol etc. which cannot be explained in terms of usual functions; in quantum mechanics Dirac introduced the concept, called then the Dirac delta function, which is not a function at all.

Though distributions as mathematical objects were introduced already by S.L. Sobolev in [13], it was Laurent Schwartz, the great French mathematician, who has developed the initial ideas in a beautiful and important, both for mathematics and physics, theory of distributions (see [12]). The theory was founded on functional analysis and this required a deep study and development of new areas in this field. Distributions (including the Dirac delta distribution δ) are linear continuous functionals on the space $\mathcal{D}$ of smooth (infinitely differentiable) functions of compact support on the real line (endowed with the topology of inductive limit of certain normed spaces), i.e. they are elements of the dual space $\mathcal{D}'$. The space $\mathcal{D}'$ of distributions embraces the class C of all continuous functions and even the wider class $L^1_{\rm loc}$ of all locally integrable functions. Distributions have a very nice property: the derivative of an arbitrary order of every distribution (in particular, every locally integrable function) always exists (in a more general, distributional sense). On the other hand, an important structural theorem says that each distribution is locally (i.e. on each open bounded interval) a distributional derivative of finite order of a continuous function. These properties of distributions had a great influence for a development of mathematical analysis and other domains of mathematics.

In contrast differentiation other operations acting on functions cannot be so easily extended to distributions in the Schwartz theory, e.g. the product $f \cdot g$ of two distributions f and g makes sense only in special cases (e.g. if f is a distribution and g is a smooth function).

The great Polish mathematician, Jan Mikusiński, has discovered (see [8], [4]) another approach to the theory of distributions, equivalent to the functional analysis approach of Schwartz. Mikusiński's approach, called elementary or sequential, is much simpler and based on elementary notions of mathematical analysis. In the frame of this approach, J. Mikusiński has proposed a uniform scheme for defining all operations on distributions. Some of them (regular) are defined and exist for all distributions, other ones (irregular) can be defined (by using the notion of a delta-sequence) for all distributions and exist for a large class of distributions, though not for all of them. The theory of regular and irregular operations is discussed in [10], [4] and is also briefly recalled in the article [14] presented in this volume.

In this paper, we will be concentrated on the operation of composition of distributions. A particular case of the composition of distributions is the one-argument operation C_1 of substitution of a fixed smooth function ω with $\omega' \neq 0$ into an arbitrary distribution f. This operation is defined first on smooth functions ψ (by $C_1(\psi) := \psi \circ \omega$) and then is extended as a regular operation to every distribution f by continuity: the result $C_1(f) := f \circ \omega$ always exists (see [4], [14]). However, the two-argument operation C_2 of composition of smooth functions (defined by $C_2(\varphi, \psi) := \psi \circ \varphi$) is not a distributionally continuous operation (see [3]).

We will consider here the composition as an irregular operation of two arguments, whose value $C_2(f, g)$ for given distributions $f, g \in \mathcal{D}'$ will be denoted by $g(f)$, called the distributional composition of f and g and defined by the formula: $g(f) := \lim_{n\to\infty} g_n \circ f_n$, where (f_n), (g_n) are so-called regular sequences of smooth functions approximating distributions f, g, the smooth functions $g_n \circ f_n$ are compositions of f_n, g_n in the usual functional sense and the limit is meant in the distributional sense (for the exact definition see section 2; see also [4]). Notice that in case f and g are locally integrable functions the usual functional composition $g \circ f$ of f and g does not coincide, in general, with the distributional composition $g(f)$ of f and g. We will show, however, that in case f and g are continuous functions both the functional $g \circ f$ and distributional $g(f)$ compositions coincide (see Theorem 1 in section 4). Such coincidence holds also under other assumptions on functions f and g (see Theorem 2 in section 4).

The distributional composition was studied by P. Antosik who gave in [3] several important results concerning the existence of the distributional composition $g(f)$ in case g is a continuous function and f is a locally integrable function, a measure or a quasi-measure. The theorems formulated in [3] may be a matter of interest for physicists, because they lead to numerous astonishing results and interesting formulas with the Dirac delta distribution, for example:

$$\sqrt[r]{\sum_{j=0}^{r} a_j \delta^j} = \sqrt[r]{a_r} + \delta, \qquad \log\left(\sum_{j=0}^{r} a_j \delta^j\right) = 0,$$

where $a_j > 0$ for $j = 0, \ldots, r$. The above formulas are new, not mentioned in [3]; notice that the powers δ^j of δ above should be meant as results of the operation of substitution of δ into a certain continuous function and not the operation of product of distributions. For other formulas see [3] and [14]. All of them follow from the main Theorem 8, formulated in section 5.

However, the proofs of Theorems 2, 4 and 7 in [3], crucial for the final result of the paper, are incorrect, since they are based on some wrong implications, e.g. that if $f_n \to f$ in measure or almost everywhere and $g_n \to g$ locally uniformly, then $g_n \circ f_n \to g \circ f$ almost everywhere. Fortunately, all the proofs can be corrected, due to using new techniques and subtle applications of the classical theorems of the measure theory; we give here the full proofs of all results from [3], which contained gaps. Consequently, the proofs of the mentioned above results as well as of the remaining ones from [3], including the main Theorem 9, are now completed. Moreover, we can generalize the results presented in [3] in a few directions. First of all, some of Antosik's proofs required the assumptions that delta-sequences used in the definition of the distributional composition of distributions have nonnegative members (condition 0°, see section 2) and we may replace this assumption by a weaker one (condition 3°_0, see section 2) and condition 3° is needed only in cases when the Łojasiewicz value of a distribution is involved. Secondly, in all results concerning the distributional composition $g(f)$ we use arbitrary delta-sequences, which can be different for f and g. Thirdly, the real line can be replaced in some theorems by its open subsets and in the main theorem by an open, finite or infinite, interval (see [14]). Finally, let us notice that our Theorem 2 is a generalization of Antosik's result from the class of $L^1_{\rm loc}$-functions to the functions of the class $L^p_{\rm loc}$, $p \in [1, \infty)$. The proof of Theorem 2 also required new techniques.

2. Basic notions and notation

The symbols $\mathbb{N}$, $\mathbb{N}_0$ and $\mathbb{R}$ denote traditionally the sets of all positive, all nonnegative integers and all real numbers. All functions and distributions will be real-valued and considered on the real line $\mathbb{R}$, though some results can be extended to open subsets of $\mathbb{R}$ (see [14]) and to multidimensional case. By *intervals* we will always mean finite (bounded) intervals: open (a, b) or closed $[a, b]$ (i.e. compact) for $a, b \in \mathbb{R}$ with $a < b$. The families of all finite open and all compact intervals will be denoted by $\mathcal{I}^\circ$ and $\mathcal{I}$, respectively. The Lebesgue measure of the (Lebesgue measurable) set $E \subseteq \mathbb{R}$ will be denoted by $|E|$. All measurable functions considered in the paper will be finite almost everywhere in the sense of the Lebesgue measure.

By L^p and $L^p_{\rm loc}$ for $p \in [1, \infty)$, we mean the spaces of all measurable real-valued functions f with the properties: $\|f\|_p := \left(\int_{\mathbb{R}} |f(x)|^p\,dx\right)^{1/p} < \infty$ and $\|f\|_{p,I} := \left(\int_I |f(x)|^p\,dx\right)^{1/p} < \infty$ for every $I \in \mathcal{I}$, respectively, where the integrals are considered with respect to the Lebesgue measure; the spaces are endowed with the topology of the norm $\|f\|_p$ and the topology of the family $\{\|f\|_{p,I} : I \in \mathcal{I}\}$ of semi-norms, respectively. By L^∞ and $L^\infty_{\rm loc}$, we mean the spaces of all measurable real-valued functions f with the properties: $\|f\|_\infty := \operatorname{ess\,sup}_{x\in\mathbb{R}}|f(x)| < \infty$ and $\|f\|_{\infty,I} := \operatorname{ess\,sup}_{x\in I}|f(x)| < \infty$ for every $I \in \mathcal{I}$, respectively; the spaces are endowed with the topology of the norm $\|f\|_\infty$ and the topology of the family $\{\|f\|_{\infty,I} : I \in \mathcal{I}\}$ of semi-norms. If $f_n, f \in L^1_{\rm loc}$ and $f_n \to f$ in $L^1_{\rm loc}$, we say that $f_n \to f$ locally in mean (and write shortly $f_n \to f$ loc.). The last convergence implies the distributional convergence: $f_n \to f$ in $\mathcal{D}'$, i.e. $< f_n, \varphi > \int_{\mathbb{R}} f_n\varphi \to \int_{\mathbb{R}} f\varphi =< f, \varphi >$ for every function $\varphi \in \mathcal{D}$ or, equivalently, for every $I \in \mathcal{I}^\circ$ there are an index $k \in \mathbb{N}$ and a sequence (F_n) of smooth functions on I such that $F_n^{(k)} = f_n$ on I and $F_n \rightrightarrows$ on I (see [4]).

Let f_n, f be given measurable functions. The uniform convergence, the locally uniform convergence (i.e. uniform convergence on each compact subset), the almost everywhere convergence and the convergence in measure of the sequence (f_n) to the function f on (a given measurable subset of) $\mathbb{R}$ will be denoted shortly by $f_n \rightrightarrows f$, $f_n \to f$ l.u., $f_n \to f$ a.e. and $f_n \to f$ i.m. on (the given subset of) $\mathbb{R}$, respectively (see [5], [6], [11]).

We say that $f_n \to f$ asymptotically (and write: $f_n \to f$ as.), if from each subsequence $(\overline{f}_n)$ of the sequence (f_n) one can select a subsequence $(\overline{\overline{f}}_n)$ convergent a.e. to f; it can be shown that, in general, $f_n \to f$ a.e. implies $f_n \to f$ as. and $f_n \to f$ i.m. implies $f_n \to f$ as., that is, the asymptotic convergence is a common generalization of the almost everywhere convergence; on the sets of finite measure the asymptotic convergence is just equivalent to the convergence in measure (see [11]).

Recall three classical theorems from measure theory in the form convenient further on.

Theorem A (of Riesz; see [6], p. 74-77; [5], p. 93-94) *Let (f_n) be a sequence of measurable functions (finite a.e.) for which Cauchy's condition in measure is satisfied. Then there exists a subsequence (f_{k_n}) of the sequence (f_n) and a measurable function f such that $f_{k_n} \to f$ a.e. and*

$f_n \to f$ i.m. In particular, if $f_n \to f$ i.m., then there is a subsequence (f_{k_n}) of (f_n) such that $f_{k_n} \to f$ a.e.

Theorem B (of Egoroff; see [6], p. 77; [5], p. 88) *Let f and f_n ($n \in \mathbb{N}$) be measurable functions (finite a.e.) defined on a measurable set $E \subset \mathbb{R}$ of a finite measure and assume that $f_n \to f$ a.e. on E. Then for every $\varepsilon > 0$ there exists a measurable subset E_0 of E such that $|E \setminus E_0| < \epsilon$ and $f_n \rightrightarrows f$ on the set E_0.*

Theorem C (of Lusin; see [6], p. 106; [5], p. 242-243) *Let f be a measurable function (finite a.e.) defined on a measurable set $E \subseteq \mathbb{R}$. Then for every $\varepsilon > 0$ there exists a measurable subset E_0 of E such that $|E \setminus E_0| < \epsilon$ and f is continuous on the set E_0.*

The fourth theorem, the following generalized theorem on dominated convergence, proved by P. Antosik in [2] (see also [11]), will be crucial in the proofs of our further results.

Theorem D (of Antosik; see [2]; [11], p. 36) *Let (f_n) be a sequence of locally integrable functions f_n such that $f_n \to f$ as. and let (g_n) be a sequence of locally integrable functions such that $|f_n| \le g_n$ and $g_n \to g$ loc. Then $f_n \to f$ loc.*

By a *delta-sequence* we mean a sequence (δ_n) of smooth functions, defined on $\mathbb{R}$, for which there exists a sequence of positive numbers (α_n), converging to 0, such that some of the following conditions are satisfied:

0° $\delta_n \ge 0$ for $n \in \mathbb{N}$;

1° $\delta_n(x) = 0$ for $x \notin O_n : (-\alpha_n, \alpha_n)$, $n \in \mathbb{N}$;

2° $\int_{\mathbb{R}} \delta_n(t)\, dt = 1$ for $n \in \mathbb{N}$;

3°_0 there is a positive constant M_0 such that

$$\int_{\mathbb{R}} |\delta_n(t)|\, dt \le M_0 \qquad n \in \mathbb{N};$$

3° for every $k \in \mathbb{N}_0$ there is a positive constant M_k such that

$$\alpha_n^k \int_{\mathbb{R}} |\delta_n^{(k)}(t)|\, dt \le M_k, \qquad n \in \mathbb{N}.$$

Various results of our paper can be proved, if we restrict ourselves to one of the following three classes of delta-sequences: a) Δ_+, b) Δ_0, c) Δ_∞, defined as the sets of all delta-sequences satisfying the conditions: a) 0°, 1°, 2°, b) 1°, 2°, 3°_0, c) 1°, 2°, 3°, respectively. But for simplicity we will assume in the whole paper that all conditions (i.e. conditions 0°, 1°, 2° and 3°) are satisfied, i.e. our delta-sequences belong to the intersection $\Delta := \Delta_+ \cap \Delta_0 \cap \Delta_\infty$ of the mentioned classes.

In the sequel, we will frequently consider, instead of one delta-sequence (δ_n) from Δ, a pair $(\delta_{1,n})$ and $(\delta_{2,n})$ of delta-sequences from Δ. We will then always denote the numerical sequence (α_n) in conditions 1° and 3°, corresponding to $(\delta_{1,n})$ and $(\delta_{2,n})$, by $(\alpha_{1,n})$ and $(\alpha_{2,n})$, respectively, and constants M_k for $k \in \mathbb{N}$ in conditions 3°_0 and 3°, corresponding to $(\delta_{1,n})$ and $(\delta_{2,n})$, by $M_{1,k}$ and $M_{2,k}$, respectively.

The sequence (f_n), $f_n := f * \delta_n$, is called a *regular sequence* for the distribution f. If f is a distribution, then elements of the sequence (f_n) are smooth functions and the sequence converges to f distributionally; if f is a continuous function, the sequence (f_n) converges to f locally uniformly, i.e. uniformly on every compact set; if f is a function of the class L^p_{loc}, $p \in [1, \infty)$, then (f_n) converges to f in L^p_{loc}, i.e. in every semi-norm $\|\cdot\|_{p,I}$ (see [4]). Denote by $f_n := f * \delta_{1,n}$, $g_n := g * \delta_{2,n}$ ($n \in \mathbb{N}$) elements of regular sequences for the distributions f, g, corresponding to delta-sequences $(\delta_{1,n}),(\delta_{2,n}) \in \Delta$, respectively. If for arbitrary $(\delta_{1,n}),(\delta_{2,n}) \in \Delta$, the sequence of compositions $(g_n \circ f_n)$ is fundamental, i.e. converges distributionally to a distribution h (see [4], [14]), then h is said to be the *distributional composition* of f and g and is denoted by $g(f)$. In other words,

$$g(f) := \lim_{n\to\infty} (g * \delta_{2,n}) \circ (f * \delta_{1,n}), \tag{1}$$

whenever the limit exists in $\mathcal{D}'$ for all delta-sequences $(\delta_{1,n})$, $(\delta_{2,n}) \in \Delta$.

A distribution f has the *Łojasiewicz value* $f_L(x_0) \in \mathbb{R}$ at a point x_0 if for every regular sequence $f_n = f * \delta_n$, where $(\delta_n) \in \Delta_\infty$, we have $\lim_{n\to\infty} f_n(x_0) = f_L(x_0)$ (see [7], [9], [4]).

By a *quasi-measure* we mean a distribution f in $\mathbb{R}$ such that f has the Łojasiewicz value $f_L(x)$ for almost all $x \in \mathbb{R}$ and the function $x \mapsto f_L(x)$ defined almost everywhere on $\mathbb{R}$, called the *regular part* f_r of f, is a locally integrable function. Denoting the *singular part* $f - f_r$ of f by f_s, we can write $f = f_r + f_s$. The above equality is called the *Łojasiewicz decomposition* of a quasi-measure of f.

3. Auxiliary results

The following three lemmas will be needed in the next section:

Lemma 1 *Let $\varphi \in L^\infty_{\mathrm{loc}}$, $\psi \in C$ on $\mathbb{R}$ and let φ_n, ψ_n be arbitrary functions on $\mathbb{R}$ for $n \in \mathbb{N}$. Assume that $\varphi_n \to \varphi$ l.u. and $\psi_n \to \psi$ l.u. Then $\psi_n \circ \varphi_n \to \psi \circ \varphi$ l.u.*

Proof. Fix an arbitrary interval $I \in \mathcal{I}$. Since $\varphi_n \rightrightarrows \varphi$ on I, there is an $n_0 \in \mathbb{N}$ such that

$$\|\varphi_n - \varphi\|_{\infty,I} = \sup_{x\in I} |\varphi_n(x) - \varphi(x)| \le 1, \qquad n \ge n_0,$$

i.e. $\|\varphi_n\|_{\infty,I} \le \beta$ for $n \in \mathbb{N}$, where

$$\beta := \max\{\|\varphi\|_{\infty,I} + 1, \max_{n\le n_0} \|\varphi_n\|_{\infty,I}\}.$$

Denote $J := [-\beta, \beta]$ and fix $\varepsilon > 0$. Since ψ is uniformly continuous on J and $\psi_n \rightrightarrows \psi$, there exist an $\eta > 0$ and an $n_1 \in \mathbb{N}$ such that $y_1, y_2 \in J, |y_1 - y_2| < \eta, n \ge n_1$ imply

$$|\psi_n(y_i) - \psi(y_i)| < \frac{\varepsilon}{6}, \qquad i = 1, 2 \tag{2}$$

and

$$\begin{aligned} |\psi_n(y_1) - \psi_n(y_2)| &\le |\psi_n(y_1) - \psi(y_1)| + |\psi(y_1) - \psi(y_2)| \\ &+ |\psi(y_2) - \psi_n(y_2)| < \frac{\varepsilon}{6} + \frac{\varepsilon}{6} + \frac{\varepsilon}{6} = \frac{\varepsilon}{2}, \end{aligned} \tag{3}$$

i.e. (ψ_n) is equicontinuous. On the other hand, since $\varphi_n \rightrightarrows \varphi$ on I, there is an $n_2 \in \mathbb{N}$ such that $\|\varphi_n - \varphi\|_{\infty,I} < \eta$ for $n \ge n_2$ and so, by equicontinuity of (ψ_n) just proved in (3), we get

$$\|\psi_n \circ \varphi_n - \psi_n \circ \varphi\|_{\infty,I} = \sup_{x\in I} |\psi_n(\varphi_n(x)) - \psi_n(\varphi(x))| \le \frac{\varepsilon}{2}, \qquad n \ge n_2.$$

Hence, again applying (2), we obtain

$$\|\psi_n \circ \varphi_n - \psi \circ \varphi\|_{\infty,I} \le \|\psi_n \circ \varphi_n - \psi_n \circ \varphi\|_{\infty,I} + \|\psi_n \circ \varphi - \psi \circ \varphi\|_{\infty,I} < \varepsilon$$

for $n > \max(n_1, n_2)$, since $\varphi(x) \in J$ for $x \in I$. We conclude, since I was an arbitrary compact interval, that $\psi_n \circ \varphi_n \to \psi \circ \varphi$ l.u.

Lemma 2 *Assume that f and f_n $(n \in \mathbb{N})$ are measurable functions, defined on a bounded measurable set $A \subset \mathbb{R}$, such that f is bounded a.e. and $f_n \to f$ i.m. on A. Moreover, assume that g and g_n $(n \in \mathbb{N})$ are continuous functions on $\mathbb{R}$ such that $g_n \to g$ l.u. Then $g_n \circ f_n \to g \circ f$ i.m. on A.*

Proof. Denote $h_n := g_n \circ f_n$. Let $(\overline{h}_n)$ be an arbitrary subsequence of the sequence (h_n), i.e. $\overline{h}_n := \overline{g}_n \circ \overline{f}_n$, where $\overline{g}_n := g_{k_n}$ and $\overline{f}_n := f_{k_n}$ with $k_n \in \mathbb{N}$, $k_n \uparrow \infty$. Clearly, $\overline{f}_n \to f$ i.m. on the set A of finite measure. By Theorem A of Riesz, there exists a subsequence $(\overline{\overline{f}}_n)$ of the sequence $(\overline{f}_n)$, i.e. $\overline{\overline{f}}_n := \overline{f}_{l_n} = f_{k_{l_n}}$ with $l_n \in \mathbb{N}, l_n \uparrow \infty$, such that $\overline{\overline{f}}_n \to f$ a.e. on A. Since f is bounded a.e., there is a measurable set $B, B \subseteq A$, $|A \setminus B| = 0$, and a bounded interval $J = [-a, a]$ such that $f(x) \in J$ and $\overline{\overline{f}}_n(x) \to f(x)$ for all $x \in B$. By Egoroff's Theorem B, there exists a measurable set $B_\varepsilon \subseteq B$ such that $\overline{\overline{f}}_n \rightrightarrows f$ on B_ε and $|B \setminus B_\varepsilon| < \varepsilon$. Hence, since $f(x) \in J$ for $x \in B \subseteq A$, there is an $n_0 \in \mathbb{N}$ such that

$$|\overline{\overline{f}}_n(x)| \leq |f(x)| + 1 \in \tilde{J} \tag{4}$$

for $x \in B_\varepsilon$ and $n \geq n_0$, where $\tilde{J} := [-a-1, a+1]$.

Notice that, since $g_n \to g$ l.u., the subsequence $(\overline{\overline{g}}_n)$ of the sequence (g_n) defined by $\overline{\overline{g}}_n := \overline{g}_{l_n} = g_{k_{l_n}}$ is uniformly convergent to g on $\tilde{J}$, so there is an $n_1 \in \mathbb{N}, n_1 \geq n_0$, such that

$$|\overline{\overline{g}}_n(\overline{\overline{f}}_n(x)) - g(\overline{\overline{f}}_n(x))| < \frac{\varepsilon}{2} \tag{5}$$

for all $x \in B_\varepsilon$ and $n \geq n_1$ by (4).

On the other hand, since g is uniformly continuous on $\tilde{J}$, there is an $\eta > 0$ such that

$$y, y' \in \tilde{J},\ |y - y'| < \eta \quad \text{imply} \quad |g(y) - g(y')| < \frac{\varepsilon}{2}. \tag{6}$$

But $\overline{\overline{f}}_n \rightrightarrows f$ on B_ε, so there is an $n_2 \in \mathbb{N}, n_2 \geq n_1$, such that

$$\overline{\overline{f}}_n(x), f(x) \in \tilde{J}, \quad |\overline{\overline{f}}_n(x) - f(x)| < \eta \tag{7}$$

for $n \geq n_2$ and $x \in B_\varepsilon$. Consequently,

$$|g(\overline{\overline{f}}_n(x)) - g(f(x))| < \frac{\varepsilon}{2} \tag{8}$$

for $x \in B_\varepsilon$ and $n \geq n_2$, in view of (6) and (7).

By (5) and (8), we see that $x \in B_\varepsilon$ implies $|\overline{\overline{g}}_n(\overline{\overline{f}}_n(x)) - g(f(x))| < \varepsilon$ for $n \geq n_2$, i.e.

$$|\{x \in A:\ |\overline{\overline{g}}_n(\overline{\overline{f}}_n(x)) - g(f(x))| \geq \varepsilon\}| \leq |A \setminus B_\varepsilon||B \setminus B_\varepsilon| < \varepsilon$$

for $n \geq n_2$, which means that $\overline{\overline{h}}_n := \overline{\overline{g}}_n \circ \overline{\overline{f}}_n \to g \circ f$ i.m. Clearly, (h_n) contains a subsequence $(\widetilde{h}_n)$, with $\widetilde{h}_n := \overline{\overline{h}}_{m_n}$, which converges to $g \circ f$ a.e. on A.

Thus for every subsequence $(\overline{h}_n)$ of the sequence (h_n) we have selected a subsequence $(\widetilde{h}_n)$ converging to h a.e. on A. This means that $h_n = g_n \circ f_n \to g \circ f$ asymptotically on A or, equivalently, in measure on A.

Lemma 3 *Assume that f, f_n $(n \in \mathbb{N})$ are measurable functions on a given compact set $K \subset \mathbb{R}$, $f \in L^1$ on K and g, g_n $(n \in \mathbb{N})$ are continuous functions on $\mathbb{R}$. If $f_n \to f$ in measure (asymptotically) on K and $g_n \to g$ l.u. in $\mathbb{R}$, then $g_n \circ f_n \to g \circ f$ in measure (asymptotically) on K.*

Proof. Notice that for every $\varepsilon > 0$ there is a symmetric interval $J = J(\varepsilon) \in \mathcal{I}$ such that

$$|K \setminus A| < \frac{\varepsilon}{2}, \qquad \text{where } A := \{x \in K \colon\ |f(x)| \in J\}. \tag{9}$$

In fact, in the converse case there would exist an $\varepsilon_0 > 0$ such that, for every $J = J(n)$ of the form $[-n, n]$, the set A defined via J in (9) would satisfy the inequality $|K \setminus A| \geq \varepsilon_0/2$. But then

$$\int_K |f(x)|\, dx \geq \int_{K\setminus A} |f(x)|\, dx \geq n|K \setminus A| \geq n\frac{\varepsilon_0}{2} \to \infty,$$

which contradicts the assumption that $f \in L^1$ on K.

Fix $\varepsilon > 0$ and select an interval $J = [-a, a]$, $a > 0$, and the respective set A such that (9) holds. Obviously, $f_n \to f$ i.m. on A. Since $g_n \to g$ l.u. on $\mathbb{R}$, we may apply Lemma 2 to conclude that $g_n \circ f_n \to g \circ f$ i.m. on A. This means that, denoting

$$C_{\varepsilon,n} := \{x \in A :\ |(g_n \circ f_n)(x) - (g \circ f)(x)| > \varepsilon\},$$

we have $|C_{\varepsilon,n}| < \varepsilon/2$ for sufficiently large n. Hence

$$|\{x \in K :\ |(g_n \circ f_n)(x) - (g \circ f)(x)| > \varepsilon\}| \leq |K \setminus A| + |C_{\varepsilon,n}| < \frac{\varepsilon}{2} + \frac{\varepsilon}{2} = \varepsilon$$

for sufficiently large n. Since $\varepsilon > 0$ was fixed arbitrarily, we have proved that $g_n \circ f_n \to g \circ f$ i.m. on K.

4. Distributional compositions

Theorem 1 *Assume that f and g are continuous functions on $\mathbb{R}$. Then the distributional composition $g(f)$ exists and equals in the distributional sense to the function $g \circ f \in L^1_{\mathrm{loc}}$.*

Proof. The assertion follows directly from Lemma 1, because f is locally finite, $\varphi_n : f * \delta_{1,n} \to f$ l.u. and $\psi_n : g * \delta_{2,n} \to g$ l.u. for arbitrary $(\delta_{1,n}), (\delta_{2,n}) \in \Delta$.

Theorem 2 *Assume that g is a continuous function on $\mathbb{R}$ such that, for some $p \in [1, +\infty)$ and $\beta > 0$,*

$$|g(y)| \leq \beta(1 + |y|)^p, \qquad y \in \mathbb{R}$$

and $f \in L^p_{\mathrm{loc}}$ for the same p as above. Then the distributional composition $g(f)$ exists and is equal in the distributional sense to the function $g \circ f \in L^1_{\mathrm{loc}}$.

Proof. Since $L^p_{\mathrm{loc}} \subseteq L^1_{\mathrm{loc}}$ for $p \in [1, \infty)$, we have $f \in L^1_{\mathrm{loc}}$, so f can be identified with the respective distribution and $g \circ f$ is a measurable function such that

$$\int_I |g(f(x))|\, dx \leq 2^p\beta \int_I (1 + |f(x)|^p)\, dx < \infty$$

for every $I \in \mathcal{I}$, i.e. $g \circ f \in L^1_{\mathrm{loc}}$.

Fix $I \in \mathcal{I}$. Since $f \in L^p_{\mathrm{loc}}$, we have

$$\|f_n - f\|^p_{p,I} = \int_I |f_n(x) - f(x)|^p\, dx \to 0, \tag{10}$$

where $f_n := f * \delta_{1,n}$ and $f_n \to f$ i.m. on I for every delta-sequence $(\delta_{1,n}) \in \Delta$. It follows from (10) that

$$\|f_n\|_{p,I} \leq 1 + \|f\|_{p,I}, \tag{11}$$

for sufficiently large n. On the other hand, $g_n := g * \delta_{2,n} \to g$ l.u. on $\mathbb{R}$ for every delta-sequence $(\delta_{2,n}) \in \Delta$. By Lemma 2, $g_n \circ f_n \to g \circ f$ i.m. on I. Since the convergence in measure is asymptotic on the interval I chosen arbitrarily in $\mathcal{I}$, we may use the standard diagonal procedure to deduce that

$$g_n \circ f_n \to g \circ f \quad \text{as. on } \mathbb{R}. \tag{12}$$

Moreover, we have

$$\begin{aligned} |g_n(y)| &\leq \beta \int_{O_n} (1 + |y - t|)^p |\delta_{2,n}(t)|\, dt \\ &\leq 3^p \beta (1 + |y|^p) M_{2,0} + 3^p \beta \int_{O_n} |t|^p |\delta_{2,n}(t)|\, dt \\ &\leq \gamma(1 + |y|^p), \qquad y \in \mathbb{R}, \end{aligned}$$

where $O_n := [-\alpha_{2,n}, \alpha_{2,n}]$ and $\gamma := 3^p \beta M_{2,0}(1 + \max_{n \in \mathbb{N}} \alpha_{2,n}^p)$. Consequently,

$$|g_n(f_n(x))| \leq \gamma(1 + |f_n(x)|^p), \qquad x \in \mathbb{R}. \tag{13}$$

The sequence of the functions $g_n \circ f_n$ on the left hand side of (13) converges asymptotically to $g \circ f$, by (12), and the assumptions of Theorem D of Antosik will be satisfied, if we prove that the right hand side of (13) converges locally to the function $x \mapsto \gamma(1 + |f(x)|^p)$. Clearly, it suffices to prove for every fixed $I \in \mathcal{I}$ that

$$\int_I \big|\, |f_n(x)|^p - |f(x)|^p \,\big|\, dx \to 0. \tag{14}$$

We begin with the proof of the following inequality:

$$a^p - b^p \leq (p + 1)(a - b)a^{p-1}, \qquad 0 < b \leq a \tag{15}$$

or, equivalently, the inequality

$$p\left(\frac{a}{b}\right)^p \geq (p + 1)\left(\frac{a}{b}\right)^{p-1} - 1, \qquad 0 < b \leq a. \tag{16}$$

Since the function ρ given by

$$\rho(t) := pt^p - (p + 1)t^{p-1} + 1, \qquad t \in [1, \infty) \tag{17}$$

is increasing on $[1, \infty)$ and $\rho(1) = 0$, so $\rho(t) \geq 0$ on $[1, \infty)$ and (16) follows from (17) for $t = \frac{a}{b}$. From (15) we have

$$\big||a|^p - |b|^p\big| \leq (p + 1)|a - b|(\max(|a|, |b|))^{p-1} \tag{18}$$

for arbitrary $a, b \in \mathbb{R}$. Consequently, denoting $\widetilde{f_n}(x) := \max(|f_n(x)|, |f(x)|)$, we get from (18) the estimate

$$\begin{aligned} \int_I \big||f_n(x)|^p - |f(x)|^p\big|\, dx &\leq (p + 1) \int_I |f_n(x) - f(x)| \cdot [\widetilde{f_n}(x)]^{p-1}\, dx \\ &\leq (p + 1)\|f_n - f\|_{p,I} \cdot \Big(\int_I [\widetilde{f_n}(x)]^{(p-1)q}\Big)^{\frac{1}{q}} \\ &= (p + 1)\|f_n - f\|_{p,I} \cdot (\|\widetilde{f_n}\|_{p,I})^{\frac{p}{q}}, \end{aligned} \tag{19}$$

where $1/p + 1/q = 1$. We have

$$\|\widetilde{f_n}\|_{p,I}^p \leq \int_I (|f_n(x)|^p + |f(x)|^p)\, dx \|f_n\|_{p,I}^p + \|f\|_{p,I}^p$$
$$\leq (1 + \|f\|_{p,I})^p + \|f\|_{p,I}^p < \infty$$

for sufficiently large n, in view of (11). Therefore (10) and (19) imply (14).

The assumptions of Theorem D are satisfied, due to (12), (13) and (14). By Theorem D, $g_n \circ f_n \to g \circ f$ loc. and thus $g_n \circ f_n \to g \circ f$ in $\mathcal{D}'$. Consequently, $g(f)$ exists and $g(f) = g \circ f \in L^1_{\text{loc}}$, as desired.

Theorem 3 *Let g be a continuous and bounded function on $\mathbb{R}$ and f be a distribution such that the Łojasiewicz value $f(x)$ at x exists for almost all $x \in \mathbb{R}$ and the regular part f_r of f is a measurable function. Then the distributional composition $g(f)$ exists in the distributional sense, the composition $g \circ f_r$ exists a.e., $g \circ f_r \in L^1_{\text{loc}}$ and $g(f) = g \circ f_r$.*

Proof. First notice that since g is continuous and f_r-measurable the composition $g \circ f_r$ is a measurable function. Moreover, if $\sup_{y \in \mathbb{R}} |g(y)| =: \gamma$, then

$$\int_I |g(f_r(x))|\, dx \leq \int_I \gamma\, dx = \gamma |I| < \infty$$

for every interval $I \in \mathcal{I}$, i.e., $g \circ f_r \in L^1_{\text{loc}}$.

Fix an interval I and $\varepsilon > 0$. By Luzin's Theorem C, there exists a measurable set $B \subseteq I$ such that $|I - B| < \varepsilon/3$ and the restriction $f_r|_B$ of f to B is a continuous function. By the known property of the Lebesgue measure, there is a closed set $C \subseteq B$ such that $|B - C| < \varepsilon/3$. Hence

$$|I \setminus C| \leq |I \setminus B| + |B \setminus C| < 2\varepsilon/3. \tag{20}$$

Since $C \subseteq B \subseteq I$, the set C is compact and the continuous function $f_r|_C$ is therefore bounded.

Fix $(\delta_{1,n}), (\delta_{2,n}) \in \Delta$ and denote $f_n := f * \delta_{1,n}$ and $g_n := g * \delta_{2,n}$. Evidently, $f_n \to f_r$ a.e. and the more $f_n \to f_r$ i.m. on C. Also, $g_n \to g$ l.u. on $\mathbb{R}$. Hence, by Lemma 3, we have $g_n \circ f_n \to g \circ f_r$ i.m. on C, i.e.

$$|C_{\varepsilon,n}| < \frac{\varepsilon}{3}, \tag{21}$$

where

$$C_{\varepsilon,n} := \{x \in C : \ |(g_n \circ f_n)(x) - (g \circ f_r)(x)| > \varepsilon\},$$

for sufficiently large n. Consequently, by (20) and (21),

$$|\{x \in I : |(g_n \circ f_n)(x) - (g \circ f_r)(x)| > \varepsilon\}| \leq |I \setminus C| + |C_{\varepsilon,n}| < \varepsilon$$

for sufficiently large n. Thus we have proved that $g_n \circ f_n \to g \circ f_r$ i.m. on I.

Since $|g(y)| \leq \gamma$ and

$$|g_n(y)| \leq \int_{-\alpha_n}^{\alpha_n} |g(y - t)| \cdot |\delta_n(t)|\, dt \leq \sup_{z \in \mathbb{R}} |g(z)| \cdot M_0 \leq \gamma \cdot M_0$$

for $y \in \mathbb{R}$, $n \in \mathbb{N}$, the functions $|g_n \circ f_n - g \circ f_r|$ are commonly bounded by the constant $\gamma \cdot (M_0 + 1)$, which is an integrable function on I. By the classical Lebesgue's dominated convergence theorem

$$\lim_{n \to \infty} \int_I |g_n(f_n(x)) - g(f_r(x))|\, dx = 0$$

for every $I \in \mathcal{I}$, i.e. $g_n \circ f_n \to g \circ f_r$ locally and the more distributionally. Thus $g(f) = \lim_{n \to \infty} g_n \circ f_n = g \circ f_r$.

5. Measures

Let m^+, m^- and m be functions on $\mathbb{R}$ given by

$$\begin{aligned} &m^+(x) : \max(x,0) = \frac{|x|+x}{2}, \qquad m^-(x) := \max(-x,0) = \frac{|x|-x}{2}, \\ &m(x) : |x| \end{aligned} \tag{22}$$

for $x \in \mathbb{R}$ and denote:

$$\varphi^+ := m^+ \circ \varphi, \qquad \varphi^- := m^- \circ \varphi, \qquad |\varphi| := m \circ \varphi,$$

for an arbitrary smooth function φ on $\mathbb{R}$. Given a distribution f on $\mathbb{R}$, consider the distributional compositions $m^+(f), m^-(f), m(f)$ in the sense, defined in (1), of two-argument irregular operation. Simultaneously, consider the compositions $f^+, f^-, |f|$ of the continuous functions m^+, m^-, m and the distribution f as a one-argument irregular operation, i.e. defined by

$$f^+ := \lim_{n\to\infty} (f * \delta_n)^+, \quad f^- := \lim_{n\to\infty} (f * \delta_n)^-, \quad |f| := \lim_{n\to\infty} |f * \delta_n|,$$

whenever the limits exist for arbitrary delta-sequences $(\delta_n) \in \Delta$.

Lemma 4 *Let m be given by (22), let (δ_n) be a delta-sequence satisfying conditions 1°, 2°, 3°_0 with the corresponding sequence (α_n) of positive reals converging to 0 satisfying condition 1° and the constant M_0 satisfying condition 3°_0 (in particular, let $(\delta_n) \in \Delta$) and let $m_n := m * \delta_n$ for $n \in \mathbb{N}$. Then*

$$\sup_{y\in\mathbb{R}} |m_n(y) - m(y)| \le \alpha_n M_0, \qquad n \in \mathbb{N}. \tag{23}$$

Consequently,

$$m_n \rightrightarrows m \qquad on \ \mathbb{R} \tag{24}$$

as $n \to \infty$.

Proof. Let $O_n := (-\alpha_n, \alpha_n)$. By the definition of the convolution and conditions 1°, 2° and 3°_0, we have

$$\begin{aligned} |m_n(y) - m(y)| & \left| \int_{-\infty}^{\infty} (|y-t| - |y|) \cdot \delta_n(t)\, dt \right| \\ &\le \int_{O_n} \big||y-t| - |y|\big| \cdot |\delta_n(t)|\, dt \\ &\le \int_{O_n} |t| \cdot |\delta_n(t)|\, dt \le \alpha_n \int_{-\infty}^{\infty} |\delta_n(t)|\, dt \le \alpha_n M_0 \end{aligned}$$

for all $y \in \mathbb{R}$. Hence

$$\sup_{y\in\mathbb{R}} |m_n(y) - m(y)| \le \alpha_n M_0 \to 0$$

as $n \to \infty$, i.e. $m_n \rightrightarrows m$ on $\mathbb{R}$.

Theorem 4 *The following conditions are equivalent:*

(i) $m(f)$ *exists;*
(ii) $m^+(f)$ *and* $m^-(f)$ *exist;*
(iii) $|f|$ *exists,*
(iv) f^+ *and* f^- *exist;*
(v) f *is a Radon measure.*

Moreover, if the conditions are satisfied, then $m(f) = |f|, m^+(f) = f^+, m^-(f) = f^-$.

Proof. For $i = 1, 2$, let $(\delta_{i,n}) \in \Delta$ and, according to the definition of a delta-sequence in section 2, let $(\alpha_{i,n})$ and $M_{i,k}$ (for $k \in \mathbb{N}_0$) be the sequence of positive reals converging to 0 and the constants from conditions 1° and 3°, corresponding to the delta-sequence $(\delta_{i,n})$. Moreover, let $O_{i,n} := (-\alpha_{i,n}, \alpha_{i,n})$ for $i = 1, 2$. Denote $f_n := f * \delta_{1,n}$, $m_n := m * \delta_{2,n}$, $m_n^+ := m^+ * \delta_{2,n}$ and $m_n^- := m^- * \delta_{2,n}$. By Lemma 4, we have

$$\sup_{x \in \mathbb{R}} |m_n(f_n(x)) - m(f_n(x))| \le \alpha_{2,n} M_{2,0} \to 0 \tag{25}$$

as $n \to \infty$, i.e. $m_n \circ f_n - |f_n| \rightrightarrows 0$ on $\mathbb{R}$. Hence conditions (i), (iii) are equivalent and $m(f) = |f|$ in case one of them holds.

Now, if $|y| > \alpha_{2,n}$, we have $|(y - t)^+ - y^+| \le |t| \le \alpha_{2,n}$ and if $|y| \le \alpha_{2,n}$, we have $|(y - t)^+ - y^+| \le 2\alpha_{2,n}$ for $t \in O_n$. Hence

$$|(y - t)^+ - y^+| \le 2\alpha_{2,n}, \qquad y \in \mathbb{R},\ t \in O_n.$$

Consequently,

$$|m_n^+(y) - m^+(y)| \left| \int_{-\infty}^{\infty} ((y - t)^+ - y^+)\, \delta_{2,n}(t)\, dt \right|$$
$$\le \int_{O_n} |(y - t)^+ - y^+| \cdot |\delta_{2,n}(t)|\, dt \le 2\alpha_{2,n} M_{2,0}, \qquad y \in \mathbb{R}.$$

Hence

$$\sup_{x \in \mathbb{R}} |m_n^+(f_n(x)) - m^+(f_n(x))| \le 2\alpha_{2,n} M_{2,0} \tag{26}$$

for $n \in \mathbb{N}$. Analogously to (26), we show that

$$\sup_{x \in \mathbb{R}} |m_n^-(f_n(x)) - m^-(f_n(x))| \le 2\alpha_{2,n} M_{2,0} \tag{27}$$

for $n \in \mathbb{N}$. By (26) and (27), conditions (ii) and (iv) are equivalent. Moreover, $m^+(f) = f^+$ and $m^-(f) = f^-$ if one of conditions (ii) and (iv) holds.

Suppose that condition (iii) holds. This means that the sequence $(|f_n|)$ converges in $\mathcal{D}'$ to a measure $|f|$, i.e. the numerical sequence $(< |f_n|, \varphi >)$ is convergent for every $\varphi \in \mathcal{D}$ or, equivalently, for every nonnegative $\varphi \in \mathcal{D}$ to $< |f|, \varphi >$ (see [1]). Fix such a nonnegative $\varphi \in \mathcal{D}$. Since $|f_n| = f_n^+ + f_n^-$, we have $|f_n| \ge f_n^+$ and $|f_n| \ge f_n^-$ and so the sequences $(< f_n^+, \varphi >)$ and $(< f_n^-, \varphi >)$ are bounded. Hence for an arbitrary increasing sequence (k_n) of positive integers there is its subsequence (k_{l_n}) such that the sequences $(< f_{k_{l_n}}^+, \varphi >)$ and $(< f_{k_{l_n}}^-, \varphi >)$ are convergent and the limits are equal to $< (|f| + f)/2, \varphi >$ and $< (|f| - f)/2, \varphi >$, respectively. Consequently, the sequences (f_n^+) and (f_n^-) converge in $\mathcal{D}'$ to $(|f| + f)/2 = f^+$ and $(|f| - f)/2 = f^-$, respectively. Thus the implication $(iii) \Rightarrow (iv)$ is proved.

If (iv) holds, then (f_n^+) and (f_n^-) converge to positive distributions f^+ and f^-. Consequently, $f = \lim_{n\to\infty}(f_n^+ - f_n^-) = f^+ - f^-$ is a Radon measure. This means, the implication $(iii) \Rightarrow (v)$ is shown.

Assume condition (v) holds, i.e. f is a measure, and let $(\delta_n) \in \Delta_+$. By Theorem 12 in [3], the limit $\lim_{n\to\infty} |f_n|$ exists in $\mathcal{D}'$ and hence, due to the equivalence $(i) \Leftrightarrow (iii)$, the limit $\lim_{n\to\infty} |m_n(f_n)|$ exists and the limits are equal, i.e. condition (v) implies conditions (iii) and (i). The proof is finished.

Using techniques used in [3], we can prove the two following result:

Theorem 5 *Suppose that g is a continuous function in $\mathbb{R}$ such that $g - I$ is a bounded function where I is the identity function and suppose that f is a quasi-measure and $f = f_r + f_s$ is the Łojasiewicz decomposition of f. Then the distributional compositions $g(f)$ and $g(f_r)$ exist, $g(f_r)$ is a loc. integrable function and $g(f) = g(f_r) + f_s$ is the Łojasiewicz of $g(f)$.*

Theorem 6 *If f is a measure and $f = f_r + f_s$ is the Łojasiewicz decomposition of f, then*

$$m(f) = m(f_r) + m(f_s) \qquad (\textit{equivalently } |f| = |f_r| + |f_s|) \tag{28}$$

is the Łojasiewicz decomposition of $m(f) = |f|$.

Proof. It will be convenient to denote $f^0 := f$, $f^1 := f_r$, $f^2 := f_s$. Fix two delta-sequences $(\delta_{1,n}), (\delta_{2,n}) \in \Delta$ and denote

$$m_n := m * \delta_{2,n}, \quad f_n^0 := f^0 * \delta_{1,n}, \quad f_n^1 := f^1 * \delta_{1,n}, \quad f_n^2 := f^2 * \delta_{1,n}$$

for $n \in \mathbb{N}$ and $y_n^1 := f_n^1(x), y_n^2 := f_n^2(x)$ for $x \in \mathbb{R}$ and $n \in \mathbb{N}$. Evidently,

$$y_n^0 := f_n^0(x) = f_n^1(x) + f_n^2(x) = y_n^1 + y_n^2 \tag{29}$$

for $x \in \mathbb{R}$ and $n \in \mathbb{N}$.

We have

$$\begin{aligned} |m_n(y_1 + y_2) - m_n(y_2)| &= \left| \int (|y_1 + y_2 - t| - |y_2 - t|)\, \delta_{2,n}(t)\, dt \right| \\ &\le \int |(y_1 + y_2 - t) - (y_2 - t)| |\delta_{2,n}(t)|\, dt \\ &\le |y_1| \int |\delta_{2,n}(t)|\, dt \le |y_1| M_{2,0} \end{aligned}$$

for arbitrary $y_1, y_2 \in \mathbb{R}$, so

$$|m_n(y_n^0) - m_n(y_n^1) - m_n(y_n^2)| \le |y_n^1| M_{2,0} + m_n(y_n^1). \tag{30}$$

By (23) in Lemma 4,

$$|m_n(y_n^1)| \le |y_n^1| + \alpha_{2,n} M_{2,0}. \tag{31}$$

Combining (30) and (31), we get

$$|m_n(y_n^0) - m_n(y_n^1) - m_n(y_n^2)| \le (1 + M_{2,0})|y_n^1| + \alpha_{2,n} M_{2,0}$$

or, equivalently,

$$\begin{aligned} &|(m_n \circ f_n^0)(x) - (m_n \circ f_n^1)(x) - (m_n \circ f_n^2)(x)| \\ &\qquad \le (1 + M_{2,0})|f_n^1(x)| + \alpha_{2,n} M_{2,0}. \end{aligned} \tag{32}$$

We are going to prove first that the left hand side of inequality (32) converges asymptotically to 0. Clearly,

$$\begin{aligned} &|m_n(y_n^0) - m_n(y_n^1) - m_n(y_n^2)| \\ &\qquad \le |m_n(y_n^0) - |y_n^1|| + ||y_n^1| - m_n(y_n^1)| + |m_n(y_n^2)|. \end{aligned} \tag{33}$$

Notice that

$$|m_n(f_n^i(x)) - |f^1(x)|| \le |m_n(f_n^i(x)) - m(f_n^i(x))| + ||f_n^i(x)| - |f^1(x)|| \tag{34}$$

for $i = 0, 1$, and

$$|m_n(f_n^2(x))| \leq |m_n(f_n^2(x)) - m(f_n^2(x))| + |f_n^2(x)| \tag{35}$$

for $i = 2$. By Lemma 4, we have

$$\sup_{x \in \mathbb{R}} |m_n(f_n^i(x)) - m(f_n^i(x))| \to 0 \tag{36}$$

i.e. $m_n \circ f_n^i - m \circ f_n^i \rightrightarrows 0$ as $n \to \infty$ for $i = 0, 1, 2$ and thus all the first addends on the right hand sides of the three inequalities in (34), (35) converge uniformly to 0.

On the other hand, since f^1 is a locally integrable function, we have $f_n^1 \to f^1$ loc., so the more $f_n^1 \to f^1$ i.m. Further, by the definition of f_r, we have $f_n^0 \to f^1$ a.e., so the more $f_n^0 \to f^1$ i.m. (since $|I| < \infty$). Hence $f_n^2 = f_n^0 - f_n^1 \to 0$ i.m. This means that the second addends on the right hand sides of the three inequalities in (34), (35) converge in measure to 0. Hence each of the three addends on the right hand side of (33) converges in measure to 0., the left hand side of the inequality (32) converges i.m. to 0 and, consequently, the left hand side of (32) converges in measure to 0.

On the other hand, we will prove that the right hand side of (32) converges locally in mean to $(1 + M_{2,0})|f^1(x)|$. In fact, for every $I \in \mathcal{I}$, we have

$$\begin{aligned} &\int_I \left| [(1 + M_{2,0})|f_n^1(x)| + \alpha_{2,n} M_{2,0} - (1 + M_{2,0})|f^1(x)|] \right| dx \\ &\quad \leq (1 + M_{2,0}) \int_I \left| |f_n^1(x)| - |f^1(x)| \right| dx + \alpha_{2,n} M_{2,0} \int_I dx \\ &\quad \leq (1 + M_{2,0}) \int_I |f_n^1(x) - f^1(x)| \, dx + \alpha_{2,n} M_{2,0} |I| \end{aligned} \tag{37}$$

and the right hand side goes to 0 as $n \to \infty$, since $f_n^1 \to f^1$ loc. as $n \to \infty$. Hence

$$(1 + M_{2,0})(m \circ f_n^1) + \alpha_{2,n} M_{2,0} \;\to\; (1 + M_{2,0})(m \circ f^1) \quad \text{loc.} \tag{38}$$

as $n \to \infty$.

Now, by the generalized theorem on dominated convergence (Theorem D in section 2), we conclude that $m_n \circ f_n^1 + m_n \circ f_n^2 - m_n \circ f_n^0 \to 0$ loc. and so

$$m_n \circ f_n^1 + m_n \circ f_n^2 - m_n \circ f_n^0 \to 0 \qquad \text{in } \mathcal{D}',$$

as $n \to \infty$. Thus the assertion is proved.

Applying Theorems 5 and 6, in a way shown in [3], we prove the two following results:

Theorem 7 *Suppose that g is a continuous function in $\mathbb{R}$, $g(x) = o(|x|)$, f is a measure in $\mathbb{R}$ and $f = f_r + f_s$ is the Łojasiewicz decomposition of f. Then the compositions $g(f)$ and $g(f_r)$ exist, $g(f_r)$ is a locally integrable function and $g(f) = g(f_r)$.*

Theorem 8 *Assume that g is a continuous function on $\mathbb{R}$ such that*

$$\lim_{x \to -\infty} \frac{g(x)}{x} = \alpha, \qquad \lim_{x \to +\infty} \frac{g(x)}{x} = \beta.$$

Assume, moreover, that f is a measure and $f = f_r + f_s$ is the Łojasiewicz decomposition of f. Then the distributional composition $g(f)$ exists and

$$g(f) = g \circ f_r - \alpha f_s^- + \beta f_s^+.$$

Acknowledgments

We would like to thank the anonymous referee for helpful remarks and stimulating comments which helped us to improve the text of the paper. We also want to thank the organizers of the conference SSPCM 2005 for giving us a possibility to present this material and for extremely nice cooperation.

References

[1] Antosik P 1967 On the modulus of a distribution *Bull. Acad. Pol. Sci., Ser. Sci. Math. Astronom. Phys.* **15** 681-3

[2] Antosik P 1968 Some conditions for mean convergence *Bull. Acad. Pol. Sci., Ser. Sci. Math. Astronom. Phys.* **16** 641-4

[3] Antosik P 1988-89 Composition of distributions *Tech. Report Series of the Department of Mathematical Sciences, Univ. of Wisconsin-Milwaukee* **53201**

[4] Antosik P, Mikusiński J and Sikorski R 1973 *Theory of Distributions. The Sequential Approach* (Elsevier-PWN, Amsterdam-Warszawa)

[5] Halmos P R 1950 *Measure Theory* (Van Nostrand Reinhold, New York)

[6] Hartman S, Mikusiński J 1961 *The Theory of Lebesgue Measure and Integration* (Pergamon Press-PWN, New York-Warszawa)

[7] Łojasiewicz S 1957 Sur la valeur et le limite d'une distribution dans un point *Studia Math.* **16** 1-36

[8] Mikusiński J 1948 Sur la méthode de généralisation de M. Laurent Schwartz et sur la convergence faible *Fund. Math.* **35** 235-9

[9] Mikusiński J 1960 On the value of a distribution at a point, *Bull. Acad. Pol. Sci., Ser. Sci. Math. Astronom. Phys.* **8** 681-3

[10] Mikusiński J 1961 Irregular operations on distributions *Studia Math.* **20** 163-9

[11] Mikusiński J 1978 *The Bochner Integral* (Birkhäuser, Basel-Stutgart)

[12] Schwartz L 1950-1 *Théorie des distributions* vol I-II (Hermann Paris)

[13] Sobolev S L 1936 Méthode nouvelle à résoudre le probléme de Cauchy pour les équations hiperboliques normales *Recueil Math. (Mat. Sbornik)* **1** 39-72

[14] Sorek S 2005 Operations on distributions: regular and irregular *this volume*

Equivariant functions in topological quantum mechanics.

Jan Milewski
Institute of Mathematics, Poznań University of Technology, ul. Piotrowo 3A
60-965 Poznań, Poland

Abstract. In the paper we deal with topological quantum mechanics. We show on the example of anyon system that the space of quantum states depends on a Hamiltonian.

1. Introduction

The aim of the paper is to show that the space of quantum states in topological quantum mechanics (TQM) depends on the Hamiltonian of the system. In quantum mechanics (QM) wave functions are complex square–integrable functions on the classical configuration space. For different Hamiltonians the systems of eigenstates are different but the Hilbert space spanned on any system does not depend on a Hamiltonian, what is a consequence of the spectral theorem. States in QM forme the Hilbert space of complex square–integrable functions on the classical configuration space, and Hamiltonians are Hermitian operators.

In TQM the situation is totally different. Wave functions can be defined on the universal covering of the classical configuration space, or as multivalued functions defined on this space. The wave functions are equivariant with respect to the standard action of the fundamental group of the classical configuration space. Hamiltonians are not necessarily hermitian operators and we cannot use the spectral theorem. Hamiltonians in TQM are Hermitian operators on their domains, but the domain depends on the Hamiltonian in contrary to QM.

The paper is organised as follows. In Section 2 we give some general information about TQM. In Section 3, we consider a special case of TQM – anyons on a plane. We show that the space of quantum states of a system of free anyons is the whole Hilbert space of square–integrable equivariant functions. Further we show that the space of quantum states for the system of free anyons and for the system of anyons in a homogeneous potential field has only the zero element as the intersection. At last we show that the space of quantum states for the anyon system in anisotropic harmonic oscillator potential is the zero space.

2. Topological quantum mechanics

Let M be a non-simply connected Riemann manifold of classical configurations of a physical system, with the fundamental group $\pi_1(M)$. Let us consider the universal covering

$$p : \widetilde{M} \longrightarrow M \tag{1}$$

consistent with the Riemann structure on M. It means that $\widetilde{M}$ is also a Riemann manifold, and p is a local isometry.

Let

$$L : \pi_1(M) \times \widetilde{M} \longrightarrow \widetilde{M} \tag{2}$$

be the standard left free action of the fundamental group on the total space $\widetilde{M}$ and

$$\chi : \pi_1(M) \longrightarrow U(1) \tag{3}$$

be a unitary one-dimensional representation. A function $f : \widetilde{M} \longrightarrow C$ is called χ–equivariant when

$$f(L([\gamma], \widetilde{m})) = \chi([\gamma]) f(\widetilde{m}) \tag{4}$$

for $[\gamma] \in \pi_1(M)$, $\widetilde{m} \in \widetilde{M}$. For the case of χ being the trivial representation a χ-equivariant function is $\pi_1(M)$–invariant and it is the lifting of a function defined on the base space M.

We can define the scalar product of equivariant functions by the formula

$$< f, g >= \int_M f g^* d\mu \tag{5}$$

because the integrand is $\pi_1(M)$-invariant ($d\mu$ denotes the volume form on M). The space of quantum space in TQM is a subspace of the Hilbert space of square–integrable χ–equivariant functions [1], [2]

$$\mathcal{H} = L^2_\chi(M, C, d\mu) \tag{6}$$

satisfying some boundary conditions.

The kinetic energy operator is defined as

$$\hat{T} = -\frac{1}{2m}\widetilde{\Delta} , \tag{7}$$

where $\widetilde{\Delta}$ is the Laplace operator on $\widetilde{M}$ (it is the lifting of the Laplace operator Δ on M) and we put $\hbar = 1$. Although Δ is a hermitian operator in $L^2(M, C, d\mu)$ the operator $\hat{T}$ is in general not an operator in $L^2_\chi(M, C, \widetilde{d\mu})$, as we will see on an example of anyons. $\hat{T}$ is defined only on a subspace of $L^2_\chi(M, C, \widetilde{d\mu})$ Moreover for Hamiltonians of the form

$$\hat{H} = \hat{T} + V \tag{8}$$

the domain of the operator $\hat{H}$ depends on the potential V.

3. Anyons on a plane.

Anyons are hard core particles, so two or more particles can not be placed at the same point. Let us consider a system of several kinds of anyons on a plane: N_1 anyons of the first kind, N_2 anyons of the second, ..., N_K anyons of the K–th. Anyons of the same kind are indistinguishable [3], [4]. We identify the plane with the complex plane C. The classical configuration space of the system is the quotient space

$$Q_{N_1,\ldots,N_K} = \frac{C^N \setminus D_N}{S_{N_1} \times \ldots \times S_{N_K}}, \; N = N_1 + \ldots N_K, \tag{9}$$

where the direct product of symmetric groups $S_{N_1} \times \ldots \times S_{N_K}$ acts on C^N in the standard way and D_N denotes the fat diagonal of C^N. The symmetric group S_{N_K} permutes positions of the k–th kind particles. The stratification (9) result with that anyons of the same kind are indistinguishable even on the classical level.

The fundamental group of the configuration space is the generalized braid group

$$B_{N_1,\dots,N_K} = \pi_1(Q_{N_1,\dots,N_K}) \ . \tag{10}$$

Let us specify two special kinds of loops in $Q_{N_1,\dots,N_K}$ at the point $[z_{11},\dots z_{KN_K}]$. We put the square bracket to denote the $S_{N_1} \times \dots \times S_{N_K}$-orbit of $(z_{11},\dots z_{KN_K})$. Let

$$\gamma : I \longrightarrow Q_{N_1,\dots,N_K},\ [\gamma_{11}(t),\dots,\gamma_{KN_K}(t)] \tag{11}$$

be a loop with $I = [0,1]$.

I. We say that the loop (11) is a proper transposition of two anyons of the k-th kind placed at z_{kp} and z_{kq} if

$$\gamma_{lr}(t) = z_{lr} \text{ for } (l,r) \neq (k,p),(kq)$$
$$\gamma_{kp}(0) = \gamma_{kq}(1) = z_{kp}\ ,\ \ \gamma_{kp}(1) = \gamma_{kq}(0) = z_{kq} \tag{12}$$

and the composition of the curves γ_{kp}, γ_{kq} is a positively oriented Jordan loop with no particles inside the area limited by the loop.

II. We say that the loop (11) is a proper encircling of a particle of the l-th kind by a particle of the k-th kind placed at z_{lq} and z_{kp} respectively, if

$$\gamma_{ir}(t) = z_{ir} \text{ for } (i,r) \neq (k,p) \tag{13}$$

and γ_{kp} is a positively directed Jordan loop with the only (l,q)-th particle inside.

Let $\chi_{\nu\mu}$ be a unitary one–dimensional representation of the fundamental group $\pi_1(Q_{N_1,\dots,N_K})$. The representation is labeled by statistical parameters $\nu = (\nu_1,\dots \nu_K)$ and encircling parameters $\mu = (\mu_{k,l}),\ \ 1 \le k < l \le K$, and is uniquely determined by two conditions:

$$\chi_{\nu\mu}(\alpha_k) = e^{\pi i \nu_k}, \tag{14}$$

where α_k is a proper transposition of anyons of the k-th kind, and

$$\chi_{\nu\mu}(\alpha_{k,l}) = e^{2\pi i \mu_{k,l}} \tag{15}$$

is a proper encircling of a particle of the l-th kind by a particle of the k-th kind.

Let $L^2_{\chi_{\nu\mu}0}(Q_{N_1,\dots,N_K}, C, \widetilde{d\mu})$ denote the subspace of all equivariant square–integrable functions satisfying the boundary conditions:

$$\lim_{z_{k,p}\longrightarrow z_{l,q}} \Psi(z_{1,1},\dots,z_{K,N_K}) = 0 \ . \tag{16}$$

The space of quantum states of the anyon system is a subspace of $L^2_{\chi_{\nu\mu}0}(Q_{N_1,\dots,N_K}, C, \widetilde{d\mu})$.

We consider hamiltonians of the form $\hat{H} = \hat{T} + V$. The kinetic energy operator is given by

$$\hat{H}_0 = \hat{T} = -\sum_{k,p} \alpha_k \frac{\partial^2}{\partial z_{kp} \partial z^*_{kp}}\ ,\ \ \alpha_k = \frac{2}{m_k}\ , \tag{17}$$

where m_k denotes the mass of an anyon of the k-th kind. We will analyse free anyon system, anyons in homogeneous potential field and anyons in anisotropic harmonic oscillator potential. For these hamiltonians solutions of the Schrödinger equation can be decomposed into two classes of equivariant functions:

$$\Phi = [\prod_{k=1}^{K} \prod_{1\le p<q\le N_k} (z_{kq} - z_{kp})^{\nu_k}] \times \tag{18}$$

$$[\prod_{1\le k<l\le K}\prod_{p=1}^{N_k}\prod_{q=1}^{N_l}(z_{lq}-z_{kp})^{\nu_{kl}}]\times$$
$$G(\ldots,z_{kp},\ldots;\ldots,z^*_{kp},\ldots)$$

and

$$\Phi=[\prod_{k=1}^{K}\prod_{1\le p<q\le N_k}(z^*_{kq}-z^*_{kp})^{2-\nu_k}]\times \tag{19}$$
$$[\prod_{1\le k<l\le K}\prod_{p=1}^{N_k}\prod_{q=1}^{N_l}(z^*_{lq}-z^*_{kp})^{1-\nu_{kl}}]\times$$
$$G(\ldots,z_{kp},\ldots;\ldots,z^*_{kp},\ldots)\,,$$

where G denotes a power series of holomorphic z_{kp} and antiholomorphic z^*_{kp} variables with nonegative integer exponents. The topological factors

$$[\prod_{k=1}^{K}\prod_{1\le p<q\le N_k}(z_{kq}-z_{kp})^{\nu_k}][\prod_{1\le k<l\le K}\prod_{p=1}^{N_k}\prod_{q=1}^{N_l}(z_{lq}-z_{kp})^{\nu_{kl}}]$$

and

$$[\prod_{k=1}^{K}\prod_{1\le p<q\le N_k}(z^*_{kq}-z^*_{kp})^{2-\nu_k}]\times[\prod_{1\le k<l\le K}\prod_{p=1}^{N_k}\prod_{q=1}^{N_l}(z^*_{lq}-z^*_{kp})^{1-\nu_{kl}}]$$

are harmonic functions (the first is a holomorphic and the second is an antiholomorphic function). Let

$$\Psi(\ldots,z_{kp},\ldots;\ldots,z^*_{kp},\ldots;t) \tag{20}$$

be an equivariant solution of the Schrödinger equation

$$\hat{T}\Psi=i\frac{\partial\Psi}{\partial t} \tag{21}$$

for the system of free anyons with an equivariant initial condition:

$$\Psi(\ldots,z_{kp},\ldots;\ldots,z^*_{kp},\ldots;t=0)= \tag{22}$$
$$\Phi(\ldots,z_{kp},\ldots;\ldots,z^*_{kp},\ldots)\,.$$

Now let us modify the initial condition

$$\Psi_a(\ldots,z_{kp},\ldots;\ldots,z^*_{kp},\ldots;t=0)= \tag{23}$$
$$\sum_{\sigma\in S_{N_1}\times\ldots S_{N_K}}\exp[i\sum_{k=1}^{K}\sum_{p=1}^{N_k}(a^*_{\sigma(kp)}z_{kp}+a_{\sigma(kp)}z^*_{kp})]$$
$$\Phi(\ldots,z_{kp},\ldots;\ldots,z^*_{kp},\ldots).$$

At [5] we have analysed a similar modification but with an unbouded factor at the all plane. Although the modified initial condition is also equivariant, the solution

$$\Psi_a(\ldots,z_{kp},\ldots;\ldots,z^*_{kp},\ldots;t)= \tag{24}$$

$$\sum_{\sigma}\exp[i\sum_{k,p}(-\alpha_k|a_{\sigma(kp)}|^2t+a^*_{\sigma(kp)}z_{kp}+a_{\sigma(kp)}z^*_{kp})]$$

$$\Psi(\ldots,z_{kp}-\frac{a_{\sigma(kp)}}{\alpha_k}t,\ldots;\ldots,z^*_{kp}-\frac{a^*_{\sigma(kp)}}{\alpha_k}t,\ldots)$$

is not equivariant. Equivariancy for transpositions of the same kind of particles is broken and equivariancy for encircling of one kind of particle by other kind is broken as well because the argument at the last factor is translated by a vector quantity proportional to time t. So the all space of square–integrable equivariant functions spanned by (18, 19) is not the space of quantum states of the free anyon system. Let us denote the space of quantum states by $\mathcal{H}_0$. One can see that the space $\mathcal{H}_0$ is infinity dimensional

$$\dim\mathcal{H}_0=\infty\ .$$

Any quantum state from the space $\mathcal{H}_0$ can be put as an initial condition (22) for the Schrödinger equation (21). The function (23) is also square–integrable equivariant function but (24) shows that (23) is not a quantum state from $\mathcal{H}_0$.

Now let us consider the anyon system in homogeneous potential field. At [6] we have considered an analogical problem for a particle on a plane with one point removed. The Hamiltonian of the system is

$$\hat{H}_f=\hat{T}-\frac{1}{2}\sum_{k,p}(f^*_k z_{k,p}+f_k z^*_{k,p})\ ,$$

where f_k is the force acting on the k-th kind particle. The evolution operator is given by the formula

$$\hat{U}_f(t)=\prod_{k,p}\exp\frac{it^3}{3m_k}\hat{\mathrm{Tr}}_{\Pi_{kp}}(\frac{-t^2}{2m_k}f)\exp i<f_k,z_{kp}>U_0(t)\ ,\qquad(25)$$

where $\hat{U}_0(t)$ is the evolution operator of the free anyon system and $\hat{\mathrm{Tr}}_{\Pi}(v)$ is the translation operator in the plane Π by a vector v. Let the initial condition for the Schrödinger equation

$$\hat{H}_f\Psi=i\frac{\partial\Psi}{\partial t}\qquad(26)$$

be any wave function Φ from the space $\mathcal{H}_0$. The solution

$$\hat{U}_f(t)\Phi$$

is not equivariant in general. Equivariancy for transpositions of the same kind of particles is conserved, but equivariancy for encircling one particle by other kind is broken. As we see from the formula (eq. 25) the singularities move with uniform accelerations

$$a_k=\frac{f_k}{m_k}$$

for the k-kind particle, which is a formula identical to the second Newton's law of motion. The solution is equivariant only for the case $a_k=a_l$ for any k,l. Although

$$\dim\mathcal{H}_f=\infty\ ,\qquad(27)$$

for $a_k\neq a_l$ we get

$$\mathcal{H}_0\cap\mathcal{H}_f=\{0\}.\qquad(28)$$

Finally let us consider the anyon system in an anisotropic harmonic oscillator potential. Let the Hamiltonian of the system be given by

$$\hat{H}_{\omega_1,\omega_2} = \hat{T} + \frac{1}{2}\sum_{kp} m_k({\omega_1}^2\hat{x}^2_{kp} + {\omega_2}^2\hat{y}^2_{kp}) \ , \tag{29}$$

where $z_{kp} = x_{kp} + iy_{kp}$. The evolution operator is

$$\hat{U}_{\omega1,\omega_2}(t) = \frac{1}{\sqrt{\cos\omega_1 t \cos\omega_1 t}}$$

$$\prod[\exp[-i\frac{m_k}{2}({\omega_1}^2\hat{x_{kp}}^2 \tan\omega_1 t + {\omega_2}^2\hat{y_{kp}}^2 \tan\omega_1 t)]$$

$$\hat{d}_{x_{kp}}\left(\sqrt{\frac{\tan\omega_1 t}{\omega_1}}\right)\hat{d}_{y_{kp}}\left(\sqrt{\frac{\tan\omega_2 t}{\omega_2}}\right)]$$

$$\hat{U}_0(1)\prod_{kp}\hat{d}_{x_{kp}}\left(\sqrt{\frac{2\omega_1}{\sin 2\omega_1 t}}\right)\hat{d}_{y_{kp}}\left(\sqrt{\frac{2\omega_2}{\sin 2\omega_2 t}}\right) \ ,$$

where $\hat{d}_x(a)$ is the dilatation operator at the direction x in the scale a. The composition $d_{x_{kp}}\left(\sqrt{\frac{2\omega_1}{\sin 2\omega_1 t}}\right)\hat{d}_{y_{kp}}\left(\sqrt{\frac{2\omega_2}{\sin 2\omega_2 t}}\right)$ of dilatations deforms the holomorphic variable z_{kp}. So deformed topological factors at the equations (18, 19) are also equivariant functions but they are not harmonic functions for $\omega_1 \neq \omega_2$. Consequently the deformed functions do not belong to the domain of the operator $\hat{U}_0(1)$. So the space of quantum states for the anisotropic oscillator is the zero space

$$\mathcal{H}_{\omega_1,\omega_2} = \{0\} \ . \tag{30}$$

4. Concluding remarks

I. We see that if the function (22) is a quantum state for free anyons system, then the function (23) (also equivariant) is not a quantum state for the system. So the quantum states of states for the system is not the all Hilbert space of equivariant square–integrable functions.

II. The spaces of quantum states for the free anyon system and for the free anyon system and for the system of anyons in homogeneous potential field are infinitely dimensional Hilbert subspaces of the space of square–integrable equivariant functional. The equation (28) shows that these quantum states of states have only the zero common vector.

III. The space of quantum states for the anyon system in anisotropic harmonic oscillator potential is the zero space.

So we see that the space of quantum states in quantum mechanics of anyons depends on the Hamiltonian.

Acknowledgments

The work has been supported by projects: grant KBN No 1 P03A 039 29 and PB-43-041/05 BW of Poznań University of Technology.

References

[1] Dowker J S 1972 *J. Phys.* **A 5** 936.
[2] Laidlaw M, DeWitt C 1971 *Phys. Rev.* **D 3** 1375
[3] Leinaas J, Myrheim J 1977 *Il Nuovo Cimento* **37** 1
[4] Forte S 1992 *Rev. Mod. Phys.* **64** No. 1 193
[5] Milewski J and Lulek T 1996 *Reports on Mathematical Physics* **38** (279–282)
[6] Milewski J 1999 *Molecular Physics Reports* **23** 171

Geometry and rigged strings in Bethe Ansatz.

D Jakubczyk[1], **T Lulek**[1], **P Jakubczyk**[2], **B Lulek**[1]

[1]Chair of Physics, Rzeszòw University of Technology, Powstancòw Warszawy 6, 35-959 Rzeszòw, Poland

[2]Institute of Physics, University of Rzeszòw, Aleja Rejtana 16c, 35-959 Rzeszòw, Poland

E-mail: djak@prz.edu.pl

Abstract. The main purpose of this report is a thorough analysis of completeness of solutions of the one-dimensional Heisenberg Hamiltonian through the hypothesis of strings. A somehow astonishing conclusion emerges from studying of the structure of the classical configuration space of this system. Namely, all allowed information concerning quantum states, which are exact solutions of the Bethe equations, encoded in quantum numbers, are predictable via a bijection between the set of the magnetic configurations and the string configurations. This startling and beautiful observation constitutes the proof of the completeness of the eigenstates of the Heisenberg Hamiltonian, deduced in a purely combinatorial way. We interpret the set of all magnetic configurations with a fixed number r of spin deviations as the classical configuration space of a hypothetic system of r Bethe pseudoparticles, which move, in a stroboscopic manner, on the magnetic ring. The geometry of this configuration space, induced by the action of Heisenberg Hamiltonian and the translation symmetry group of the ring, implies the structure of a locally r-dimensional hypercubic lattice with well defined F-dimensional boundaries, $1 \leq F \leq r$. We demonstrate that rigged string configurations originate from these boundaries, depending upon the island structure of spin deviations. We show that a relatively simple combinatoric definition of rigged strings reproduces completely exact results of Bethe Ansatz. It is expressed in terms of a combined bijection: Robinson-Schensted with Kerov-Kirillov-Reshetikhin (RSKKR) which produces a geography of exact Bethe Ansatz solutions on the classical configuration space.

1. Introduction

The famous Bethe Ansatz (BA) [1] provides the exact solution of the eigenproblem for the Heisenberg Hamiltonian for a finite magnetic ring of N spins $s = 1/2$. Eigenfunctions are classified by combinatorial objects called rigged string configurations [2], which are a refined version of Takahashi integers [3]. Kerov, Kirillov and Reshetikhin (KKR)[2] introduced a bijection between the irreducible basis of the Weyl duality [4] (cf. also [5], [6]) and the set of exact BA eigenfunctions. Another bijection, known as the Robinson - Schensted (RS) algorithm [7], [8] (cf. also [9]-[11]), associates each magnetic configuration, treated as a word of the length N in the alphabet $\{+, -\}$ of single-node spins, with the irreducible basis of the Weyl duality, that is a pair of standard Weyl and Young tableaux with the same shape.

The set of all magnetic configurations with a given number r of spin deviations from the magnetic saturation (all spins "+") can be equipped with a geometry imposed by the action of the Heisenberg Hamiltonian and the cyclic group C_N - the translational symmetry group of the magnet [12]. This set can be also interpreted along the Schrodinger picture of Quantum Mechanics, as the "classical" configuration space for the system of r spin deviations (Bethe pseudoparticles), moving in a stroboscopic way on the magnetic ring. In the present paper we aim to consider some poperties of a *composed* bijection, $KKR \circ RS$, which maps magnetic configurations immediately to exact BA solutions, omitting thus the basis of duality of Weyl. We aim to point out that strings of the length $l > 1$, interpreted usually as bound states of l magnons, originate indeed from *boundaries* of the classical configuration space, i. e. from those magnetic configurations in which some Bethe pseudoparticles gather into islands. In this way, the $RSKKR$ bijection establishes a "geography" of distribution of rigged string configurations over the classical configuration space.

2. Preliminaries

Let $\tilde{N} = \{j = 1, 2, \ldots, N\}$ be the set of nodes of a magnetic ring, and $\tilde{2} = \{+, -\}$ - the set of single-node spin projections. Then the set of all magnetic configurations is

$$\tilde{2}^{\tilde{N}} = \{f : \tilde{N} \longrightarrow \tilde{2}\}, \tag{1}$$

and the space $\mathcal{H}$ of all quantum states of the magnetic ring is spanned on this set, i. e.

$$\mathcal{H} = lc_{\mathbb{C}}\, \tilde{2}^{\tilde{N}}. \tag{2}$$

Moreover, the set $\tilde{2}^{\tilde{N}}$ provides an *orthonormal* basis in $\mathcal{H}$. We recall briefly that the space $\mathcal{H}$ of all quantum states of the Heisenberg magnet, with $\dim\mathcal{H} = 2^N$, decomposes as

$$\mathcal{H} = \sum_{r=0}^{N} \oplus\, \mathcal{H}^r, \tag{3}$$

into subspaces $\mathcal{H}^r$, $\dim\mathcal{H}^r = \binom{N}{r}$, with the fixed number r of Bethe pseudoparticles (spin deviations). Stated otherwise, $\mathcal{H}^r$ is the space of all states of the magnet with the projection

$$M = \frac{N}{2} - r \tag{4}$$

of the total spin, or of all states with the *weight*

$$\mu = \{N - r, r\}. \tag{5}$$

Clearly, $\mathcal{H}^r$ and $\mathcal{H}^{N-r}$ are particle-hole counterparts, so we restrict in the sequel to $r \leq N/2$. Then, each space $\mathcal{H}^r$ decomposes as

$$\mathcal{H}^r = \sum_{S=N/2-r}^{N/2} \oplus\, \mathcal{H}^{rS} \tag{6}$$

into subspaces $\mathcal{H}^{rS}$ with the total spin S (and $M = N/2 - r$). By the duality of Weyl [4] between the actions of the symmetric group Σ_N and the unitary group $U(n)$, $n = 2s + 1 = 2$, in the space $\mathcal{H}$ (cf., e.g., [5], [6] for detail), it defines the partition λ of the integer N, of the *shape*

$$\lambda = \{N - r', r'\}, \tag{7}$$

such that λ labels the irrep Δ^λ of Σ_N,

$$\dim \Delta^\lambda = \dim \mathcal{H}^{rS}, \tag{8}$$

and

$$S = \frac{N}{2} - r', \;\; 0 \leq r' \leq r. \tag{9}$$

For $r \leq N/2$, orbits of the symmetric group Σ_N are classical configuration spaces $Q^{(r)}$ for the system of r spin deviations, such that

$$\mathcal{H}^r = lc_{\mathbb{C}}\mathcal{O}_\mu \equiv lc_{\mathbb{C}}Q^{(r)}, \tag{10}$$

and

$$\dim \mathcal{H} = |Q^{(r)}| = \begin{pmatrix} N \\ r \end{pmatrix}. \tag{11}$$

Let

$$\mathbf{j} = (j_1, j_2, \ldots, j_r), \text{ for some set } 1 \leq j_1 < j_2 < \ldots < j_r \leq N, \tag{12}$$

be the distribution of Bethe pseudoparticles which specifies a magnetic configuration $f \in Q^{(r)}$, and let

$$\mathbf{t} = (t_1, t_2, \ldots, t_r), \; \sum_{\alpha \in \tilde{r}} t_\alpha = N. \tag{13}$$

be the *relative* distribution of Bethe pseudoparticles [13] within an orbit of the translational symmetry group of the ring C_N (C_N-orbit), being subgroup of the symmetric group Σ_N. The vector $\mathbf{t}$ provides a unique label for classification of C_N-orbits in $Q^{(r)}$, so that

$$Q^{(r)} = \bigcup_{\mathbf{t} \in T} \mathcal{O}_{\mathbf{t}}, \tag{14}$$

where $T = Q^{(r)}/C_N$ is the set of orbits, which we refer to, after [13], as a *skeleton* of the configuration space $Q^{(r)}$ (cf. [13] for detail definitions).

Let

$$f^{\mathbf{t}} = (\mathbf{t}, 1) = |\underbrace{- + + \ldots +}_{t_1}\underbrace{- + + \ldots +}_{t_2} - \ldots \underbrace{- + + \ldots +}_{t_r} > \tag{15}$$

be, by convention, the *initial* magnetic configuration in the orbit $\mathcal{O}_{\mathbf{t}}$. The distribution $\mathbf{j}^{\mathbf{t}}$ of Bethe pseudoparticles for the initial magnetic configuration $f^{\mathbf{t}}$ is given by

$$j_\alpha^{(\mathbf{t},1)} = 1 + \sum_{\alpha' < \alpha} t_{\alpha'}, \;\; \alpha \in \tilde{r}. \tag{16}$$

An arbitrary configuration in this orbit has the form of composition of mappings

$$(\mathbf{t}, j) = (\mathbf{t}, 1) \circ C_N^{1-j}, \;\; j \in \tilde{N}, \tag{17}$$

where the mapping $C_N : \tilde{N} \longrightarrow \tilde{N}$ is defined by

$$C_N = \begin{pmatrix} j \\ (j+1) \bmod N \end{pmatrix} \in \Sigma_N, \;\; j \in \tilde{N} \tag{18}$$

(we use the same symbol for the group $C_N \subset \Sigma_N$, and for its generating element C_N).

3. Structure of islands of spin deviations

We recall that the action of the Heisenberg Hamiltonian $\hat{H}$ in $\mathcal{H}^r$ is given by

$$2\hat{H}|\mathbf{j}>= \sum_{\mathbf{j}'\in Q_{\mathbf{j}}^{(r)}} (|\mathbf{j}'> - |\mathbf{j}>), \ \ \mathbf{j} \in Q^{(r)}, \tag{19}$$

where $|\mathbf{j}> \in \mathcal{H}^r$ is the quantum state of the model, corresponding to the magnetic configuration determined by the distribution $\mathbf{j} \in Q^{(r)}$ of Bethe pseudoparticles, and $Q_{\mathbf{j}}^{(r)} \subset Q^{(r)}$ is the set of all *nearest neighbours* of the magnetic configuration $\mathbf{j}$ within $Q^{(r)}$.

In this way, the dynamics (19) of BA imposes, in a local manner, the structure of a hypercubic lattice in the classical configuration space $Q^{(r)}$: for a generic position $\mathbf{j} \in Q^{(r)}$, its nearest neighbours $\mathbf{j}' \in Q_{\mathbf{j}}^{(r)}$ define a local coordinate system in a real r-dimensional linear space $\mathbb{R}^r$ into which $Q_{\mathbf{j}}^{(r)}$ can be embedded, with the origin at $\mathbf{j}$. Such a description reflects the r degrees of freedom for stroboscopic movements of the system $\tilde{r}$ of Bethe pseudoparticles over the ring $\tilde{N}$: a pseudoparticle $\alpha \in \tilde{r}$ can jump forward or back by one step on the crystal $\tilde{N}$, independently on other pseudoparticles $\alpha' \neq \alpha$.

The classical configuration space $Q^{(r)}$, however, is not homogeneous with respect to the degree of freedom. A look at the form (15) of a magnetic configuration reveals that some pseudoparticles can be the nearest neighbours each of other, forming thus "islands" of spin deviations in the "sea" of nodes with the spin $+ \in \tilde{2}$. Clearly, each island contributes to a single degree of freedom, since its leftmost pseudoparticle can only jump to the left, its rightmost - to the right, and all internal pseudoparticles of this island are *kinematically frozen*.

The structure of islands of spin deviations is defined as a sequence of lengths

$$\xi = (\xi_1, \xi_2, \ldots, \xi_F), \ \sum_{\gamma=1}^{F} \xi_\gamma = r. \tag{20}$$

It thus imposes a decomposition

$$Q^{(r)} = \bigcup_{F=1}^{r} Q^{(r,F)} \tag{21}$$

of the classical configuration space into subsets

$$Q^{(r,F)} = \{\mathbf{j} \in Q^{(r)} | \, |Q_{\mathbf{j}}^{(r)}| = 2F\}, \tag{22}$$

consisting of C_N-orbits with exactly $F \leq r$ islands of spin deviations. The case $F = r$ is *generic*, whereas subsets $Q^{(r,F)}$ with $F < r$ correspond to *boundaries* of the classical configuration space with a lower dimension. Thus the integer F, defined in Eq. (20) as the number of islands of spin deviations for a magnetic configuration within an orbit $\mathcal{O}_{\mathbf{t}}$, plays also the role of *local dimension* for all elements of the boundary $Q^{(r,F)}$ (such that $\mathcal{O}_{\mathbf{t}} \subset Q^{(r,F)}$) of the classical configuration space $Q^{(r)}$. The local dimension F varies within the range

$$1 \leq F \leq r. \tag{23}$$

It will follow from the sequel (Sect. 5-6) that

$$r' \leq F, \tag{24}$$

where r' determines the total spin $S = N/2 - r'$ of the RSKKR image of $\mathbf{j} \in Q^{(r,F)}$.

4. The combinatorics of strings.

A *string configuration* [2], [14], [15] ν is, by the definition, a partition of the integer r', $\nu \vdash r'$. Each part of the partition ν is referred to as a *string*, and its size (the number of boxes in the Young diagram of ν) - the *length* of the string. Denoting by m_l the number of strings of the length l in ν, we have

$$\sum_l l m_l - r'. \tag{25}$$

Each string (lv), $v \in \tilde{m}_l = \{1, 2, \ldots, m_l\}$, is equipped with a nonnegative integer L, called its *rigging*. The rigging L varies within the range

$$0 \leq L \leq P_l, \tag{26}$$

where

$$P_l = N - 2Q_l, \tag{27}$$

with Q_l being the number of boxes in the first l *columns* of the Young diagram of the string configuration ν. The integer P_l is referred to as the number of *holes* for strings of the length l. The pair $\nu\mathcal{L}$, where

$$\mathcal{L} = \{L_{lv} | l = 1, 2, \ldots; \ v \in \tilde{m}_l\}, \tag{28}$$

is the set of all riggings of ν, is referred to as a *rigged string configuration*. All m_l strings of the length l are distinguished by their riggings only, so that they can be arranged in $\nu\mathcal{L}$ in the nondecreasing order (from the top to the bottom of the $l \times m_l$ - rectangle incorporated in the Young diagram $\nu \vdash r'$), that is

$$0 \leq L_{lv} \leq L_{lv'} \leq P_l \text{ for } 1 \leq v < v' \leq m_l, \ l = 1, 2, \ldots \tag{29}$$

Such a counting implies that the set $z(\nu) = \{\nu\mathcal{L}\}$ of all rigged string configurations corresponding to a given $\nu \vdash r'$ has the cardinality

$$|z(\nu)| = \prod_l \binom{P_l \ + \ m_l}{m_l}, \tag{30}$$

and, moreover,

$$\dim \Delta^\lambda = \sum_{\nu \vdash r'} |z(\nu)|. \tag{31}$$

Each rigged string configuration $\nu\mathcal{L}$ classifies an exact eigenstate of the Heisenberg Hamiltonian $\hat{H}$ in each space $\mathcal{H}^r$, $r' \leq r \leq N/2$ (cf. Eq. (19). The eigenstate $|\nu\mathcal{L}>$ is a superposition [6] of magnetic configurations $|f>$, $f \in Q^{(r)}$, and can be obtained as a solution of highly non-linear Bethe equations. It is amasing that this state is fully characterised by the combinatoric data $\nu\mathcal{L}$, which can be simply interpreted in terms of strings. In this combinatoric interpretation, a string of the length l is a collection of l *consecutive* nodes $j, j+1, \ldots, j+l-1$, each occupied by a spin deviation, followed by next l nodes $j+l, j+l+1, \ldots, j+2l-1$, occupied by a spin $''+'' \in \tilde{2}$. In the case when there are no other strings, then the remaining $P_l = N - 2l$ nodes of the crystal $\tilde{N}$ play the role of the "sea of holes" for the string configuration $\nu = \{l\}$. The rigging of this l-string is $L = j - 1$, which coincides with the number of holes to the left of the molecule representing this string [16]. All possible riggings of this l-string correspond to various locations of the molecule inside the *open* chain $\tilde{N}$, with distinguished initial and final node. In other words, the left boundary node j of such a string should satisfy $1 \leq j \leq N - 2l$. Clearly, the number of such locations is $|z(\{l\})| = P_l + 1 = N - 2l + 1$, in accordance with Eq. (30). More complex string configurations require some combinatoric extension which we describe in the next paragraph.

5. The RSKKR bijection

In fact, such an apparently simplified, but surprisingly adequate picture of an exact BA solution, emerges by virtue of existence of two bijections. The first is Robinson-Schensted correspondence [7], [8] between the set $\tilde{2}^{\tilde{N}}$ of all magnetic configurations and the set b_{irr} of an irreducible basis of the duality of Weyl, and the second - the bijection of Kerov, Kirillov and Reshetikhin [2] between the set b_{irr} and appropriate rigged string configurations. The irreducible basis of the duality of Weyl, written in the form

$$b_{irr} = \bigcup_{0 \leq r' \leq N/2} SYT(\{N-r', r'\}) \times WT(\{N-r', r'\}, \tilde{2}), \tag{32}$$

where $SYT(\lambda)$ is the set of all standard Young tableaux of the shape λ in the alphabet $\tilde{N}$ of nodes, and $WT(\lambda, \tilde{2})$ is the set of all Weyl tableaux (that is, semistandard Young tableaux) of the same shape λ in the alphabet $\tilde{2}$ of spins, with $\lambda = \{N-r', r'\}$, mediates thus between the initial basis $\tilde{2}^{\tilde{N}}$ for quantum computations, and the final basis of rigged string configurations which classify the exact BA eigenstates. Here we consider the *composition* of these two bijections, which runs immediately from magnetic configurations to rigged string configurations, omitting thus the pairs of Young and Weyl tableaux of the duality of Weyl. The classical configuration space $Q^{(r)}$ serves as the source of the bijection, and the target is the basis of those exact BA eigenstates which span the space $\mathcal{H}^r = lc_{\mathbb{C}} Q^{(r)}$. We thus describe the composite RSKKR bijection

$$\rho : Q^{(r)} \longrightarrow RC(N, r), \tag{33}$$

where

$$RC(N, r) = \bigcup_{r'=0}^{r} RC(N, r, r'), \text{ and } RC(N, r, r') = \bigcup_{\nu \vdash r'} z(\nu). \tag{34}$$

Here $RC(N, r)$, $RC(N, r, r')$, and $z(\nu)$ denotes the set of all rigged string configurations on the magnetic ring $\tilde{N}$ which correspond to the system of r Bethe pseudoparticles, all those with the total spin $S = N/2 - r'$ and its projection $M = N/2 - r$, and all riggings of the string configuration $\nu \vdash r'$, respectively. The completeness of exact BA solutions reads

$$lc_{\mathbb{C}}\, RC(N, r, r') = \mathcal{H}^{rS}, \text{ and } lc_{\mathbb{C}}\, RC(N, r) = \mathcal{H}^r, \; S = N/2 - r'. \tag{35}$$

We proceed to describe the RSKKR bijection in detail. To this aim, we introduce the notion of a *minimal compensated subword* of a magnetic configuration $f \in Q^{(r)}$. Let

$$v = f(j)f(j+1)\ldots f(j+2A-1), \; 2A \leq N - j + 1, \tag{36}$$

be a subword of f, consisting of 2A letters of consecutive nodes. Let F^{α} be the subword of v, consisting of its first α letters, i. e.

$$F^{\alpha} = f(j)f(j+1)\ldots f(j+\alpha-1), \; \alpha = 1, 2\ldots, 2A, \tag{37}$$

and let $w(F^{\alpha}) = \{w_1, w_2\}$ be the weight of the word F^{α}, so that w_i is the number of occurences of the letter $i \in \tilde{2}$ in F^{α}, and thus $w_1 + w_2 = \alpha$. By the definition, v is called a minimal compensated subword of f if

$$w_1 \begin{cases} < w_2 & \text{for} \quad \alpha < 2A, \\ = w_2 = A & \text{for} \quad \alpha = 2A. \end{cases} \tag{38}$$

A minimal compensated word of the length 2A corresponds to an ensemble of A pairs of nodes (in general *not* adjacent), each pair consisting of a Bethe pseudoparticle, followed by a compensating node occupied by the spin $''+''$.

It is evident that each magnetic configuration f can be decomposed, in a unique way, into consecutive subwords

$$f = u_1 v_1 u_2 v_2 \dots v_a u_{a+1} \tag{39}$$

such that $v_1, v_2, \dots, v_a$ are minimal compensated subwords, whereas $u_1, u_2, \dots, u_{a+1}$ are formed consecutively from the remaining letters of f (some u_i might have zero length, when it happens that v_{i-1} and v_i are adjacent). The main algorithm which determines the RSKKR bijection can be expressed transparently in terms of paths. The path [14] $p(f)$, corresponding to a magnetic configuration f in the RSKKR bijection is given by

$$p_j(f) = \left\{ \begin{array}{lll} \text{up} & \text{if} & f(j) = ''-'', \\ & \text{or if} & f(j) = ''+'' \quad \text{and } f(j) \in u(f), \\ \text{down} & \text{if} & f(j) = ''+'' \quad \text{and } f(j) \in v(f), \end{array} \right\}, j \in \tilde{N}. \tag{40}$$

Thus all Bethe pseudoparticles in a configuration f correspond to steps up in the path $p(f)$, whereas the nodes with $f(j) = ''+''$ correspond to the step either up or down, depending if they belong to the subword $u(f)$ or $v(f)$, respectively. Each straight line u_α belongs to the "sea of holes" for all strings, whereas a multipyramid v_α constitutes a collection of overlapping strings. We proceed to interpret such a multipyramid.

Each of two or more pyramids can move within its range of riggings, obeying two simple *rules of navigation* [16]. Firstly, two l-strings are mutually inpenetrable, that is, they behave like hard-core objects. It is due to the fact that the strings cannot exchange their positions in the one-dimensional ring $\tilde{N}$ since the order of the queue of Bethe pseudoparticles is fixed. Secondly, for $l < l'$, the l-string, treated as a pyramid, can slide over both slopes of the l'-string until the heights of their tops coincide. In fact, the first rule can be looked at as a special case of the second.

As a result, each multipyramid $v(f)$ represents a collection of overlapping strings.

6. Behaviour of RSKKR preimages of strings under translations

We proceed to consider some properties of the mapping $\mathcal{K}_N : RC(N,r) \longrightarrow RC(N,r)$, defined by

$$\mathcal{K}_N = \rho \circ A(C_N) \circ \rho^{-1}. \tag{41}$$

In other words, $\mathcal{K}_N$ is the push-forward of the shift $A(C_N) : Q^{(r)} \longrightarrow Q^{(r)}$ from the classical configuration space $Q^{(r)}$ to the corresponding set $RC(N,r)$ of exact BA solutions along the RSKKR bijection $\rho : Q^{(r)} \longrightarrow RC(N,r)$.

The Fig. 1 presents the travel of the coupled system of a 1-string and a 3-string over the ring of $N = 10$ nodes along the orbit $\mathbf{t} = (1216)$. The corresponding multipyramid

$$v = - - + - - + ++ \tag{42}$$

has the size $2(1+3) = 8$, and can be thus located on the open chain of 10 nodes in three different positions. RSKKR preimages of these positions are given by magnetic configurations $(\mathbf{t}, j)$ with $j \in \{1,2,3\}$. For the next three nodes, $j \in \{4,5,6\}$, the right slope of the 3-string disappears at the terminal node of the chain, so this string is shortened to 2-string ($j = 4$), 1-string ($j = 5$), and vanishes completely at $j = 6$. Each such shortening results in extension of the sea of unbounded nodes by 2: one "+" which arises at the beginning of the chain, and one "-", which is no longer bound to the shortened string. Then, at the nodes $j \in \{7,8\}$ a string arises at the initial node (1-string and 2-string for $j = 7$ and 8, respectively). In the meantime, the 1-string reaches the terminal node for $j = 8$, then disappears at $j = 9$, and arises again at $j = 10$ as the neighbour of the 2-string. The next step, corresponding to $j = 1$, creates the multipiramid (42) at its leftmost position.

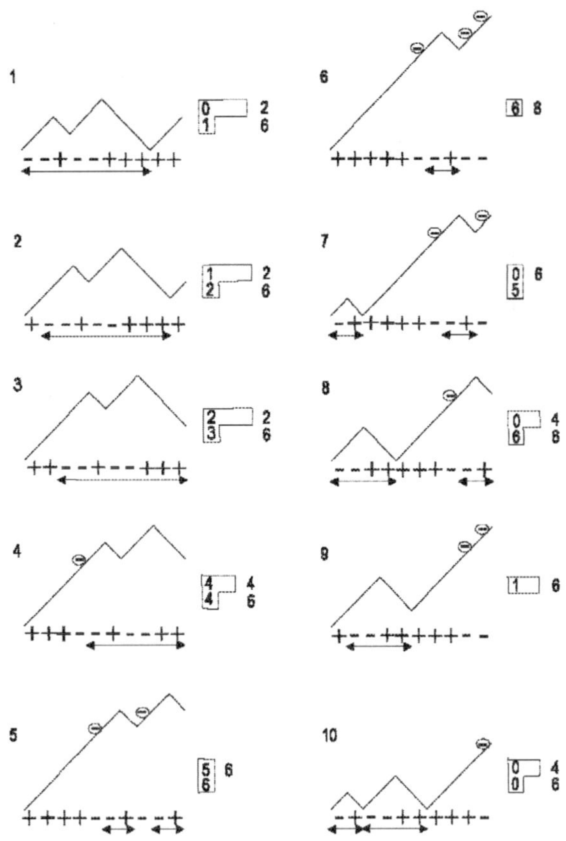

Figure 1. The RSKKR bijection for the C_{10}-orbit $\mathbf{t} = (1216)$ ($N = 10$, $r = 4$).

To describe this process in detail, one has to distinguish three cases: (i) $f(N)$="+" and $f(N) \in u(f) \equiv$ the last node is occupied by the spin "+", and belongs to the sea of holes; (ii) $f(N)$="+" and $f(N) \in v(f) \equiv$ the last node is "+" and belongs to the (last) minimal compensated subword $v_a(f) \subset v(f)$; (iii) $f(N)$="-". The total spin $S' = \frac{N}{2} - r'$ changes in the cases described above according to the formula

$$S' = \begin{cases} S & \text{in the case (i)}, \\ S-1 & \text{in the case (ii)}, \\ S+1 & \text{in the case (iii)}. \end{cases} \tag{43}$$

It is worth to observe that the total number of strings,

$$q(f) = \sum_{l \geq 1} m_l, \tag{44}$$

can be deduced immediately from the shape of the path $p(f)$ - it is just equal to the number of cuspidal maxima *inside* the open chain $\tilde{N}$. Clearly the variation of $q(f)$ along an orbit $\mathcal{O}_{\mathbf{t}}$ is much smaller than that of S, namely

$$q(\mathbf{t}, j) \in \{F, F-1\} \text{ for } \mathcal{O}_{\mathbf{t}} \subset Q^{(r,F)}. \tag{45}$$

It follows from the observation that at most one island can approach the rightmost end of the open chain, and thus not to contribute to a string, at each step along the orbit $\mathcal{O}_{\mathbf{t}}$.

7. Final remarks and conclusions

The main purpose of this report was a thorough analysis of completeness of the exact solutions of the eigenproblem for the Heisenberg Hamiltonian for a finite magnetic ring of N spins $s = 1/2$. We have shown, that this result can be obtained in a purely combinatorial way, merely by studying the geometry of the classical configuration space of this system, by means of the RSKKR bijection. This composite bijection maps magnetic configurations immediately to exact BA solutions, omitting thus the basis of the duality of Weyl. We also pointed to the fact that the number of uncompensated spin deviations in a magnetic configuration are responsible for variation of the total spin S.

References

[1] Bethe H 1931 Z. Physik **71** 205 (in German; English translation in: Mattis D C 1993 The Many-Body Problem (World Sci. Singapore 689-716)

[2] Kerov S V, Kirillov A N and Reshetikhin N Yu 1986 *LOMI* **155** 50 (in Russian; English translation: 1988 *J. Sov. Math.* **41** 916)

[3] Takahashi M 1971 *Progr. Theor. Phys.* **46** 401

[4] Weyl H 1931 *Gruppentheorie und Quantummechanik*(Leipzig, Hirzel) (English translation: 1950 *The Theory of Groups and Quantum Mechanics* (New York, Dover)

[5] Lulek B Jakubczyk D Jakubczyk P 2004 Kinematics of the Heisenberg chain and irreducible bases of the Weyl duality *Mol. Phys.* **102** 1269-1278

[6] Lulek T Jakubczyk P Jakubczyk D 2004 Magnetic interpretation of the Robinson - Schensted - Knuth algorithm *Mol. Phys.* **102** 1279-1284

[7] Robinson G 1938 *Amer. J. Math.* **60** 745-60

[8] Schested G *Canad. J. Math.* **13** 179-91

[9] Knuth G 1970 *Pacific. J. Math.* **34** 709-27

[10] Schutzenberger M P 1977 Combinatoire et Representation du Groupe Symetrique (edited by D. Foata) *Lect. Notes in Math.* **579** 59

[11] Kerber A 1999 *Applied Finite Group Actions* (Berlin: Springer)

[12] Lulek B Jakubczyk D 2003 *CEJP* **1** 132-144

[13] Lulek B, Lulek T, Wal A and Jakubczyk P 2003 The basis of wavelets for a finite Heisenberg magnet *Physica B - Condensed Matter* **337/1** 375-387

[14] Dasmahapatra S and Foda O *Int. J. Mod. Phys.* **38** 1041

[15] Schilling A 2002 *Symmetry and Structural Properties of Condensed Matter* (World Sci., Singapore 2002) 201

[16] Warnaar S O 1996 *J. Stat. Phys.* **82** 657

Operations on distributions: regular and irregular

S Sorek

Institute of Mathematics, University of Rzeszów, Rejtana 16 A, 35-310 Rzeszów, Poland

E-mail: `ssorek@univ.rzeszow.pl`

Abstract. In the frame of the Mikusiński sequential theory of L. Schwartz's distributions, regular and irregular operations on distributions are considered with applications to some formulas which may be of interest to physicists, e.g. $\sqrt{\delta} = 0$, $\sqrt{1+\delta^2} = 1+\delta$, $\log(1+\delta) = 0$, $\sin\delta = 0$.

1. Introduction

J. Mikusiński in his sequential theory of L. Schwartz's distributions (see [6], [4], [2]) introduced a general method of extending operations defined for (smooth, continuous, locally integrable) functions to distributions. Operations which are *regular* (see the definition in section 2) can easily be extended to the class of all distributions. That the operation of substitution of a fixed smooth diffeomorphism into a given function (distribution) is regular is not so easy to prove as the regularity of other operations. We will show a proof of the regularity of this operation in the multidimensional case (see section 2). If a given operation is *irregular*, J. Mikusiński has also proposed (see [5], [2]) a general method of extending it to distributions, but not to all of them, in general. The method is based on the notion of a *delta-sequence*, i.e. a sequence (δ_n) of smooth functions convergent to the distribution δ, satisfying special conditions, and the notion of *regular sequence* for a given distribution f, i.e. a sequence (f_n) of smooth functions of the form $f_n := f * \delta_n$, which is fundamental and distributionally convergent to f, where the symbol $f * \delta_n$ denotes the convolution of f with n-th element of a certain delta-sequence (δ_n) and is meant as the result of a regular operation (see [2], p. 153). The class of all regular sequences for a distribution f is narrower than the class of all fundamental sequences for f. Consequently, irregular operations are feasible for a quite large class of distributions.

Irregular operations, e.g. the two argument operations of product and composition of distributions, and the one argument operation of the substitution of distributions into a given function, may be particularly interesting to physicists. We are going to justify mathematically some formulas concerning the distribution δ which cannot be explained on the base of the classical approach to the theory of distributions. P. Antosik obtained in [1] some important results on composition of distributions in $\mathbb{R}$ in case the external distribution is a continuous function and the internal one is a measure or a quasi-measure. The results enabled him to receive several interesting formulas, e.g. $\sqrt{1+\delta^2} = 1+\delta$. However, the proofs of a few main theorems in [1] are incorrect. The incorrectness is entirely removed in [3], due to some results concerning measure theory.

In this note we present without proofs the generalizations of certain theorems in [1] and [3] from the real line $\mathbb{R}$ to an arbitrary open subset or an arbitrary open (bounded or not) interval in $\mathbb{R}$. This makes possible to extend applications of the main theorem (e.g. to prove the formula $\log(1+\delta)=0$).

2. Regular operations

In the whole paper distributions will be considered on open subsets of a given multidimensional Euclidean space. By intervals in $\mathbb{R}^d$, we always mean d-dimensional finite (open or closed) intervals. Let us briefly recall the notation and definitions concerning the sequential approach to the theory distributions (see [2] and [3]). If the closure $[a,b]$ of a given interval $I=(a,b)$, with $a,b \in \mathbb{R}^d$ and $a<b$, is contained in a given open set $U \subset \mathbb{R}^d$, we write $I \sqsubset U$. By $C^\infty(U)$ we denote the space of smooth (i.e. C^∞-) functions on U. A sequence (φ_n) of functions in $C^\infty(U)$ is said to be *fundamental* on U if for every $I \sqsubset U$ there exist a multi-index $k \in \mathbb{N}_0^d$ and functions $\Phi_n \in C^\infty(U)$ $(n \in \mathbb{N})$ such that $\Phi_n^{(k)} = \varphi_n$ on I and $\Phi_n \rightarrow$ *locally uniformly* (shortly: l.u.) on I, i.e. uniformly on each compact set $K \subset I$ to a function on I (see [2] and [3]). It is worth noting that locally uniformly convergent sequences are fundamental (for $k=0$), but in general, fundamental are sequences of smooth functions which are locally derivatives of a finite order of locally uniformly convergent sequences. Obviously, the multi-index k can be always increased. We call two fundamental sequences (φ_n) and (ψ_n) *equivalent* on U, $(\varphi_n) \sim (\psi_n)$, if the sequence $\varphi_1, \psi_1, \varphi_2, \psi_2, \ldots$ is fundamental on U. Clearly, $\sim$ is an equivalence relation. Equivalence classes of the relation $\sim$ are called *distributions* on U. The distributions represented by the fundamental sequences $(\varphi_n), (\psi_n), \ldots$ will be denoted by $f, g, \ldots$, respectively, i.e. $f=[(\varphi_n)]$, $g=[(\psi_n)]$, ... The space of all distributions on U will be denoted by $\mathcal{D}'(U)$.

Definition 1. Let $R: C^\infty(U)^j \rightarrow C^\infty(U)$, $j \in \mathbb{N}$, be an operation defined on all j-tuples of smooth functions whose values are smooth functions (or numbers treated as constant functions on U). The operation R is called *regular* if the sequence $(R(\varphi_n^1, \ldots, \varphi_n^j))$ is fundamental on U for arbitrary fundamental sequences $(\varphi_n^1), \ldots, (\varphi_n^j)$ on U. Every regular operation R can be defined for arbitrary distributions $f_1 = [(\varphi_n^1)], \ldots, f_j = [(\varphi_n^j)]$ on U by means of the formula $R(f_1, \ldots, f_j) := [(R(\varphi_n^1, \ldots, \varphi_n^j))]$. Clearly, the result of every regular operation for given distributions does not depend on the choice of fundamental sequences.

Examples of regular operations are: addition, differentiation, multiplication by a fixed smooth function, tensor product, convolution with a given smooth function of bounded support and multidimensional substitution.

Definition 2. Let U and V be open sets in $\mathbb{R}^d$ and $\mathbb{R}^m$, respectively, where $m \in \mathbb{N}, m \leq d$. Let $\sigma_j \in C^\infty(U)$ for $j \in \{1, \ldots, m\}$, $\sigma := (\sigma_1, \ldots, \sigma_m)\colon U \rightarrow V$. Assume that for every $x \in U$ at least one of the Jacobians:

$$(J_{j_1,\ldots,j_m}\sigma)(x) := \frac{\partial(\sigma_1, \ldots, \sigma_m)(x)}{\partial(\xi_{j_1}, \ldots, \xi_{j_m})} = \det A(x) \tag{1}$$

is not equal to 0, where $j_1, \ldots, j_m \in \mathbb{N}, j_1 < \ldots < j_m \leq d$ and the elements of the matrix $A(x)$ are $\frac{\partial \sigma_i}{\partial \xi_{j_k}}(x)$ for $i,k \in \{1, \ldots, m\}$. The operation S of the multidimensional substitution of the function σ into $\varphi \in C^\infty(V)$ is now defined by the formula: $S(\varphi) := \varphi \circ \sigma$ on U, i.e. $S(\varphi)(x) := \varphi(\sigma(x))$, $x \in U$.

To prove that S is a regular operation we need three lemmas, of which the first is very well known.

Lemma 1 *Let U and V be open sets in $\mathbb{R}^d$ and $\mathbb{R}^m$, respectively, where $m \in \mathbb{N}, m \leq d$. Let $\sigma_j \in C^\infty(U)$ for $j \in \{1, \ldots, m\}$ and $\sigma := (\sigma_1, \ldots, \sigma_m): U \to V$. If $\varphi \in C^\infty(V)$, then $\varphi \circ \sigma \in C^\infty(U)$ and*

$$(\varphi \circ \sigma)^{(e_i)}(x) = \sum_{j=1}^{m} (\varphi^{(e_j)} \circ \sigma)(x) \frac{\partial \sigma_j}{\partial \xi_i}(x), \qquad i \in \{1, \ldots, d\}, \tag{2}$$

for $x = (\xi_1, \ldots, \xi_d) \in \mathbb{R}^d$, where $\psi^{(e_i)}(x) := \frac{\partial}{\partial \xi_i} \psi(x)$ for a given function ψ and $x \in \mathbb{R}^d$.

Lemma 2 *Under the assumption of Definition 2, we have*

$$(\varphi^{(e_j)} \circ \sigma) = \frac{1}{J} \sum_{j_1 < \ldots < j_m} J_{j_1, \ldots, j_m}\sigma \cdot J_{\varphi, j; j_1, \ldots, j_m}\sigma, \qquad j \in \{1, \ldots, m\},$$

where

$$J := \sum_{j_1 < \ldots < j_m} (J_{j_1, \ldots, j_m}\sigma)^2 \tag{3}$$

and

$$(J_{\varphi, j; j_1, \ldots, j_m}\sigma)(x) := \frac{\partial(\sigma_1, \ldots, \sigma_{j-1}, \varphi \circ \sigma, \sigma_{j+1}, \ldots, \sigma_m)(x)}{\partial(\xi_{j_1}, \ldots, \xi_{j_m})}. \tag{4}$$

Proof. Fix $j \in \{1, \ldots, m\}$ and indices $j_1, \ldots, j_m \in \{1, \ldots, d\}$ such that $j_1 < \ldots < j_m$. To prove that

$$(J_{j_1, \ldots, j_m}\sigma)(x) \cdot (\varphi^{(e_j)} \circ \sigma)(x) = (J_{\varphi, j; j_1, \ldots, j_m}\sigma)(x), \qquad x \in U. \tag{5}$$

Consider the matrices $A(x) = (a_{i,k}(x))$ and $B(x) = (b_{i,k}(x))$, where

$$a_{i,k}(x) := b_{i,k}(x) := \frac{\partial \sigma_i}{\partial \xi_{j_k}}(x), \qquad i \in \{1, \ldots, m\},\ i \neq j;$$

$$a_{j,k}(x) := (\frac{\partial \varphi}{\partial \sigma_j} \circ \sigma)(x) \frac{\partial \sigma_j}{\partial \xi_{j_k}}(x); \qquad b_{j,k}(x) := \sum_{n=1}^{m} (\frac{\partial \varphi}{\partial \sigma_n} \circ \sigma)(x) \frac{\partial \sigma_n}{\partial \xi_{j_k}}(x)$$

for $k \in \{1, \ldots, m\}$. Clearly,

$$(J_{j_1, \ldots, j_m}\sigma)(x) \cdot (\varphi^{(e_j)} \circ \sigma)(x) = \det A(x).$$

Multiplying each i-th (for $i \neq j$) row of the matrix $A(x)$ by $(\frac{\partial \varphi}{\partial \sigma_i} \circ \sigma)(x)$ and adding the results to the j-th row, we get the matrix $B(x)$, i.e.

$$(J_{j_1, \ldots, j_m}\sigma)(x) \cdot (\varphi^{(e_j)} \circ \sigma)(x) = \det A(x) = \det B(x).$$

But $b_{j,k}(x) = \frac{\partial(\varphi \circ \sigma)}{\partial \xi_{j_k}}(x)$, so $\det B(x) = (J_{\varphi, j; j_1, \ldots, j_m}\sigma)(x)$ and equality (5) is shown. Multiplying both sides of (5) by $(J_{j_1, \ldots, j_m}\sigma)(x)$, we get

$$(J_{j_1, \ldots, j_m}\sigma)^2 \cdot (\varphi^{(e_j)} \circ \sigma) = J_{j_1, \ldots, j_m}\sigma \cdot J_{\varphi, j; j_1, \ldots, j_m}\sigma$$

for all m-tuples $(j_1, \ldots, j_m) \in \mathbb{N}^m$ such that $j_1 < \ldots < j_m$. Summing the above equalities over all such m-tuples $(j_1, \ldots, j_m)$, we obtain

$$\sum_{j_1 < \ldots < j_m} (J_{j_1, \ldots, j_m}\sigma)^2 \cdot (\varphi^{(e_j)} \circ \sigma) = \sum_{j_1 < \ldots < j_m} J_{j_1, \ldots, j_m}\sigma \cdot J_{\varphi, j; j_1, \ldots, j_m}\sigma$$

for every $j \in \{1, \ldots, m\}$. Since $J > 0$ on U, our assertion follows, by (3).

Lemma 3 *Let σ be a given function satisfying the assumption of Definition 2. If $\Phi_n \in C^\infty(V)$ $(n \in \mathbb{N})$ and the sequence $(\Phi_n \circ \sigma)$ is fundamental on U, then the sequence $(J_{\Phi_n,j;j_1,\dots,j_m}\sigma)$ is fundamental on U.*

Proof. Assume that the sequence $(\Phi_n \circ \sigma)$ is fundamental on U. By (4), due to the Laplace theorem, we have

$$(J_{\Phi_n,j;j_1\dots,j_m}\sigma)(x) = \sum_{k=1}^{m} (-1)^{j+k} \frac{\partial(\Phi_n \circ \sigma)}{\partial \xi_{j_k}}(x) \cdot \det\left(\frac{\partial \sigma_s}{\partial \xi_{j_t}}(x)\right),$$

where $s \in \{1,\dots,j-1,j+1,\dots,m\}$ and $t \in \{1,\dots,k-1,k+1,\dots,m\}$.

Since $\sigma_i \in C^\infty(U)$ for $i \in \{1,\dots,m\}$, all elements of the above determinants are smooth functions on U. Consequently, the determinants themselves are smooth functions on U. But differentiation is a regular operation and we infer from the assumptions that the sequence $(\frac{\partial(\Phi_n\circ\sigma)}{\partial \xi_{j_k}})$ is fundamental on U for $k \in \{1,\dots,m\}$. Finally, the sequence $(J_{\Phi_n,j;j_1,\dots,j_m}\sigma)$ is fundamental on U, as a finite iteration of the regular operations.

Theorem 1 *Multidimensional substitution is a regular operation.*

Proof. Suppose that σ satisfies the assumptions of Definition 2 and the sequence (φ_n) is fundamental on $V \subset \mathbb{R}^m$, i.e. for each interval $I' \sqsubset V$ there exist a multi-index $k \in \mathbb{N}_0^m$ and functions $\Phi_n \in C^\infty(V)$ $(n \in \mathbb{N})$ such that $\Phi_n^{(k)} = \varphi_n$ and $\Phi_n \to$ l.u. to a certain function on I'. Fix an interval $I \sqsubset U$ and a compact set $K \subset I$. Since the map σ is continuous, the image $\sigma(K)$ of the compact set K is a compact set contained in $V \subset \mathbb{R}^m$. Hence $\Phi_n \rightrightarrows$ on $\sigma(K)$ and thus $\Phi_n \circ \sigma \rightrightarrows$ on K, which implies that $\Phi_n \circ \sigma \to$ l.u. on U.

We will prove, by induction, that the sequence $(\Phi_n^{(k)} \circ \sigma)$ is fundamental on U for each $k \in \mathbb{N}_0^m$. For $k = 0$ our assertion is obvious, because $\Phi_n \circ \sigma \to$ l.u. on U and every locally uniformly convergent sequence is fundamental. Suppose that the sequence $(\Phi_n^{(k)} \circ \sigma)$ is fundamental on U for $k \in \mathbb{N}_0^m$. We will show that the sequence $(\Phi_n^{(k+e_j)} \circ \sigma)$ is fundamental on U for $j \in \{1,\dots,m\}$. By Lemma 2, we have

$$(\Phi_n^{(k+e_j)} \circ \sigma) = (\Phi_n^{(k)})^{(e_j)} \circ \sigma = \frac{1}{J} \sum_{j_1<\dots<j_m} J_{j_1,\dots,j_m}\sigma \cdot J_{\Phi_n^{(k)},j;j_1,\dots,j_m}\sigma \tag{6}$$

for $j \in \{1,\dots,m\}$. Since $\sigma_j \in C^\infty(U)$ for $j = 1,\dots,m$ and $J > 0$, we see that $J \in C^\infty(U)$, $J_{j_1,\dots,j_m}\sigma \in C^\infty(U)$ for $j_1 < \dots < j_m$ and $\frac{1}{J} \in C^\infty(U)$. By Lemma 3 and the induction hypothesis, the sequence $(J_{\Phi_n^{(k)},j;j_1,\dots,j_m}\sigma)$ is fundamental on U. By (6), since any finite iteration of regular operations is regular, we conclude that a sequence $(\Phi_n^{(k+e_j)} \circ \sigma)$ is fundamental on U.

Consequently, the sequence $(\Phi_n^{(k)} \circ \sigma)$ is fundamental on U for each $k \in \mathbb{N}_0^m$ and this implies the assertion.

Corollary 1 *Let σ be a given function satisfying the assumption of Definition 2. The operation S of the multidimensional substitution of the function σ into a distribution on an open set $V \subset \mathbb{R}^m$ exists for every $f = [(\varphi_n)]$ on V in the sense of the formula: $S(f) = f \circ \sigma := [(\varphi_n \circ \sigma)]$.*

3. Irregular operations

We discuss now irregular operations based on the notion of a delta-sequence (see [5] and [2]). We consider the class Δ (see [3]) of delta-sequences (δ_n), whose elements are smooth functions defined in a neighborhood of 0, which satisfy the following conditions: 0° $\delta_n \geq 0$ for $n \in \mathbb{N}$; 1° there is a sequence of positive numbers (α_n), converging to 0, such that $\delta_n(x) = 0$ for $|x| \geqslant \alpha_n$, $n \in \mathbb{N}$; 2° $\int \delta_n = 1$ for $n \in \mathbb{N}$; 3° for every multi-index $k \in \mathbb{N}_0^d$ there is a positive constant M_k such that $\alpha_n^k \int |\delta_n^{(k)}| \leq M_k$ for $n \in \mathbb{N}$.

By a regular sequence for a given distribution f, we mean every sequence (f_n) of the form $f_n := f * \delta_n$ for $(\delta_n) \in \Delta$ and $n \in \mathbb{N}$. Clearly, every regular sequence is a fundamental sequence.

Let $R : C^\infty(U)^j \to C^\infty(U)$, $j \in \mathbb{N}$, be an arbitrary operation as in Definition 1, but not necessarily regular. We say that the operation R is feasible for given distributions $f_1, \ldots, f_k$ in the sense of irregular operation if for arbitrary regular sequences $(f_n^1), \ldots, (f_n^k)$ corresponding to $f_1, \ldots, f_k$ the sequence $(R(f_n^1, \ldots, f_n^k))$ is fundamental and we put then

$$R(f_1, \ldots, f_k) := [(R(f_n^1, \ldots, f_n^k))] = \lim_{n\to\infty} R(f_n^1, \ldots, f_n^k),$$

where the limit is meant in the distributional sense.

The examples of irregular operations are: the one argument operations of the Łojasiewicz value of a distribution at a point, the integral of a distribution, the Fourier transform of a distribution, the substitution of distributions into a fixed distribution (in particular, a continuous function), and the two argument operations of product, convolution and composition of distributions (see [2]).

In particular, the last operation, the composition $g(f)$ of distributions f and g, is meant in the sense of the distributional limit:

$$g(f) := \lim_{n\to\infty} g_n \circ f_n,$$

where the symbol $\circ$ denotes the composition of functions, $f_n := f * \delta_{1,n}$, $g_n := g * \delta_{2,n}$ and the limit is assumed to exist for all $(\delta_{1,n}), (\delta_{2,n}) \in \Delta$. This operation is studied in [1] and [3] and interesting results are obtained for f and g from certain classes of functions and distributions on the real line $\mathbb{R}$.

Let us remark that some of these results can be generalized, due to certain modifications in the proofs, to the case of an arbitrary open subset of $\mathbb{R}$. Namely, Theorems 2, 3 and 7 from [3] can be proved in the following more general form:

Theorem 2 *Assume that g is a continuous function on V such that, for some $p \in [1, +\infty)$ and $A > 0$,*

$$|g(y)| \leq A(1 + |y|)^p, \qquad y \in V$$

and $f \in L^p_{\text{loc}}(U)$ for the same p as above. Then the composition $g(f)$ exists and is equal to the function $g \circ f \in L^1_{\text{loc}}(U)$ in the distributional sense.

Theorem 3 *Let g be a continuous and bounded function on V and f be a distribution such that the Łojasiewicz value $f(x)$ at x exists for almost all $x \in U$ and the regular part f_r of f is a measurable function. Then the composition $g(f)$ exists in the distributional sense, the composition $g \circ f_r$ exists a.e., $g \circ f_r \in L^1_{\text{loc}}(U)$ and $g(f) = g \circ f_r$.*

Theorem 4 *Suppose that g is a continuous function on (a,b), where $-\infty \leq a < b \leq \infty$, $g(x) = o(|x|)$ as $x \to a+$ and $x \to b-$, f is a measure on U, $f(U) \subset (a,b)$ and $f = f_r + f_s$ is the Łojasiewicz decomposition of f. Then the compositions $g(f)$ and $g(f_r)$ exist, $g(f_r)$ is a locally integrable function and $g(f) = g(f_r)$ on U.*

The main Theorem 8 of the paper [3] is also true in a more general form, with $\mathbb{R}$ replaced by an arbitrary open (bounded or not) interval in $\mathbb{R}$. Namely,

Theorem 5 *Assume that g is a continuous function on (a,b), where $-\infty \le a < b \le \infty$ such that*

$$\lim_{x\to a^+} \frac{g(x)}{x} = \alpha, \qquad \lim_{x\to b^-} \frac{g(x)}{x} = \beta.$$

Moreover, assume that f is a measure on U, $f(U) \subset (a,b)$ and $f = f_r + f_s$ is the Łojasiewicz decomposition of f. Then the composition $g(f)$ exists and

$$g(f) = g \circ f_r - \alpha f_s^- + \beta f_s^+ \qquad \text{on } U.$$

By the Łojasiewicz decomposition of a distribution (a measure) f, in the formulation of Theorems 3, 4 and 5 above, we understand the equality $f = f_r + f_s$, where f_r is the regular part of the distribution f (see [1]), which is a locally integrable function and $f_s := f - f_r$. By f^+ and f^- for a measure f, we mean $f^+ := \frac{|f|+f}{2}$ and $f^- := \frac{|f|-f}{2}$ (see [1]).

Applying Theorem 5, the following formulas can be derived: $1°\ \sqrt{\delta} = 0$; $2°\ \sin\delta = 0$; $3°\ \sqrt{1+\delta^2} = 1+\delta$; $4°\ \log(1+\delta) = 0$. To obtain them it suffices to put $(a,b) := (-\infty,\infty)$ in $1° - 3°$, $(a,b) := (-\frac{1}{2},\infty)$ in $4°$, $f := \delta$ in all cases $1° - 4°$, and

$$g_1(x) = \begin{cases} \sqrt{x} & \text{for } x \geqslant 0, \\ 0 & \text{for } x < 0; \end{cases}$$

$g_2(x) := \sin x$; $g_3(x) := \sqrt{1+x^2}$; $g_4(x) := \log(1+x)$ for $x \in (a,b)$ in cases $1° - 4°$, respectively. The mentioned formulas follow from Theorem 5, because its assumptions are obviously fulfilled by f and g_i $(i = 1,2,3,4)$ defined above.

Acknowledgment

I would to thank the referee for his useful remarks which allowed me to remove mistakes and to improve the text of my paper.

References

[1] Antosik P 1988-89 Composition of distributions *Tech. Report Series of the Department of Mathematical Sciences, Univ. of Wisconsin-Milwaukee* **53201**

[2] Antosik P Mikusiński J and Sikorski R 1973 *Theory of Distributions. The Sequential Approach* (Amsterdam-Warszawa: Elsevier-PWN)

[3] Kamiński A and Sorek S 2005 On the operation of composition on distributions *this volume*

[4] Mikusiński J 1948 Sur la méthode de généralisation de M. Laurent Schwartz et sur la convergence faible *Fund. Math.* **35** 235-39

[5] Mikusiński J 1961 Irregular operations on distributions *Studia Math.* **20** 163-9

[6] Schwartz L 1950-1 *Théorie des distributions* vol I-II, (Hermann Paris)

The duality of Weyl and linear extension of Kostka matrices

P Jakubczyk[1], T Lulek[2], D Jakubczyk[2], B Lulek[2]

[1]Institute of Physics, University of Rzeszòw, Aleja Rejtana 16c, 35-959 Rzeszòw, Poland
[2]Chair of Physics, Rzeszòw University of Technology, Powstancòw Warszawy 6, 35-959 Rzeszòw, Poland

E-mail: pjakub@univ.rzeszow.pl

Abstract. Application of the Robinson-Schensted algorithm to the basis of magnetic configurations of the one-dimensional Heisenberg magnet with an arbitrary spin gives an efficient way for a classification of the irreducible basis of the Weyl duality. The plactic monoid is shown to be an adequate tool for describing this irreducible basis in a way consistent with the Schensted insertion procedure, i.e. the creation of a new single-particle state (a letter of the single-node spin) in already constructed Young and Weyl tableaux. Schensted insertion is interpreted in terms of Gelfand triangles - combinatoric analogues of Weyl tableaux with exposed occupation numbers, consistent with canonical chains of subgroups of both the symmetric and the unitary group. A transition matrix between these two bases should exist due to the linear structure of the Hilbert space. This matrix can be looked at as the linear extension of the famous Kostka matrix. We show how to obtain this matrix and give an interpretation of its elements as coefficients of certain wave packet with exactly defined symmetry.

1. Introduction

The one-dimensional Heisenberg magnet is a paradigmatic example of use of the Weyl duality [1] between the actions of the symmetric group Σ_N - the group of all permutations of nodes of a magnet, and the unitary group $U(n)$ of the highest quantum symmetry of each constituent part of the magnet - the node with the spin s. The irreducible basis of the Weyl duality is given by the combinatoric algorithm of Robinson and Schensted (RS) [2, 3] in terms of a pair (P, Q) of Weyl and Young (semi)standard tableaux. In this paper we aim to show an explicit construction of this irreducible basis. It can be interpreted as a generalization of famous Kostka matrices $\{K_{\lambda,\mu}\}$ at the level of bases, spanned on an orbit of action of the symmetric group with the weight μ, corresponding to an Σ-irrep of the shape λ.

Within the paper, we use interchangeably semistandard tableaux and Gelfand patterns. The former are perhaps better adapted to RS algorithm of insertion of a letter into a tableau, whereas the latter are more convenient in physical interpretation as collections of appropriate occupation numbers.

2. Heisenberg magnetic ring

We consider a one dimensional Heisenberg magnet consisting of N nodes with single-node spin s. A state of the magnet is given by ascribing a projection of the single node spin to each node of a magnet. Mathematically, this is formulated in terms of two sets:

$$\tilde{N} = \{j = 1, 2, ..., N\} \quad \text{and} \quad \tilde{n} = \{i = 1, 2, ..., n\}, \;\; n = 2s + 1, \tag{1}$$

the set of N nodes of the crystal - *the alphabet of nodes*, and the set of the single node states - *the alphabet of spins*, respectively. The state of the whole magnet is a mapping $f : \tilde{N} \to \tilde{n}$ and can be presented as

$$|f\rangle = |f(1)f(2)\cdots f(N)\rangle \equiv |i_1, i_2, ..., i_N>, \;\; i_j \in \tilde{n}, j \in \tilde{N}. \tag{2}$$

We call this state *a magnetic configuration* and treat "it" as a word of the length N in the alphabet of spins.

The set $\tilde{n}^{\tilde{N}} = \{f : \tilde{N} \longrightarrow \tilde{n}\}$ of all magnetic configurations spans the Hilbert space

$$\mathcal{H} = lc_{\mathbb{C}}\, \tilde{n}^{\tilde{N}}, \tag{3}$$

with the scalar product $\langle f|f'\rangle = \delta_{ff'}, \;\; f, f' \in \tilde{n}^{\tilde{N}}$.

The dynamics of the Heisenberg magnet is defined by the Hamiltonian - a hermitian operator $\hat{H} \in End\, \mathcal{H}$. In general a setting, this Hamiltonian is constructed from generators $e_{ii'}$, $i, i' \in \tilde{n}$ of the Lie algebra of the unitary group $U(n)$. For example, for the case $n = 3$ (or $s = 1$) for the isotropic XXX model we have

$$H = (J/4) \sum_{j \in \tilde{N}} \left(\hat{\mathbf{s}}_j \cdot \hat{\mathbf{s}}_{j+1} - (\hat{\mathbf{s}}_j \cdot \hat{\mathbf{s}}_{j+1})^2 \right), \tag{4}$$

where $\hat{\mathbf{s}}_j = (\hat{s}^x_j, \hat{s}^y_j, \hat{s}^z_j)$ is the vector operator for the spin $s = 1$.

3. The Duality of Weyl

We recall that the space $\mathcal{H}$ is a scene of two dual actions, $A : \Sigma_N \times \mathcal{H} \to \mathcal{H}$ and $B : U(n) \times \mathcal{H} \to \mathcal{H}$. The most important quantum -mechanical observation is that the two actions mutually commute, that is,

$$[A(\sigma), B(u)] = 0, \;\; \sigma \in \Sigma_N, \;\; u \in U(n), \tag{5}$$

despite the fact that both dual groups are, for $N > 2, n > 1$, highly noncommutative. It implies compatibility of appropriate quantities related to both groups in the spirit of the Heisenberg uncertainty principle: these quantities "can be measured simultaneously". A maximal system of such compatible (commuting) observables is realized in an irreducible basis in the space $\mathcal{H}$, adapted to the symmetry of both dual groups. We proceed to describe this scheme in more detail.

We introduce the set of partitions of the integer N with no more than n parts

$$\mathcal{D}_W(N, n) = \{\lambda \vdash N, \;\; |\lambda| \leq n\} \tag{6}$$

where $|\lambda|$ denotes the number of non zero parts of the partition λ of N. Each element λ of this set of partitions serves as the label of the irreducible representation (irrep) Δ^λ of the symmetric group Σ_N and, at the same time, as the label of the irrep D^λ of the unitary group $U(n)$.

The corresponding decompositions of both dual actions into irreps read

$$A = \sum_{\lambda \in \mathcal{D}_W(N,n)} m(A, \Delta^\lambda)\; \Delta^\lambda \quad \text{and} \quad B = \sum_{\lambda \in \mathcal{D}_W(N,n)} m(B, D^\lambda)\; D^\lambda, \tag{7}$$

where appropriate multiplicities, on the strength of duality of Weyl [1], satisfy relations

$$m(A, \Delta^\lambda) = dim\, D^\lambda, \quad m(B, D^\lambda) = dim\, \Delta^\lambda. \tag{8}$$

In this way, the duality of Weyl decomposes the whole space $\mathcal{H}$ of quantum states of the composite system into *sectors* $\mathcal{H}^\lambda$,

$$\mathcal{H} = \sum_{\lambda \in \mathcal{D}_W(N,n)} \oplus \mathcal{H}^\lambda \quad \text{such that} \quad A|_{\mathcal{H}^\lambda} = (dim\, D^\lambda)\, \Delta^\lambda, \quad \text{and} \quad B|_{\mathcal{H}^\lambda} = (dim\, \Delta^\lambda)\, D^\lambda. \tag{9}$$

Thus each sector $\mathcal{H}^\lambda$ carries $m(A, \lambda) = dim\, D^\lambda$ copies of Δ^λ-irreducible subspaces, and, at the same time, $m(B, \lambda) = dim\, \Delta^\lambda$ copies of D^λ-irreducible ones.

The duality of Weyl admits therefore the irreducible basis of the form

$$b_{irr} = \{|\lambda\, t\, y\rangle | \lambda \in \mathcal{D}_W(N, n), \;\; t \in \tilde{D}^\lambda, \;\; y \in \tilde{\Delta}^\lambda\}, \tag{10}$$

where $\tilde{D}^\lambda$ and $\tilde{\Delta}^\lambda$ are some standard bases for the irrep D^λ and Δ^λ, respectively. It is convenient to take standard bases in the form $\tilde{\Delta}^\lambda = SYT(\lambda)$, $\tilde{D}^\lambda = WT(\lambda, \tilde{n})$, where $SYT(\lambda)$ denotes the set of all standard Young tableaux of the shape λ in the alphabet $\tilde{N}$ of nodes, and $WT(\lambda, \tilde{n})$ is the set of all Weyl tableaux of the shape λ in the alphabet $\tilde{n}$ of spins. In this way, the duality of Weyl provides a complete classification of basis states in the space $\mathcal{H}$.

4. The Robinson-Schensted combinatoric bijection

Now we describe a way of labelling scheme of Weyl duality, namely the RS algorithm. Let us take the set of words

$$\tilde{n}^* = \bigcup_{N=0}^{\infty} \tilde{n}^{\tilde{N}}, \tag{11}$$

i.e. the free monoid [4] with juxtaposition of words as monomial multiplication and the empty word $\emptyset$ as the unit. Physically, the free monoid is the disjoint union of sets $\tilde{n}^{\tilde{N}}$ of all magnetic configurations for the rings $\tilde{N}$, $N = 0, 1, \ldots$, with $N = 0$ corresponding to the empty set. Let

$$b^* = \bigcup_{N=0}^{\infty} b_{irr}(N) \tag{12}$$

be the disjoint union of irreducible bases of the duality of Weyl for the rings $\tilde{N}$, $N = 0, 1, \ldots$, with $b_{irr}(0) = \emptyset$. The Robinson-Schensted algorithm [2, 3] defines a bijection $RS : \tilde{n}^* \to b^*$ between magnetic configurations and appropriate irreducible bases of the duality of Weyl, by putting

$$RS(f) = (P(f), Q(f)), \;\; f \in \tilde{n}^*, \tag{13}$$

where $P(f)$ and $Q(f)$ denotes the Weyl and Young tableau, respectively.

The construction of these tableaux is given by the following rules [5, 11]:

- The magnetic configuration $f \in \tilde{n}^{\tilde{N}}$ can be presented in two-line notation: $f = \binom{1\,2\ldots N}{i_1 i_2 \ldots i_N}$ where the upper row contains the labels of nodes and the lower one the corresponding single node spin states.
- Next we construct a sequence of pairs of tableaux: $(\emptyset, \emptyset) = (P_0, Q_0)$, (P_1, Q_1), (P_2, Q_2), $\ldots$, $(P_N, Q_N) = (P, Q)$ where one has to insert letters $i_1, i_2, \ldots, i_N$ into the P's and the numbers $1, 2, \ldots, N$ into the Q's so that $sh\, P_k = sh\, Q_k$ for all k (sh denotes the shape of a tableau).

In order to insert the letter x into the intermediate tableau P, we proceed as follows:

```
set r:=the first row of P
while x is less than some element of row r do
  let y be the smallest element of r greater than x and replace y by x in row r
  set x:=y and r:=the next row down
now x is greater than every element of r, so place x at the end of row r and
stop
```

Performing the RS algorithm on all magnetic configurations we obtain the complete set of labels of the basis of duality of Weyl.

5. The decomposition of Kostka at the level of bases

Consider the action A of the symmetric group Σ_N as a purely permutational representation $A : \Sigma_N \times \tilde{n}^{\tilde{N}} \to \tilde{n}^{\tilde{N}}$. This action decomposes the set $\tilde{n}^{\tilde{N}}$ of all magnetic configurations of the ring into orbits

$$\mathcal{O}_\mu = \{f \circ \sigma^{-1} | \sigma \in \Sigma_N\} \tag{14}$$

labelled by *weights*. The weight is a composition $\mu = (\mu_1, \mu_2, \ldots, \mu_n)$, $\sum_{i \in \tilde{n}} \mu_i = N$, where $\mu_i = |\{i_j = i \,|\, j \in \tilde{N}\}|$, $i \in \tilde{n}$ is the occupation number for the single-node state $i \in \tilde{n}$, for any $f \in \mathcal{O}_\mu$. The restriction

$$A|_{\mathcal{O}_\mu} \equiv R^{\Sigma_N : \Sigma^\mu} \tag{15}$$

of the action A to the orbit $\mathcal{O}_\mu$ is a *transitive representation* of Σ_N, with the stabiliser being the Young subgroup $\Sigma^\mu = \Sigma_{\mu_1} \times \Sigma_{\mu_2} \times \cdots \times \Sigma_{\mu_n}$.

Let us consider next a (non-empty) intersection $lc_{\mathbb{C}}\mathcal{O}_\mu \cap \mathcal{H}^\lambda$, $\lambda \trianglerighteq \mu$, of the carrier space of the action R^μ with the sector $\mathcal{H}^\lambda$ of the Weyl duality. $\lambda \trianglerighteq \mu$, means that λ is greater than or equal to μ in the dominance order. This intersection carries some copies of the irrep Δ^λ of the symmetric group Σ_N, according to *the Kostka decomposition*

$$R^{\Sigma_N : \Sigma^\mu} = \sum_{\lambda \trianglerighteq \mu} K_{\lambda\mu} \Delta^\lambda, \tag{16}$$

with the multiplicities $K_{\lambda\mu}$ being Kostka numbers. The Kostka decomposition can be specified at the level of bases as

$$|\mu\, \lambda\, t\, y\rangle = \sum_{f \in \mathcal{O}_\mu} \begin{bmatrix} \mu & \lambda & t \\ f & y & \end{bmatrix} |f\rangle, \tag{17}$$

where the symbol in the rectangular brackets forms a unitary matrix in the space $lc_{\mathbb{C}}\mathcal{O}_\mu$, which transforms the initial basis $\mathcal{O}_\mu$ of magnetic configurations to the irreducible basis of the Weyl duality, t denotes the Weyl tableau, and y the Young tableau.

In the sequel we identify the Weyl and Young tableaux with combinatorially equivalent Gelfand patterns along [6, 7]. The elements of this matrix are calculated on the basis of a ladder construction following the combinatorial growth of a Weyl tableau t by attachement of consecutive nodes $j = 1, 2, \ldots N$ to already constructed state of $j-1$ nodes [10, 12]. Thus we have

$$\begin{bmatrix} \mu & \lambda & t \\ f & y & \end{bmatrix} = \sum_{\{t_{12}\, t_{123} \cdots t_{1..N-1}\, t\}} \begin{bmatrix} \{1\} & \{1\} & \lambda_{12} \\ f(1) & f(2) & t_{12} \end{bmatrix} \begin{bmatrix} \lambda_{12} & \{1\} & \lambda_{123} \\ t_{12} & f(3) & t_{123} \end{bmatrix} \cdots \begin{bmatrix} \lambda_{1...N-1} & \{1\} & \lambda \\ t_{1...N-1} & f(N) & t \end{bmatrix}, \tag{18}$$

where the sum is over all possible chains of Gelfand patterns $\{t_{12} \;\; t_{123} \;\; \cdots \;\; t_{1..N-1} \;\; t\}$, adjusted to the configuration f and successive steps of RS algorithm along the configuration $f' = RS^{-1}(\lambda\, t\, y)$.

In more detail, a chain $\{t_{12}\ t_{123}\ \cdots\ t_{1..N-1}\ t\}$ is constructed recursively, namely: $t_{1..j}$ is formed from $t_{1..j-1}$ by adding (in all possible ways) a letter $f(j)$ of magnetic configuration to $t_{1..j-1}$ such that: the n-th row of the Gelfand pattern $t_{1..j}$ is equal to $\lambda_{1..j}$ and the standardness of Gelfand pattern has to be conserved.

Note that adding a letter a into a Weyl tableau implies that the appropriate entries in all rows j, $(a \leq j \leq n)$ of the corresponding Gelfand pattern increase by one. Procedure of insertion of a letter a yields that the row $\lambda_{1i}\lambda_{2i}\ldots\lambda_{ii}$ of Gelfand pattern is given by:

$$\lambda_{1i}\lambda_{2i}\ldots\lambda_{ii} \xrightarrow{a} \lambda_{1i}\lambda_{2i}\ldots\lambda_{li}+1\ldots\lambda_{ii}, \quad 1 \leq l \leq i,$$

what means that this row can be changed in various ways, depending on the choice of l. The symbol

$$\begin{bmatrix} \lambda_{1\ldots j-1} & \{1\} & \lambda_{1\ldots j} \\ t_{1\ldots j-1} & f(j) & t_{1\ldots j} \end{bmatrix} \tag{19}$$

in (18) denotes the coupling coefficient (Wigner-Clebsch-Gordan coefficient) at the j-th stage of growth, corresponding to the Littlewood-Richardson decomposition

$$D^{\lambda_{1\ldots j-1}} \otimes D^{\{1\}} = \sum_{\lambda_{1\ldots j}} \oplus\ D^{\lambda_{1\ldots j}} \tag{20}$$

for the unitary group $U(n)$. The bottom row of the coupling coefficient (19) exhibits the standard bases corresponding to irreps presented in the first row. Intermediate irrep $\lambda_{1\ldots j}$ is defined in terms of quantum numbers $(\lambda\, t\, y)$ by means of the Robinson-Schensted correspondence, whereas the corresponding bases $t_{1\ldots j}$, $j \in \tilde{N}$ $(\lambda_{1\ldots N} = \lambda,\ t_{1\ldots N} = t)$ are defined by the insertion procedure, adjusted to the shapes from RS correspondence.

The Wigner-Clebsch-Gordan coefficient (19) can be calculated using a technique called *Pattern Calculus* [7, 8, 9] by interpreting it in terms of fundamental tensor operators. A fundamental tensor operator $t_{i_1 i_2}$ (double Gelfand pattern) is defines by the second column $\left(\begin{smallmatrix}\{1\}\\ f(j)\end{smallmatrix}\right)$ of (19), by assigning: $i_1 = f(j)$, $i_2 =$ the row of $\lambda_{1..j}$ in which the new box is added at the j-th step. Then we have

$$\begin{bmatrix} \lambda_{1\ldots j-1} & \{1\} & \lambda_{1\ldots j} \\ t_{1\ldots j-1} & f(j) & t_{1\ldots j} \end{bmatrix} = \langle t_{1..j} | t_{i_1 i_2} | t_{1..j-1} \rangle, \tag{21}$$

this allow us to calculate this coefficient. Examples of matrix (17) are given in Table 1.

6. Conclusions

We have shown that the Robinson-Schensted algorithm plays two roles in the context of the Heisenberg magnet. Firstly, it serves for labelling the irreducible basis of the Weyl duality, so that $|\lambda\, t\, y\rangle = |RS(f)\rangle$. Secondly, it provides a complete information for evaluation of elements $\langle f | RS(f) \rangle$ of the Kostka matrix at the level of bases. This allow us to construct a wave packet with exactly defined symmetry.

Table 1. Kostka matrix at a level of bases for $N = 3$, $n = 2$ ($S = 1/2$), and $N = 3$, $n = 3$ ($S = 1$).

	λ	$\{3\}$	$\{2,1\}$	
	t	aab	aa/b	aa/b
	y	123	$13/2$	$12/3$
f				
aab		$\frac{1}{\sqrt{3}}$	0	$\frac{2}{\sqrt{6}}$
aba		$\frac{1}{\sqrt{3}}$	$\frac{1}{\sqrt{2}}$	$-\frac{1}{\sqrt{6}}$
baa		$\frac{1}{\sqrt{3}}$	$-\frac{1}{\sqrt{2}}$	$-\frac{1}{\sqrt{6}}$

	λ	$\{3\}$	$\{2,1\}$				$\{1^3\}$
	t	abc	ab/c	ab/c	ac/b	ac/b	$a/b/c$
	y	123	$12/3$	$13/2$	$12/3$	$13/2$	$1/2/3$
f							
abc		$\frac{1}{\sqrt{6}}$	$\frac{2}{\sqrt{12}}$	0	0	$\frac{2}{\sqrt{12}}$	$\frac{1}{\sqrt{6}}$
bac		$\frac{1}{\sqrt{6}}$	$\frac{2}{\sqrt{12}}$	0	0	$-\frac{2}{\sqrt{12}}$	$-\frac{1}{\sqrt{6}}$
cba		$\frac{1}{\sqrt{6}}$	$-\frac{1}{\sqrt{12}}$	$-\frac{1}{2}$	$-\frac{1}{2}$	$\frac{1}{\sqrt{12}}$	$-\frac{1}{\sqrt{6}}$
acb		$\frac{1}{\sqrt{6}}$	$-\frac{1}{\sqrt{12}}$	$\frac{1}{2}$	$\frac{1}{2}$	$\frac{1}{\sqrt{12}}$	$-\frac{1}{\sqrt{6}}$
bca		$\frac{1}{\sqrt{6}}$	$-\frac{1}{\sqrt{12}}$	$\frac{1}{2}$	$-\frac{1}{2}$	$-\frac{1}{\sqrt{12}}$	$\frac{1}{\sqrt{6}}$
cab		$\frac{1}{\sqrt{6}}$	$-\frac{1}{\sqrt{12}}$	$-\frac{1}{2}$	$\frac{1}{2}$	$-\frac{1}{\sqrt{12}}$	$\frac{1}{\sqrt{6}}$

References

[1] Weyl H 1931 *Gruppentheorie und Quantummechanik* (Leipzig: Hirzel) (English translation: 1950 *The Theory of Groups and Quantum Mechanics* (New York: Dover))

[2] Robinson G 1938 *Amer. J. Math.* **60** 745

[3] Schested G 1961 *Canad. J. Math.* **13** 179

[4] Lascoux A Lecrerc B and Thibon J -Y 2001 *The Plactic Monoid* (Algebraic Combinatorcs on Words) ed M Lothaire (Cambridge: Univ. Press)

[5] Sagan B 2001 *The symmetric group: Representations, Combinatorial Algorithms, and Symmetric Functions* (New York: Springer-Verlag) 2nd edition

[6] Louck J D and Biedenharn L C 1991 (Symmetry and Structural Properties of Condensed Matter) ed W Florek, T Lulek and M Mucha (Singapore: World Sci.) p 207

[7] Louck J D 1997 (Symmetry and Structural Properties of Condensed Matter) ed T.Lulek W Florek and B Lulek (Singapore: World Sci.) p 231

[8] Biedenharn L C, Chen W Y C, Lohe M A and Louck J D 1995 (Symmetry and Structural Properties of Condensed Matter) ed T Lulek, W Florek and S Walcerz (Singapore: World Sci.) p 150

[9] Louck J D, Chen W Y and Galbraith W H 1999 (Symmetry and Structural Properties of Condensed Matter) ed T Lulek, B Lulek and A Wal (Singapore: World Sci.) p 112

[10] Lulek B, Jakubczyk D, Jakubczyk P 2004 *Mol. Phys.* **102** 1269

[11] Lulek T, Jakubczyk P and Jakubczyk D 2004 *Mol. Phys.* **102** 1279

[12] Jakubczyk P and Lulek B 2005 *Phys. Stat. Sol.* (b) **242** 327

Quasimomenta of string configurations

T Lulek[1], B Lulek[2], P Jakubczyk[3] and D Jakubczyk[4]

[1]Chair of Physics, Rzeszòw University of Technology, Powstancòw Warszawy 6, 35-959 Rzeszòw, Poland

[3]Institute of Physics, University of Rzeszòw, Aleja Rejtana 16c, 35-959 Rzeszòw, Poland

E-mail: tadlulek@prz.edu.pl

Abstract. We sketch a reciprocal space analogue of the combinatorial bijection of Robinson-Schensted and Kerov-Kirillov-Reshetikhin (RSKKR) between magnetic configurations (the initial basis for quantum calculations of the eigenproblem of the Heisenberg Hamiltonian for a one-dim finite Heisenberg chain), and rigged string configurations (the classification labels for the exact results of Bethe Ansatz). Existence of such a bijection admits an interpretation of the exact quantum numbers of riggings as quasimomenta of l-strings. The extended size of an l-string results in selection rules for these quasimomenta, and thus in a division of the Brillouin zone into compact subzones of forbidden and allowed states of the system of coupled Bethe pseudoparticles. The forbidden Brillouin subzone for a particular l-string is evidently the effect of kinematical restrictions for motions of constituent Bethe pseudoparticles. These restrictions can be easily predicted in a combinatorially unique way due to completness of the RSKKR bijection.

1. Introduction

The famous Bethe Ansatz (BA, [1]) provides exact solutions to the eigenproblem of the Hamiltonian for the one-dimensional Heisenberg magnetic ring with N nodes, each occupied by a spin 1/2 object. A complete set of exact eigenfunctions is classified by rigged string configurations [2]-[9] - combinatoric objects introduced by Kerov, Kirillov and Reshetikhin (KKR, [3]). Each rigged string configuration represents a correlated wave function - a superposition of magnetic configurations, the latter being the initial basis for quantum calculations. We argue in a parallel paper [9] that the set of all magnetic configurations with r reversed spins (and thus with the total magnetization $M = N/2 - r$), denoted by

$$Q^{(r)} = \{\mathbf{j} = (j_1, j_2, \ldots, j_r) | 1 \leq j_1 < j_2 < \ldots < j_r \leq N\}, \tag{1}$$

acquires the natural interpretation of a classical configuration space for the system of r Bethe pseudoparticles, each moving on the chain $\tilde{N} = \{j = 1, 2, \ldots, N\}$. Thus a magnetic configuration presented by $\mathbf{j} \in Q^{(r)}$ is a quantum state of the system $\tilde{r} = \{\alpha = 1, 2, \ldots, r\}$ of Bethe pseudoparticles, in which the pseudoparticle $\alpha \in \tilde{r}$ is localized at the node $j_\alpha \in \tilde{N}$. The dynamics of the system $\tilde{r}$, imposed by the standard Heisenberg Hamiltonian, is such that the pseudoparticle $\alpha \in \tilde{r}$ can move (or "jump") to a nearest neighbor $(j_\alpha \pm 1)$ mod N in the ring $\tilde{N}$ provided that the "hard-core" condition $(j_\alpha \pm 1)$ mod $N \neq j_{(\alpha \pm 1) \text{mod } r}$ is satisfied. Moreover, a single quantum transition involves exactly one such a jump, with all other pseudoparticles at fixed positions. The hard-core condition distinguishes some elements $\mathbf{j} \in Q^{(r)}$ in which all r pseudoparticles are free to make any jump, from those in which some of them are "frozen" by the proximity of others. In this way, the classical configuration space $Q^{(r)}$ acquires the geometry of locally r-dimensional hypercubic lattice, with F-dimensional boundaries, $1 \leq F < r$, corresponding to F islands of consecutive nodes occupied by Bethe pseudoparticles (the case $F = r$ is generic).

We also consider there the RSKKR bijection (the composition of KKR bijection [3] and the well known combinatoric correspondence of Robinson [10] and Schensted [11]) which maps the classical configuration space $Q^{(r)}$ onto the set $RC(N, r)$ of all rigged string configurations for fixed N and r. We demonstrate there that the geometry of $Q^{(r)}$ is naturally reflected - by the RSKKR bijection - in the set $RC(N, r)$ of exact BA solutions.

In the present paper we aim to propose a combinatoric assignment of definite quasimomenta to each string entering a rigged string configuration $(\nu\mathcal{L}) \in RC(N, r)$. The procedure bases on an observation that a simple combinatoric description of rigged strings in terms of some paths in the (l, j) plane should be representation independent, and allows to replace the set $\tilde{N}$ of nodes of the ring (interpreted as the "real lattice" for a single Bethe pseudoparticle) by the Brillouin zone B, i.e. by the set of all quasimomenta, which are irreps of the translation group C_N (seen as the "reciprocal lattice"). In short, we replace the positional representation of the system $\tilde{r}$ of Bethe pseudoparticles by the momentum one.

2. Strings within the momentum representation

Each magnetic configuration $f \in Q^{(r)}$ is a mapping $F : \tilde{N} \to \{+, -\}$, i.e. a word of the length N in the alphabet $\{+, -\}$ of single-node spins, with the weight $wt(f) \equiv (w_1, w_2) = (N - r, r)$, where w_1 and w_2 is the number of occurrences of the letter "+" and "-", respectively, in the word f. We recall after [9, 12] that the RSKKR bijection $\rho : Q^{(r)} \to RC(N, r)$ is specified, for each $f \in Q^{(r)}$, by a *path* in the (l, j)-plane.

A path [4], [5] of the length N is defined recursively with respect to $j \in \tilde{N}$ as a continuous line, consisting of consecutive segments such that (i) the segment at the step $j - 1$ joins the point $(j - 1, l)$ with one of the two points: either $(j, l + 1)$ for the step up, or $(j, l - 1)$ for the step down, (ii) the path starts at $(0, 0)$. Thus a path p can be presented as

$$p = (p_1, p_2, \ldots, p_N), \quad p_j \in \{\text{up}, \text{down}\}, \quad j \in \tilde{N}. \tag{2}$$

In order to define the path $p(f)$, we write down the magnetic configuration f, treated as the word of the length N in the alphabet $\tilde{2}$ of single - node spins, in the following form

$$f = u_1 v_1 u_2 v_2 \ldots u_a v_a u_{a+1}. \tag{3}$$

We thus distinguish here, within the word f, some subwords v_i of consecutive $2A_i = |v_i|$ letters, $i = 1, 2, \ldots, a$. A subword v_i, referred by us to as a *minimal compensated subword*, is defined by the following conditions: (i) $wt(v_i) = (A_i, A_i)$, i.e. all letters "-" are compensated by some next "+" within v_i, (ii) the length $2A_i$ is *minimal* with respect to compensation property, which means that for each proper subword $v_i^{(j)}$ of v_i, consisting of its *first* j letters, one has $w_1 > w_2$ for $wt(v_i^{(j)}) = (w_1, w_2), i = 1, 2, \ldots 2A_i - 1$ (so that the compensation is reached at the end of v_i only). We refer to all letters $f(j), i \in \tilde{N}$, which enter any minimal compensated subword v_i, as to nodes *bounded* into strings, and all the remaining letters of f as *unbounded* ones. The unbounded letters between v_{i-1} and v_i form the (possibly empty) subword $u_i, i = 1, 2, \ldots, a + 1$, with u_1 and u_{a+1} formed from the leftmost and rightmost part of f, respectively.

The path $p(f)$ is now specified by the formula

$$p_j(f) = \begin{cases} \text{up} & f(j) = " - ", \\ & \text{or if } f(j) = " + ", \text{ and } f(j) \in u(f), \\ \text{down} & \text{if } f(j) = " + ", \text{ and } f(j) \in v(f), \end{cases} \tag{4}$$

where

$$u(f) = u_1 u_2 \dots u_a u_{a+1} \tag{5}$$

and

$$v(f) = v_1 v_2 \dots v_a \tag{6}$$

are subwords of f. The decomposition (3) is unique, and the corresponding path $p(f)$ looks like a mountain range, with cuspidal local maxima and minima. Each internal maximum (for $1 < j < N$) corresponds to a string, each minimal compensated subword v_i represents a multipyramid, i.e. a segment of the mountain range with the same initial and final heights and all intermediate heights larger, and each subword u_i represents a segment of the "sea of holes" for all multipyramids, in a form of a straight line parallel to the diagonal $j = L$. The subword $v(f)$, with the weight $wt(v(f)) = (r', r')$, defines the string configuration $\nu \vdash r'$, or $\nu = (m_1, m_2, \dots)$, with m_l being the number of l-strings, and

$$\sum_{i=1}^{a} A_i = \sum_{l \geq 1} m_l = r', \tag{7}$$

whereas the word $u(f)$, taken together with all words $v_i, i = 1, 2, \dots, a$ defines riggings L_{lv}, $l = 1, 2, \dots, v \in \tilde{m}_l \equiv \{1, 2, \dots, m_l\}$ for all l-strings. The collection $\mathcal{L} = \{L_{lv}\}$ and string configuration $\nu \vdash r'$ determine the rigged string configuration $\nu\mathcal{L} = \rho(f)$.

Within this combinatoric picture of strings, based essentially on the positional representation, each l-string is an extended object on the path $p(f)$, which occupies the size $2l$, and can move by jumps over that part of the ring $\tilde{N}$ which is not occupied by this string, nor by other strings. Moreover, this string can also move by sliding over both sides of *larger* l'-strings ($l' > l$), until their relative heights on the path $p(f)$ coincide (cf. [9] for detail).

This picture should be representation independent, i.e. it should work in momentum representation (e.g. in the basis of wavelets [13]), with the "real space" $\tilde{N}$ replaced by the Brillouin zone

$$B = \{k = 0, \pm 1, \pm 2, \dots, \left\{ \begin{array}{ll} \pm(N/2 - 1), N/2 & \text{for } N \text{ even} \\ \pm(N-1)/2, & \text{for } N \text{ odd} \end{array} \right\}, \tag{8}$$

counting irreps

$$\Gamma_k(j) = exp\,(2\pi i k j / N), \quad j \in \tilde{N} \tag{9}$$

of the translation symmetry group C_N [13] by quasimomenta k.

In the momentum representation, the kinematical restrictions resulting from the size of an l-string define the forbidden region in the Brillouin zone, concentrated symmetrically over its center $k = 0$. All quasimomenta outside this region are admissible for l-strings.

3. Allowed and forbidden Brillouin subzones for l-strings

Let us consider first an isolated, single l-string, that is the string configuration $\nu = \{l\}$. Then the rigged string configuration is

$$\nu\mathcal{L} = \underbrace{\boxed{L}\boxed{}\boxed{}\boxed{}\boxed{}}_{l} \quad P_l = N - 2l, \tag{10}$$

with the rigging L varying in the range

$$0 \leq L \leq P_l, \tag{11}$$

such that P_l is the total number of l-holes, and L - the number of holes to the left of the l-string on the corresponding path $p(f)$. In this simple case, one has a single minimal compensated subword $v = - - \cdots - + + \cdots +, |v| = 2l$, and $f = u_1 v u_2$, such that $|u_1| = L$ and $|u| = P_l$ for $u = u_1 u_2$. The number P_l of holes is here equal to the total length of the word u, whereas the rigging is the length of the subword u_1, i.e. the number of the letters to the left of the minimal compensated subword v. The total number of riggings is

$$|z(\nu)| = P_l + 1 = N - 2l + 1. \qquad (12)$$

In the momentum representation, the Brillouin zone is divided as

$$B = B_f \cup B_a, \quad B_f \cap B_a = \emptyset, \qquad (13)$$

with

$$B_f = \{|k| \leq k_e\} \qquad (14)$$

being the region forbidden for the l-string, and

$$B_a = \{|k| > k_e\} \qquad (15)$$

being the set of quasimomenta which are admissible. The quasimomentum

$$k_e = l - 1 \qquad (16)$$

defines the edge of the forbidden Brillouin zone, and is the quantity which replaces the number P_l of l-holes in the momentum representation. The admissible value k of quasimomentum is related to the rigging L by the formula

$$k = (L + l) \bmod B, \qquad (17)$$

where $k \bmod B$ means reduction of the integer k to the Brillouin zone B, i.e. reduction mod N to the range indicated by Eq. (8). These results for single l-strings are readily extended to an arbitrary rigged string configuration $\nu\mathcal{L}$. Each l-string (l, v), $l = 1, 2, \ldots,\quad v \in \tilde{m}_l$, corresponds to quasimomentum

$$k(\nu\mathcal{L}; lv) = (L_{lv} + Q_l) \bmod B, \qquad (18)$$

where Q_l is the number of boxes in the first l columns of the Young diagram $\nu \vdash r'$. The integer Q_l determines precisely the kinematical restrictions imposed on the quasimomentum of a string, resulting from the hard-core property of constituent Bethe pseudoparticles. These restrictions define the forbidden Brillouin zone for l-strings, of the size

$$|B_f(\nu, l)| = 2Q_l - 1, \qquad (19)$$

so that the allowed quasimomenta belong to $B_a(\nu, l)$, of the size

$$|B_a(\nu, l)| = P_l + 1 = N - 2Q_l + 1. \qquad (20)$$

In this way, the total quasimomentum of an exact BA eigenstate reads

$$k(\nu\mathcal{L}) = \sum_{l \geq 1} (m_l Q_l + \sum_{v \in \tilde{m}_l} L_{lv}) \bmod B. \qquad (21)$$

We consider in some detail an example of distribution of rigged string configurations over the Brillouin zone for the case $N = 10$, $\nu = \{21^2\}$. A rigged string configuration $\nu\mathcal{L} = \rho(f)$ has now the form

$$\nu\mathcal{L} = \begin{array}{|c|c|}\hline L & \\ \hline L_1 \\ \cline{1-1} L_2 \\ \cline{1-1}\end{array} \begin{array}{c} 2 \\ 4 \end{array} \quad \text{with} \quad \begin{array}{c} 0 \leq L \leq 2, \\ 0 \leq L_1 \leq L_2 \leq 4, \end{array} \tag{22}$$

so that L is the rigging of the 2-string, and $L_1 \leq L_2$ are the riggings of the two 1-strings, the latter counted from the left to the right in the open chain $\tilde{N} = \tilde{10}$. The quasimomentum for this case is

$$k(\nu\mathcal{L}) = (L + L_1 + L_2 + 10) \bmod B \tag{23}$$

This case corresponds to $r' = 4$ Bethe pseudoparticles, coupled into a 2-string ($m_2 = 1$) of the form

$$v_2 = --++ \tag{24}$$

and two ($m_1 = 2$) 1-strings, each of the form

$$v_1 = -+\,. \tag{25}$$

The total number of holes for the 2-string is $P_2 = 10 - 2 \cdot 4 = 2$, whereas that for each of the 1-strings is $P_1 = 10 - 2 \cdot 3 = 4$. Thus the total number of different riggings is

$$|z(\{21^2\})| = \binom{P_1 + m_1}{m_1}\binom{P_2 + m_2}{m_2} = \binom{2+1}{1}\binom{4+2}{2} = 45. \tag{26}$$

The Brillouin zone for this case with appropriate distribution of strings is

$$\begin{array}{rcccccccccccl} B = \{ & & -4 & -3 & -2 & -1 & 0 & 1 & 2 & 3 & 4 & 5 & \} \\ \text{number of riggings} & & 7 & 5 & 4 & 2 & 2 & 2 & 4 & 5 & 7 & 7 & \text{45 altogether} \end{array} \tag{27}$$

and consists of the center $\{0\}$, the boundary $\{5\}$, and the interior $\{\pm 1, \pm 2, \pm 3 \pm 4\}$.

4. Conclusions

Existence of RSKKR bijection between magnetic configurations (the initial basis for quantum calculations), and rigged string configurations (the exact solutions of BA), can be exploited not only in the positional, but also in the momentum representation. As a result, each string acquires its own quasimomentum. Kinematically, an l-string in a string configuration ν can take on any quasimomentum outside a forbidden region of the Brillouin zone. This forbidden region is concentrated symmetrically around the center $k = 0$, and its size is determined by the combinatoric parameter dependent on ν. The total quasimomentum of a rigged string configuration $\nu\mathcal{L}$ can be readily evaluated by merely combinatoric data od $\nu\mathcal{L}$.

It is well known that each exact solution $\nu\mathcal{L}$ of BA equations has a definite quasimomentum $k(\nu\mathcal{L})$ - the sum of pseudomomenta of Bethe pseudoparticles. The fact that each string has its own conserved quasimomentum is a manifestation of integrability of the one - dimensional Heisenberg magnet.

The distribution of exact BA solutions of a set $z(\nu)$ of all riggings of a particular string configuration ν is considerably inhomogeneous. An interesting question arises what is the band structure of BA solutions, and, in particular, what is the relation between energy bands (admittedly composite) and string configurations.

We point out here that the assignment of a definite quasimomentum to each string in an exact BA solution has only a combinatoric meaning, similar to RSKKR bijection, and resulting just from the *existence* of this bijection. Clearly, exact BA solutions are much more complicated wave packets than naive pictures stemming from shapes of paths, overlapped multipyramids, etc. Still, it is amasing for us that a merely combinatoric bijection is able to reproduce precisely all quantum numbers of exact BA eigenstates, in simple and transparent terms of quasimomenta of strings.

References

[1] Bethe H 1931 *Z. Physik* **71** 205 (in German: English translation in: Mattis D C 1993 *The Many-Body Problem* (Singapore: World Sci.) pp 689-716

[2] Takahashi M 1971 *Progr. Theor. Phys.* **46** 401-15

[3] Kerov S V, Kirillov A N and Reshetikhin N Yu 1986 *LOMI* **155** 50 (English translation: 1988 *J. Sov. Math.* **41** 916-24)

[4] Dasmahapatra S and Foda O 1997 *Int. J. Mod. Phys.* **38** 1041-68

[5] Warnaar S O 1996 *J. Stat. Phys.* **82** 657-85

[6] Baxter R J 2002 *J. Stat. Phys.* **108** 1-48

[7] Lulek B, Jakubczyk D and Jakubczyk P 2004 *Mol. Phys.* **102** pp 1269-78

[8] Lulek T, Jakubczyk P and Jakubczyk D 2004 *Mol. Phys.* **102** pp 1279-84

[9] Jakubczyk D, Lulek B, Jakubczyk P and Lulek T, This volume

[10] Robinson G de B 1938 *Amer. J. Math.* **60** 745-60

[11] Schested C 1961 *Canad. J. Math.* **13** 179-91

[12] Jakubczyk P, Lulek B, Jakubczyk D and Lulek T, This volume

[13] Lulek B, Lulek T, Wal A and Jakubczyk P 2003 *Physica B - Condensed Matter* **337/1** pp 375-87

Unexpected magnetism in low dimensional systems: the role of symmetry

MC Muñoz[1], L Chico[2], MP López-Sancho[1], JI Beltrán[1], S Gallego[1] and J Cerdá[1]

[1] Instituto de Ciencia de Materiales de Madrid, Consejo Superior de Investigaciones Científicas, Cantoblanco, 28049 Madrid, Spain
[2] Departamento de Física Aplicada, Facultad de Ciencias del Medio Ambiente, Universidad de Castilla-La Mancha, 45071 Toledo, Spain

E-mail: mcarmen@icmm.csic.es

Abstract.
The symmetry underlying the geometric structure of materials determines most of their physical properties. In low dimensional systems the role of symmetry is enhanced and can give rise to new phenomena. Here, we report on unexpected magnetism in carbon nanotubes and O-rich surfaces of ionic oxides, to show how its existence is closely related to the symmetry conditions. First, based on tight-binding models, we demonstrate that chiral carbon nanotubes present spin splitting at the Fermi level in the absence of a magnetic field, whereas achiral tubes preserve spin degeneracy. These remarkably different behaviors of chiral and non-chiral nanotubes are due to the intrinsic symmetry dependence of the spin-orbit interaction. Second, the occurence of spin-polarization at ZrO_2, Al_2O_3 and MgO surfaces is proved by means of *ab-initio* calculations within the density functional theory. Large spin moments develop at O-ended polar terminations, transforming the non-magnetic insulator into a half-metal. The magnetic moments mainly reside in the surface oxygen atoms, and their origin is related to the existence of $2p$ holes of well-defined spin polarization at the valence band of the ionic oxide. The direct relation between magnetization and local loss of donor charge shows that at the origin of these phenomena is the reduced surface symmetry.

1. Introduction

The underlying symmetry of a material is probably its most fundamental characteristic, since it determines not only the structure but also the macroscopic properties. Pure geometrical arguments are usually sufficient to understand qualitatively many physical properties of matter. When dimensions are reduced, the role of geometry and symmetry is enhanced and can give rise to new and unexpected phenomena. Together with its inherent fundamental interest, this has important technological consequences, auguring the birth of new technologies [1, 2].

Carbon-based materials are a unique example of how different spatial distributions of atoms result in materials with different properties. Carbon is the only element of the Periodic Table which presents stable structures of any dimensionality, from the three dimensional (3D) diamond to the zero-dimensional (0D) fullerenes. Even though the only difference between carbon-based materials is the relative position of the C atoms, their physical properties are dissimilar. To put an example, diamond -the hardest material in nature- is an insulator, while two-dimensional (2D) graphite is a semimetal. Carbon nanotubes (CNTs) are formed by rolling up a graphene sheet into a cylinder, so the structure is one-dimensional (1D). Some tubes have a spiral conformation of atoms in the unit cell, they are called chiral tubes. They constitute a new class of mesoscopic 1D quantum systems [3] in which physical properties are intimately related to geometry. They can either be metals or semiconductors depending on their diameter and chirality [4, 5].

Ceramic oxides as ZrO_2, Al_2O_3 and MgO are non-magnetic ionic insulators. Their electronic structure can be roughly described as a valence band formed by the filled O $2p$ orbitals and a conduction band formed by the empty metal levels. The crystal can be described as a set of charged ions bonded by strong electrostatic forces, due to the large charge transfer from the metal cation to O. The resulting electronic charge distribution is highly localized in the neighborhood of the ion cores, with the transferred electrons closely bound to oxygen. Surfaces of ionic compounds can be polar or non-polar. In polar surfaces each atomic plane is formed solely by either oxygen anions or metals cations, resulting in an alternated distribution of negative and positive charged layers. Contrary, in the non-polar surfaces, atomic layers are neutral due to the presence of both anions and cations at the same plane. Formation of polar surfaces requires the modification of the charge density via faceting, reconstruction, changes of the surface stoichiometry or any alternative mechanism which disrupts their polar divergent character.

In this paper, we explore the close connection between magnetism and symmetry in low dimensional systems. We demonstrate that chiral carbon nanotubes and O-rich polar surfaces of binary insulating oxides present a magnetic ground state whose origin lies in the reduced symmetry of the structures [6, 7].

2. Model and method

2.1. Metallic carbon nanotubes

We model CNTs by a Slater-Koster [8] empirical tight-binding (ETB) hamiltonian including sp^3 orbitals, employing the Tománek-Louie parametrization for graphite [9] to take into account the actual discrete nature of the lattice, and including the spin-orbit interaction (SOI). The unit cell is generated by rolling up a portion of graphene layer, thus including curvature effects [10]. Metallic tubes present a band crossing at the Fermi level either at Γ or at two thirds of the distance from Γ to the zone boundary [6]; here we focus on 2/3-metals CNTs.

SOI is caused by the coupling of the spin of a moving electron with an electric field which acts as a magnetic field in the rest frame of the electron. In a crystalline environment, the major internal contribution arises from the electron orbital motion in the crystal potential V, and thus its effects are related to the crystal symmetry. The SOI is given by $H_{SO} = \frac{\hbar}{4mc^2}\sigma(\nabla V \times \mathbf{p})$, where σ represents the Pauli matrices and $\mathbf{p}$ the electron momentum. The relevant contribution of the crystal potential is that close to the atomic cores, so assuming spherical symmetry, the above hamiltonian can be expressed as $H_{SO} = \lambda \mathbf{L} \cdot \mathbf{S}$, where the spin-orbit (SO) coupling constant (λ) depends on the atomic orbital and $\mathbf{L}$ is the angular momentum of the electron. Within the ETB framework, H_{SO} couples p orbitals on the same atom, and λ can be either greater or smaller than the atomic value [11, 12]. The present calculations have been performed with $\lambda = 0.2$ eV [6]. SO effects have been customarily neglected in graphite-like materials, and the exact value of λ is not known. Although there are no recent measurements, low dimensionality, electron-electron interactions and magnetic fields are shown to induce an enhancement of SO coupling [13, 14, 15]. Nevertheless, the SO strength can be controlled by a proper choice of gate potentials [16], as was first proposed for semiconductor heterostructures.

2.2. Surfaces of ionic oxides

We have studied low-index surfaces of binary insulating oxides of different crystal structures: MgO, Al_2O_3 and ZrO_2. MgO adopts the rock-salt lattice. α-Al_2O_3 can be viewed as an hexagonal close-packed array of O atoms with the Al atoms occupying two thirds of the interstitial octahedral sites. ZrO_2 displays three polymorphs at room temperature [18], either in its pure form (m-ZrO_2, monoclinic), or by dopage with substitutional cations (t-ZrO_2, tetragonal, and c-ZrO_2, cubic). The cubic structure corresponds to the CaF_2 lattice. t-ZrO_2 can be obtained from the cubic unit cell elongating one of the equivalent edges of the cube and introducing small opposite shifts at the O sublattice positions. The m-ZrO_2 lattice is more complex and has lower symmetry; it is formed by further distorting the tetragonal structure, with non-orthogonal lattice vectors. Depending on the crystallographic orientation, the stacking of atomic planes of the above lattices leads to very different surface structures. Perhaps the best example is c-ZrO_2(111), which shows three polar terminations labelled according to the composition of the two topmost layers: O-O-, O-Zr- and Zr-O-. Here we have considered all low-index terminations of c-ZrO_2, and selected t-ZrO_2, m-ZrO_2, MgO and Al_2O_3 surfaces: MgO(111) consists of a stacking of alternated pure Mg and O planes; in α-Al_2O_3(0001) consecutive O planes are separated by two Al layers; m-ZrO_2(001) defines a layered structure of chemical stacking sequence O-O-Zr, while c-ZrO_2(001) shows an alternance of O and Zr planes.

Our results are obtained from first-principles spin-polarized calculations within the DFT (density functional theory) under the GGA (generalized gradient approximation) for exchange and correlation [19]. We use the SIESTA package [24, 25] with basis sets formed by double-zeta polarized localized numerical atomic orbitals. More details about the conditions of the calculations can be found elsewhere [7, 26]. Each surface is modelled as a periodic slab containing a vacuum region of 10 Å. The width of the vacuum region is enough to inhibit interaction between neighbouring surfaces. The number of M_xO_y units in a slab -M being the metal donor and x,y accounting for the particular metal to oxygen ratio- depends on the particular structure, and ranges from 5 to 7. The slabs are symmetric about the central plane, to avoid an artificial divergent Madelung energy, although we have verified that all the conclusions of this work remain unaltered if asymmetric slabs are considered instead. Bulk-like behaviour is attained at the innermost central layers. The atomic positions are allowed to relax until the forces on the atoms are less than 0.06 eV/Å. For comparison, a non-spin polarized (NSP) calculation is also performed for each case.

3. Results and Discussion

3.1. Metallic carbon nanotubes

From symmetry considerations, the bands of any nanotube are at most fourfold degenerate. Considering spin, the bands crossing at E_F in 2/3-metals are only doubly degenerate. Thus, neglecting curvature effects, there is an accidental fourfold degeneracy at E_F in all 2/3-metals CNTs. First, we focus on achiral armchair tubes which possess inversion symmetry. Figure 1a presents the atomic structure and the spin-resolved density of states (DOS) of an armchair (5,5) CNT calculated at $\mathbf{k_F}$ for energies around the Fermi level, with and without SO coupling. It clearly shows that, even though armchair tubes remain metallic after including curvature effects (DOS calculated without SOI), the inclusion of SO interaction leads to a small gap opening as predicted by Ando [27]. The experimental measurement of the SO-induced gap could provide a way to estimate the SO-interaction strength. However, the spin degeneracy is not lifted by the SO interaction, although the states are no longer spin eigenstates. The eigenfunctions (not shown here) do not correspond to pure spin-states, although they have a dominant weight of one spin orientation. Thus, as was previously pointed out [27], due to the SOI in achiral CNTs, spin scattering is induced even by spin-independent scatterers and impurities can cause spin-relaxation.

Figure 1b shows the atomic structure and the corresponding DOS of a chiral (7,1) tube. Chiral NTs do not have an inversion center and only possess spiral symmetry operations. In contrast to the results shown for the (5,5) armchair CNT, curvature effects open a small gap at the Fermi energy (see DOS calculated without SOI). Further, the SO interaction lifts all degeneracies producing an energy splitting between states with different spin orientation in

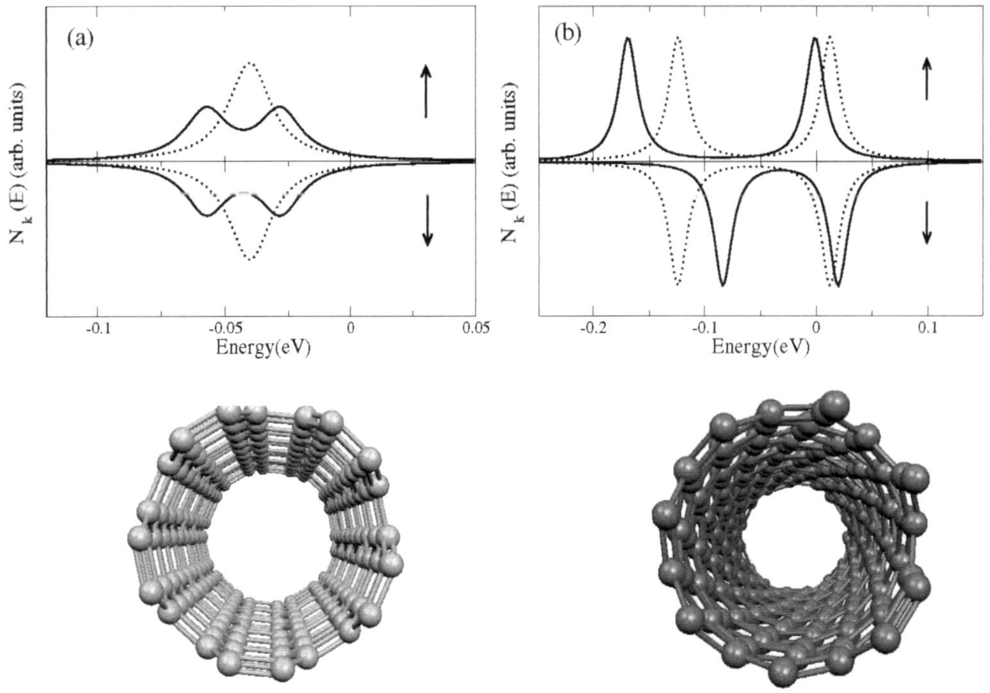

Figure 1. Atomic structure and spin-resolved DOS at $\mathbf{k_F}$ for (a) the (5,5) nanotube and (b) the (7,1) nanotube, calculated with (full line) and without (dotted line) SO interaction.

chiral nanotubes. The bands correspond to spin eigenstates. The motion of electrons with spin up and spin down are completely decoupled. Thus, in chiral CNTs there is not spin scattering from scatterers with a spin-independent potential. In fact, an electron with spin up cannot be scattered into a state with spin down and viceversa. This would explain the huge spin-flip scattering length (≈ 130 nm) observed for spin-polarized electrons injected into multiwall CNTs [28]. In all tubules the band-crossing at E_F is between energy bands of different symmetry; then they respond differently to the SOI interaction, as can be seen in the figure since the SO splitting is band dependent.

SO-related effects in CNTs have a fundamental symmetry dependence: while in achiral tubes they do not affect the spin degenerate states, in chiral tubes spin degeneracy is completely removed. The origin of the asymmetric behavior of achiral and chiral CNTs under the SOI is due to the interplay of time reversal and spatial inversion symmetries. Kramers' theorem on time reversal symmetry states that $E_{\mathbf{k},\sigma} = E_{-\mathbf{k},-\sigma}$, where $E_{\mathbf{k},\sigma}$ is the energy of the eigenstate with wavevector $\mathbf{k}$ and spin σ. In achiral tubes, because there is an inversion center: $E_{\mathbf{k},\sigma} = E_{-\mathbf{k},\sigma}$, therefore, $E_{\mathbf{k},\sigma} = E_{\mathbf{k},-\sigma}$ and spin degeneracy can not be removed. For chiral tubes only the condition imposed by the time reversal invariance holds and thus spin degeneracy is lifted. In that case, electron velocities become dependent on both the spin and the direction of the motion.

The different response to the SOI of chiral CNT bands of different symmetry can lead to sequences of states with parallel spin. Such effect occurs at energies around a band gap and whenever there are two bands close in energy with different symmetry. This is illustrated in Figure 2, which shows the spin-resolved DOS of a chiral (9,3) CNT for energies around the Fermi level. Recently, spin-dependent transport experiments have been reported which were in apparent contradiction, suggesting the existence of different spin ground states in CNTs. On one

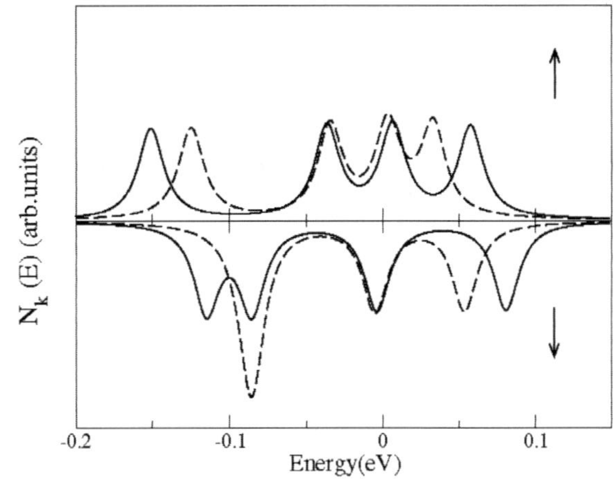

Figure 2. Spin-resolved DOS for the (9,3) tube at $\mathbf{k} = 0.025$ (full line), and $\mathbf{k} = 0.0125$ (dashed line). Wavevector values are given with respect to the BZ length.

hand, measurements performed at low magnetic fields in a bundle of CNTs [29] and in quantum dots [30, 31] found the expected spin alternation and could be explained by a simple shell-filling model. On the other, successive electrons entering isolated CNTs showed that consecutive electrons could have the same spin-direction, indicating a possible spin polarization of the nanotube [32]. This experimental observation could be explained by the sequential ordering of states with parallel spin [6]. We have focused in metallic tubes, but SO-induced spin splitting is also present in chiral semiconductor tubes, where SO-related effects are analogous.

3.2. Surfaces of ionic oxides

Table 1 summarizes our main result. It shows the total magnetic moments at the surface oxygen plane for all the structures which present a magnetic ground state, together with their in-plane (||) and normal ($\perp$) p projections onto the surface. All spin moments arise almost exclusively from the p electrons at the valence band of the oxide. They reach large values, ranging from 0.8 to 1.6 μ_B. Notice that only the O-rich oxide surfaces are magnetic. Both the cation-ended and the non-polar O surfaces not in the table do not show any spin polarization.

The distinct characteristic of all these magnetic structures is that they are polar divergent surfaces [33] with a loss of cationic coordination for the outermost O atoms. The loss of the surface O counterparts at the M_xO_y coordination unit leads to an asymmetric charge distribution and to the creation of *2p* holes at the oxide valence band. This already suggests that the O spin polarization is intimately related to the decrease of the oxygen ionic charge at the surface with respect to the bulk, that is, to the existence of *2p* holes at the valence band. The correlation is evident after inspection of Table 1, where the rightmost columns provide the oxygen Mulliken populations at the surface layer and the bulk. In all these cases the surface oxygen loses ionic charge approaching a charge neutral state. Moreover, the smaller the oxygen ionic charge, the larger its associated magnetic moment.

The calculated surfaces with a non-magnetic ground state have oxygen charges close to the bulk ones, which corroborates the correlation between the reduction of the ionic charge and the formation of the magnetic moment. As an example, we can compare the surface O Mulliken populations of the two O-ended c-ZrO_2(111) surfaces. O atoms at the O-O- surface -which presents a high magnetic moment of 1.6 μ_B- have an ionic charge of 6.02, while that corresponding to the non-magnetic O-Zr- surface is 6.68, closer to the 6.79 obtained for bulk oxygen. This means that the spin polarization is a response of the system to the loss of

Table 1. Spin moments (in μ_B) at the topmost layer of all the oxide surfaces with a magnetic ground state, together with the decomposition of the p orbital contribution parallel and normal to the surface plane. The rightmost columns provide the corresponding Mulliken charge populations (Q) compared to the inner bulk value (Q_b). The O_1 and O_2 entries for the non-cubic ZrO_2 structures refer to the two inequivalent in-plane positions.

Surface	m_{tot}	$m^p{}_{\parallel}$	$m^p{}_{\perp}$	Q	Q_b
c-$ZrO_2(111)_{O-O-}$	**1.56**	1.50	0.04	**6.02**	6.79
m-ZrO_2(001)-O_1	**1.43**	0.81	0.61	**6.10**	6.66
-O_2	**1.43**	0.81	0.60	**6.11**	
t-ZrO_2(001)-O_1	**1.19**	0.54	0.64	**6.13**	6.76
-O_2	**0.59**	0.55	0.03	**6.17**	
c-ZrO_2(001)	**0.83**	0.62	0.21	**6.29**	6.79
Al_2O_3(0001)	**0.97**	0.74	0.23	**6.10**	6.49
MgO(111)	**0.83**	0.06	0.76	**6.37**	6.72

transferred charge.

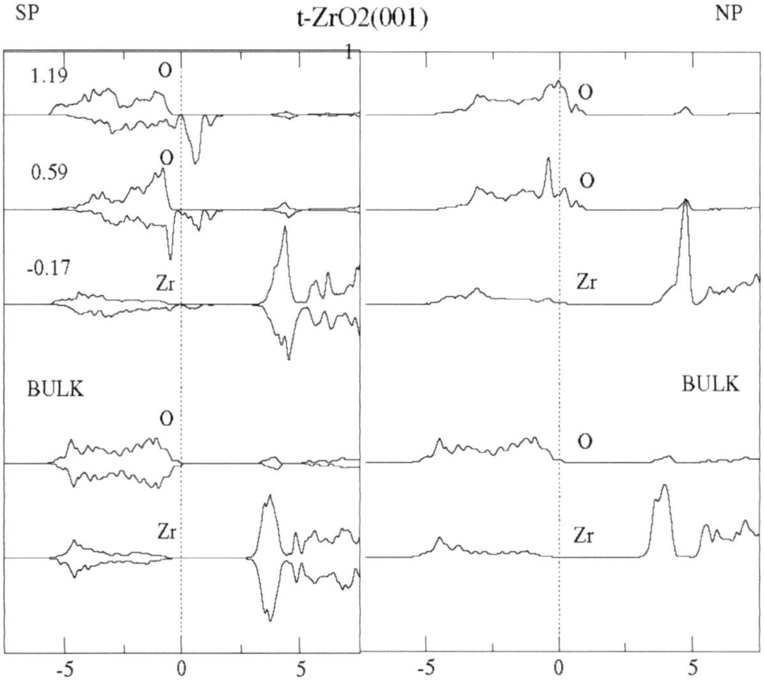

Figure 3. DOS for the three topmost planes and the bulk of t-ZrO_2(001) after SP (left) and NSP (right) calculations. Positive (negative) values correspond to majority (minority) spin states. Energies are referred to the Fermi level. The total spin moments at each layer are indicated for the SP case.

All the magnetic surfaces are half-metallic. Figure 3 shows the spin-resolved density of states (DOS) at the three uppermost layers and the bulk for the t-ZrO_2(001) surface, both for SP (left) and NSP (right) calculations. As in all the magnetic structures investigated, the Fermi level (E_F)

crosses the valence band of the topmost layer. Remarkably, the bands crossing E_F have always a well defined spin polarization, the majority spin states being full while the minority spin band is partially filled. The result is a half-metallic system. The most significant effect of the spin exchange is the shift of the minority spin DOS and the reduction of the DOS at E_F. However, a deep redistribution of charge occurs, as manifests the shape of the surface oxygen DOS; the changes of the electronic distribution affect mostly to the minority spin electrons. Even if the magnetization is a local effect, rooted in the lack of donor electrons for the outermost O atoms, the spin polarization at the surface layer induces a magnetic moment at the neighbouring planes, which decays as we deepen into the bulk. This can be observed in Figure 3, which provides the magnetization of the three surfacemost layers for t-ZrO_2(001). A similar trend corresponds to all magnetic surfaces calculated here.

The major changes observed in the DOS are also present in the spatial charge distributions. This can be directly seen in the left and central panels of Figure 4, which depict the spin-resolved charge density differences (CDD, total charge density minus the superposition of atomic charge densities) of the [0001] α-Al_2O_3 surface, within a plane normal to the surface along the bond direction between a surface O atom and a nearest neighbour at the consecutive O layer. The majority spin band being completely filled, as was shown in the DOS, the corresponding surface CDD adopts the spherical symmetry of the bulk. On the contrary, anisotropic lobular shapes appear in the minority spin charge. This asymmetry between the spatial distributions of the spin states is a common feature to all the magnetic structures studied here. The rightmost panels of the figure show the corresponding spin density difference (SDD), CDD of majority spin minus CDD of minority spin. The crystal field induces directionality to the spin moments, involving orbitals along specific directions. For the [0001] α-Al_2O_3 surface the dominant contribution comes from orbitals contained within the surface plane, although there is a significant contribution to the magnetic charge from orbitals normal to the surface.

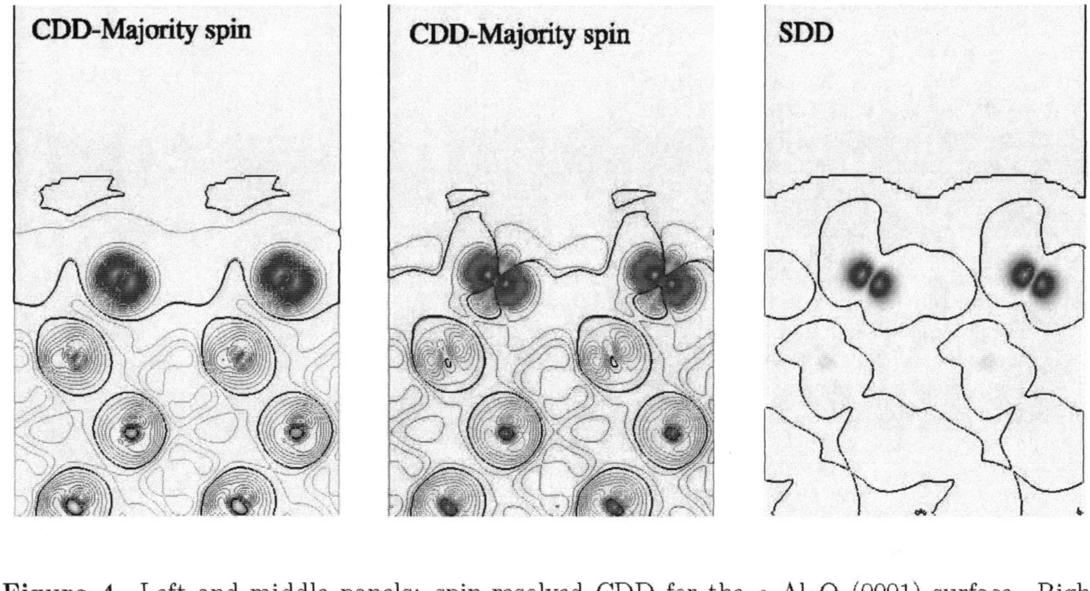

Figure 4. Left and middle panels: spin-resolved CDD for the α-Al_2O_3(0001) surface. Right panel: corresponding SDD.

Very recently, unexpected ferromagnetism has been measured in thin films of undoped non-magnetic oxides, like HfO_2 and ZrO_2, and assigned to the presence of lattice defects concentrated at the film interface. [34, 35]. The mechanism presented here provides an explanation for the origin of the magnetic moments of these so called d^0 ferromagnets. Furthermore, we predict that this kind of magnetism can be extended to a wider class of non-magnetic oxides, like MgO and Al_2O_3, where the cation is not necessarily a d metal.

4. Conclusions

In summary, we have shown that low dimensional systems as carbon nanotubes and surfaces of ionic oxides present new and unexpected magnetism directly related to their reduced symmetry. The SO interaction removes spin degeneracy in chiral CNTs due to the lack of inversion symmetry of the crystal potential, while in achiral CNTs spin degeneracy is not lifted. Thus, SO interaction in CNTs presents an intrinsic symmetry dependence. In addition, we have proved the existence of local magnetism at the ideal O-ended polar surfaces of ceramic oxides. Large spin moments are formed at the outermost surface plane due to the lack of donor charge and the subsequent creation of $2p$ holes in the valence band of the oxide. This leads to a half-metallic ground state, with the majority spin band completely filled and only minority spin states at the Fermi level. The fact that low dimensional materials exhibit a 100 % spin polarization at the Fermi level opens a new route to the manipulation of spin currents.

Acknowledgments

This work has been partially financed by the Spanish Ministerio de Educación y Ciencia and the DGES under contracts MAT2002-04095-C02-01, MAT2002-04540-C05-03 and MAT2003-04278, and by the Comunidad Autónoma de Madrid under CAM2004-0440. S.G. acknowledges financial support from the Ramón y Cajal program of the Spanish Ministerio de Educación y Ciencia, and J.I.B. from the I3P program of the CSIC.

5. References

[1] Wolf S A *et al* 2001 *Science* **294** 1488
[2] Prinz G A 1998 *Science* **282** 1660
[3] Saito R, Dresselhaus G and Dresselhaus M 1998 *Physical Properties of Carbon Nanotubes*, London (Imperial College Press)
[4] Wildöer J W G *et al.* 1998 *Nature* **391** 59
[5] Odom T W *et al.* 1998 *Nature* **391** 62
[6] Chico L, López-Sancho M P and Muñoz M C 2004 *Phys. Rev. Lett.* **93** 176402
[7] Gallego S, Beltran J I, Cerdá J and Muñoz M C 2005 *J. Phys.: Condens. Matter* **17** L451
[8] Slater J C and Koster J F 1954 *Phys. Rev.* **94** 1498
[9] Tománek D and Louie S G 1988 *Phys. Rev. B* **37** 8327
[10] Blase X *et al.* 1994 *Phys. Rev. Lett.* **72** 1878
[11] Wang C S and Callaway J 1974 *Phys. Rev. B* **9** 4897
[12] Gallego S and Muñoz M C 1999 *Surf. Sci.* **423** 324
[13] Moroz A V and Barnes C H W 1999 *Phys. Rev. B* **60** 14272
[14] Chen G H and Raikh M E 1999 *Phys. Rev. B* **60** 4826
[15] Valín-Rodríguez M 2004 *Phys. Rev. B* **70** 033306
[16] Rashba E I and Efros A L 2003 *Phys. Rev. Lett.* **91** 126405
[17] Kleiner A and Eggert S 2001 *Phys. Rev. B* **63** 73408
[18] Howard C J, Hill R J and Reichert B E 1988 *Acta Crystallogr. Sect. B: Struct. Sci.* **44** 116
[19] Perdew J P *et al.* 1992 *Phys. Rev. B* **46** 6671
[20] Kralik B, Chang E K and Louie S G 1998 *Phys. Rev. B* **57** 7027
[21] Jansen H J F 1991 *Phys. Rev. B* **43** 7267
[22] Eichler A 2001 *Phys. Rev. B* **64** 174103
[23] Zhao X and Vanderblit D 2002 *Phys. Rev. B* **65** 075105
[24] Ordejón P, Artacho E and Soler J M 1996 *Phys. Rev. B* **53** R10441
[25] Soler J M *et al.* 2002 *J. Phys.: Condens. Matter* **14** 2745
[26] J.I. Beltrán *et al.*, *Phys. Rev. B* **68**, 075401 (2003).
[27] Ando T 2000 *J. Phys. Soc. Jpn.* **69** 1757
[28] Tsukagoshi K, Alphenaar B W and Ago H 1999 *Nature* **401** 572
[29] Cobden D H *et al.* 1998 *Phys. Rev. Lett.* **81** 681

[30] Cobden D H and Nygård J 2002 *Phys. Rev. Lett.* **89** 046803
[31] Liang W, Bockrath M and Park H 2002 *Phys. Rev. Lett.* **88** 126801
[32] Tans S J *et al.* 1998 *Nature* **394** 761
[33] Tasker P W 1979 *J. Phys. C: Sol. State Phys.* **12** 4977
[34] Venkatesan M, Fitzgerald C B and Coey J M D 2004 *Nature* **430** 630
[35] M. Coey, Abstract U5.00005 of 2005 APS March Meeting

Coherent quantum dynamics of mesoscopic metallic ring with a barrier

M Szopa, E Zipper, M Szelag and M Kurpas
University of Silesia, Katowice, Poland

Abstract. In this paper we consider a mesoscopic 1D, ballistic, metallic ring with a potential barrier. We show that the coherent coupling between two distinct quantum states with different winding numbers can lead to a formation of a qubit. We discuss the possible realizations of such a ring, the adjustment of a potential barrier parameters and the possible decoherence sources.

Recently a number of systems, which can be effectively reduced to a two level systems [1], [2], [3], [4], have been examined as candidates for quantum computing hardware. In the present paper we propose a candidate for the quantum qubit, which is based on a mesoscopic metallic ring with a barrier.

Persistent currents in small nonsuperconducting rings threaded by a magnetic flux are a manifestation of quantum coherence in a submicron system. If the ring circumference L is smaller than the phase coherence L_ϕ, the electron wave function may extend coherently over L even in the presence of elastic scatterers [5], [6]. In other words a normal loop with $L < L_\phi$ has a nontrivial ground state with a circulating persistent current (PC). In this paper we perform same model calculations on mesoscopic metallic ring which show that quantum tunneling between states with nearly equal energy and opposite PC can lead to a formation of a qubit.

Let us consider the mesoscopic metallic quasi $1D$ ring of radius R ($2\pi R < L_\phi$) in the presence of static magnetic flux ϕ_e, $\phi_e = B_e \cdot \pi R^2$, B_e is the applied magnetic field perpendicular to the plane of the ring.

The Hamiltonian of the system in the free electron model is

$$H = \frac{1}{2m_e}(p - eA)^2 + V(x),$$

where $x = R\varphi$, $x \in (0, L)$, $A = \frac{\phi}{L}$ and m_e is the electron mass. $V(x)$ is the potential energy in the presence of an energy barrier. We assume that a ring is made from a very clean material i.e. we are in the ballistic regime. In the absence of the energy barrier $V(x) = 0$ the formulas for the energy levels and the wave functions in a quasi $1D$ ring are [7]:

$$E_n = \frac{\hbar^2}{2m_e R^2}\left(n - \phi'\right)^2, \quad \psi_n(x) = \frac{1}{\sqrt{2\pi}} e^{in\frac{x}{R}}, \quad n = 0, \pm 1..., \tag{1}$$

where $\phi' = \frac{\phi}{\phi_0}$, n is the orbital quantum number (winding number) for an electron going around the ring. For very thin ring the selfinductance effect can be neglected and $\phi = \phi_e$, $\phi_0 = \frac{h}{e}$.

The system has a set of quantum size energy gaps, the gap at the Fermi surface (FS) at $\phi' = 0.5$ is $\Delta = \frac{hv_F}{L}$. Assuming a ring made of a metal (e.g. copper) with $R = 400\AA$, we get $\Delta \sim 290K$. The energy spectrum as a function of ϕ' is shown in Fig. 1 by the dashed lines. At integer and half integer ϕ' the states E_n are doubly degenerate, therefore neglecting the spin, each level E_n can be occupied by one or two electrons.

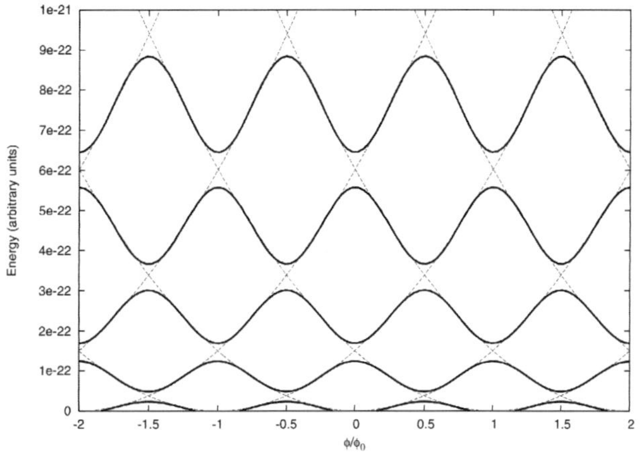

Figure 1. The energy spectrum of a 1D ring as a function of the applied flux ϕ'.

We can see from Fig. 1, that for $\phi' = 0.5$ and the number of electrons in the ring $N = N_{even} = 2n_F$ the last occupied level, i.e. the level below the FS, is doubly occupied, whereas for $N = N_{odd} = 2n_F + 1$ the last level below the FS is occupied by a single electron only. The energy spectrum is periodic with period $\phi' = 1$. We restrict our considerations to $\phi' \in [0, 1]$.

With each energy level we can associate a microscopic current

$$I_n = -\frac{\partial E_n}{\partial \phi} = \frac{e\hbar}{2\pi m_e R^2}\left(n - \phi'\right), \quad n = 0, \pm 1, ... \tag{2}$$

The current is persistent at $kT \ll \Delta$.

In the presence of the energy barrier of finite length a and height V_0, the tunneling occurs which mixes the states from the two potential wells. Effectively this is connected with the possibility of an electron wave function phase slip at the barrier. The phase slip rate increases with the height and width of the barrier. The phase slip is more likely to occur close to the degeneracy points of the energy spectrum and then the eigenstates, which are the superposition of flux states with different winding numbers, can be formed. This leads to the level splitting of the initial energy levels, which depends on the barrier parameters (the solid line at Fig. 1). Quantum tunneling should thus lead to a qubit i.e. a quantum superposition of the two opposed current states. The Hamiltonian in the second quantization is

$$H = \sum_{m \neq n}\left[\frac{\hbar^2}{2m_e R^2}\left(n - \phi'\right)^2 |n\rangle\langle n| - \frac{1}{2}\delta_{m,n}\left(|n\rangle\langle m| + |m\rangle\langle n|\right)\right], \tag{3}$$

where the summation is over all pairs of different occupied electronic states, $\delta_{m,n}$ is the phase slip rate between states $|m\rangle$ and $|n\rangle$.

The energy states with $|n| < n_F$ are fully occupied and form the "Fermi sea". The energy states for $|n| > n_{F+1}$ are separated by large energy gaps from the FS and at $kT \ll \Delta$ are fully empty. Thus the only states which can take part in the tunnelling are the states in the immediate neighbourhood of the FS and we can consider a mesoscopic ring as a two-state quantum system.

In this case the summation in (3) can be restricted only to two states closest to the FS. Throughout this paper we assume e.g. that $N = N_{odd}$ and $\phi^{'}$ is close to 0.5. Then these states are $|n_F + 1\rangle = |\beta\rangle = \binom{1}{0}$ and $|-n_F\rangle = |\alpha\rangle = \binom{0}{1}$ and the Hamiltonian (3) becomes

$$H = \begin{bmatrix} E_\alpha & -\frac{1}{2}\delta_{\alpha,\beta} \\ -\frac{1}{2}\delta_{\alpha,\beta} & E_\beta \end{bmatrix}, \tag{4}$$

where $E_\alpha = E_{-n_F}$, $E_\beta = E_{n_F+1}$ and $\delta_{\alpha,\beta} = \delta_{-n_F, n_F+1}$.

For most values of $\phi^{'}$ the $\delta_{\alpha,\beta}$ are small compared with the energy of orbital motion of an electron in the ring. However close to the degeneracy points the $\delta_{\alpha,\beta}$ term mixes them strongly.

At $\phi^{'} = 0.5$ the states $|\beta\rangle$ and $|\alpha\rangle$ have exactly the same energies

$$E_\alpha = E_\beta, \tag{5}$$

but opposite currents

$$I_\alpha = -I_\beta. \tag{6}$$

In this case the phase slip rate $\delta_{\alpha,\beta}/\hbar$ increases at all half integer $\phi^{'}$.

The quantum tunneling can provide coherent coupling between these two distinct quantum states. The energy after quantum tunneling is the same, but the persistent current is reversed. Quantum tunneling should thus lead to a qubit i.e. a quantum superposition of the two opposed current states. The amplitude of these currents is $I_0 = \frac{ev_F}{L}$. Assuming e.g. $R = 400\mathring{A}$, for metallic ring $I_0 \sim 1\ \mu A$.

In a pseudospin notation (4) can be written as

$$H = -\frac{1}{2}B_z\widehat{\sigma}_z - \frac{1}{2}B_x\widehat{\sigma}_x, \tag{7}$$

where $\widehat{\sigma}_z$ and $\widehat{\sigma}_x$ are Pauli spin matrices.

The energy splitting, which is controlled by the magnetic flux corresponds here to the z component of the magnetic field

$$B_z = E_\beta - E_\alpha = \Delta\left[1 - 2\frac{\phi}{\phi_0}\right]. \tag{8}$$

The x component of the effective magnetic field B_x describes the tunneling amplitude $\delta_{\alpha,\beta}$ between the two potential wells. It depends on the height and width of the barrier. The phase slip may occur as a result of a reflection of an electron on a single potential barrier encountered along the ring.

The potential barrier in a mesoscopic ring can be realized in a number of ways. One of them is the ring in which the small fragment of a convex arc have been replaced with a concave one [8]. In this case the resulting potential barrier has a rectangular shape. The potential barrier produces forward and backward scattered electrons with equal energies and opposite currents (Eq. (5), (6)). Another possibility to form the barrier is the local point-like electric gate close to the ring. The electric potential of the gate forms the barrier, the height of which can be modified by the gate voltage. The electric potential barrier can be combined with the geometric one described above. In this case the barrier height and therefore the mixing term $\delta_{\alpha,\beta}$ can be easily adjusted to the required value. Finally, to obtain a system described by Eq. (7) one can apply a real magnetic field B_x close to the degeneracy point.

Diagonalizing the Hamiltonian (4) we obtain two energy bands (see Fig. 2)

$$E_{I,II} = \pm\frac{1}{2}\sqrt{(E_\beta - E_\alpha)^2 + \delta^2_{\alpha,\beta}}. \tag{9}$$

Introducing the "mixing angle" [2], $\eta = \tan^{-1}\frac{\delta_{\alpha,\beta}}{E_\beta - E_\alpha}$ the eigenstates of Eq. (7) are

$$|I\rangle = \cos\frac{\eta}{2}|\beta\rangle + \sin\frac{\eta}{2}|\alpha\rangle, \tag{10}$$

$$|II\rangle = -\sin\frac{\eta}{2}|\beta\rangle + \cos\frac{\eta}{2}|\alpha\rangle. \tag{11}$$

At the degeneracy point $E_\beta = E_\alpha$, $\eta = \frac{\pi}{2}$ the energy splitting is

$$E_I - E_{II} = \delta_{\alpha,\beta} \tag{12}$$

and the respective energy eigenstates are

$$|I\rangle = \frac{1}{\sqrt{2}}(|\beta\rangle + |\alpha\rangle),$$

$$|II\rangle = \frac{1}{\sqrt{2}}(-|\beta\rangle + |\alpha\rangle). \tag{13}$$

We performed the numerical calculations of the energy levels $E_{I,II}$ for different sets of the barrier parameters and extracted the magnitude of $\delta_{\alpha,\beta}$ from it.

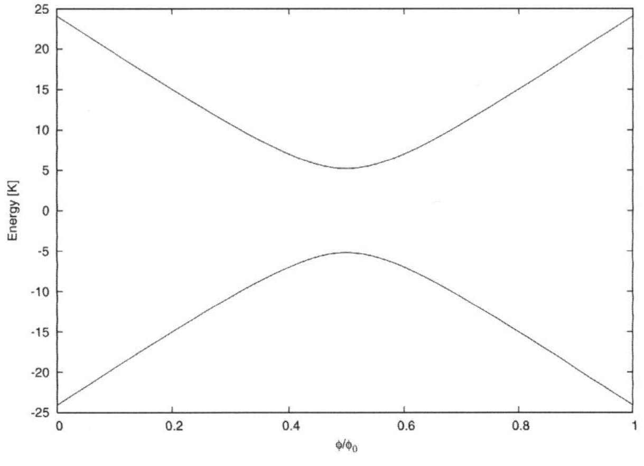

Figure 2. The energy bands $E_{I,II}$ as a function of an applied flux ϕ'.

Let us estimate the energy splitting $\delta_{\alpha,\beta}$ of the degenerate states due to the tunneling through the potential barrier of width $a \ll L$ and height V_0 [9]. By making use of a transfer matrix method one obtains

$$\delta_{\alpha,\beta} = \frac{\Delta}{\pi}\arccos\sqrt{T}, \tag{14}$$

where T is the transmission probability of an electron at the FS. The transmission probability T as a function of V_0 and a is presented in Fig. 3.

The tunneling amplitude depends on the height of the barrier which can be raised or lowered by electronic gating. Assuming e.g. $E_F \sim 1eV$ one obtains the $\delta_{\alpha,\beta}$ in the range of a few K. The two energy levels given by Eq. (9) for a ring with $V_0 = 1.015eV$ and $a = 20\mathring{A}$ are shown in Fig. 2.

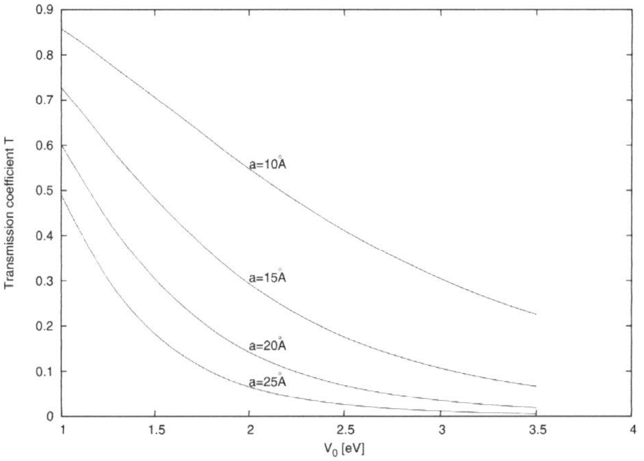

Figure 3. The transmission coefficient T as a function of the potential barrier height V_0 and width a.

Our model calculations has been performed for a quasi $1D$ ring. However they are also valid with some modifications for the mesoscopic metallic ring with very small thickness d, $d \ll R$ i.e. with a few transverse channels (M) [9]. Sharp energy levels E_n will then be replaced by narrow energy bands with M sublevels. However the gaps between these bands will still be large and at low temperature we can approximate thin ring by a quasi $1D$ ring. In the area of present development of nanotechnology there is a hope that such rings will be produced in the near future.

In this paper we discussed the possibility of making a flux qubit on a nonsuperconducting, mesoscopic, metallic, ballistic quasi $1D$ ring. Such ring can be effectively reduced to a two-state system and described in the pseudo-spin notation. The term B_z in Eq. (7) can be tuned by the applied flux. The coherent coupling between two distinct quantum states having different winding numbers n is via coherent tunneling through the potential barrier which can be raised or lowered by electrical gating. The off-diagonal term in Eq. (7) can be also accomplished by the real magnetic field B_x perpendicular to the z axis (or in the plane of the ring). With these two external control parameters the elementary single-bit operations i.e. z and x rotations can be performed.

The system discussed in the paper can behave quantum mechanically provided that it is sufficiently decoupled from its environment. Thus the important constraints on the device are dephasing effects due to various decoherence sources. They are mainly caused by material problems such as defects. The magnetic noise in practical currents is smaller than the electrical noise and random fluctuations generated by charged defects. The fluctuations in the barrier and in the magnetic field, necessary to get the degeneracy of the states for $N = N_{odd}$, can be the source of decoherence. We can estimate the influence of known sources of decoherence, but it is impossible to determine the real decoherence time with certainty, except by measurements. Till now we discussed a single qubit design. Two or more qubits can be coupled by means of the flux that the circulating currents generate [1]. The flux state of the sample can be measured by a separate magnetometer inductively coupled to the sample. The proposed qubits should be of considerable interest for fundamental studies of quantum coherence in mesoscopic systems.

Acknowledgments
The authors acknowledge the support of the Polish Ministry of Scientific Research and Information Technology under the (solicited) grant No PBZ-MIN-008/P03/2003.

References

[1] Mooij J E, Orlando T P, Levitov L, Lin Tian, van der Wal Caspar H and Lloyd Seth 1999 *Science* **285** 1036
Mooij J E *CJPM Harmans*, cond-mat/0508440
[2] Makhalin Y, Schön G and Schnirman A 2001 *Rev. Mod. Phys.* **73** 357
[3] Loss J D and Vincenzo D P 1998 *Phys. Rev.* A **57** 120
[4] Cirac J I and Zoller P 2000 *Nature* **404** 579
[5] Büttiker M, Imry Y and Landauer R 1993 *Phys. Lett.* A **96** 365
[6] Levy L P, Dolan G, Dunsmuir J and Bouchiat H 1990 *Phys. Rev. Lett.* **64** 2074
[7] Cheung H F, Gefen Y, Riedel E K and Shih W H 1988 *Phys.Rev.* B **37** 6050
[8] Pershin Y V and Piermarocchi C 2005 *Phys. Rev.* B **72** 125348
[9] Gogolin A O and Prokofev N V 1994 *Phys. Rev.* B **50** 4921

Symmetry and structural properties of carbon nanotube quantum dots and superlattices

W Jaskólski[1] **and L Chico**[2]

[1] Instytut Fizyki UMK, Grudziądzka 5, 87-100 Toruń, Poland

[2] Facultad de Ciencias del Medio Ambiente, Universidad de Castilla-La Mancha, 54071 Toledo, Spain

Abstract. The electronic structure and quantum conductance of metallic carbon nanotube quantum dots and superlattices are investigated. We study two kinds of quantum dots and superlattices: rotationally symmetric and rotationally asymmetric ones. It is shown that rotational invariance significantly influences the conductance of quantum dots. In contrast, quantum conductance of superlattices (multiple quantum dots) does not critically depend on their rotational symmetry. All the calculations are performed within the tight-binding model and π-electron nearest-neighbors approximation.

Carbon nanotubes (CN) are quasi-one-dimensional systems made by rolling-up a graphite sheet into a cylinder of nanosize diameter. They were discovered by S. Iijima in 1991[1], and since then, the interest in their fundamental properties and potential applications has been rapidly growing. One of the focus of current interest of nanotubes is their electronic behavior: depending on the rolling direction and on the tube diameter they may have metallic or semiconducting character. Carbon nanotubes are identified by two integers, (n, m), which are the coordinates of the circumference vector $\boldsymbol{C_h}$ in the graphene primitive basis $\{\boldsymbol{a_1}, \boldsymbol{a_2}\}$, chosen to form an angle[3] of 60°. If $\boldsymbol{C_h}$ goes along two neighboring carbon atoms, then $\boldsymbol{C_h} = n\boldsymbol{a_1} + n\boldsymbol{a_2}$, so the tube is labeled (n, n). This kind of nanotube is called *armchair*; it has $2n$ atoms around the circumference, and it is invariant under C_n rotations around the nanotube axis. If $\boldsymbol{C_h}$ goes along a lattice vector, e.g. $\boldsymbol{a_1}$, the resulting tube, $(n, 0)$ has n atoms at the circumference and it is invariant under C_n rotations. Because of the arrangement of atoms around its circumference, it is called *zig-zag*. All the other tubes, labeled (n, m), with $n \neq m$ and $n, m \neq 0$, are *chiral*: their atoms are arranged in a helical way around the circumference. Any smallest section of the CN cylinder, periodically repeatable along the tube axis, defines the carbon nanotube unit cell. In general, when n and m have a common divisor p, the (n, m) nanotube is invariant under C_p rotations[3]. The geometrical information in (n, m) is directly related to the electronic properties: if $(n - m)/3$ is integer the tube is metallic, otherwise it is semiconducting.

Introducing topological defects in the hexagonal carbon network allows to connect different tubes, which can have different electronic character, e.g., metallic and semiconductor. The junctions between different nanotubes are most usually realized by the introduction of pentagon/heptagon defects at the interface between them.[4] In the same fashion, one can design

[3] Initially, the basis with vectors at 120° was also used [2]. With this choice of basis, the labeling of nanotubes changes.

[4] Pentagons and heptagons are the weakest perturbations of the ideal hexagonal lattice. They naturally appear

a double junction, $(n,m)/N(n',m')/(n,m)$, formed by sandwiching a short section of N unit cells of one nanotube in between of two other (semiinfinite or long enough) tubes. Due to finite size of the central section, the bands of (n',m') quantize into discrete energy levels, so it can be considered as a carbon nanotube quantum dot (QD). And finally, a periodic sequence of junctions, $M(n,m)/N(n',m')$, is a carbon nanotube superlattice (SL). Such junctions, as well as other types of T-like or Y-like connections, have already been achieved experimentally[4, 5, 6, 7]. Together with the already demonstrated applications as e.g., field-effect and single-electron transistors or rectifying diodes, it opens up a route to an all-carbon based nanoelectronics.

Motivated by such experimental findings and possible future applications, different kinds of carbon nanotube junctions have been theoretically studied showing their unusual electronic and transport properties. Chico *et al.* [9] first found that when joining two different metallic tubes, a wide conductance gap can open up at the Fermi energy, when the entire system preserves some rotational symmetries and the bands of the joined tubes have different values of the discrete angular momentum[8] in a common energy interval. We use the term 'discrete angular momentum' in the meaning explained in Ref. [9]. Because of these apparently unusual behavior of all-metallic junctions, we explored the properties of CN double junctions, and we showed[10] that in quantum dots fulfilling the above-mentioned conditions, the conductance gap is covered by a series of non-conductive localized quantum dots states. In contrast, when the quantum dot does not preserve any rotational symmetry, the gap is covered by a series of conducting resonances, the widths of which depend on the wavevector mismatch[10, 11] between the (band) states of the metallic tubes forming the quantum dot. Finally, we demonstrated that in superlattices, built as periodic sequences of rotationally invariant quantum dots, some localized states transform into conducting minibands[12].

In this work we perform a comparative study of the electronic structure and quantum conductance of rotationally symmetric and rotationally non-symmetric quantum dots and superlattices. Knowing how the rotational invariance affects the conductance of quantum dots (carbon nanotube double junctions), we investigate effect of the rotational invariance on the conductance of superlattices (carbon nanotube multiple junctions). In particular we study: (A) rotationally symmetric $(12,0)/4(6,6)/(12,0)$ quantum dot and $4(12,0)/4(6,6)$ superlattice, (B) rotationally non-symmetric $(8,2)/4(5,5)/(8,2)$ quantum dot and $4(8,2)/4(5,5)$ superlattice, (C) rotationally symmetric $(9,3)/4(6,6)/(9,3)$ quantum dot and $4(9,3)/4(6,6)$ superlattice, (D) rotationally non-symmetric $(9,3)/4(6,6)/(9,3)$ quantum dot and $4(9,3)/4(6,6)$ superlattice. (A) and (B) are structurally different systems, while (C) and (D) differs only by the symmetry of the junction between $(9,3)$ and $(6,6)$ nanotubes. In case (C) the junction has a common rotation axis with both tubes, while in case (D) the junction is not rotationally invariant.

We work within the tight-binding model with one π-orbital per atom. The hopping parameter for nearest neighbors is fixed to $V_{pp\pi} = t = -2.66$ eV and second-neighbor interactions are neglected. The local density of states (LDOS) of quantum dots is calculated using the Green function matching approach[13]. The band structure and density of states (DOS) of superlattices is calculated by direct diagonalization of the tight-binding Hamiltonian matrix. The conductance of quantum dots is calculated employing the multichannel Landauer formula[14], while for superlattices it is calculated by counting the number of available channels (minibands) at each energy.

Let us consider first the systems built of $(12,0)$ and $(6,6)$ tubes. Although the considered tubes have slightly different diameters, they can be easily connected by means of 6 pairs of 5,7 rings. Such a connection is invariant under C_6 rotations around the common axis. Since $(12,0)$ and $(6,6)$ undergo the same rotational symmetry C_6, thus any system built of such nanotubes connected along the tube axes is invariant under C_6 rotations. However, the value of the discrete

in fullerenes and have also been experimentally detected in carbon nanotubes.

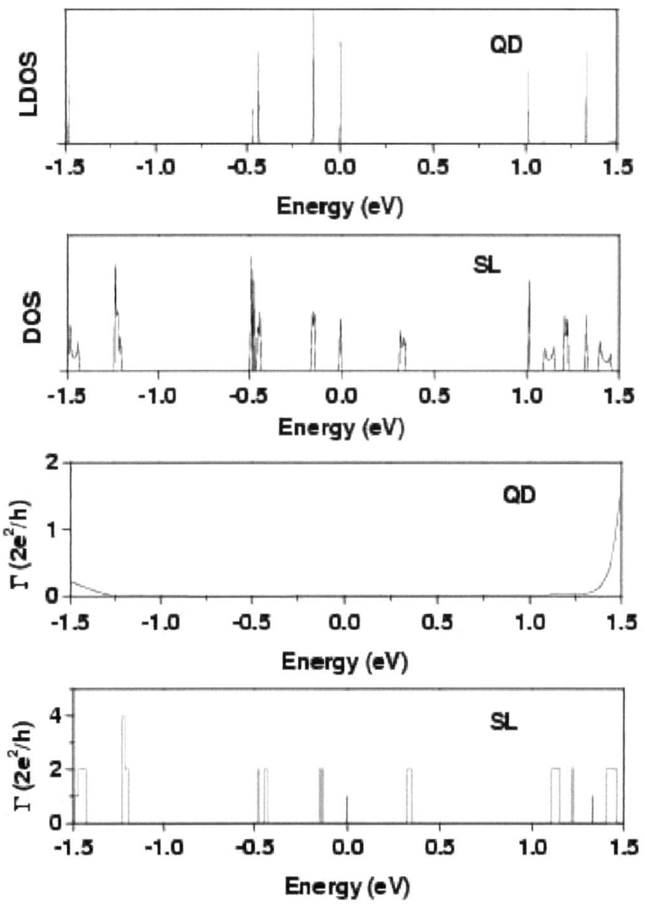

Figure 1. Density of states (local density of states in case of quantum dot) and conductance of rotationally symmetric $(12,0)/4(6,6)/(12,0)$ quantum dot and $(12,0)/4(6,6)/(12,0)$ superlattice. LDOS is taken at the $4(6,6)$ side of the interface.

angular momentum of the bands[15] close the Fermi level E_F in the $(12,0)$ tube is equal $L = 4$, while the corresponding value for the $(6,6)$ tube is $L' = 6$. Therefore, the conductance gap, as wide as ≈ 2.5 eV, opens up at E_F in the system built by connecting two such tubes[9]. The conductance gap persists in the $(12,0)/N(6,6)/(12,0)$ quantum dots[10] if $N > 2$. For $N = 4$ the gap is visualised in Fig. 1. The figure shows also LDOS at the $4(6,6)$ side of the interface. The δ-like peaks of LDOS correspond to a series of non-conducting states. Some of these states originate from the quantized bands of $(6,6)$ and from dispersionless bands of $(12,0)$ tube[12]. They are localized in the $4(6,6)$ quantum dot. The peaks at ≈ 0.15 and ≈ 0.43 eV correspond to the interface states. The corresponding wavefunctions are not invariant under any rotation.

In Fig. 1 the DOS and conductance of $4(12,0)/4(6,6)$ superlattice is also presented. In this case the localized QD states transform into narrow minibands. Some additional minibands appear also at -1.25, 0.3, 1.1, 1.2 and 1.4 eV; they originate from the quantized bands of the short sections of $(12,0)$ nanotube. All these minibands, except the dispersionless states (e. g., at 1.0 eV), are conductive, as confirmed in the bottom panel of Fig. 1.

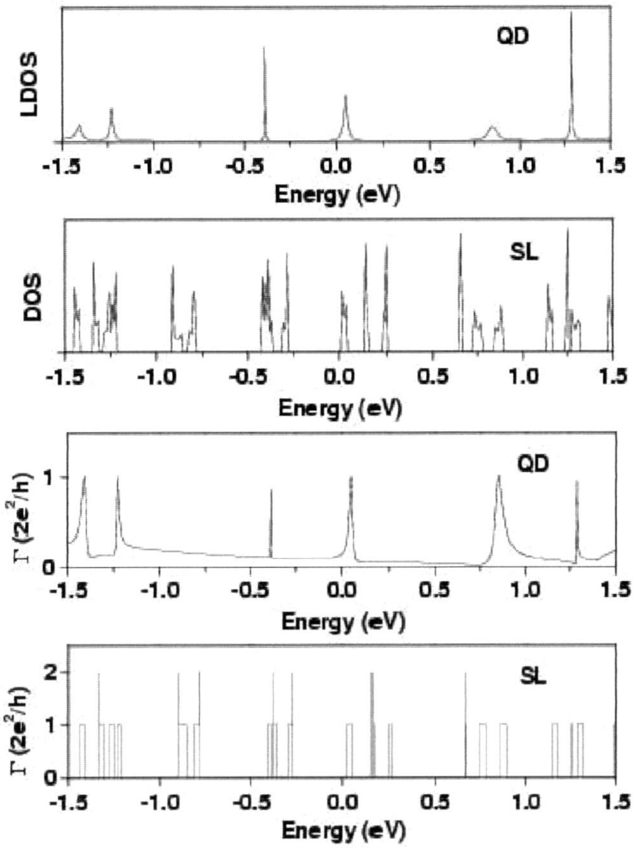

Figure 2. Density of states (local density of states in case of quantum dot) and conductance of rotationally asymmetric $(8,2)/4(5,5)/(8,2)$ quantum dot and $(8,2)/4(5,5)/(8,2)$ superlattice.

Now, let us consider systems that do not preserve any rotational symmetry, namely QD and SL built of metallic $(8,2)$ and $(5,5)$ tubes. The pure rotation groups of $(8,2)$ and $(5,5)$ are C_2 and C_5, respectively, so they have no common rotational symmetry. Furthermore, one can connect these tubes by a sequence of 6,6,5,7,5,7,6,6,5,7 rings, which is not invariant under any rotation. Therefore, no conductance gap opens up when two such tubes are connected to form a single junction. In the case of $(8,2)/N(5,5)/(8,2)$ quantum dot (double junction) the quantized bands of $(5,5)$ as well as interface states yield a series of resonance states, quasi-localized in the $N(5,5)$ dot[5], but coupled to the bands of $(12,0)$ leads. For $N=4$ these states are visualized in Fig. 2 as wide LDOS peaks. Fig. 2 shows also that these resonances lead to non-zero conductance. The detailed explanation of different widths of these states is given in Refs.[10, 11]. Similarly to the previous case, when a superlattice is formed as an infinite sequence of multiple $4(8,2)/4(5,5)$ quantum dots, the resonance states transform into minibands. Again, some additional minibands appear; they originate from quantized bands of the $(8,2)$ tube. As shown in Fig. 2, all the minibands are obviously conducting.

The presented results show that, although the rotational symmetry does critically influence

[5] The interface state are quasilocalized at the interfaces

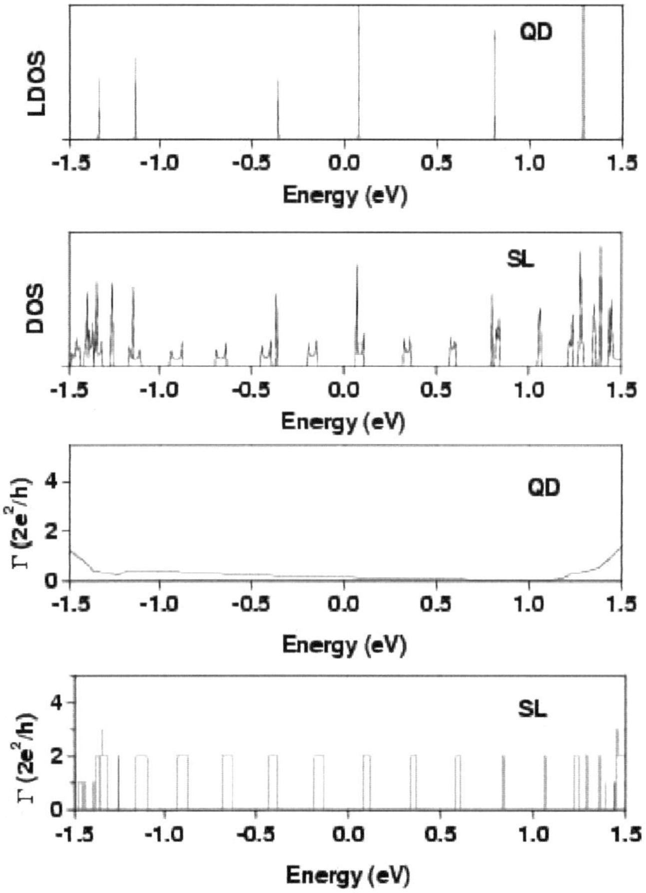

Figure 3. Density of states (local density of states in case of quantum dot) and conductance of rotationally symmetric (9, 3)/4(6, 6)/(9, 3) quantum dot and (9, 3)/4(6, 6)/(9, 3) superlattice.

the transport properties of quantum dots, leading even to wide conducting gaps in QDs built of all-metallic nanotubes, it does not significantly influence the conductance of short-period superlattices. In both kinds of superlattices, rotationally symmetric and asymmetric ones, the conductance gap around E_F is always supplemented by a series of narrow conducting minibands.

In the preceding examples we have explored the role of symmetry by comparing systems built of different pairs of nanotubes, so that the non-symmetric case consisted of tubes that had no common rotational symmetries themselves. But the importance of symmetry can be more strikingly shown by choosing one particular pair of tubes with a common rotational symmetry and joining them in different ways, with rotationally symmetric junctions or asymmetric junctions. As an example, let us consider now quantum dots and superlattices built of (9, 3) and (6, 6) nanotubes. As in the (12, 0)/(6, 6) case, close to E_F the bands of the joined tubes have different values of the discrete angular momentum. The (9, 3) tube is invariant under rotations of C_3 around its axis while the rotational symmetry of (6, 6) is C_6. Both tubes can be connected by a the sequence of 6,6,7,5,6,6,7,5,6,6,7,5 rings. Such a junction assures that the entire system is invariant under C_3 rotations. However, one can also connect these tubes by

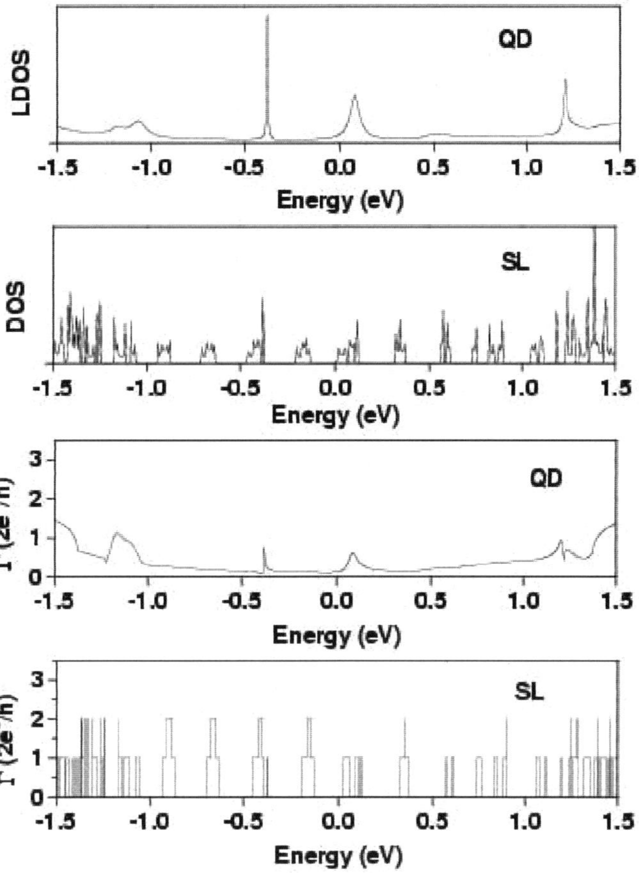

Figure 4. Density of states (local density of states in case of quantum dot) and conductance of (9, 3)/4(6, 6)/(9, 3) quantum dot and (9, 3)/4(6, 6)/(9, 3) superlattice with asymmetric junction.

rotationally asymmetric junction of 5,7,5,7,6,6,5,7,6,6,6,6 rings. Such an asymmetric connection makes the entire system rotationally non-invariant. The asymmetric junction can be consider as the perturbation of the symmetric one by a simple transposition of one 5,7 pair with the 6,6 pair. It is interesting to check the influence of such a perturbation on the transport properties of quantum dots and superlattices built of (9, 3) and (6, 6) nanotubes.

In Fig. 3 the DOS/LDOS and conductance of rotationally symmetric (9, 3)/4(6, 6)/(9, 3) QD and 4(9, 3)/4(6, 6) SL are presented. The results for rotationally asymmetric dot and superlattice are shown in Fig. 4. By comparison to Figs. 1 and 2 one can draw the same conclusions as before: the rotational symmetry of quantum dots built of metallic nanotubes can lead to a wide conductance gap opening at the Fermi energy.[6] If the rotational symmetry is broken, the gap is substituted by a series of conducting minibands. The conductance of the superlattice does not change significantly. In both, symmetric and asymmetric cases the conductance gap is covered by a series of conducting minibands. The main effect of the symmetry breaking on the

[6] The non-totally-zero conductance of (9, 3)/4(6, 6)/(9, 3) QD is due to the residual tunneling conductance between the (9, 3) leads through the finite 4(6, 6) section.

4(9, 3)/4(6, 6) superlattice minibands is lifting of their degeneracy.

Acknowledgments

This work has been supported by the Polish KBN(MNI) Grants 3T1104326 and PZB-MIN-008/P03/2003 and Spanish Grants MAT2002-04540-C05-03 (DGES) and GR/MAT/0440/2004 (Comunidad Autónoma de Madrid).

References

[1] Iijima S 1991 *Nature* (London) **354** 56
[2] Hamada N *et al* 1992 *Phys. Rev. Lett.* **68** 1579
[3] Saito R Dresselhaus G and Dresselhaus M S 1998 *Physical Properties of Carbon Nanotubes* (London: Imperial College Press)
[4] Tans S J, Verschueren A R M and Dekker C 1998 *Nature* (London) **393** 49
[5] Stadermann M *et al* 2004 *Phys. Rev.* B **69** 201402
[6] Menon M and Srivastava S 1997 *Phys. Rev. Lett.* **79** 4453
[7] Jing L, Papadopoulos C and Xu J 1999 *Nature* (London) **402** 253
[8] We use the term 'discrete angular momentum' in the meaning explained in Ref. [9]
[9] Chico L, Benedict L X, Louie S G and Cohen M L 1996 *Phys. Rev.* B **54** 2600
[10] Chico L and Jaskólski W 2004 *Phys. Rev.* B **69** 089406
[11] Chico L, Jaskólski W, Lopez-Sancho M P and Munoz M C 2005 *Int. J. Nanotechnology* **2** 103
[12] Jaskólski W and Chico L 2005 *Phys. Rev.* B **71** 155405
[13] García-Moliner F and Velasco V R 1992 *Theory of Single and Multiple Interfaces* (Singapore: World Scientific)
[14] Datta S 1995 *Electronic Transport in Mesoscopic Systems* (Cambridge: Cambridge University Press)
[15] Jaskólski W, Stachów A and Chico L *Acta Phys. Pol.* in press

Symmetry analysis of quadrupolar and structural ordering in single-q phases of UPd3

W Sikora[1], F Białas[2], L Pytlik[1] and J Malinowski[1]

[1] Faculty of Physics and Applied Computer Science, AGH University of Science and Technology, al.Mickiewicza 30, 30-059 Krakow, Poland

[2] Nowy Sacz School of Business - National Louis University, Nowy Sacz, Poland

E-mail: sikora@novell.ftj.agh.edu.pl

Abstract. The paper describes the application of symmetry analysis to construction of possible quadrupolar ordering modes in a crystal with a given space group symmetry. The matrix representing a given tensor property (e.g. quadrupolar moment) is converted to a vector form, for which the standard symmetry analysis algorithms can be applied. The possible ordering modes are obtained as linear combinations of basis vectors of irreducible representations of a given space group, but taking into account the constraints, specific for a given physical quantity (e.g. symmetric matrix form and zero trace for quadrupolar tensors). The results are applied to the single-q phases of UPd_3 compound. For completeness the possible structural ordering modes are also calculated, which are consistent with the symmetry of a given quadrupolar ordering.

1. Introduction

In the present paper the ordering of quadrupolar tensor components and possible deformation of the initial potential shape in a quadrupolar phase transition are investigated by the symmetry analysis method [1]. The components of quadrupolar moment tensor (QMT) are calculated from the charge density distribution by the respective textbook formula. The matrix of QMT is always symmetrical and additionally in orthogonal coordinate systems traceless. The QMT matrix can be brought to the principal axes, which may be treated as a local symmetry-adapted coordinate system. Then the matrix of QMT takes the simplest form: only three values on the diagonal can be nonzero and only two of them are independent. For an initial spherical potential shape the QMT form defines the directional deviations from the spherical form, resulting from the presence of nonzero quadrupolar moment.

The quadrupolar ordering concept is intended to characterize, at all atomic positions in the crystal, the spatial shape of the surface associated with the quadrupolar potential (or the surface associated with the deformed total potential) and the orientation of QMT's principal axes in relation to the crystallographic system. The signs of QMT coefficients characterize the type of deformation and their values characterize the magnitude of the potential surface deformation. By the symmetry analysis method all types of quadrupolar ordering, allowed for a given parent structure, are calculated, together with their resulting final symmetry groups. An additional analysis can be also carried out to determine the possible patterns of atomic displacements of the neighboring atoms, which are consistent with the symmetry imposed by the quadrupolar ordering.

In the present work the method is applied to quadrupolar moment ordering of U atoms and possible patterns of atomic displacements of the neighboring Pd atoms in the hexagonal UPd_3 compound. The high-temperature structure of UPd_3 - the parent structure - is described by the hexagonal space group $P6_3/mmc$ (D_{6h}^4) and it is isostructural to Ni_3Ti structure [11]. The atomic positions found from the diffraction experiments are as follows: four uranium atoms are located in quasi-cubic 2(a) sites: U1(0,0,0), U2(0,0,1/2) and hexagonal 2(c) sites: U3(1/3,2/3,1/4), U4(2/3,1/3,3/4) while the twelve palladium atoms are located in 6(h) and 6(g) sites [5,12]. The experimental data from neutron and X-ray diffraction, [2,3] as well as measurements of heat capacity, thermal expansion, ultrasound attenuation [4], μSR [5], magnetic susceptibility [6] indicate that a sequence of phase transitions is encountered at temperatures T_0=7.6K, T_{+1}=6.9K, T_{-1}=6.7K and T_2=4.4K. The analysis of diffraction data (neutrons, X-rays) indicates that below T_0 various forms of antiferroquadrupolar ordering are present, together with the accompanying structural changes, and that some form of antiferroquadrupolar ordering coexists with antiferromagnetic ordering below T_2 [2,6]. The non-zero quadrupolar moments are found mainly for the U atoms located in the quasi-cubic 2(a) sites. In the quadrupolar phase the ordering vector is found to be $\boldsymbol{k}=(1/2,0,0)$ and principal axes of the quadrupolar moments coincide with orthorhombic coordinate system axes. Below T_{+1} a deflection of the quadrupolar moments from the coordinate system axes is observed [5]. Additionally there is a suggestion [4,7] that both phases below T_{-1} exhibit the so called triple-q structure in order to avoid anti-phase domain walls and that the associated phase transitions at T_{-1} and T_2 are of the first order, what possibly breaks the group-subgroup relation with the parent hexagonal phase.

The present paper deals with the possible types of quadrupolar ordering modes in the two single-q phases ($T_{-1}< T < T_0$), the resulting final symmetry groups, and the possible patterns of atomic displacements of the neighboring Pd atoms, which are consistent with the symmetry imposed by the quadrupolar ordering. The models of ordered structure and discussion of corresponding sets of free parameters calculated from symmetry considerations are compared with the results of experimental investigations.

2. Elements of the symmetry analysis

2.1. *Description of the ordering pattern.* The symmetry analysis method, used for construction of the ordering pattern, is known for a long time and it works properly if some conditions are fulfilled [1]. The symmetry change should preserve the group – subgroup relation between the initial and the final phase and the described property should be localized on some set of equivalent positions in the unit cell of the initial phase. The property can be represented by a scalar, vector (polar or axial) or tensor quantity.

The method itself has been derived from the theory of group and representations and is based on the decomposition of full (permutational, , mechanical, magnetic, or quadrupolar) representation of the crystal space group, calculated for a given wave vector **k** and a given orbit of positions, into its irreducible representations (IR-s). Such a decomposition allows calculation of the so called basis vectors (BV-s) attributed to each IR, i.e. vectors which under the action of symmetry operations transform according to the respective IR. The final ordering pattern of the investigated property is built up as a linear combination of basis vectors attributed to one or more irreducible representations. Thus the global vector $\vec{S}$, describing the considered local property $\hat{u}$ in the whole crystal can be written as:

$$\vec{S}^{\{k_L\}} = \begin{bmatrix} \begin{bmatrix} \hat{u}(\vec{r}_1) \\ \dots \\ \hat{u}(\vec{r}_n) \end{bmatrix} \\ \dots \\ \begin{bmatrix} \hat{u}(\vec{r}_1 + \vec{t}_{p,q,s}) \\ \dots \\ \hat{u}(\vec{r}_n + \vec{t}_{p,q,s}) \end{bmatrix} \end{bmatrix} = \sum_{k_L,\nu,\lambda} C_{\nu,\lambda}^{(k_L)} \begin{bmatrix} \begin{bmatrix} \Psi_{\nu,\lambda}^{(k_L)}(\vec{r}_1) \\ \dots \\ \Psi_{\nu,\lambda}^{(k_L)}(\vec{r}_n) \end{bmatrix} \\ \dots \\ \begin{bmatrix} \Psi_{\nu,\lambda}^{(k_L)}(\vec{r}_1)\, e^{i\vec{k}_L \vec{t}_{p,q,s}} \\ \dots \\ \Psi_{\nu,\lambda}^{(k_L)}(\vec{r}_n)\, e^{i\vec{k}_L \vec{t}_{p,q,s}} \end{bmatrix} \end{bmatrix} \tag{1}$$

where the sum on the right-hand side runs over $\{\boldsymbol{k}_L\}$, denoting one or more wave vectors describing the ordering. The other summation indices ν, λ, denote respectively the IR label and an additional index, numbering the basis vectors belonging to the given ν-th IR. The symbols $\Psi_{\nu,\lambda}^{(k_L)}$ denote the parts of the respective basis vectors, calculated for atom i, for a given wave vector $\boldsymbol{k}_L$.. The translation vector: $\boldsymbol{t}_{p,q,s}=p\boldsymbol{a}_1+q\boldsymbol{a}_2+s\boldsymbol{a}_3$ runs over all unit cells of the crystal, when the indices p,q,s scan the set of integer numbers. As can be noticed the translation of the individual site $\boldsymbol{r}_n$ by the lattice vector $\vec{t}_{p,q,s}$ is equivalent to multiplication of the local part of the respective basis vector by a factor $\exp(i\vec{k}_L \cdot \vec{t}_{p,q,s})$. Taking this into account one can focus attention on the single-site description of the property $\hat{u}$:

$$\hat{u}_\alpha(\vec{r}_i) = \hat{u}_{i,\alpha}^{(k_L)} = \sum_{k_L}\sum_m C_m^{(k_L)} \Psi_{m,a}^{(k_L)}(\vec{r}_i) \tag{2}$$

where α labels the components of $\hat{u}$ at the local site and the new sum index $m \equiv \nu,\lambda$ numbers the consecutive basis vectors used in the linear combination.

As shown above construction of ordering mode requires the exact knowledge of basis vectors attributed to individual irreducible representations of a given space group for all the wave vectors contributing to the ordering pattern. There are several software packages that can be used for such calculations. The basic application written and used by the authors of the present paper is the MODY program [8].

The local description of a given property, denoted here as $\hat{u}$, can be a vector (displacement, magnetic moment) or a scalar (electron density, probability) , which is in fact a reduced, one element vector. Therefore the regular, well-known symmetry analysis has been elaborated for vector quantities. In order to apply the whole algorithm to tensor quantities (e.g. electric or magnetic quarupolar moment) a special procedure has to be applied, transforming both the matrix representing the local tensor quantity $\hat{u}\ (r_n)$ and the transfomation matrices. The tensor 3x3 matrix is transformed into a nine element vector with the new vector element index r calculated as: $r=3(i-1)+j$. If the same index recalculation procedure is applied to the transformation matrices the well-known vector algorithm can be applied directly [9].

2.2. *The constraints imposed on the solution.* The coefficients of the linear combination $C_m^{(k_L)}$ are the objects, which in fact specify the exact form of the global ordering pattern. In principle they should be arbitrary, as they explore the freedom left after imposing the symmetry requirements. However there are additional conditions that have to be fulfilled, if the solution for $\vec{S}^{\{k_L\}}$ is to represent a given physical quantity. For most vector quantities (displacement, magnetic moment) the basic requirement is that all the components of $\hat{u}$ should be real, even if the basis vector used for construction of the solution are complex. The respective equation takes the form:

$$\mathrm{Im}\{\hat{u}_{i,\alpha}^{(k_L)}\} = 0 \qquad \forall i,\alpha \tag{3}$$

For tensor quantities the detailed constraints depend on the exact nature of the physical quantity. For moment of inertia tensor the only requirement would be the symmetric form of the matrix, for electric quadrupolar moment tensor an additional condition is the zero trace of the matrix. The respective equations look as follows:

$$u_{i,12}^{(k_L)} = u_{i,21}^{(k_L)} \quad u_{i,23}^{(k_L)} = u_{i,32}^{(k_L)} \quad u_{i,13}^{(k_L)} = u_{i,31}^{(k_L)} \; u_{i,11}^{(k_L)} + u_{i,22}^{(k_L)} + u_{i,33}^{(k_L)} = 0 \qquad \forall i \tag{4}$$

The above requirements generate a set of equation for the unknown $C_m^{(k_L)}$ coefficients, and after solving this equation set the final solution can be written. It usually happens that even after imposing the constraints there are still some free parameters left, like the amplitude or phase of the ordering mode, the number of which depends on the irreducible representations used for construction of the solution.

3. Quadrupolar ordering concept

3.1. *Quadrupolar potential contribution and its visualisation.* The quadrupolar ordering concept is intended to characterise the electric potential and electron density distribution at individual atomic positions in the crystal. In particular it is intended to describe:

- orientation of principal axes of quadrupolar moment tensor (QMT) in relation to the crystallographic system axes
- the spatial shape of quadrupolar potential, related to the values of QMT components

The expansion of electric potential from any local atomic charge distribution may be written as:

$$\Phi(\vec{r}) = \Phi_m(\vec{r}) + \Phi_d(\vec{r}) + \Phi_q(\vec{r}) + \ldots \tag{5}$$

where the m,d and q indices denote the monopolar, dipolar and quadrupolar contributions to the total potential. The shape of the potential and charge distribution around the local crystal site and depends on the relative contributions from the individual terms of the expansion. It is often assumed that for zero external field conditions the dipolar contribution is negligible and the shape of local charge and potential distributions is controlled by the quadrupolar contribution, as the monopolar contribution is fully symmetric. The quadrupolar contribution can be characterized by its expansion into spherical functions [10]. Such expansion has been used for visualization of the angular dependence of both the quadrupolar potential contribution and the total potential. The analyzed potential values are calculated for a fixed distance from the local center (r_0=const). In the program Tensor_Vis by J.Malinowski [9] the potential value for a given set of spherical angles (θ, φ) is translated into the radial distance from the center, thus graphically reflecting the spatial shape of the potential distribution. All the potential shape pictures found below have been produced using that program.

3.2. *Properties of quadrupolar moment tensor matrix* The components of quadrupolar moment tensor (QMT) in orthogonal coordinate systems are usually defined as [10]:

$$Q_{ij} = \int (3x_i x_j - \rho^2 \delta_{ij}) q(\vec{\rho}) d^3x \tag{6}$$

The definition implies two properties for QMT matrix: it is always symmetric i.e. $Q_{ij} = Q_{ji}$ and its the trace is always zero i.e. $Q_{33} = -Q_{11} - Q_{22}$. The general form of the QMT matrix can always be transformed to the diagonal form by transformation to the local coordinate system oriented along the so called principal axes of QMT tensor. The general and diagonal forms of QMT look respectively as below:

$$\hat{Q}_{gen} = \begin{pmatrix} Q_{11} & Q_{12} & Q_{13} \\ Q_{12} & Q_{22} & Q_{23} \\ Q_{13} & Q_{23} & (-Q_{11}-Q_{22}) \end{pmatrix} \quad \hat{Q}_{diag} = \begin{pmatrix} Q_a & 0 & 0 \\ 0 & Q_b & 0 \\ 0 & 0 & (-Q_a-Q_b) \end{pmatrix} \tag{7}$$

The general form of QMT suggests that there should be five arbitrary parameters, but the transformation to diagonal form shows that there are only two parameters characterizing the shape of the potential itself: Q_a and Q_b. The other three parameters characterize the spatial orientation of the potential (electron density) shape. It should be stressed that the presence of two free parameters is a direct proof that not all shapes of the electron density cloud (potential) can be described by a prolate or oblate spheroid.

$$\begin{bmatrix} 4 & 0 & 0 \\ 0 & -4 & 2 \\ 0 & 2 & 0 \end{bmatrix}$$

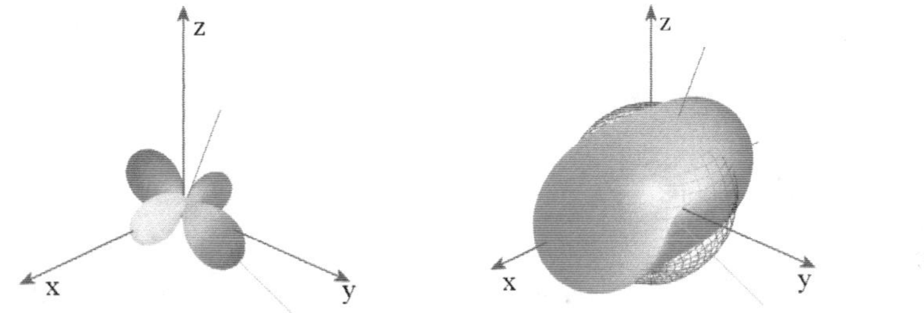

Figure 1. An off-diagonal QMT form (*left*) and the respective shapes of the quadrupolar contribution (*center*) and the total potential (*right*). The Q_{xx} and Q_{yy} elements control mainly the shape and the nonzero Q_{yz} elements control the orientation of principal axes.

The QMT matrix components can be used for calculation of the expansion coefficients in the potential expansion into spherical functions. The necessary relations can be found in [10] and they have been used to produce the values of quadrupolar contribution to the potential, later used for its visualization. A direct illustration of such visualization can be found in Figure 1. The diagonal forms of QMT can be classified with respect to the signs and relative values of the Q_a and Q_b parameters, and the resulting shapes can be attributed to the well-known classification quadrupolar operators $Q_{\alpha,\beta}$. For example the QMT matrix in Figure 1 and the corresponding shapes of the quadrupolar and total potential surfaces can be attributed to the operator combination labeled as $(Q_{x^2-y^2} + Q_{yz})$.

4. Results of the symmetry analysis of quadrupolar ordering

Symmetry analysis of the quadrupolar ordering applied to the parent structure with the *P 6₃/mmc* symmetry group and U quadrupolar moments localized on 2(a) positions indicates, that these positions belong to one orbit of the wave vector group G(k) for the ordering vector ***k***=*(1/2,0,0)*. All irreducible representations for that ***k*** vector are real and one-dimensional. IR-s that are potentially active in the quadrupolar phase transition are contained in the full representation decomposition for U(2a) positions: τ_1 and τ_5 - five times each, τ_3 and τ_7 – four times each. It means, that we have five different basis vectors, which transform according to the τ_1 representation, and which may be used in the linear combination, to produce a quadrupolar structure solution leading to he final orthorhombic *P mma* symmetry, corresponding to this representation. The similar situation occurs for τ_5, leading to quadrupolar structure with the *P nnm* symmetry group, which differs from the τ_1 structure by the sign of the QMT at the U2 site. For IR τ_3 we have four different basis vectors, which may be used in the linear combination, leading to the structure with *P nma* symmetry group. The same situation occurs for τ_7, leading to a structure for which the QMT signs at U(2) sites are opposite to the respective signs in τ_3 structure, and which exhibits the *P bcn* symmetry group.

The full representation decomposition for U(2c) positions contains: IR-s τ_1 and τ_8 - three times each, and IR-s τ_2 ,τ_3 ,τ_4 ,τ_5 ,τ_6 and τ_7 – two times each. The representations in QMT ordering at U(2c), which are possibly present also for QMT ordering at U(2a) are: τ_1 ,τ_3 ,τ_5 and τ_7 .

The BV functions calculated by the MODY program for *P 6₃/mmc* group are given in the non-orthogonal, hexagonal setting. Therefore the linear combinations of basis vectors are also calculated in the hexagonal system. For discussion of the minimal number of relevant free parameters in the final solution, i.e. after applying the conditions imposed on the coefficients of the linear combination and for drawing the shape of potential it is convenient to transform these functions to the orthogonal system. Therefore for the discussion of solutions the orthorhombic system is naturally chosen, in the setting related to the orthorhombic cell of the structure after phase transition.

For U(2a) IR-s τ_1 and τ_5 the five unknown coefficients C_m $(m=1,5)$ of the linear combination in the hexagonal system are treated as free parameters *(A,B,C,D,E)*. The conditions imposed on the solution require that the parameters have to be related by the following equations: $A-B/2+E=0$ and $C=D$. As can be seen from these equations two groups of parameters may be regarded independently: *(A,B,0,0,E)* and *(0,0,C,D,0)*. The same types of QMT structures, which correspond to these two groups of parameters, are found for QMT solutions at U(2c) sites, but in the (2c) case the respective parameter sets are attributed to different IR-s.

Table 1a. The characteristic types of U(2a) QMT solutions for IR τ_1 and parameter set *(A,B,0,0,E)*, (valid also for U(2c) solutions for IR τ_8 and parameter set *(A',B',E')*)

Coefficients of linear combinations	QMT matrix in hexagonal setting Q(1)=Q(2)	QMT matrix in orthogonal setting Q(1)=Q(2)	Angular distribution of Φ_q	Directions of principal axes
$A=-2Q$ $B=-2.5Q$ $E=0.75Q$	$\begin{pmatrix} -2Q & -Q & 0 \\ -Q & 0.25Q & 0 \\ 0 & 0 & 0.75Q \end{pmatrix}$	$\begin{pmatrix} -1.5Q & 0 & 0 \\ 0 & 0.75Q & 0 \\ 0 & 0 & 0.75Q \end{pmatrix}$	Fig.2a	**[2 1 0]$_h$** [0 1 0]$_h$ [0 0 1]$_h$
$E=Q$ $B=2Q$ $E=0$	$\begin{pmatrix} Q & 0.5Q & 0 \\ 0.5Q & -0.5Q & 0 \\ 0 & 0 & 0 \end{pmatrix}$	$\begin{pmatrix} 0.75Q & 0 & 0 \\ 0 & -0.75Q & 0 \\ 0 & 0 & 0 \end{pmatrix}$	Fig.2b	[2 1 0]$_h$ [0 1 0]$_h$ **[0 0 1]$_h$**

The characteristic types of quadrupolar structures for these types of parameter sets are given in the Table 1a and Table 1b correspondingly. The axis shown in boldface in the last column denotes that QMT principal axis, which is also the full rotation symmetry axis of the deformed potential, or the direction where deformation of spherical potential is zero. The amplitude parameter Q changes only the amplitude of deformation, without changing the type of potential shape or directions of QMT principal axes. As shown in Table 1a, essentially different choices of *A,B,E* parameters, which all fulfill the equation (4) condition, lead to quadrupolar orderings with QMT principal axes directed along the other two axes of orthogonal system. The respective solutions at U(2a) for τ_5 are different by the QMT sign at the U2 site. The same types of solutions are obtained for U(2c) sites, but the one which corresponds to IR τ_5 and *(A,B,0,0,E)* at U(2a) is attributed to τ_1 and *(A',B',E')* at U(2c), while the one corresponding to τ_5 and *(0,0,C,D,0)* at U(2a) is attributed to τ_4 and *(C',D')* at U(2c).

Table 1b. The characteristic types of U(2a) QMT solutions for IR τ_1 and parameter set *(0,0,C,D,0)* (valid also for U(2c) solutions for IR τ_5 and parameter set *(C',D')*)

Atom	QMT matrix in hexagonal setting Q(1)= -Q(2)	QMT matrix in orthogonal setting Q(1)= -Q(2)	Angular distribution of Φ_q	Directions of principal axes
U1 (0,0,0) *C=D=Q*	$\begin{pmatrix} 0 & 0 & Q \\ 0 & 0 & 0.5Q \\ Q & 0.5Q & 0 \end{pmatrix}$	$\begin{pmatrix} 0 & 0 & \sqrt{3}Q/2 \\ 0 & 0 & 0 \\ \sqrt{3}Q/2 & 0 & 0 \end{pmatrix}$	Fig.2c	**$[0\ 1\ 0]_h$ $[0\ 1\ 0]_o$** $[1\ 0\ 1]_o$ $[1\ 0\ \text{-}1]_o$

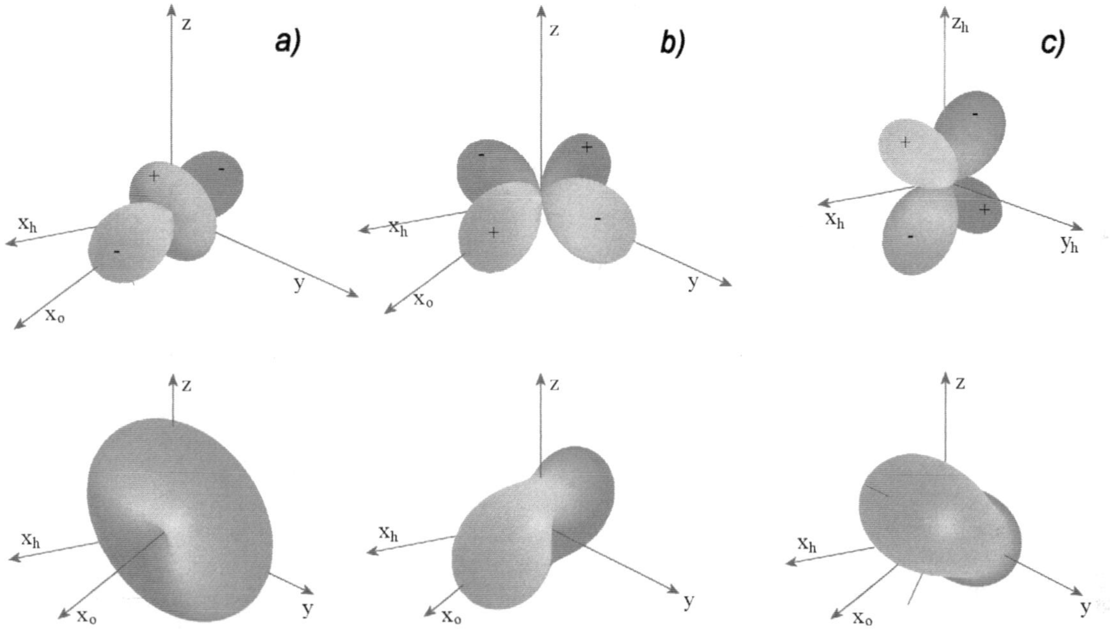

Figure 2. The angular distributions of Φ_q and Φ for IR τ_1 and the following parameter sets: a) *(-2Q,-2.5Q,,0.75Q);* b) *(Q,,2Q,0,0,0)*; c) (*0,0,Q,,Q,*0),

For U(2a) QMT solutions for IR τ_3 and τ_7 the sets of four parameters *(A',B',C' D')* in the hexagonal system, which fulfill the Eq.(4) conditions, have to be obey the following equations: *A'=C'* and *B'=D'*. Now it can be seen that the following parameter sets may be considered independently: *(A',0,A',0)* and *(0,B',0,B')*. Also in this case the types of U(2a) QMT structures, which correspond to these two groups of parameters, are similar to QMT solutions at U(2c) sites, but again in (2c) both cases are attributed to different IR-s.

The characteristic types of QMT structures for these parameter sets and for their combination *(A',B',A',B')* are given in the table 2. The respective solutions for IR τ_7 are different only by the QMT sign at the U2 site. The same type of solutions are obtained for U(2c) sites, but the one corresponding to IR τ_7 and *(A',0,A',0)* at U(2a) sites is attributed to IR τ_3 .and *(A',A')* at U(2c), and the solution type which belongs to τ_7 and *(0,B',0,B')* at U(2a) sites is attributed to τ_2 and *(B',B')* at U(2c) sites.

Table 2. The characteristic types of U(2a) QMT solutions for IR τ_3, and parameter set (A,B,A,B) (valid also for U(2c) QMT solutions for IR-s τ_6 and τ_7) and independent parameter sets $(A,0,A,0)$ and $(0,B,0,B)$

Coefficients of linear combinations	QMT matrix in hexagonal setting	QMT matrix in orthogonal setting	Angular distribution of Φ_q	Direction of principal axes
$p=(A;B;A;B)$ U(2a): τ_3 U(2c): $\tau_6+\tau_7$	$\begin{pmatrix} 0 & A & 0 \\ A & A & -B \\ 0 & -B & 0 \end{pmatrix}$	$\begin{pmatrix} 0 & \sqrt{3}A/2 & 0 \\ \sqrt{3}A/2 & 0 & -B \\ 0 & -B & 0 \end{pmatrix}$	Fig.3a	**$[k\ 0\ l]_o$**
$p=(A;0;A;0)$ Q(1)=Q(2) U(2a): τ_3 U(2c): τ_6	$\begin{pmatrix} 0 & A & 0 \\ A & A & 0 \\ 0 & 0 & 0 \end{pmatrix}$	$\begin{pmatrix} 0 & \sqrt{3}A/2 & 0 \\ \sqrt{3}A/2 & 0 & 0 \\ 0 & 0 & 0 \end{pmatrix}$	Fig.3b	**$[0\ 0\ 1]_o$** $[1\ \text{-}1\ 0]_o$ $[1\ 1\ 0]_o$
$p=(0;B;0;B)$ Q(1)= -Q(2) U(2a): τ_3 U(2c): τ_7	$\begin{pmatrix} 0 & 0 & 0 \\ 0 & 0 & -B \\ 0 & -B & 0 \end{pmatrix}$	$\begin{pmatrix} 0 & 0 & 0 \\ 0 & 0 & -B \\ 0 & -B & 0 \end{pmatrix}$	Fig.3c	**$[1\ 0\ 0]_o$** $[0\ 1\ 1]_o$ $[0\ 1\ \text{-}1]_o$

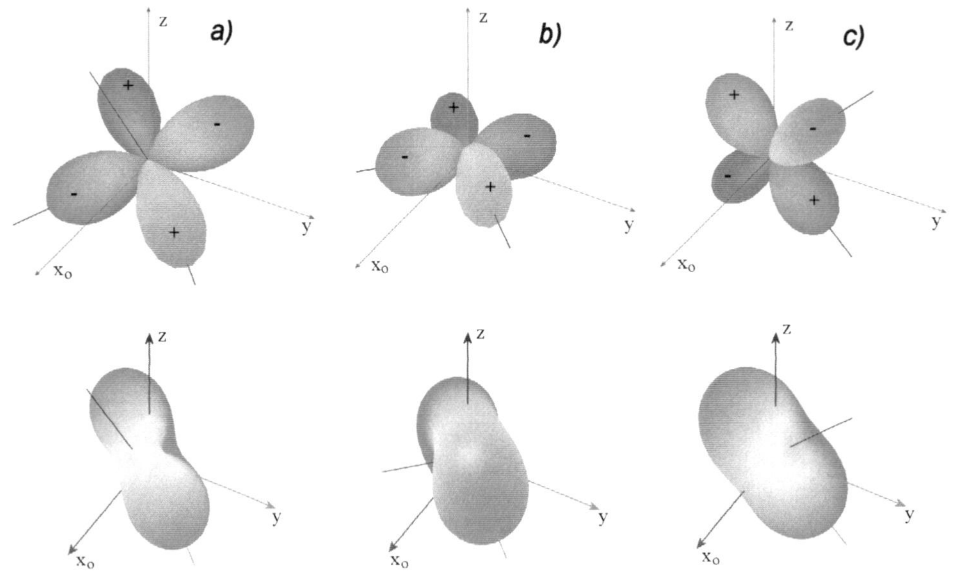

Figure 3. The angular distribution of Φ_q and Φ for τ_3 and sets of coefficients: a) (A,B,A,B); b) $(A,0,A,0)$, c) $(0,B,0,B)$

The symmetry analysis calculations carried out for different arms of the k-vector star ($\boldsymbol{k_2}=(0,1/2,0)$ and $\boldsymbol{k_3}=(-1/2,1/2,0)$) lead to quadrupolar structures, which are simply different domain orientations of the structures proposed for the $\boldsymbol{k_1}=(1/2,0,0)$ and described above.

5. Symmetry analysis of the accompanying structural deformations

The structural deformations, accompanying the quadrupolar ordering, should exhibit the same final symmetry, i.e. they should be described by the same active IR-s. The decomposition of reducible mechanical (crystal) representation carried out for U atoms localized at 2(a) positions of *P 6₃/mmc* group, with the same ***k***-vector as for quadrupolar ordering indicates, that the IR-s active in these structural changes are τ_2, τ_4, τ_6 and τ_8. None of them appears as representation possibly active in the quadrupolar ordering. It indicates, that the displacements of U(2a) atoms are not directly associated with quadrupolar phase transition. For U(2c) sites the possible active representations are τ_1, τ_2, τ_3, but only τ_1 and τ_3 are shared with the quadrupolar ordering. The respective displacements of U(2c) atoms are shown in Table 4c.

Table 3a. Splitting of 6(h) positions on two orbits in G(k), and corresponding IR-s active in the structural deformations

Orbit	Atoms position				Active representations
1	Pd1	(x,2x,0.25)	Pd2	(1-x,1-2x,0.75)	$2\tau_1, \tau_2, 2\tau_3, \tau_4,$
	Pd3	(1-x,x,0.75)	Pd4	(x,1-x,0.25)	$\tau_5, 2\tau_6, \tau_7, 2\tau_8$
2	Pd5	(1-2x,1-x,0.25)	Pd6	(2x,x,0.75)	$\tau_1, \tau_3, \tau_4, \tau_5, \tau_6, \tau_8$

Table 3b. Splitting of 6(g) positions on two orbits in G(k), and corresponding IR-s active in the structural deformations

Orbit	Atoms position				Active representations
1	Pd1	(0.5,0,0)	Pd2	(0.5,0,0.5)	$3\tau_1,, 3\tau_3,$
	Pd3	(0.5,0.5,0)	Pd4	(0.5,0.5,0.5)	$3\tau_5, 3\tau_7$
2	Pd5	(0,0.5,0)	Pd6	(0,0.5,0.5)	$\tau_2, 2\tau_4, \tau_6, 2\tau_8$

The atoms of Pd located at 6(h) and 6(g) positions in the parent UPd_3 structure split into two orbits in the G(k_1) group. They are given in Table 3a and Table 3b, together with corresponding active IR-s. The active IR-s common for Pd displacements and quadrupolar ordering are τ_1, τ_3, τ_5 and τ_7. Because these representations are real and one dimensional, the BV-s calculated by the MODY program give directly the possible displacements of Pd atoms leading to the structures with the symmetry groups discussed earlier for the quadrupolar ordering. They are listed in the Table 4a and 4b.

Table 4a. Possible displacements of Pd atoms at 6(h) sites for ***k***=*(1/2,0,0)* accompanying the quadrupolar ordering

IR	Version	Orbit 1				Orbit 2	
		Pd1	Pd2	Pd3	Pd4	Pd5	Pd6
τ_1	1	*(A,0,0)*	*(A,0,0)*	*(A,A,0)*	*(A,A,0)*	*(2C,C,0)*	*(2C,C,0)*
	2	*(0,B,0)*	*(0,B,0)*	*(0,-B,0)*	*(0,-B,0)*		
τ_3	1	*(A,0,0)*	*(A,0,0)*	*(-A,-A,0)*	*(-A,-A,0)*	*(0,C,0)*	*(0,C,0)*
	2	*(0,B,0)*	*(0,B,0)*	*(0,B,0)*	*(0,B,0)*		
τ_5	1	*(0,0,A)*	*(0,0,A)*	*(0,0,A)*	*(0,0,A)*	*(0,0,C)*	*(0,0,C)*
τ_7	1	*(0,0,A)*	*(0,0,A)*	*(0,0,-A)*	*(0,0,-A)*		

Table 4b. Possible displacements of Pd atoms at 6(g) sites for $\boldsymbol{k}=(1/2,0,0)$ accompanying the quadrupolar ordering

IR	Version	Orbit 1				Orbit 2	
		Pd1	Pd2	Pd3	Pd4	Pd5	Pd6
τ_1	1	*(A,0,0)*	*(A,0,0)*	*(A,A,0)*	*(A,A,0)*	*non-related*	
	2	*(0,B,0)*	*(0,B,0)*	*(0,-B,0)*	*(0,-B,0)*		
	3	*(0,0,A)*	*(0,0,-A)*	*(0,0,A)*	*(0,0,-A)*		
τ_3	1	*(A,0,0)*	*(A,0,0)*	*(-A,-A,0)*	*(-A,-A,0)*	*non-related*	
	2	*(0,B,0)*	*(0,B,0)*	*(0,B,0)*	*(0,B,0)*		
	3	*(0,0,A)*	*(0,0,-A)*	*(0,0,-A)*	*(0,0,A)*		
τ_5	1	*(A,0,0)*	*(-A,0,0)*	*(A,A,0)*	*(-A,-A,0)*	*non-related*	
	2	*(0,B,0)*	*(0,-B,0)*	*(0,-B,0)*	*(0,B,0)*		
	3	*(0,0,A)*	*(0,0,A)*	*(0,0,A)*	*(0,0,A)*		
τ_7	1	*(A,0,0)*	*(-A,0,0)*	*(-A,-A,0)*	*(A,A,0)*	*non-related*	
	2	*(0,B,0)*	*(0,-B,0)*	*(0,B,0)*	*(0,-B,0)*		
	3	*(0,0,A)*	*(0,0,-A)*	*(0,0,A)*	*(0,0,-A)*		

Table 4c. Possible displacements of uranium atoms in 2(c) positions for $\boldsymbol{k}=(1/2,0,0)$ accompanying quadrupolar ordering with active IR-s τ_1 , τ_3 and τ_5

IR	U3	U4
τ_1	*(2A,A,0)*	(2A,A,0)
τ_2	*(0,A,0)*	*(0,A,0)*
τ_3	*(0,0,A)*	*(0,0,A)*

6. Quadrupolar ordering and structural deformations in UPd_3

Below the first transition temperature T_0=7.6K the ordering of quadrupolar moments localised on U positions is found, with wave vector $\boldsymbol{k}=(1/2,0,0)$, The quadrupolar moments at U1 and U2 sites form an antiquadrupolar arrangement, which is probably accompanied by structural deformations (atomic displacements) leading to orthorhombic symmetry. The principal axes of QMT are parallel to the orthorhombic system axes. The symmetry of the AFQ ordering mode can be described by the $(Q_{x^2-y^2})$ type of quadrupolar operator at U(2a) sites. For $T_{1+} < T < T_0$ the experimental data are consistent with the model described by τ_5 , leading to the *P nnm* symmetry group, and the set of parameters given by $p=(Q;2Q;0,0,0)$ for U(2a) sites. Such quadrupolar ordering may be accompanied by displacements parallel to the z axis for Pd atoms at (6h) sites and parallel to the z axis or located in the basal plane for Pd atoms at (6g) sites (see Table4a and Table 4b). Magnitudes and signs of displacements for atoms belonging to the first orbit are independent to the magnitudes and signs of displacements of the atoms belonging to the second orbit. Within each orbit in one hexagonal elementary cell the displacements are parallel. One of the possible models of common quadrupolar and structural ordering in this temperature range is shown in Figure 4.

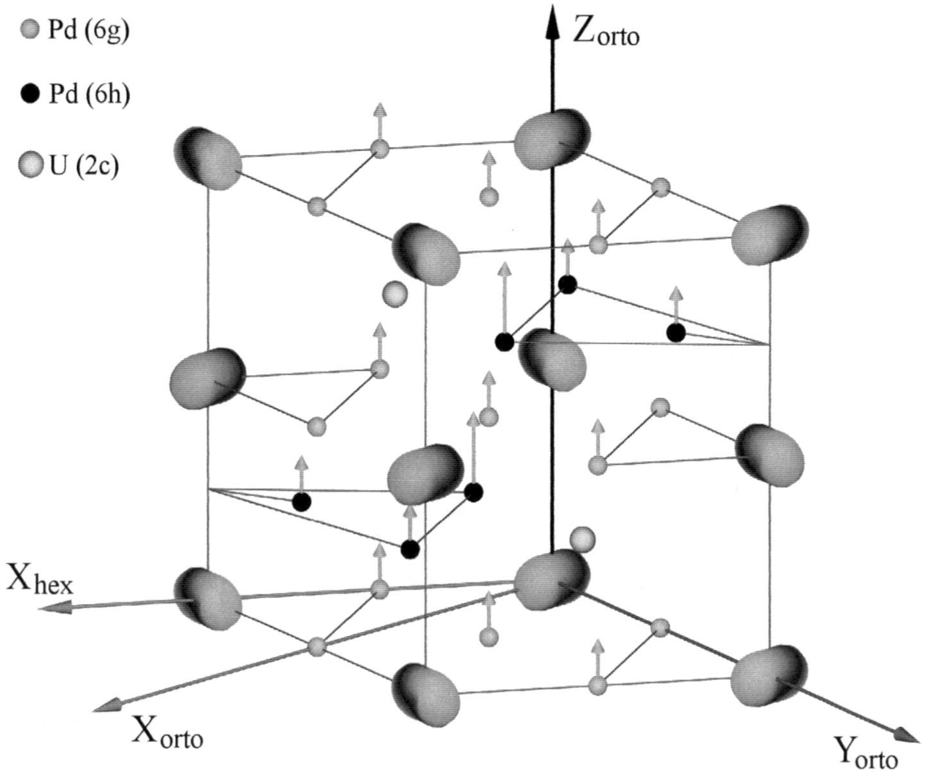

Figure 4. One of the possible models of common quadrupolar and structural ordering for IR τ_5, quadrupolar parameter set *(Q,2Q,0,0,0)* and independent amplitudes of atomic displacements for various orbits of Pd atoms.

Below T_{+1}=6.9K and T_{-1}=6.7K temperatures further deviations of the QMT principal axes directions are observed, leading to further symmetry reductions of AFQ ordering at U(2a) sites, which are respectively described by $(Q_{x^2-y^2}, Q_{yz})$ and $(Q_{x^2-y^2}, Q_{yz}, Q_{zx})$ quadrupolar operators. It corresponds to the symmetry analysis result described by the τ_3 IR, with the set of parameters given by *p=(0,B,0,B)*. The addition of emergence of τ_3 solution leads to the rotation of the τ_5 QMT shape along the orthorhombic ***x*** axis (Q1 and Q2 in antiphase) and change of the QMT shape. Next the τ_1 solution is added, with the set of parameters given by *p=(0,0,Q',Q',0)*, leading to the rotation of the τ_5 QMT shape along the ***y*** axis (Q1 and Q2 also in antiphase) and further change of the QMT shape. Such quadrupolar ordering may be also accompanied by additional displacements of Pd atoms in the basal plane at (6h) and (6g) sites, which are given in the Table 4a and 4b. Below T_1 the possible (Q_{zx}) deviations at U(2c) sites [2] correspond to the τ_5 IR and parameter set *p'=(C',C')* The resulting type of structure belongs to the monoclinic symmetry. The last stage described above (for $T<T_{-1}$) is probably not a valid description of the situation in UPd_3 as the experimental data [6] strongly indicate, that a triple-q phase is formed below T_{-1} temperature in a discontinuous phase transition. The symmetry analysis of the triple-q phase with corresponding quadrupolar, structural and magnetic ordering will be discussed in the next paper.

References:

[1] Izyumov Yu A, Syromyatnikov V N 1990 *Phase Transitions and Crystal Symmetry* (Dordrecht: Kluver Academic Publishers).

[2] McEwen K A, Steigenberger U, Clausen K N, Kulda J, Park J-G and Walker M B 1998 *J. Magn. Magn. Mat.* **177-181** 37

[3] McMorrow D F, McEwen K A, Steigenberger U, Ronnow H M and Yakhou F 2001 *Phys. Rev. Lett.* **87** 057201

[4] Lingg N, Maurer D, Mueller V and McEwen K A 1999 *Phys. Rev.* B **60** R8430

[5] Schenck A, Gynax F N and McEwen K A 2002 *J. Phys.: Condensed Matter* **14** 4595

[6] McEwen K A, Park J-G, Gipson A J, Gehring G A (2003) *J. Phys.: Condensed Matter* **15** S1923

[7] Walker M B, Kappler C, McEwen K A, Steigenberger U and Clausen K N 1994 *J. Phys.: Condensed Matter* **6** 7365

[8] Sikora W et al. 2004 *J. Appl. Crystall.* **37** 1015-1019 (program available from: http://novell.ftj.agh.edu.pl/~sikora/modyopis.htm)

[9] Sikora W, Bialas F, Pytlik L and Malinowski J 2005 *Solid State Sciences* **7** 645-655 (Tensor_Vis program available at: http://novell.ftj.agh.edu.pl/~malinowski/opis/Tensor_Vis_en.html)

[10] Jackson D 1975 *Classical Electrodynamics* (John Wiley & Sons, Inc.) Chapter 4

[11] Heathman S et al. 2003 *Phys. Rev* B **67** 180101

[12] Pasturel A, Colinet C, Nguyen Manh D, Paxton A T and van Schilfgaarde M 1995 *Phys. Rev.* B **52** 15176-15190

Symmetry preserving optimised effective potential theory (application to atoms)

Andreas K Theophilou[1], Petros G Papaconstantinou[1] and Vitaly Glushkov[2]

[1] Institute of Materials Science, DEMOCRITOS NCSR, TT15310, Athens, Greece

[2] Department of Physics, State University, Dniepropetrovsk 320625, Nauchny 13, Ukraine

Abstract. In this paper, general symmetry properties of physical systems are used in order to produce a mapping of the external potential of a many electron system to its optimized effective potential (OEP). The so derived effective potential is used to calculate the OEP ground state energies and spin orbitals. Applications are made to atoms and ions and the results are compared to those of the exact Hartree-Fock approximation. The relative deviations from the exact HF theory ($\Delta E/E$) are of the order of 10^{-4}. The same holds for molecules. One of the features of the present theory is that the many electron wave functions derived by the present method, transform according to the irreducible representations of the exact states.

1. Introduction

Symmetry is important in the many-electron problem as it gives the main features of the many-electron wave function, or in more technical terms, the energy eigenstates of the physical system under consideration, transform according to the irreducible representations (Irreps) of the symmetry group of the Hamiltonian. For this reason, it is an advantage for a perturbation theory to conform to this requirement. This means that the approximate Hamiltonian should be invariant under the same symmetry group as the exact one. Nevertheless, this fundamental aspect is ignored in most approximation theories. In fact, there is a good reason for this symmetry breaking: The approximate ground state energy is lower when symmetry breaking is allowed. We shall deal with this matter in more detail.

Among the most popular approximations for electronic structure calculations of atoms, molecules and solids are the Hartree-Fock approximation (HF) and Density Functional Theory (DFT) in its Kohn and Sham (KS) formulation. The first one, HF, is based on a well known variational principle and gives approximate solutions and energies, whereas the second one, DFT, in principle gives the exact ground state energy and density. The first method is a standard topic in Quantum Mechanics books, whereas on the second, there are already many special textbooks[1]. The advantage of the KS theory over HF is that it includes only local terms, whereas in HF the exchange operator is nonlocal. However, as recent calculations have shown, the optimized effective potential approximation (OEP) to HF gives states and energies very close to those of the exact HF. The advantage of OEP is that it involves only local potentials. The difficulty with this approximation is that the effective potential $V_{OEP}(\mathbf{r})$ of this theory cannot be found in a direct way. In fact one puts the constraint that the N-electron Slater determinant $|\Phi\rangle = |\phi_1, ...\phi_N\rangle$ is constructed by spin orbitals $|\phi_i\rangle$ obeying a Schroedinger type equation with effective potential $V_{eff}(\mathbf{r})$ and therefore $|\Phi\rangle = |\Phi_{veff}\rangle$. Then one has to find the minimum of $E(V_{eff}) = \langle\Phi_{veff}|H|\Phi_{veff}\rangle$ with respect to V_{eff}. This procedure however is time consuming.

The original DFT theory was developed for ground state calculations by Kohn and his coworkers[2, 3], but it was later extended to include excited states [4, 5, 6, 7] and time-dependent (TDFT) problems[8]. Other recent developments in this field, especially in the context of the Optimized Effective Potential method(OEP) [9] increased the accuracy and speed of DFT calculations. However, the accuracy of DFT is still limited, because the explicit form of the exchange and correlation energy as functional of the density are sufficiently accurate, although a lot of research work has been done in this field.

In both theories the Hamiltonian involves terms expressed as mappings of the electron density $\rho_\Phi(\mathbf{r}) = \langle\Phi|\widehat{\rho}(\mathbf{r})|\Phi\rangle$ and the density matrix $\rho_\Phi(\mathbf{r},\mathbf{r}') = \langle\Phi|\widehat{\rho}(\mathbf{r},\mathbf{r}')|\Phi\rangle$, where $|\Phi\rangle$ is the lower energy eigenstate of the Hartree-Fock (see Apendix 1 for details) and KS theory (see Section 2). But electron densities and density matrices are bilinear forms of the wavefunction $|\Phi\rangle$ and therefore if one takes $|\Phi\rangle$ to belong to a multidimensional Irrep of the symmetry group G of the exact Hamiltonian, then $\rho_\Phi(\mathbf{r})$ and $\rho_\Phi(\mathbf{r},\mathbf{r}')$ in general, do not have the symmetry of the external potential $V(r)$ and neither does the effective Hamiltoniann. This is obvious for the HF theory as the Hartree potential, $V_H(r)$,

$$V_H(\mathbf{r};\Phi) = \int d^3r' \frac{\rho_\Phi(\mathbf{r}')}{|\mathbf{r}'-\mathbf{r}|} \tag{1}$$

is a linear mapping of the density whereas the exchange operator $\widehat{V}_{xc}(\Phi)$ is a linear mapping of the density matrix. Thus, all asymmetries of the densities are transferred to those of the potentials.

We shall prove in Apendix 1 **Theorem 1:** There are no eigenstates of the Kohn and Sham (KS) and Hartree-Fock (HF) operators belonging to the multidimensional Irreps of the symmetry group of the Hamiltonian.

From the above theorem we conclude that the lower energy states of these approximation theories cannot have the proper transformation properties. Thus, formally we can make the statement that symmetry breaking lowers the ground state energy of the HF and KS theories. However, one can restrict the search for the energy minimum in the space of Slater determinants (Sladets) belonging to an Irrep Γ of the symmetry group G of the external potential[4, 5]. Thus, in the case of degenerate states one can talk about a minimizing subspace and not a single state. For this reason it is necessary to apply the corresponding subspace DFT theory [4, 5], in which the various functionals are defined as traces in a subspace of certain dimension. In this case the subspace density is

$$\rho_\Gamma(\mathbf{r}) = \frac{1}{M_\Gamma}\sum_{m=1}^{M_\Gamma}\langle\Phi_m^\Gamma|\widehat{\rho}(\mathbf{r})|\Phi_m^\Gamma\rangle$$

where the summation is over the orthonormal basis of Irrep Γ. Note, that a change of the orthonormal basis of the subspace M_Γ does not affect the left hand side[10] as the above expression is the Trace of the density operator $\widehat{\rho}(\mathbf{r})$ divided by the dimension of the subspace. Thus,

$$\rho_\Gamma(\mathbf{r}) = \frac{1}{M_\Gamma}Tr\{\widehat{\rho}(\mathbf{r}), S_\Gamma\}$$

and as the trace is a mapping which depends only on the subspace, the density is invariant under transformations which map the subspace S_Γ to itself, i.e., the above density belongs to the Identity Irrep of the symmetry group of the external potential. Thus the effective KS potential in this case will have the symmetry of the external potential and therefore one can develop HF and DFT theories where the effective potential preserves symmetry. The advantage is that the minimizing states that one gets have the same transformation properties as the exact eigenstates of the physical system. One would argue that the problem of degeneracy can be overcome, by applying a small symmetry breaking perturbation. In this way the conventional ground state DFT is applicable and one can take the limit when the perturbation tends to zero. This means that one cannot use anymore only the symmetric part of the density in order to determine the effective potential. Then, the problem becomes extremely difficult for the numerical calculations as the perturbed effective potential does not regain its symmetry when the perturbation tends to zero. This is a consequence of the fact that the single state density does not tend to a density having the symmetry of the external potential. For this reason, in atoms, the practice was to use spherically symmetric potentials for both the HF and KS theories, long before the subspace theory was developed.

Another reason justifying the need of a symmetry preserving theory is that in order to use the HK or KS many electron states for higher order corrections, it is essential for the approximate states to have the same symmetry i.e. to belong to the same Irrep as the exact states, otherwise, many selection rules and other simplifications resulting from symmetry, like e.g. the Wigner-Eckhardt theorem [11], are not applicable.

Recently, we developed a symmetry preserving DFT theory where the KS potential is given as a direct mapping of the external potential [12, 13]. The same is true for the OEP theory. In fact the first order approximations of these theories coincide. The direct mapping theory was applied to molecules where the eigenstate energies obtained were very close to those of the exact HF as the relative deviation from HF ($\Delta E/E$) was of the order of 10^{-4}. In this paper we shall apply this theory to atoms. An advantage of this theory is that the input KS potential in the calculations is known, whereas in standard DFT, it is expressed as a mapping of the electron density, which has to be determined self-consistently. In the present state of the theory, the parameters of the mapping of the external potential to the KS potential were determined variationally. Thus in a certain sense our method in its present implementation is closely related to the optimized effective potential approximation.

It is useful here to give a brief outline of DFT. The basic theorem of DFT, put forward by Hohenberg and Kohn [2], states that the ground state density of any system, $\rho(\mathbf{r})$, determines uniquely the external potential $V(\mathbf{r})$ and consequently the ground state $|\Psi_0\rangle$, provided that no degeneracy is present. A direct consequence of this fundamental theorem is that we can, in principle, express the total energy as a functional of $\rho(\mathbf{r})$ and find its minimum, bypassing the solution of the many-particle Schroedinger equation. A difficulty of this approach is that the space of densities is not well defined and for finding the energy minimum, it is essential to define the space where these minima are sought. A further difficulty was the derivation of sufficiently accurate density functionals and in particular that of the kinetic energy. All efforts for this functional had limited success [15, 16]. This difficulty was overcome by Kohn and Sham [3] who, instead of energy density functionals, considered the energy as a functional of a noninteracting state having the same electronic density as that of the ground state of the real interacting system. The minimization of this functional leads to a set of one-particle Schroedinger equations, the KS equations, with potential: $V_{KS}(\mathbf{r}) = V(\mathbf{r}) + V_K(\mathbf{r})$, where $V_K(\mathbf{r})$, to be referred to as the Kohn potential, is a function of the electronic density. Still, this theory needs a better explicit form of the exchange and correlation functional, E_{xc}. Although, considerable effort was put in this field [17], there are not yet any such functionals which provide sufficient accuracy.

Closely related to the present theory is the work of the model potential for atoms developed by Green and coworkers [18] which was done in relation to the HF approximation. However, the attempts for parametrized model potentials was restricted to atoms and not general mappings were sought. Some applications concerning molecules were also attempted, but to our knowledge they had no continuation [19]. The essential difference with previous work is that we search for a general mapping of the external potential V to the effective KS potential V_{KS} or to the optimized effective potential V_{OEP} and not one limited to a single point charge. In fact the initial scope of applicability of the independent particle model was more ambitious as it aimed to include not only energy eigenstates but also scattering states.

In section 2 we give an outline of an alternative formulation of DFT [20] valid also for degenerate states. (The first attempts for this formulation appeared in [21]). The advantage of this formulation is that all densities involved are single state representable and no functional derivatives are needed.

In Section 3, we derive a first approximation of the effective potential as a mapping of the external potential. This mapping is based on symmetry properties and on some physical considerations regarding some simple many electron systems, like atoms, molecules and the finite radius jellium model.

In Section 4, we implement our proposed method to the lower energy states of atoms of the three first rows of the periodic table (Z=2 to Z=18). Some ionization energies are also given. A brief discussion about the accuracy of the present method is given.

2. The Formulation of DFT without the use of functional derivatives

The Hamiltonian considered is the N-electron Hamiltonian

$$H = T + H_{int} + \hat{V}, \tag{2}$$

where T is the kinetic energy operator

$$T = \frac{1}{2}\int d^3r \sum_s \nabla\psi^{s\dagger}(\mathbf{r})\nabla\psi^s(\mathbf{r}), \tag{3}$$

H_{int} is the interaction energy,

$$H_{int} = \frac{1}{2}\int d^3r \int d^3r' \sum_{st} \frac{\psi^{s\dagger}(\mathbf{r})\psi^{t\dagger}(\mathbf{r}')\psi^t(\mathbf{r}')\psi^s(\mathbf{r})}{|\mathbf{r}-\mathbf{r}'|}, \tag{4}$$

$$\hat{V} = \int d^3r\hat{\rho}(\mathbf{r})V(\mathbf{r}) \tag{5}$$

$\hat{\rho}(\mathbf{r})$ is the density operator

$$\hat{\rho}(\mathbf{r}) = \sum_s \psi^{s\dagger}(\mathbf{r})\psi^s(\mathbf{r}) \tag{6}$$

and $\psi^{s\dagger}(\mathbf{r})$, $\psi^s(\mathbf{r})$ are the fermion field creation and annihilation operators. The summation over s and t indicates summation with respect to the spin indices.

As the initial DFT theory uses functional derivatives, the existence of which recently has been questioned, we shall give a brief summary of a different formulation of the KS theory in terms of a minimum principle [20], in which no use is made of functional derivatives and the Hohenberg and Kohn theorem, i.e. the one to one correspondence between states and densities [2] is not needed. Further, instead of the energy minimum, in our formulation one has to determine the minima of a new functional denoted by $G(\Psi)$, the definition of which will be given below.

We define first the space S_Ψ as the space of all states $|\Phi\rangle$ having the same density as the state $|\Psi\rangle$. The following functionals are well defined for any state $|\Psi\rangle \in \mathcal{H}$, where $\mathcal{H}$ is the Hilbert space of the N-electron wave functions having finite kinetic energy and normalized to unity.

$$K(\Psi) = \min\{\langle\Phi|T|\Phi\rangle : \rho_\Phi(\mathbf{r}) = \rho_\Psi(\mathbf{r})\} \tag{7}$$

This functional assigns a kinetic energy to every state $|\Psi\rangle$ belonging to the Hilbert space $\mathcal{H}$. However the kinetic energy *is not* that of $|\Psi\rangle$, but of the state $|\Phi\rangle$ which has the minimum kinetic energy in the space S_Ψ. Note that the value of this functional is the same in the whole space S_Ψ, so in fact its value depends on the density. We next define the functional

$$L(\Psi) = \min\{\langle\Phi|T + H_{int}|\Phi\rangle : \rho_\Phi(\mathbf{r}) = \rho_\Psi(\mathbf{r})\} \quad (8)$$

This functional is related to the kinetic plus interaction energy and its value is the same for all states with the same density. Finally we define

$$G(\Psi; V) = L(\Psi) + \langle\Psi|T|\Psi\rangle - K(\Psi) + \int d^3r\rho_\Psi(\mathbf{r}) > V(\mathbf{r}) \quad (9)$$

Note that all these functional are defined for the whole Hilbert space of the N-particle states $|\Psi\rangle$ and not in the space of densities, although densities play a significant role in these definitions. In [20] we proved

Theorem 2. The ground state energy $E(V)$ for given potential V is

$$E(V) = minG(\Psi; V) = G(\Phi; V) \quad (10)$$

We note here that since $\langle\Psi|T|\Psi\rangle - K(\Psi) \geq 0$ it follows that the minimum of $G(\Psi; V)$ is obtained when equality holds, i.e. when the minimising $|\Phi\rangle$ gives the minimum of the kinetic energy under the condition that $\rho_\Phi(\mathbf{r}) = \rho_\Theta(\mathbf{r})$, where here $|\Theta\rangle$ is the exact ground state. This implies that there is an effective potential $V_{eff}(\mathbf{r})$ such that the noninteracting state obeys the KS equation

$$T|\Phi\rangle + \widehat{V}_{eff}|\Phi\rangle = E|\Phi\rangle \quad (11)$$

for some effective potential $V_{eff}(\mathbf{r}) = V(\mathbf{r}) + V_K(\mathbf{r})$, where $V_K(\mathbf{r})$ depends on the exact ground state density $\rho_\Theta(\mathbf{r})$. But this implies that the effective potential $V_{eff}(\mathbf{r})$ is a unique mapping of the external potential $V(\mathbf{r})$, since by the Hohenberg and Kohn theorem there is one to one correspondence between density and external potential. A detailed treatment concerning the existence of this potential is given in [13]. Obviouslly equation (11) is an equation corresponding to a noninteracting system and therefore $|\Phi\rangle$, to be referred in the following as the *KS state,* is not restricted to a single Sladet as in the initial KS theory, but a linear combination of such determinants is allowed when demanded by symmetry. Note that the present formulation does not exclude degenerate states as no use is made of the Hohenberg and Kohn theorem. However, for each Irrep, the minimization has to be obtained in the subspace of states transforming according to this Irrep. This means that the Kohn potential will have the symmetry of the external potential, otherwise, no minimization is possible. This is the reason that we use the label eK instead of KS. The spin orbitals (one-particle wave functions) $\phi_i(\mathbf{r})$ of $|\Phi\rangle$ which minimizes $G(\Psi; V)$ satisfy the equation

$$-\frac{1}{2}\nabla^2\phi_i(\mathbf{r}) + V_{eff}(\mathbf{r})\phi_i(\mathbf{r}) = \varepsilon_i\phi_i(\mathbf{r}). \quad (12)$$

As in the present formulation linear combinations of Sladets are allowed, the number of spin orbitals needed in some cases is larger than the number of particles N. Although according to Harriman and Gilbert [22], for each density, one can find a single Sladet having this density, this Sladet may not fulfill at all the symmetry requirements. Such an example is the density of the $S = 0$, $l = 2$, $m = 1$ state $|\Phi_1^2\rangle = \frac{1}{\sqrt{2}}\{|\phi_0^{1\uparrow}, \phi_1^{1\downarrow}\rangle + |\phi_1^{1\uparrow}, \phi_0^{1\downarrow}\rangle$. The density of this state is the same as that of the Sladet $|\phi_0^{1\uparrow}, \phi_1^{1\downarrow}\rangle$, or $|\phi_1^{1\uparrow}, \phi_0^{1\downarrow}\rangle$ but the spin symmetry is lost, as this is not anymore an eigenstate of S^2. One might argue that either of the above could represent $|\Phi_1^2\rangle$, but then this could represent equally well the state $|\Phi_1^1\rangle = \frac{1}{\sqrt{2}}\{|\phi_0^{1\uparrow}, \phi_1^{1\downarrow}\rangle - |\phi_1^{1\uparrow}, \phi_0^{1\downarrow}\rangle$ which is again an $S = 0$ state, with $l = 1$, $m = 1$. The exact eigenstates of an atom corresponding to the above quantum numbers correspond to very different energies and densities. Further, the state

$|\Phi_1^1\rangle = |\phi_0^{1\uparrow}, \phi_1^{1\uparrow}\rangle$ has the same density as $|\Phi_1^2\rangle$ but its spin eigenvalues are $S = 1$, $S_z = 1$. Thus if one neglects symmetry, one can have many states having the same density but as our examples show, it is not clear to which exact eigenstate they are assigned.

We conclude that in the present formulation one is not forced to use a single Sladet as in the initial formulation of Kohn and Sham. The basic philosophy of the present DFT theory is that the noninteracting state, must have the same symmetry (transform according to the same Irrep) as the exact ground state. The present formulation allows linear combinations of Sladets until the proper symmetry is obtained. This implies that there are many Sladets, built from orbitals satisfying equation (12), satisfying the noninteracting state equation (11). However, in order to get the correct Irrep state, one may have to use linear combinations of such determinants.

As we discussed in the introduction there are many advantages when we express the effective potential V_{eff} of the present theory as a mapping of the external potential. This mapping is much more convenient since one maps a physical quantity to another one with the same symmetry and of the same physical nature. In this formulation the total energy is also a functional of the density, because the Hohenberg and Kohn theorem is also valid, but the noninteracting system equations are not derived by minimizing this functional and thus no functional derivatives are used. Beyond the symmetry considerations, the new methodology, has the advantage that one can determine the self-consistent eK potential at the first stage of the calculations as the external potential of a physical system is a function that one knows, whereas in conventional DFT, one has to perform a series of iterations to find the self-consistent density in terms of which the self-consistent KS potential is determined. Thus, finally one expects a significant increase of the speed of calculations. In our present treatment we preferred to calculate the ground state energy in terms of the Hartree-Fock energy using the KS state in place of the HF one and adding a correlation term, i.e. we used the formula

$$E[\Phi_{KS}] = \langle\Phi_{KS}|T|\Phi_{KS}\rangle + \int d^3r \langle\Phi_{KS}|\hat{\rho}(\mathbf{r})|\Phi_{KS}\rangle V(\mathbf{r}) +$$

$$\frac{1}{2}\int\int d^3r d^3r' \{\frac{\rho(\mathbf{r})\rho(\mathbf{r}') - \sum_s \rho_s(\mathbf{r},\mathbf{r}')\rho_s(\mathbf{r}',\mathbf{r})}{|\ \mathbf{r}-\mathbf{r}'\ |}\} + E_c[\Phi_{KS}] \tag{13}$$

where $\rho_s(\mathbf{r}',\mathbf{r}) = \langle\Phi_{KS}|\psi^{s\dagger}(\mathbf{r})\psi^s(\mathbf{r})|\Phi_{KS}\rangle$ and as usually, the successive terms in the second side of the equation stand for the kinetic, the potential, the Coulomb (Hartree) and the exchange energy, respectively. Finally the correlation energy $E_c[\Phi_{KS}]$ is the only term about which there is no exact functional form. Note that $E_c[\Phi_{KS}]$ does not coincide with the one corresponding to Hartree-Fock, as $|\Phi_{KS}\rangle$ and $|\Phi_{HF}\rangle$ do not coincide. In practice we found that, for atoms and molecules that we tried up to now, they are not very different.

So far we refer to the KS theory, but the mapping of the external potential to the effective potential is also valid for the Optimized Effective Potential Theory.

3. A first approximation of the mapping of the external potential $V(r)$ to the effective potential $V_{eff}(r)$

According to the previous section, the external plus Kohn potential of the present KS theory, as well as the optimized effective potential V_{OEP}, which in the following will both be denoted by $V_{eff,}$ must be invariant under the symmetry group G of the external potential V. This will be a consequence of covariance. This means that, if the external potential of a physical system is subject to a rotation or translation all physical properties, like e.g. the density, must be transformed accordingly. Then the mapping of V to $V_{eff,}$ must belong to the identity Irrep of I_3^3, the group which contains all rotation and translations in the three dimensional Euclidean space. Such a mapping will map the translated potential $V_a(\mathbf{r}) = V(\mathbf{r}-\mathbf{a})$ to $V_{eff,a}(\mathbf{r}) = V_{eff}(\mathbf{r}-\mathbf{a})$. Thus e.g. if all nuclei of a molecule in zero external field are translated by **a**, the external potential $V_a(\mathbf{r})$ becomes $V(\mathbf{r}-\mathbf{a})$ and then as one expects the relation $V_{KS,a}(\mathbf{r}) = V_{KS}(\mathbf{r}-\mathbf{a})$ must hold. This mapping implies that all physical properties of the initial system must be translated by **a** when the external potential is translated. Thus in the case of molecules, where the external potential is due to the Coulomb forces of the atomic nuclei, transformation of the all nuclei, imply transformation of all electronic properties and therefore of the effective potential. The same relation must hold for rotations.

The simplest forms compatible with the symmetry requirements are:

(a) $V_{eff}(\mathbf{r}) = \lambda(N)V(\mathbf{r})$, (b) $V_{eff}(\mathbf{r}) = \lambda_{2n}(N)\nabla^{2n}V(\mathbf{r})$, $n \geq 2$, or linear combinations of the above. (Note that $\nabla^2 = \nabla.\nabla$ transforms according to the identity irreducible representation of the I_3^3 group, i.e. this operator is invariant under all rotations and translations).

These mappings were rejected as they did not comply with the uniform electron density of the finite radius jellium model [12]. Note that because of the universality of the mapping, if for a single case it gives unphysical results, it is not acceptable.

The next family of mappings of $V(\mathbf{r})$ to $V_{eff}(\mathbf{r})$ complying with covariance includes nonlocal integral forms like:

$$V_{eff}(\mathbf{r}) = \int V(\mathbf{r}')Q(|\mathbf{r}'-\mathbf{r}|;N)d^3r' \tag{14}$$

As it is shown in the Apenxix of [12] we can write this formula in an alternative way by which we can have more physical insight, namely:

$$V_{eff}(\mathbf{r}) = \int d^3r' \frac{K(|\mathbf{r}'-\mathbf{r}|;N)}{|\mathbf{r}'-\mathbf{r}|}\nabla'^2 V(\mathbf{r}'). \tag{15}$$

Since $4\pi\rho^+(\mathbf{r}) = -\nabla^2 V(\mathbf{r})$ we have effectively a mapping of the external charge density to the effective potential.

Next, we consider a non extended system, with total charge $\int \rho^+(\mathbf{r}')d^3r' = Q$ in a finite volume. The asymptotic behavior of the external potential is $-\frac{Q}{r}$ for $r \to \infty$ and that of the KS potential should be:

$\lim_{r\to\infty} V_{eff}(\mathbf{r}) = -\frac{Q-N+1}{r}$. It then follows that $K(r;N) = \frac{Q-N+1}{Q}$ for $r \to \infty$.

One can readily verify that this is compatible with the previous demand by considering the case of point charges.

We now consider the positive density $\rho^+(\mathbf{r}) = \frac{1}{4\pi}\sum_k Z_k\delta(\mathbf{r}-\mathbf{R}_k)$ of a molecule or solid, due to the point charges Z_k of atomic nuclei at positions $\mathbf{R}_k$. Then, because of the $\delta-$functions, the integration of the right hand side of the previous equation can be carried out and we find

$$V_{eff}(\mathbf{r}) = -\sum_k Z_k \frac{K(|\mathbf{R}_k-\mathbf{r}|;N)}{|\mathbf{R}_k-\mathbf{r}|}. \tag{16}$$

The effective potential in the vicinity of $\mathbf{R}_k$, in a generalized Taylor expansion, must not include powers smaller than $|\mathbf{r}-\mathbf{R}_k|^{-1}$, i.e. it must be of the form $V_{eff}(\mathbf{r}) \to Z_k(\frac{-1}{|\mathbf{r}-\mathbf{R}_k|} + c_0 + c_1|\mathbf{r}-\mathbf{R}_k| + ...)$ for $\mathbf{r} \to \mathbf{R}_k$. It then follows that $K(|\mathbf{R}_k-\mathbf{r}|) = \frac{1}{|\mathbf{r}-\mathbf{R}_k|} + c_0 + c_1|\mathbf{r}-\mathbf{R}_k| + ...$when $\mathbf{r} \to \mathbf{R}_k$.

We can construct an approximate formula of the KS potential by a proper interpolation of these two asymptotic forms. The simplest way to do it is described by the formula:

$$V_{KS}(\mathbf{r}) = \frac{-Z+N-1}{r} - \frac{(N-1)}{r}e^{-r/a}. \tag{17}$$

One can easily verify that the above expression has the proper asymptotic behavior by expanding the exponential in powers of r. In addition the KS potential is shallower in the vicinity of the nucleus relative to the external potential to take into account the flow of charge away from areas of high density, when the repulsive forces between the electrons are switched on. The constant a, which could depend on Z and N, is expected to be of the order of magnitude of the atomic unit as it is a measure of the extension of the electronic cloud around the nucleus.

We tried this parametric form of $V_{KS}(\mathbf{r})$, equation (17), in many atoms and ions, in combination with equation (13) which gives the total ground state energy. Since in the present state of development of this theory, the parameter a is free, differing from atom to atom, we had to try a series of values of the parameter a in order to find the one leading to the minimum of the energy as mentioned above. The number of values of a for which we had to solve the KS equations was dramatically diminished by plotting E v.s. a and noticing that the graph was a smooth parabolic curve. The results were quite encouraging, although this is the lowest order approximation of the KS potential. The calculation of the total ground state energies for the first two periods of the periodic system gave relative errors between 0.5% and 3% with a tendency to increase with Z, while the optimum a was found to vary between 0.3 and 0.6 atomic units, with a tendency to decrease with Z.

We now return to the general form of the kernel K. By equating the right hand sides of Eqs(16) and (17) for a single point charge at the origin we find

$$K(\mid \mathbf{r}' - \mathbf{r} \mid; N) = 1 - \frac{N-1}{Z} + \frac{N-1}{Z} e^{-|\mathbf{r}'-\mathbf{r}|/a}. \tag{18}$$

Searching for a better approximation, we tried the two- parameter formula:

$$V_{KS}(\mathbf{r}) = \frac{-Z + N - 1 + C}{r} - \frac{N - 1 + C}{r} e^{-r/a} \tag{19}$$

with the second parameter C, depending like a, on Z and N. This time, we had to find the minimum of the function $E = E(a, C)$, varying both a and C. The results were very promising, as the relative errors were reduced by one or even two orders of magnitude, as we will see in detail in the next section.

The problem with the above expression is that, introducing the constant C, we change the asymptotic behavior of $V_{KS}(\mathbf{r})$ at infinity, since we have $V_{KS}(r \to \infty) = -\frac{Z-N+1-C}{r}$ which deviates from the correct asymptotic form at infinity but preserving that at zero. Although this does not play an essential role in the calculations one can remedy the situation by taking C a function of $\mathbf{r}$, which is practically a constant in the vicinity of the nucleus and tends to zero in the asymptotic limit $r \to \infty$. The simplest way to achieve such a behavior is through a Fermi-type function,

$$C(\mathbf{r}, \mathbf{Z}, \mathbf{N}) = \frac{c(Z, N)}{1 + e^{|\mathbf{r}|)/b}} \tag{20}$$

Assuming that $b >> 1$, $C(\mathbf{r}, \mathbf{Z}, \mathbf{N})$ tends to a step function: $C(\mathbf{r}, \mathbf{Z}, \mathbf{N}) = c(Z, N)$ if $r \leq b$ and 0 otherwise.

Then one has to modify the kernel as follows

$$\begin{aligned} K(\mid \mathbf{r}' - \mathbf{r} \mid; N) &= 1 - \frac{N - 1 + C(|\mathbf{r}'-\mathbf{r}|; N, Q)}{Q} + \\ & \frac{N - 1 + C(|\mathbf{r}'-\mathbf{r}|, N, Q)}{Q} e^{-|\mathbf{r}'-\mathbf{r}|/a} \end{aligned} \tag{21}$$

where again Q is the total positive charge $Q = -\int d^3r \nabla^2 V(\mathbf{r})$.

Then it is easy to verify that our formula for large N gives: $V_{KS}(\mathbf{r}) \approx -\frac{Ze^{-r/a}}{r} - V_0$, which is the familiar Thomas- Fermi formula for the electronic screening in metals. We made application of our present effective potential to atoms the results of which will be discussed in the next section.

Table 1. The ground state energies of neutral atoms from Z=2 to 18. The percent error of the presents result (denoted by E_{TP} is relative to the exact energy E_{Exact}. $E_{LDA-SIC}$ denotes the energies obtained by the linear density approximation with the self-interaction correction. E_{OEP} are the results of the Optimised Effective Potential approximation. The parameters a and c are those of the effective potential used as defined in the text.

At.	Z	$-E_{exact}$	$-E_{LDA-SIC}$	$-E_{OEP}$	$-E_{TP}$	Error(%)	a	c
He	2	2.904	2.920	2.903	2.903	0.03	0.500	0.000
Li	3	7.478	7.506	7.483	7.482	0.05	0.555	0.000
Be	4	14.667	14.695	14.665	14.665	0.01	0.497	0.317
B	5	24.654	24.702	24.658	24.655	0.00	0.440	0.729
C	6	37.845	37.933	37.849	37.846	0.00	0.416	0.990
N	7	54.589	54.729	54.591	54.584	0.01	0.398	1.150
O	8	75.067	75.259	75.072	75.064	0.00	0.377	1.258
F	9	99.734	99.999	99.731	99.724	0.01	0.365	1.214
Ne	10	128.939	129.287	128.921	128.914	0.02	0.348	1.200
Na	11	162.257	162.672	162.256	162.208	0.03	0.366	0.752
Mg	12	200.059	200.536	200.063	200.006	0.02	0.353	0.867
Al	13	242.356	242.891	242.362	242.301	0.02	0.329	1.333
Si	14	289.374	289.969	289.376	289.302	0.02	0.302	1.900
P	15	341.272	341.930	341.273	341.195	0.02	0.291	2.303
S	16	398.139	398.852	398.129	398.030	0.03	0.280	2.684
Cl	17	460.196	460.967	460.129	460.043	0.03	0.270	3.055
Ar	18	527.604	528.432	527.555	527.406	0.04	0.259	3.488

4. Numerical results and conclusions

The ground state energies of the atoms which were calculated with the present method compare in accuracy with those of the optimized effective potential method and are better than other familiar methods like the local density approximation self-interaction correction LDA-SIC, as one can realize from tables 1 and 2. The relative errors with respect to the exact energies, had a tendency to increase with Z and were smaller than 0.05%, i.e. $\frac{\Delta E}{E} < 5 \cdot 10^{-4}$. Calculations of ground states of ions exhibited the same or even better accuracy than the ones of the corresponding neutral atoms (See table 2). The relative errors were bigger in excited state calculations, with the maximum error, for the first excited state of Helium, being 0.14%. As

Table 2. The energies of several ions are given for the configurations denoted. The notation used is the same as that of Table 1.

Ion	Z	N	State	$-E_{exact}$.	$-E_{LDA}$	$-E_{OEP}$	$-E_{TP}$	Err%	a	c
O^{2+}	8	6	$\{1s^2 2s^2 2p^2\}^3P$	73.224	73.319	73.219	73.218	0.01	0.303	0.700
C^{4+}	6	2	$\{1s^2\}^1S$	32.406	32.450	32.404	32.404	0.01	0.148	0.000
He	2	2	$\{1s2s\}^3S$	2.174	2.176	2.174	2.168	0.27	1.160	0.000
C^{4+}	6	2	$\{1s2s\}^3S$	21.431	21.423	21.418	21.424	0.08	0.357	0.000
Li	3	3	$\{1s^2 2p\}^2P$	7.410	7.441	7.414	7.413	0.04	0.555	0.000
Be	4	4	$\{1s^2 2s2p\}^3P$	14.567	14.608	14.573	14.571	0.03	0.466	0.470

expected, the order of magnitude of the optimum values of $a(Z, N)$ were one atomic unit. They ranged between 0.244 a.u. and 0.588 a.u. in ground state calculations of neutral atoms, with a tendency to decrease with Z. The values of $a(Z, N)$ were always smaller for ionized than for neutral atoms, something also expected since a is a measure of the mean radius of the electronic cloud, which tends to shrink when N decreases in relation to Z because of the stronger attraction by the nucleus. On the contrary, the parameter $c(Z, N)$ tends to increase with Z. This should also be expected since c is a measure of the deviation of $V_{KS}(\mathbf{r})$ from the simple interpolation of the asymptotic forms in $Z = 1$ and Z tending to ∞.

The relative errors of the first ionization potential calculations is one or even two orders of magnitude bigger than the ones in total energy calculations. This is natural, as the ionization potentials are calculated either as differences of two very large numbers (the total ground state energies of the N electron and the N-1 electron system) either by the energy of the highest occupied orbital, according to the approximation which uses Koopaman's theorem. In any case, the errors are systematically lower than 4%, and almost identical to the ones of the OEP method.

One of the most crucial tests of the proposed method was the comparison of our potential with the KS potential calculated by the OEP method in its most sophisticated version [9], the most accurate up to date method in the framework of DFT. In atomic systems, the two KS potentials were found almost identical, something very important since it means that, at least in the atoms that we tested, it is possible to find a pretty accurate effective potential without iterations.

Of course, the problem of the determination of the kernel K with higher accuracy remains open and should be one of the main priorities for the future development of the proposed method. Higher order nonlinear corrections of the mapping of the external potential on V_{KS} might also be necessary. The present method has also been tested for molecules giving quite accurate results as its relative deviation from the exact HF is less than 10^{-4} [12]. The Moeller Plesset perturbation theory has also been applied for molecules. The results which will be published elsewhere, compare well relative to other methods which do not take into account the symmetry of the basis wave functions.

Finally we remind the main advantage of the method from the point of view of numerical calculations which is the speed of computation which increases dramatically relative to conventional DFT, since one does not have to make a guess of the input KS potential or the corresponding density, but starts from the correct one, although in the form presented in this paper it has two free parameters.

Apendix 1.
Proof of Theorem 1.

Theorem: There is no solution of the HF equation which transforms according to an Irrep of the symmetry group of the exact Hamiltonian other than the identity Irrep.

Theorem 1 is also valid for the solutions of the KS equations

Proof: Assume that such a solution exists and let it belong to Irrep Γ different from the identity Irrep. Then the many particle HF equation in the second quantization representation is

$$(T+V) \mid \Phi_\Gamma\rangle + \int d^3r \int d^3r' \frac{\rho_\Phi(r')}{\mid r'-r \mid} \widehat{\rho}(r) \mid \Phi_\Gamma\rangle - \sum_s \int d^3r \int d^3r' \frac{\rho_\Phi^s(r,r')}{\mid r'-r \mid} \widehat{\rho}^s(r',r) \mid \Phi_\Gamma\rangle = E \mid \Phi_\Gamma\rangle. \quad (22)$$

We can write now the density matrix as

$$\rho_\Phi(r,r') = \rho_0(r,r') + \sum_{B\neq 0} \rho_B(r,r'). \tag{23}$$

Then the Hartree and the Exchange operator will have terms other than those belonging to the identity Irrep. Let us call them V_H' and V_{xc}'. Then

$$(V_H' + V_{xc}') \mid \Phi_\Gamma\rangle = \sum_{A\neq\Gamma} \mid \Phi_A\rangle = 0. \tag{24}$$

But the operator $(V_H' + V_{xc}')$ is positive definite and has no zero eigenstates. Hence it is not possible for $\mid \Phi_\Gamma\rangle$ to be an eigenstate of the HF operator.

The proof for the case of DFT is much easier as we have only the local potential and the local potentials have no eigenstates in the space of square integrable functions.

For a better understanding of the above theorem we consider the HF equation for the spin orbitals $\mid \varphi_i\rangle$ of the HF Sladets $|\Phi\rangle = |\varphi_1, ...\varphi_i, ...\varphi_N\rangle$. This equation has the form

$$[-\frac{1}{2}\nabla^2 + V(r)]\varphi_i^s(r) + \int d^3r' \frac{\rho_\Phi(r')}{\mid r'-r\mid}\varphi_i^s(r) - \int d^3r' \frac{\rho_\Phi^s(r,r')}{\mid r'-r\mid}\varphi_i^s(r') = \epsilon_i\varphi_i^s(r) \tag{25}$$

where s is the spin up and spin down index. It is easy to verify the above theorem by using the spin-orbital HF equation. In order to construct a Sladet belonging to a certain Irrep of the symmetry group G of the Hamiltonian, one chooses $|\varphi_i\rangle$ to belong to a certain Irrep of G. Assume now that one of the $\varphi_i(r)$ transforms according to the identity Irrep. of G and let us call it. Then in separating $\rho_\Phi(r,r')$ into $\rho^0(r,r')+ \sum \rho^A(r,r')$ we find that $\varphi_i^s(r)$ satisfies the equations

$$\int d^3r' \frac{\rho^A(r')}{\mid r'-r\mid}\varphi_i^s(r) - \int d^3r' \frac{\rho^{sA}(r,r')}{\mid r'-r\mid}\varphi_i^s(r) = 0.$$

The above equation is an eigenstate equation with eigenvalue 0. But this is not possible since this is an eigenstate of a positive definite operator.

One can verify easily the above theorem by considering a Sladet of 2M+1 electrons, where M orbitals are doubly filled and is single filled with spin up. By taking into account that the corresponding state has $s = \frac{1}{2}$ and $s_z = \frac{1}{2}$ we find that the spin up density matrix differs from the one with spin down. Then the spacial parts of the spin up orbitals are different from those of spin down. But this is contrary to our symmetry preserving Sladet, where M orbitals are doubly filled.

The proof for the KS equation is similar but simpler since we have only local potentials. By using the same procedure we reach an eigenvalue equation as above, but with only a local part, namely $V^A(\mathbf{r})\varphi_i^s(r) = 0$. But this is not possible as the $\varphi_i^s(r)$ must be square integrable.

Acknowledgments

The authors would like to thank N. Gidopoulos for constructive discussions.

References

[1] Parr R G and Yang W 1989 *Density Functional Theory in Atoms and Molecules* (New York: Oxford University Press)

Dreizler R M and Gross E K U 1990 *Density Functional Theory* (Berlin:Springer)

March N H and Deb W Eds. 1987 *The Single Particle-density in Physics and Chemistry* (London: Academic Press)

Nalewajsky R F 1996 *Density Functional Theory, Springer Series: Topics in Current Chemistry* (Berlin, Heidelberg: Springer-Verlag)

[2] Hohenberg P and Kohn W *Phys. Rev.* B **136** 864
[3] Kohn W and Sham L J 1965 *Phys. Rev.* A **140** 1133
[4] Theophilou A K 1979 *J. Phys.* C **12** 5419
Gidopoulos N I and Theophilou A K 1994 *Philosophical Magazine* B **69** 1067
[5] Theophilou A K and Papaconstantinou P 2000 *Phys. Rev* A **61** 022502
[6] Kohn W 1986 *Phys. Rev.* A **34** 737
Gross E K U, Oliveira L N and Kohn W 1988 *Phys. Rev.* A **37** 2805
Gross E K U, Oliveira L N and Kohn W 1988 *Phys. Rev.* A **37** 2809
Gross E K U, Oliveira L N and Kohn W 1988 *Phys. Rev.* A **37** 2821
[7] Levy M 1982 *Phys. Rev.* A **26** 1200
Levy M and Nagy A 1999 *Phys. Rev. Lett.* **83** 4361
Nagy A and Levy M 2001 *Phys. Rev.* A **63** 052502
Nagy A 1999 *J. Phys.* B **32** 2841
Nagy A 1995 *Int. J. Quantum. Chem. S.* **29** 297
Nagy A 1998 *Phys. Rev.* A **57** 1672
Nagy A 1997 *Phys. Rev.* A **55** 3465
Nagy A 1999 *J. Phys.* B **32** 2841
Nagy A 1995 *Int. J. Quantum. Chem. S.* **29** 297
Nagy A 1998 *Phys. Rev.* A **57** 1672
Nagy A 1997 *Phys. Rev.* A **55** 3465
[8] Runge Er and Gross E K U 1984 *Phys. Rev. Lett.* **52** 997
[9] Sharp R T and Horton G K 1953 *Phys. Rev.* **90** 317
Talman J D and Shadwick W F 1976 *Phys. Rev.* A **14** 36
Grabo T, Kreibich T, Kurth S and Gross E K U 1998 *Strong Correlations in Electronic Structure* (Tokyo:Gordon and Breach)
Chen J, Krieger J B, Li Y and Iafrate G J 1996 *Phys. Rev.* A **54** 3939
Krieger J B, Li Y and Iafrate G J 1990 *Phys. Lett.* A **146** 256
Krieger J B, Li Y and Iafrate G J 1992 *Phys. Rev.* A **45** 101
Krieger J B, Li Y and Iafrate G J 1992 *Phys. Rev.* A **46** 5453
Krieger J B, Li Y and Iafrate G J 1992 *Int. J. Quantum. Chem.* **41** 489
[10] Theophilou A K and Gidopoulos N I 1995 *Int. J. Quant. Chem.* **56** 333
[11] Tinkham M 1964 *Group Theory and Quantum Mechanics* (Toronto-London: McGraw-Hill)
[12] Theophilou A K and Glushkov V N 2005 *Int. J. Quant. Chem.* **104** 538
[13] Theophilou A K 2003 *The fundamentals of Electron Density, Density Matrix and Densiy Functional Theory in Atoms, Molecules and the Solid State* eds N I Gidopoulos and S Wilson (Dodrecht, The Netherlands: Kluwer Academic) p 115
[14] Theophilou A K and Papaconstantinou P 2000 *Proc. Int. Conf. Symmetry and Structural Properties of Condensed Matter* eds T Lulek, B Lulek *at al* (Singapore: World Scientific) pp 196-204
[15] March N H 1997 *Phys. Rev.* A **56** 1025
March N H and Pucci R 1981 *J. Chem. Phys.* **75** 496
March N H and Nalewajski R F 1987 *Phys. Rev.* A **35** 525
Nagy A and March N H 1989 *Phys. Rev.* A **39** 5512
Holas A and March N H 1991 *Phys. Rev.* A **44** 5521
Almbladh C O and Pedroza A C 1984 *Phys. Rev.* A **29** 2322
[16] Zhao Q and Parr R G 1992 *Phys. Rev.* A **46** 2337
[17] Gidopoulos N, Papaconstantinou P G and Gross E K U 2002 *Phys. Rev. Lett.* **88** 033003
Higuchi M and Higuchi K 2002 *Phys. Rev.* B **65** 195122
Rushton P P, Tozer D J and Clark S J 2002 *Phys. Rev.* B **65** 193106
Fuchs M and Gonze X 2002 *Phys. Rev.* B **65** 235109
Nekkovee M, Foulkes W M C and Needs R J 2001 *Phys. Rev. Lett.* **87** 036401
Bonetti A F, Engel E, Schmid R N and Dreizler R M 2001 *Phys. Rev. Lett.* **86** 2241
Gritsenko Q V and Baerends E J 2001 *Phys. Rev.* A **64** 042506
Balbas L C, Martins J L and Soler J M 2001 *Phys. Rev.* B **64** 165110
Tsuneda T and Hirao K 2000 *Phys. Rev.* B **62** 15527
Perdew J P, Kurth S, Zupan A and Blaha P 2001 *Phys. Rev. Lett.* **82** 2544
Goerling A 1999 *Phys. Rev. Lett.* **83** 5459
[18] Green A E S, Sellin D L and Zachor A S 1969 *Phys. Rev.* **184** 1
Ganas P S and Green A E S 1970 *Phys. Rev.* A **2** 111
Ganas P S and Green A E S 1971 *Phys. Rev.* A **4** 182

Berg R A, Purcell J E and Green A E S 1971 *Phys. Rev.* A **3** 508
Kazaks P A, Ganas P S and Green A E S 1972 *Phys. Rev.* A **6** 2169
Wallace S J, Berg R A and Green A E S 1973 *Phys. Rev.* A **7** 1616
Riewe F E, Wallace S J and Green A E S 1973 *Phys. Rev.* A **7** 880
Szydlik P P and Green A E S 1974 *Phys. Rev.* A **9** 1885
Green A E S, Garvey R H and Jackman C H 1975 *Int. J. Quantum Chem. Symp.* **9** 43
Green A E S, Garvey R H and Doda D D 1977 *Phys. Rev.* A **15** 1322
Ganas P S Talman J D and Green A E S 1980 *Phys. Rev.* A **22** 336
[19] Whaleen J E and Green A E S 1972 *American J. Phys.* **20** 1484
[20] Theophilou A K 1998 *Int. J. Quant. Chem.* **69** 461
[21] Hadjisavvas N and Theophilou A K 1984 *Phys. Rev.* A **30** 2183
Hadjisavvas N and Theophilou A K 1985 *Phys. Rev.* A **32** 720
[22] Gilbert T L 1975 *Phys. Rev.* B **12** 2111
Harriman J E 1981 *Phys. Rev.* A **24** 680

On LCAO positron wavefunctions in crystals

W Obermayr[1] **and M Sznajder**[2]

[1] Institute for Electronic Engineering, University of Applied Sciences, FH Joanneum, Werk-VI-Straße 46, A-8605 Kapfenberg, Austria

[2] Institute of Physics, University of Rzeszów, Rejtana 16a, 35-310 Rzeszów, Poland

E-mail: werner.obermayr@fh-joanneum.at

Abstract. In this work we deal with the construction of delocalized positron wavefunctions in crystalline solids within the linear combination of atomic orbitals (LCAO) scheme. The present method leads to an accurate wavefunction of a positron, both in the core and in the interstitial region of a crystal. Furthermore, its representation by a superposition of atomic orbitals is simple enough for the application to calculations of expectation values, like e.g. of accurate electron-positron annihilation rates (high momentum components). The representation of the positron wavefunction can be optimized with respect to the two points of view: firstly, the number of orbitals in the LCAO representation can be minimized (e.g. within the localized spherical orbitals (LSO) method) or, secondly, the orbitals can be chosen to be well-localized around their atoms and vanishing near all others: this leads to the computational advantage, that three- or multicenter integrals arising during numerical calculations can be replaced by one- and two-center integrals which are treated using standard numerical techniques in spherical polar or elliptic coordinates. The transformation from the first into the second representation can be performed via a "fuzzy" cellular partitioning of the crystal volume via an analytically continuous shape-function. Finally we demonstrate our method in the case of a positron in lithium.

1. Introduction

Measurements of positron annihilation radiation are a valuable method to obtain information about the electronic and atomic structures of solids. For a thorough understanding and for reliable interpretations of experimental results an accompanying theory is needed [1].

For the determination of the electronic structure and of the spatial distribution of the positron the two-component density-functional theory (2C-DFT) [2, 3] is widely used. For the electronic part many methods of band structure calculations have been developed, like for example the expansion of the electron wavefunction in terms of symmetrized plane waves, the pseudopotential theory, the augmented-plane-wave approach, Green-function techniques in the KKR method, their full-potential linearized versions, the linear combination of atomic orbitals (LCAO) method, numerical methods using finite-element techniques and others. Many of these methods have been adapted also to the determination of positron wavefunctions [4, 5, 6]. For these methods two aspects are of importance: they should lead to accurate wavefunctions, so that precise expectation values as – for example – high momentum components of annihlation rates can be derived and secondly, the mathematical representation of the positron wavefunction should be simple enough for a direct use in the calculation of expectation values.

2. Basic equations as derived via the 2C-DFT and method of calculation

The 2C-DFT and the application of both, the Kohn-Sham method [7] and the local density approximation (LDA) in the case of *one* positron in a delocalized state, leads to the set of equations

$$H^{\sigma}\,\psi_i^{\sigma}(\mathbf{r}) = \varepsilon_i^{\sigma}\,\psi_i^{\sigma}(\mathbf{r}) \qquad (\sigma = +, -) \tag{1}$$

with

$$H^{+} = -\frac{1}{2}\nabla^2 + V^{+}(\mathbf{r}) - \int \frac{n^{-}(\mathbf{r}')}{|\mathbf{r}-\mathbf{r}'|}\,d^3r' + \epsilon^{\pm}_{c;\,n^{+}\to 0}(n^{-}(\mathbf{r})) \tag{2}$$

$$H^{-} = -\frac{1}{2}\nabla^2 + V^{-}(\mathbf{r}) + \int \frac{n^{-}(\mathbf{r}')}{|\mathbf{r}-\mathbf{r}'|}\,d^3r' + \frac{\delta}{\delta n^{-}(\mathbf{r})}\{n^{-}(\mathbf{r})[\epsilon_{ex}(n^{-}(\mathbf{r})) + \epsilon_{c}(n^{-}(\mathbf{r}))]\} \tag{3}$$

$$n^{-}(\mathbf{r}) = \sum_{i=1}^{N^{-}} |\psi_i^{-}(\mathbf{r})|^2, \qquad n^{+}(\mathbf{r}) = |\psi_1^{+}(\mathbf{r})|^2.$$

We use atomic units, $\sigma = -$ or $+$ denotes electrons and positrons, respectively, H^{σ} is the Hamiltonian and ψ_i^{σ} the crystal wavefunction associated with the energy ε_i^{σ}. $n^{\sigma}(\mathbf{r})$ is the particle density, N^{-} is the number of occupied electron states, $V^{\sigma}(\mathbf{r})$ is the external potential, $\int n^{-}(\mathbf{r}')/|\mathbf{r}-\mathbf{r}'|\,d^3r'$ contains the Hartree part of the Coulomb energy, $\epsilon_{ex}(n^{-}(\mathbf{r}))$ and $\epsilon_{c}(n^{-}(\mathbf{r}))$ are the exchange and correlation energy of an homogeneous electron gas of the density $n^{-}(\mathbf{r})$ and $\epsilon^{\pm}_{c;\,n^{+}\to 0}(n^{-}(\mathbf{r}))$ is the effective electron-positron correlation potential. The exchange energy is given by $-3/4(3n/\pi)^{1/3}$ a.u. in energy per particle. For the correlation energy $\epsilon_{c}(n^{-}(\mathbf{r}))$ the results of Arponen and Pajanne [8] as interpolated by Boroński and Nieminen [3] can be used.

3. Application of the LCAO method to the calculation of positron wavefunctions

3.1. Optimization of the LCAO positron wavefunction with respect to a minimum number of orbitals

For the following discussion we choose, for the sake of clarity, a metal which contains one atom per unit cell; but this is not a restriction on the method described, it can be generalized easily to more complex solids. In this case the external potential $V^{+}(\mathbf{r})$ of the Hamiltonian (2) can be written as $V^{+}(\mathbf{r}) = \sum_{\mathbf{R}} Z/|\,\mathbf{r}-\mathbf{R}\,|$, where Z is the atomic number of the atoms and the $\mathbf{R}'s$ are the vectors of the translation lattice of the metal and, in our case, also the positions of the atomic nuclei.

Furthermore we assume, that the positron is thermalized [11] and thus in the ground-state at $\mathbf{k} = \Gamma$ with energy $\varepsilon^{+}(\Gamma_1)$. For this special case it follows from the Schrödinger equation that the positron wavefunction ψ^{+} has the symmetry of the crystal lattice. Then the positron wavefunction can be written within the LCAO scheme as

$$\psi^{+}(\mathbf{r}) = \sum_{\mathbf{R}} \chi(\mathbf{r}-\mathbf{R}) \tag{4}$$

and a suitable function $\chi(\mathbf{r})$ has to be determined. $\chi(\mathbf{r})$ is not uniquely defined [12] and it can be thus optimized with respect to several points of view.

In the following we expand $\chi(\mathbf{r})$ in spherical harmonics $Y_{lm}(\mathbf{r})$

$$\chi(\mathbf{r}) = \sum_{l=0}^{\infty} \sum_{m=-l}^{+l} c_{lm}(r)\, Y_{lm}(\mathbf{r}) \tag{5}$$

where the radial functions $c_{lm}(r)$ are the expansion coefficients. Applying the symmetry operations of the crystal lattice to the wavefunction, the symmetry conditions can easily be derived: for example, in the case of a cubic crystal lattice only orbitals with $l = 0, 4, 6, \ldots$ contribute to the expansion.

In the next step we determine an appropraite $\chi(\mathbf{r})$ in such a way, that a minimum number of functions $c_{lm}(r)$ is necessary in the expansion (5) for the representation of the positron wavefunction. For this purpose we expand it in terms of functions $\varphi_{ln}(r)$

$$c_{lm}(r) = \sum_n c_{lmn}\, \varphi_{ln}(r). \tag{6}$$

Theoretically, each function set $\{\varphi_{ln}(r)\}$ that is complete in $(0, \infty)$ may be used. Like in electron band-structure methods analytical functions like Gaussian and Slater-type orbitals may be used, in the present work we determine the $\varphi_{ln}(r)$ numerically [13, 6]: they are chosen as the solutions of a central-field one-particle atomic-like Schrödinger equation. The coefficients c_{lmn} are found by the minimum principle for the positron ground-state energy. As an example we apply the method to a positron in BCC lithium: the function $c_{00}(r)$ is already sufficient for a very accurate representation of the positron wave function in both, the core and interstitial region of the crystal (LSO approximation). The numerical results are shown in figure 1.

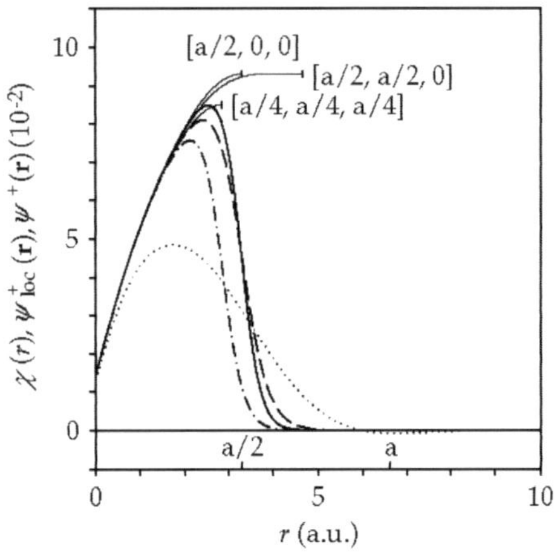

Figure 1. The functions $\chi(r)$ $(\cdots\cdots)$, $\psi^+_{\rm loc}(\mathbf{r})$ along the smmetry directions [1,0,0] (——), [1,1,0] (– – –), and [1,1,1] (— · —), and the resulting positron wavefunction $\psi^+(\mathbf{r})$ in BCC lithium along these symmetry directions (thin solid lines). The positron wavefunction is normalized to one in the unit cell, a is the lattice constant.

3.2. Optimization of the LCAO positron wavefunction with respect to well-localized orbitals

For practical calculations it is often very useful to choose in the *ansatz* (4) the orbitals $\chi(\mathbf{r})$ such that their tails are negligible or exactly zero for r larger than the distance to the nearest neighbouring atomic site. An increase of the complexity of $\chi(\mathbf{r})$ in the representation (5) is then rewarded by a reduction of the numerical effort e.g. in the calculation of expectation values, since multi-centre integrals can be avoided by replacing them by one- or two-center integrals which are treated accurately [9] by standard integration techniques in spherical polar or elliptic coordinates, respectively.

In the following we describe the transformation of the positron wavefunction into a representation with well-localized orbitals $\chi(\mathbf{r})$. For this purpose we first perform a partition of unity into a sum of localized functions $f(\mathbf{r})$ centered at the atomic sites $\mathbf{R}$

$$1 \equiv \sum_{\mathbf{R}} f(\mathbf{r} - \mathbf{R}). \tag{7}$$

Thereby the shape function $f(\mathbf{r})$ has the value *unity* in the vicinity of its center and decreases in a continuous and well-behaved manner with increasing r: $1 \geq f(r) \geq 0$ and $f(r) \sim 0$ for r larger than the distance to the nearest neighbours. $f(\mathbf{r})$ is not uniquely defined [12, 14, 15, 16];

as an example we show a smooth shape function for a BCC lattice in figure 2. In this case the crystal volume is divided not into conventional discrete cells, but into *fuzzy, overlapping,* and *analytically continuous* ones.

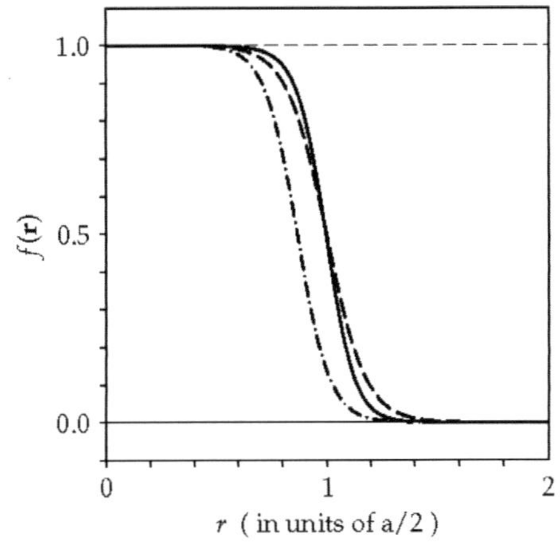

Figure 2. A shape function $f(\mathbf{r})$ for a BCC lattice along the symmetry directions: [1,0,0] (——), [1,1,0] (- - -), and [1,1,1] (— · —). a is the lattice constant.

Next we apply the function $f(\mathbf{r})$ with its property (7) to the partition of a periodic function $g(\mathbf{r})$ into well-localized orbitals $g_{\text{loc}}(\mathbf{r})$:

$$g(\mathbf{r}) = g(\mathbf{r}) \sum_{\mathbf{R}} f(\mathbf{r}-\mathbf{R}) = \sum_{\mathbf{R}} g_{\text{loc}}(\mathbf{r}-\mathbf{R}) \qquad \text{with} \qquad g_{\text{loc}}(\mathbf{r}) = g(\mathbf{r})\, f(\mathbf{r}). \tag{8}$$

For further calculations it is useful to expand $g_{\text{loc}}(\mathbf{r})$ in terms of spherical harmonics. As an example for $g(\mathbf{r})$ we choose the positron wavefunction $\psi^+(\mathbf{r})$ for a positron in BCC lithium. We show the results obtained for the localized orbitals $\psi^+_{\text{loc}}(\mathbf{r})$ in figure 1.

This procedure (8) can be applied to any periodic function: for example, to the electron and positron densities, to effective potentials in the Hamiltonians (2) and (3), etc. The representations may then be well suited for an efficient calculation of expectation values like, for example, electron positron annihilation rates [17].

References

[1] Puska M J and Nieminen R M 1994 *Rev. Mod. Phys.* **66** 841
[2] Chakraborty B and Siegel R W 1983 *Phys. Rev.* B **27** 4535
[3] Boroński E and Nieminen R M 1986 *Phys. Rev.* B **34** 3820
[4] Mijnarends P E 1979 *Positrons in Solids* ed P Hautoärvi (Berlin: Springer) pp 25-88
[5] Puska M J 1987 *Phys. Status Solidi* a **102** 11-29
[6] Obermayr W 1991 *J. Phys.: Condens. Matter* **3** 889-98
[7] Kohn W and Sham L J 1965 *Phys. Rev.* A **140** 1133-8
[8] Arponen J and Pajanne E 1979 *Ann. Phys., NY* **121** 343-89
[9] Obermayr W 2000 *Int. J. Quant. Chem.* **78** 212
[10] Obermayr W 2002 *Symmetries and Structural Properties of Condensed Matter* ed T Lulek, B Lulek and A Wal (World Scientific: Singapore) pp 426-40
[11] Lee-Whiting G E 1955 *Phys. Rev.* **97** 1557-8
[12] Obermayr W 1992 *Mater. Sci. Forum* **105-110** 795-8
[13] Zunger A and Freeman A 1977 *Phys. Rev.* B **15** 4716-37
[14] Becke A D 1988 *J. Chem. Phys.* **88** 2547
[15] Stefanou N and Zeller R 1991 *J. Phys.: Condens. Matter* **3** 7599
[16] Koepernik K and Eschrig H 1999 *Phys. Rev.* B **59** 1743
[17] Obermayr W 2001 *Phys. Status Solidi* B **225**, 63

Noether's theorem and low symmetry aspects concerning the crystal (ligand) field Hamiltonians and spin Hamiltonians

C Rudowicz[1,3] and J Qin[2]

[1] Institute of Physics, Szczecin University of Technology, Al. Piastów 17, 70-310 Szczecin, Poland
[2] Department of Physics and Materials Science, City University, Hong Kong SAR

E-mail: crudowicz@ps.pl

Abstract. This review presents a concise summary of major findings arising from our recent studies concerning the symmetry properties of crystal/ligand field (CF/LF) Hamiltonians and spin Hamiltonians (SHs). First we provide a bird's-eye view of these studies. Then we overview (i) the pertinent basic concepts and notations, (ii) the algebraic symmetry (AS) of Hamiltonians for continuous rotational symmetry (CRS) cases, and (iii) the concepts of the rotational invariants and moments of CF Hamiltonians. This enables a new look from the point of view of the Noether's theorem on the properties of CF/LF Hamiltonians and SHs invariant under CRS, i.e. hexagonal II, tetragonal II, trigonal II, monoclinic, and triclinic ones. An important theorem and a conjecture on the conserved quantities stipulated by Noether's theorem for the Hamiltonians in question formulated by us helps to elucidate the interrelationships and deeper meaning of the concepts involved. Implications of the existence of the conserved quantities for interpretation of experimental CF parameter (CFP) datasets are encapsulated in five corollaries. These considerations reveal that the feasibility of determination of CFPs from fitting experimental spectra and the reduction of the existing higher-order rotational invariants for hexagonal type II and cubic symmetry require reinterpretation. This novel approach enables adoption of better fitting strategies utilizing well-defined conserved quantities, which are invariant under CRS. The advantages of this approach are illustrated using the CFP datasets reported in literature for $RE^{3+}(4f^N)$ ions in $LiYF_4$. This review deals also with the fundamental intricate aspects, hitherto not fully understood, concerning the CF Hamiltonians for the 'low symmetry' cases, including the CRS cases as well as orthorhombic ones. This includes: (1) selection of the axis systems, (2) types of CF parameters and their properties, (3) introduction of a new notion of a *nominal* axis system for the fitted CFP datasets, (4) implications of the Noether's theorem and the AS of CF Hamiltonians, (5) correlation properties among CFP datasets, (6) the rotational degrees of freedom and the reduction of the number of independent CFPs, and (7) extension of the multiple correlated fitting technique. Clarification of these intricate aspects enables us to provide a general framework aimed at achieving an increased compatibility and reliability of CFP datasets for transition ions at low symmetry sites in crystals. The usefulness of this framework is illustrated by reanalysis of the CFP datasets for Nd^{3+} (Pr^{3+}) in $NdGaO_3$ ($PrGaO_3$) and $R$$NiO_3$.

[3] Corresponding author: phone: +91-449-4286; fax: +91-449-4181

1. Introduction

Spectroscopic properties of the transition ions, especially iron-group ions and lanthanides, at low symmetry sites in crystals play important role for various technologically important materials (for references, see [1-5]). The trivalent rare-earth (RE) $4f^N$ ions have been most extensively studied due to applications in laser host and optoelectronic materials. The experimental techniques employed to investigate the CF/LF effects, majority of which are 'non-directional' techniques - unlike those used in the EMR (electron magnetic resonance) studies [6,7], were discussed in [3]. During our recent CF/LF studies of transition ions in crystals [1-5] we have encountered several intricate low symmetry aspects. The motivation to clarify these aspects has prompted us to embark on a comprehensive project aimed at increasing the compatibility and reliability of the crystal field parameter (CFP) datasets. This review presents a concise summary of the major findings arising from our studies concerning the intricate low symmetry properties of CF/LF Hamiltonians and spin Hamiltonians (SHs). To facilitate broad understanding, a bird's-eye view of these studies is given in Section 2, where the recent results [1-5] are briefly outlined, whereas our earlier pertinent studies [8-12] are put in perspective. To provide theoretical background for the underlying considerations, the pertinent basic concepts and notations are overviewed in Section 3. Our earlier studies on the algebraic symmetry (AS) of Hamiltonians for continuous rotational symmetry (CRS) cases as well as the existing concepts of the rotational invariants and moments of CF Hamiltonians are outlined in Section 4 and 5, respectively. Brief presentation of the Noether's theorem applicable to CRS cases follows in Section 6. The various intricate 'low symmetry' aspects bearing on the compatibility and reliability of CFPs are dealt with in Section 7. The results presented in Sections 6 and 7 enable a new look at the properties of CF/LF Hamiltonians and SHs for CRS cases and consideration of practical applications in Section 8. Since this review is based on our published papers, especially the recent ones [1-5], instead of a full list of references and as a space saving measure, we provide here only pointers to the authors' papers. Hence, these references shall be considered as sources of specific references to pertinent literature and not as mere self-citations. For completeness, the most recent relevant papers not cited in [1-5] are briefly commented on in Conclusions (Section 9).

2. Bird's-eye view of our studies of low symmetry aspects in the CF/LF and EMR area

For the purpose of these considerations the term 'low symmetry' includes hexagonal II, tetragonal II, trigonal II, monoclinic, and triclinic symmetry cases [8-10], which collectively can be termed as CRS cases, as well as orthorhombic [11] ones. The former symmetries have been studied by one of us (CZR) earlier from the point of view of the AS of CF Hamiltonians [8-10]. The fact that these symmetries fall in the realm of the Noether's theorem and the arising consequences have only recently been realized [1]. The study [1] provides first consideration of the Noether's theorem in application to CF Hamiltonians and has profound implications. As discussed in Section 6, the conserved quantities stipulated by the Noether's theorem [1] are equivalent to the invariant combinations revealed earlier by the AS considerations [8-10]. Putting the studies [8-10] in perspective [1,3] provides a deeper meaning of the seemingly unrelated concepts: (i) the Noether's theorem, (ii) the AS of Hamiltonians, and (iii) the rotational invariants and moments of CF Hamiltonians. Considerations of the Noether's theorem [1] reveal hitherto unrealized implications for the properties of CFP datasets obtained from fitting the theoretical energy levels to the experimental data [3]. Utilizing the results [1], the CFP datasets for RE^{3+} ions in $LiYF_4$ were thoroughly reviewed and systematized in [2] (see, Section 8).

An important question in CF studies is to what extent the CFPs obtained for the same ion/host system by various authors are compatible and reliable. The quantity, which bears on these aspects, is the *'rhombicity'* ratio [10,11] defined as $\lambda = B_2^2 / B_2^0$ in the extended Stevens (ES) notation [6,12] and $\kappa = ReB_{22} / B_{20}$ in the Wybourne notation [6,7]. Due to the intrinsic property of orthorhombic [11] and lower symmetry CF Hamiltonians [10], this ratio can be confined to the *standard* range {0, 1} and

{0, 0.4082}, respectively, by choosing a proper axis system [10,11]. The standardization idea [11] was further extended to the higher-order CF terms and zero-field splitting (ZFS) terms [6,7] terms for orthorhombic [11] and monoclinic [10] symmetry. The computer package CST [13,14] was developed to facilitate (unit and notation) *conversions* [6,7,12], *standardization* [10,11], and *transformations* [12]. Since for any meaningful comparison the CFP and ZFSP datasets must be expressed in the same axis system, this package provides a useful tool for comparison of apparently different but physically equivalent datasets expressed in various notations and axis systems. Pertinent references for extensive applications of the standardization idea to ZFSP and CFP datasets can be found in [3].

Most recently, seeking an answer to the question: *"Can the low symmetry crystal (ligand) field parameters be considered compatible and reliable?"* posed in [3] we considered several fundamental aspects important for interpretation of the experimental and theoretical CFP datasets as well as their compatibility and reliability. The considerations [3] provide framework for re-consideration and clarification of various intricate 'low symmetry' aspects in CF studies, which have not been fully understood so far. These aspects are discussed in details in Section 7. The framework developed in [3] was utilized to reanalyse the CFP datasets for Nd^{3+} in $NdGaO_3$ and Pr^{3+} in $PrGaO_3$ [4] and rare-earth nickelates RNiO$_3$ [5]. Work is currently ongoing on the non-standard CFPs for the high temperature superconductors and related systems.

3. Overview of basic concepts and notations

For clarity, in this Section we define in the nutshell the basic concepts and notations pertinent for the considerations in the following Sections. Detailed discussion of the general properties of CF (ZFS) Hamiltonians and CF (ZFS) parameters and pertinent references can be found in [1,3] ([6,7]).

The concept of ***crystal-field*** was introduced to describe the effect of the electric field due to the surrounding ligands acting on the paramagnetic ion [1,3]. Within a given J- (or L-) multiplet the general form of the CF Hamiltonians can be expressed in terms of the ES operators [6,12] as:

$$H_{CF} = \sum_{kq} B_k^q O_k^q(\mathbf{J_x}, \mathbf{J_y}, \mathbf{J_z}) \tag{1}$$

where the nature of the ES operators $O_k^q(\mathbf{J_x}, \mathbf{J_y}, \mathbf{J_z})$ in equation (1) is explicitly indicated as being the functions of the total angular **J** (or total orbital **L**) momentum operators. The Wybourne notation utilizing the spherical tensor operators C_{kq} [1,3,6] is more widely used in the optical spectroscopy area. Within a given nl^m configuration H_{CF} in the ***compact*** form (as defined in Ref. 6) is given by:

$$H_{CF} = \sum_{kq} B_k^q O_k^q \tag{2}$$

whereas in the ***expanded*** form [6] using explicitly the pairs of the operators with $\pm$ q, H_{CF} is given by:

$$H_{CF} = \sum_k \left[B_{k0}C_{k0} + \sum_{q=1}^{k} \left\{ \mathrm{Re}\, B_{kq}\left[C_{kq} + (-1)^q C_{k-q} \right] + \mathrm{i}\,\mathrm{Im}\, B_{kq}\left[C_{kq} - (-1)^q C_{k-q} \right] \right\} \right] \tag{3}$$

Note that the CFPs B_k^q in the ES notation are real, whereas the CFPs B_{kq} in the Wybourne notation are complex, i.e. involve both real (Re) and imaginary (Im) parts. Conversion relations between the ES, Wybourne, and several other operator notations existing in the literature are provided in the comprehensive review [6].

The concept of *effective* spin Hamiltonian (SH) has been introduced to describe the splitting to the lowest orbital singlet ground state with the electronic spin S of a paramagnetic ion [6,7]. The derivation procedure and the *effective* nature of SHs were thoroughly discussed in [6,7]. Various meanings of 'spin' may apply for a particular 'spin' system: electronic (*real*), *effective,* or *fictitious* 'spin'. Within a given *effective* spin $\tilde{\mathbf{S}}$ multiplet the SH form [6], including the second-order *conventional* ZFS term (**D**-tensor) and the Zeeman electronic term (**g**-tensor) is given by:

$$\tilde{H}_{spin} = \tilde{S}.\mathbf{D}.\tilde{S} + \mu_B \mathbf{B}.\mathbf{g}.\tilde{S} \equiv \tilde{H}_{ZFS} + \tilde{H}_{Ze} \quad (4)$$

In terms of the ES operators [12] the general form of the ZFS term is nearly identical with H_{CF} in equation (1), except for the odd-order terms admissible only for H_{CF} [6,7] and replacement of B_k^q (CF) by B_k^q (ZFS) and $\mathbf{J}_i$ by appropriate 'spin' components $\tilde{S}_i$. Hence, the physical nature of the CF and ZFS quantities is completely different. However, there is a wide-spread confusion in the literature between the CF and spin Hamiltonians as well as the respective quantities. Various degrees of this confusion were discussed in the review [6] and more recently in [7,15].

Several theoretical models to compute the CFPs for transition ions in crystals exist in the literature [3]. A formula relating various CF components has often been employed to reduce the number of free parameters for fittings [4,5]:

$$B_n^{m(\alpha)} = \left(\gamma_n^{m(\alpha)} / \gamma_n^0\right) B_n^0, \quad (5)$$

where α= c and s denotes CFP components B_k^q (ES) and B_k^{-q} (ES) [12], respectively. The geometrical coordination factors $\gamma\gamma_n^{m(\alpha)}$ (α = c, s), equivalent to Hutchings' γ_{nm}^{α} [4], are to be calculated from the crystallographic data using, e.g. point charge model (PCM) or equivalent models reviewed in [3].

4. Algebraic symmetry (AS) of Hamiltonians for continuous rotational symmetry (CRS) cases

CRS cases comprise the following symmetry classes: hexagonal II (C_6, C_{3h}, C_{6h}), tetragonal II (C_4, S_4, C_{4h}), trigonal II (C_3, S_6), and monoclinic (C_2, C_{1h}, C_{2h}), i.e. the so-called uni-axial symmetries, as well as triclinic (C_i, C_1) symmetry. The concept of AS of Hamiltonians has been introduced by Clark in 1971 (Ref. 57 in [1]) to describe the symmetry of the characteristic equation of Hamiltonians [8-10,1]. Regrettably, the concept of the AS has been underutilized - apart from the applications [8-10,1] the Clark's paper [8,1] remains largely forgotten as revealed by a SCI citation search.

Considerations of the AS of CF Hamiltonians were carried out for tetragonal II [8], hexagonal II and trigonal II [9], and monoclinic [10] symmetry. These considerations revealed the existence of some conserved quantities (CQs) invariant under rotations around the axis specifying the given CRS. CQs are in fact invariant combinations (ICs) of the CFPs. The following ICs were identified explicitly [8-10]: the first-order ICs, which comprise the CFPs with q = 0:

$$B_2^0, B_4^0 \text{ and } B_6^0; \quad (6)$$

the second-order ICs of two types:

$$\text{type (i): } (\{B_k^{+q}\}^2 + \{B_k^{-q}\}^2)^{1/2} \equiv |\mathrm{v}_{kq}|, \quad (7)$$

$$\text{type (ii): } B_k^{+q} B_{k'}^{+q} + B_k^{-q} B_{k'}^{-q} \equiv \mathbf{v_{kq}} \bullet \mathbf{v_{k'q}} = |\mathrm{v}_{kq}||\mathrm{v}_{k'q}|\cos(\beta_{kq}^{k'q}). \quad (8)$$

The higher n^{th}-order (n>2) ICs exist for the cases which require solution of the characteristic equations of the order greater than 2. The third- (n=3) and fourth-order (n=4) ICs were explicitly derived for trigonal II [9] and monoclinic [10] symmetry and their basic types were discussed in [1]. The 2nd -order CQs have a geometrical meaning [8-10,1]: type (i) - the length of a 'vector' $\mathbf{v_{kq}}$, the type (ii) - the angle between any two possible vectors $\mathbf{v_{kq}}$ and $\mathbf{v_{k'q}}$ (see, also Section 6.1).

The fact that the ICs, and not separate CFPs involved in the ICs, appear in the coefficients of the characteristic equation of CF Hamiltonian has an important consequence for fitting experimental spectra to the theoretical energy levels. Three approaches to fitting the experimental spectra for RE ions at sites exhibiting one of the five symmetry types in question exist in the literature [8-10,1]. In the **approximated (A)** approach the actual point-symmetry group is approximated by a higher group, thus omitting the CFPs with negative q in Eq. (1) (or equivalently the Im parts of CFPs in Eqs (2) and (3)). In the **reduced (R)** approach one component of a pair of CFPs with +q and -q is set to zero. In the **complete (C)** approach all CFPs admissible by group theory for a given site symmetry are allowed in the fittings. Various conventions and implications regarding each approach were discussed in [1]. Critical examination of the literature reveals various inconsistencies in the experimental determination of the low symmetry CFPs [8-10,1-5]. The studies [8-10] established that determination from fitting experimental spectra of the magnitudes and relative signs of all CFPs predicted by group theory is impossible for CRS cases. However, the relationship between the Noether's theorem [1] and the AS of CF Hamiltonians has only recently been noticed (see, Section 6).

5. Rotational invariants and moments of CF Hamiltonians

Since various authors often use different axis systems for the same ion/host system, the rotational invariants S_k have been introduced by several authors [1,3] to measure the strength of the CF terms; in the Wybourne notation:

$$(S_k)^2 = \frac{1}{2k+1}\sum_q \left|B_{kq}\right|^2. \tag{9}$$

Equivalent expressions in the ES notation exist [1,3,10,11]. The quantities S_k are the 2nd-order rotational invariants [1] and provide an additional check of compatibility of the original and transformed CFPs [12-14]. The 3rd- and 4th-order rotational invariants, i.e. combinations of three and four CFPs, respectively, were also introduced [1]. However, as shown in [1] such hexagonal and cubic invariants carry no new information, since they are reducible to the first-order ICs in Eq. (6).

The nth-order CFPs rotational invariants are related to the nth-order (spectral) moments $\sigma_n(\alpha LSJ)$ of H_{CF} as follows [1]:

$$(\sigma_n(\alpha LSJ))^n = (1/(2J+1))\mathrm{Tr}(H_{CF})^n. \tag{10}$$

It appears that the invariant properties of the nth-order moments as well as their relationships with the CFPs and the CQ predicted by the AS of CF Hamiltonians and stipulated by the Noether's theorem have not been explicitly considered so far [1]. For a more detailed discussion of the rotational invariants and moments of CF Hamiltonians we refer to [1].

6. New look at the CRS cases in CF studies from the point of view of the Noether's theorem

6.1. Noether's theorem and continuous rotational symmetries

In [1] we considered generalization of CF and ZFS Hamiltonians in the form of a generic Hamiltonian:

$$H = \sum A_{lm}\chi_{lm}(\mathbf{X}), \tag{11}$$

where A_{lm} denote the parameters associated with a given generic type of the tensor operators $\chi_{lm}(\mathbf{X})$ of the specific angular momentum $\mathbf{X}(X_x, X_y, X_z)$ operators, expressed in a given axis system (x, y, z). Note that the generic symbols χ_{lm} and A_{lm} used in Eq. (11) do not define yet another tensor-operator notation, an abundance of which already exists in literature [6,7] but may represent any of the various sets of tensor operators and associated parameters. Since the arguments [1] are based on symmetry requirements, which apply to all types of Hamiltonians (11), the general conclusions concerning the implications of the Noether's theorem are valid for all such Hamiltonians.

The symmetry aspects inherent in each of the CRS cases were discussed in details in [1]. Various formulations of Noether's theorems, which utilize the geometrical symmetries, from introductory ones to advanced mathematical ones, available in the literature were discussed in [1]. Most pertinent for our consideration is the formulation given by Hill & Lederman (2001; Ref. 2 in [1]): *"For every continuous symmetry of the laws of physics, there must exist a conservation law. For every conservation law, there must exist a continuous symmetry."* Hence, the canonical Noether's theorem implies, for given symmetry transformations of the coordinates, the existence of the associated CQs, which are functions of the parameters describing a given classical or quantum system. This theorem should, in principle, apply also to any CF Hamiltonian invariant under CRS.

Here we can only afford to recap the major findings achieved in [1]. As argued in [1] for the CRS cases there must exist associated **conserved quantities**, i.e. polynomial functions of the CFPs; in general, the generic A_{lm} ones in Eq. (11). Existence of such quantities ensures the physical equivalence of the CFP datasets interrelated by the CRS transformations applicable to the cases in question. The

central idea put forward in [1] and arising from the canonical Noether's theorem for the CF and equivalent Hamiltonians invariant under CRS is encapsulated in **Theorem 1**:

> **If the CF Hamiltonian defined in Eq. (1) (and (2) & (3); in general, any Hamiltonian of the type in Eq. (11)) is invariant under CRS, then several quantities of various order in the CF parameters B_k^q (and equivalent ones; in general, the associated A_{lm} ones) are conserved.**

By 'conservation' we mean the invariance of such quantities under the CRS operations. Analytical rigorous derivations of the CQs in question may be appealing to algebraists and quantum mechanical theorists. Instead, we have exploited [1] the AS results [8-10], which provide a direct proof of Theorem 1 for all symmetry cases in question, except of triclinic symmetry not dealt with in a similar way yet. The studies [8-10] have, in fact, revealed the existence of several *"invariant combinations"* (ICs) in the characteristic equations of monoclinic or higher CRS CF Hamiltonians. ICs are polynomial functions of CFPs of the form: $f\{(B_k^q)^i(B_k^q)^j(B_k^q)^m(B_k^q)^l\}$, with various values of (*k*, *q*) in each term and the power coefficients (*i*, *j*, *m*, *l*). Theorem 1 and the AS considerations enable us [1] to put forward **Conjecture 1:**

> **The "invariant combinations" (ICs) revealed by the AS considerations and categorized above are the CQs stipulated by the canonical Noether's theorem applied to the CF Hamiltonians in question (Theorem 1).**

The ICs (CQs) identified by the explicit AS considerations [8-10] as specific combinations of CFPs arise naturally from the Noether's theorem. Theorem 1 and Conjecture 1 establish a direct link between the Noether's theorem and the AS of CF Hamiltonians invariant under CRS. The geometrical meaning of the 2nd-order ICs in Eqs (7) and (8) was illustrated in [1] using a two- and three-dimensional representation in the CFP space, respectively. The vexing question if a single conservation law can sum up all the ICs has also been discussed in [1]. The ramifications of Theorem 1 and Conjecture 1 are considered in Section 6.2.

6.2. Implications of the canonical Noether's theorem and the conserved quantities

The existence of the CQs stipulated by Theorem 1 and Conjecture 1 has profound practical implications for interpretation of experimental CFP datasets for the CRS cases. For example, for tetragonal II symmetry [8] only the 'length of the vector' $|\mathbf{v}_{44}| = (\{B_4^4\}^2 + \{B_4^{-4}\}^2)^{1/2}$, equation (7), can be experimentally determined. Hence, the approach C, which attempts to determine independently the values of both B_4^4 and B_4^{-4}, appears inappropriate. Since $|\mathbf{v}_{44}|$ is nothing else but one of the CQs stipulated by the Noether's theorem, this limitation is a direct implication of this theorem.

The properties of the CQs and their implications were summarized in Corollaries [1] listed below. Corollaries 1 to 3 generalize the conclusions arising from the AS considerations [8-10] and bear on interpretations of the experimental CFPs and fitting procedures. Implications specific for interpretation of the approach C and R were summarized in Corollaries 4 and 5 thus providing new deeper meaning of these approaches.

Corollary 1. The coefficients of the characteristic equations of the CF Hamiltonians described by the Theorem 1 are functions of the admissible CQs defined by Conjecture 1 only.

Corollary 2. Only the CQs can be directly and independently determined from fitting the experimental energy level structure to the theoretical energy levels.

Corollary 3. The CQs of the order n = 1 determine directly the magnitudes and the signs of the CFPs B_k^q with q = 0, whereas those of the order $n \geq 2$ determine the magnitudes of the allowed combinations of the CFPs and the relative signs of B_k^q with $q \neq 0$.

Corollary 4. There exists **one to many** correspondence between the CQs and the magnitudes of the CFPs with $q \neq 0$ determined in the approach **C**.

Corollary 5. There exist **one to one** correspondence between the CQs and the magnitudes of the CFPs with $q \neq 0$ determined in the approach **R** for each specific choice of the CF parameter reduced to zero.

Corollary 1 establishes the link between the coefficients of the characteristic equations of the CF Hamiltonians and the CQs. The consequences of the invariance of CF Hamiltonians under CRS stipulated by the Noether's theorem and encapsulated in Corollaries 2 to 5 provide a new insight into the symmetry properties of such Hamiltonians and CFPs. The ramifications of the present considerations provide a solid basis for interpretation of experimental CFP datasets appearing in the literature. Our considerations reveal that various aspects not fully realized in the literature as yet require reinterpretation, e.g. the feasibility of determination of CF parameters from fitting experimental spectra and the reduction of the existing higher-order rotational invariants for hexagonal type II symmetry and cubic symmetry to combinations of primary lower-order invariants. The novel approach presented in [1] enables adoption of better fitting strategies utilizing the well-defined CQs, which are invariant under CRS. The considerations enable also to solve the long-standing problem, i.e. which CF parameters can be determined from fitting the experimental energy level structure to the theoretical energy levels [1].

7. Re-consideration and clarification of various 'low symmetry' aspects in CF studies

As continuation of the studies [1,2] we considered the fundamental, yet hitherto not fully understood, intricate properties of CF Hamiltonians. The studies [3-5] are briefly summarized below. Clarification of the intricate aspects in question has enabled us to provide a general framework aimed at achieving an increased compatibility and reliability of CFP datasets for transition ions at low symmetry sites in crystals. This framework may be especially useful for trivalent lanthanides in technologically important low symmetry hosts for which only limited energy-level data are available.

7.1. Selection of the axis systems for various symmetry cases and their implications

The crystallographic axes (**a**, **b**, **c**), the local site symmetry axes, and the '*natural*' choices of the axis system (x, y, z) for the CF Hamiltonian were discussed for various symmetries in [3]. In short, for given site symmetry: **hexagonal type II** (C_6, C_{3h}, C_{6h}) - there exist only one sixfold axis (C_6); **axial type II**, i.e. tetragonal II (C_4, S_4, C_{4h}) and trigonal II (C_3, S_6) - there exist only one fourfold and one threefold local site symmetry axis, respectively; **orthorhombic** (D_2, C_{2v}, D_{2h}) - there exist three C_2 mutually perpendicular local site symmetry axes, which may or may not coincide with the axes (**a**, **b**, **c**). This leads to additional complications discussed in [3].

For **monoclinic** site symmetry (C_2, C_{1h}, C_{2h}) three complications occur [3]: (i) the vectors (**a**, **b**, **c**) may not form a Cartesian coordinate frame - then an approximate **c*** axis is chosen perpendicular to the plane (**a**, **b**); (ii) only one local site symmetry (C_2) axis (or plane) exists defining the unique monoclinic direction ($\mathbf{C_M}$) – this yields three choices of labeling the $\mathbf{C_M}$ direction: $\mathbf{C_M} \| z$, $\mathbf{C_M} \| y$, or $\mathbf{C_M} \| x$, each corresponding to different yet physically equivalent form of the CF (ZFS) Hamiltonian; (iii) once a specific axis is fixed along the $\mathbf{C_M}$ direction, a problem arises how to select the other two axes, e.g., x and y in the case of $\mathbf{C_M} \| z$ - consequently, the remaining two axes may be selected arbitrarily and hence for monoclinic symmetry an infinite number of distinct yet physically equivalent CFP datasets may be generated in two independent ways.

For **triclinic** site symmetry (C_i, C_1) there exist no single local site symmetry axis, whereas for triclinic space groups the crystallographic axes (**a**, **b**, **c**) do not form a mutually perpendicular set. This renders the choice of the axis system (x, y, z) completely arbitrary. Due to the lowest symmetry, a number of specific intricacies concern triclinic CF Hamiltonians. They are dealt with in a separate section in [3] and are too extensive to cover in details here. The major point is that there exist an infinite number of identical and physically equivalent forms of triclinic CF Hamiltonians each associated with a distinct yet physically equivalent CFP dataset.

7.2. Types of CF parameters and their properties

In [3] we identified three types of CFPs occurring in the CF/LF studies: (i) **symbolic CFPs** and two types of **numerical** CFPs: (ii) **theoretical CFPs** and (iii) **fitted CFPs**. Each type has specific intrinsic properties and general ones in regard to the axis system in which the CFPs are (presumably) expressed [3]. This novel classification helps to clarify the related intricate aspects.

The **symbolic CFPs** are defined by the CF Hamiltonians employed for a given ion-host system and hence underlie both the theoretical calculations and fitting procedures. Most often the symmetry-adapted (x, y, z) axis systems are adopted explicitly or implicitly in CF/LF studies since they yield the smallest number of the non-zero symbolic CFPs. The axis systems defined for CF Hamiltonians and hence the symbolic CFPs serve as "input" necessary to set up the CF calculations. The **theoretical CFPs** are computed using model calculations. Clear definitions of the axis system (x, y, z) in which the positions of ligands are given, e.g. w.r.t. to the axes (**a**, **b**, **c**), are indispensable for any model calculations. The existence of equivalent theoretical CFP datasets in the literature arise from 'different (implicit) choices of coordinate system' [3].

The **fitted CFPs** are obtained from fitting the experimental CF energy levels and/or intensity data to the theoretical predictions based on the symmetry-adapted CF Hamiltonians. Our extensive survey of CF data and their analysis have enabled us to make the following statements [3]. **(A)** It is feasible to specify clear definitions of the axis system (x, y, z) **only** for the CF Hamiltonians and thus the symbolic and theoretical CFPs. **(B)** The conventional techniques employed to determine CFPs for the transition ions in crystals provide information primarily about the experimental CF energy levels and in some cases also about the intensity of the transitions. These techniques provide no 'directional' information, unlike most of EMR techniques [6,7]. **(C)** Our survey reveals two kinds of attitudes concerning the axes for the fitted CFP datasets for a given ion/host system. By implication, the more theoretically inclined researchers seem to share the presumption that such axes are just identical with the axes (x, y, z) adopted initially for the CF Hamiltonian. Another attitude manifested by majority of researchers, especially experimentalists, consists in simply avoiding any explicit mention of the axis system adopted for their fitted CFP datasets. It turns out that the interpretation of the axis systems for the fitted CFP datasets amounting to their tacit equivalence with the axis systems pertinent for the symbolic or theoretical CFPs (see the statement (A) above) does not stand up to the truth. In fact, no such equivalence exists, since no particular axis system can be obtained from fitting the experimental CF energy levels determined by 'non-directional' techniques (see the statement (B) above).

7.3. Introduction of a new notion of a nominal axis system for the fitted CFP datasets

The apparent dilemma that there is no prescription how to assign an axis system to the fitted CFPs bears on the compatibility of CFP datasets since the knowledge of the axes is essential for any meaningful comparison. To solve this dilemma we have introduced [3] a **novel** notion of a ***'nominal'*** axis system defined by the statement: any CFP dataset **fitted using non-directional techniques** must initially be considered as expressed in an undefined *'nominal'* axis system denoted, e.g. by (x_n, y_n, z_n).

Ample justification for this statement coming from the intrinsic properties of the CF Hamiltonians in question and the nature of the CF calculations carried out using the least-squares fitting procedures was provided in [3]. The orientation of a *'nominal'* axis system w.r.t. the crystallographic axes is not known and cannot be assigned based on the axes used for H_{CF}. In conclusion, the fitted CFP datasets reported in the literature cannot be automatically assigned to any, even well defined, specific axis systems. Conversely, one cannot determine any axis system from fittings. These findings represent a marked difference with the wide-spread perception that the axis system once defined as "input" can be carried over and assigned to the fitted CFPs. The notion of the *'nominal'* axis system solves the existing confusion concerning the fitted CFPs.

7.4. Implications of the Noether's theorem and the AS of CF Hamiltonians

The existence of several CQs stipulated by the Noether's theorem and AS of CF Hamiltonians invariant under CRS bears on the feasibility of determination of CFPs from fitting experimental spectra. The novel approach [1] enables adoption of better fitting strategies utilizing the well-defined CQs for these symmetry cases [2,3]. The constraints inherent in any least-squares fitting of CFPs can be classified as (i) '**computational**' - related to the sufficient number of experimental CF energy levels available for meaningful fitting, and (ii) '**physical**' - applicable to CF Hamiltonians invariant under CRS. The considerations of the AS [8-10,1] reveal that for such Hamiltonians not all the non-zero CFPs predicted by group theory for a given symmetry, of total number $N_G = \sum_k N_G(k)$, where $N_G(k)$ is the number of independent k-th order CFPs, can be determined independently from least-squares fittings of experimental data. Only certain combinations of CFPs (see Section 4) can be independently determined for such symmetry cases due to the intrinsic CF Hamiltonian properties. In the set of all vectors $\mathbf{v}_{kq}$ formed by the possible pairs (+q and -q) of the non-zero k-th order CFPs admissible for a given symmetry, only one component of one vector $\mathbf{v}_{kq}$ at a time may be set to zero. This '(N_G -1)' constraint limits the number N_D of the independently determinable CFPs to (N_G -1) for the uni-axial symmetry as well as for triclinic symmetry cases [3]. The approach utilizing (N_G -1) CFPs for fitting correspond to the approach **R.** The (N_G -1) physical constraint holds irrespective of the computational constraint. Hence, even if a sufficiently large number of experimental CF energy levels is available for fittings, one cannot meaningfully determine all N_G CFPs for these symmetry cases. This may render the CFP datasets obtained using the approach **C** potentially unreliable [3]. Various approximations concerning the non-zero CFPs adopted to reduce the number of CFPs involved in fitting procedures have also been discussed in [3].

7.5. Correlation properties among CFP datasets and the standardization idea

An infinite number of physically equivalent CFP datasets of any type can be obtained by applying arbitrary transformations to the original CFPs. Such transformed and original CFPs are mutually 'correlated'. However, the symmetry-adapted form of CF Hamiltonian is preserved only for specific transformations. The transformations preserving CF Hamiltonian form and yielding equivalent CF Hamiltonian forms [3] have special importance since they generate easily identifiable physically equivalent CFP datasets. Correlation properties of such CFP datasets can be utilized to improve their overall compatibility and reliability [3]. Three types of such transformations were identified for low symmetry cases [3]: (I) the standardization transformations (see Section 2) for the orthorhombic and monoclinic symmetry cases [10,11] as well as triclinic ones [3], (II) transformations applicable to H_{CF} for continuous uni-axial rotational symmetry, and (III) transformations applicable to the three forms of monoclinic H_{CF}. In relation to the Noether's theorem [1] we note that the standardization [10,11] of the orthorhombic and lower symmetry Hamiltonians (including CF, ZFS, and related ones) involves continuous permutational symmetry, and not rotational symmetry, whereas the transformations type (II) and (III) represents implications of the Noether's theorem and AS of CF Hamiltonians.

As a consequence of the transformations type (I) to (III), there exist several equivalent CF Hamiltonians with numerically different values of CFPs, while each CFP set must yield identical energy levels. Hence fittings may yield several equivalent CFP datasets in various regions of the multidimensional CF parameter space. The question how to identify such distinct yet correlated CFP datasets was discussed in details in [3] for each type of transformations (I) – (III). Keeping in mind the possibility that the least-squares fittings may return both global and local minima, the problem arises how to correlate the various data points resulting from independent fittings. The notion of a 'common' *nominal* axis system meant as any transformed *nominal* axis system required to yield, after transformation of an original non-standard CFP dataset, the transformed standardized dataset located in the standard region in the CF parameter space, plays an important role in these considerations.

The interrelationships between physically equivalent CFP datasets located in different regions in the CF parameter space were graphically visualized [3]. The advantages of the standardization [10,11] call for a wider acceptance of unified guidelines for presentation of CFP datasets proposed in [3].

7.6. The rotational degrees of freedom and the reduction of the number of independent CFPs

These aspects are of special importance for triclinic symmetry cases and have aroused some controversy in the literature. To summarize the extensive discussion [3], for each CRS, including triclinic symmetry, the number of independent rotational degrees of freedom W = 1. The physical constraint discussed above for these cases yields a reduction of the number (X) of CFPs which can be independently fitted from experimental data. Even if there is sufficient number of experimental CF energy levels, i.e. when the computational constraint is satisfied, we have the relation: $X = N_G - W = N_G - 1$ [3].

7.7. Extension of the multiple correlated fitting technique

The idea of utilization of additional information contained in the intrinsically correlated CFP datasets underlies the multiple correlated fitting technique (MCFT) originally proposed in [10]. MCFT has, for the first time, been practically implemented using the experimental CF energy levels for Nd^{3+} in YAG and LaF_3 [16]. The study [16] has also revealed misinterpretations of the original CFPs due to inadvertent choices of different 'representations', i.e. different *nominal* axis systems [3] corresponding to different regions in the CF parameter space. MCFT provides also guidance on the selection of most appropriate 'representation' for the final fitted CFP dataset. The three types of physically equivalent CFP datasets discussed in Section 7.5 enable further extension of the capabilities of MCFT [3]. The extension includes a quantitative method for CFP datasets comparison based on the closeness parameters defined for the CFP "vectors" [3]. Hence the degree of correlation between independently fitted CFP datasets and expressed in the same *nominal* axis system may be quantitatively measured. The basic steps involved the MCFT procedure [16] are summarized in [3].

8. Application of the Noether's theorem and low symmetry aspects to CFP datasets

The CQs stipulated by the Noether's theorem were utilized in [2] for comparative analysis of the CFP datasets for the RE^{3+} ions at tetragonal type II S_4 symmetry sites in $LiYF_4$, for which sufficiently large amount of data is available across the whole $4f^N$ series. Five CQs stipulated by the Noether's theorem for S_4 symmetry, namely, B_{20}, B_{40}, B_{60}, $|\mathbf{v}_{44}|$, $|\mathbf{v}_{64}|$, and the angle β_{44}^{64} as well as the ratio $\mathrm{Im}B_{64}/\mathrm{Re}B_{64}$ and the 2nd-order rotational invariants S_k were considered. Pertinent CFP datasets together with the calculated associated quantities were tabularized and the variations of B_2^0, B_4^0, B_6^0, $|\mathbf{v}_{44}|$, and $|\mathbf{v}_{64}|$ vs N were presented in graphs [2].

This systematics facilitated consideration of trends across the $4f^N$ series among the CFPs and derived quantities. Knowledge of such trends enables identification of unreliable CFPs and prediction of CFPs for ions for which no or little data are available. The major observations for RE^{3+}:$LiYF_4$ are as follows [2]. The available data show for most ions a very large spread of the original CFP values for the same $4f^N$ ion, thus indicating inconsistency of several CFP datasets. Since the ratio $\mathrm{Im}B_{64}/\mathrm{Re}B_{64}$ used earlier in the literature is not a CQ it is inadequate for comparative purpose, whereas the angle β_{44}^{64} as a CQ is more convenient to deal with than the 'scalar product' $\mathbf{v}_{44}\cdot\mathbf{v}_{64}$. The values averaged over the available data points for each ion show no well-defined trends across the RE^{3+} series in B_k^0, whereas a weak diminishing trend with increasing N in $|\mathbf{v}_{44}|$ and $|\mathbf{v}_{64}|$. Various aspects bearing on the erratic CFP behavior were discussed [2]. The intricacies concerning the approach R and C were also clarified providing guidelines for identification of the CFPs and associated quantities reliably determinable from fitting the experimental CF spectra to the theoretical energy levels.

The applications [2] of proposed method of analysis and presentation of experimental and theoretical data [1] amply illustrate the usefulness of the CQs stipulated by the Noether's theorem and their advantages over other derived quantities used in the literature as well as the advantages of the approach R and the disadvantages of the approach C and A. The systematics presented in [2] has implications for interpretation of the published CFP datasets and their reliability and thus may bring about an increased reliability of the CFP datasets in future.

Concerning other practical applications of the results presented in Sections 6 and 7, better understanding of the intricate properties of low symmetry CF Hamiltonians has enabled consideration of specific important aspects bearing on compatibility and reliability of the CFP datasets for Nd^{3+} in $NdGaO_3$ and Pr^{3+} in $PrGaO_3$ [4] and rare-earth nickelates RNiO$_3$ (R = Pr, Nd, Sm, Eu, and $Pr_{1-x}La_x$, $0 \leq x \leq 0.7$) [5]. This includes: (i) clear definition of the axis systems, (ii) relevant coordination factors used in the point charge model, (iii) multiplicative factors missing in some CFP datsets, (iv) reduction of the number of the independent CF parameters, and (v) conserved quantities existing for monoclinic symmetry [4,5].

9. Conclusions and further readings

The summary of major findings arising from our recent studies concerning the symmetry properties of crystal/ligand field Hamiltonians and spin Hamiltonians presented in this review offers several advantages. This review provides all-in-one concise yet comprehensive treatment of the important problems dealt previously with in separate papers. The aspects considered here are crucial for better understanding of the procedures used to obtain the experimental CFP values, especially for low symmetry cases, and hence interpretation of the CFP datasets available in the literature. Clarification of these intricate aspects has enabled us to provide a general framework aimed at achieving an increased compatibility and reliability of CFP datasets for transition ions at low symmetry sites in crystals. It may be hoped that by making this general framework more accessible to general practitioners more meaningful interpretation and comparison of the CFP datasets may be carried out in future studies.

Concerning recent pertinent literature we note the following papers. Optical absorption spectroscopy in geosciences has been reviewed with focus on the basic concepts of crystal field theory [17] and quantitative aspects of crystal fields [18]. Generalization of the extended Stevens operators to higher ranks and spins and systematic review of the tables of the tensor operators and their matrix elements are provided in [19].

The paper by Mulak [20] proposing a parameterization scheme in which the number of the k-even triclinic CF parameters $N_G(k)$ = 5, 9, and 13 can be reduced by 4, 8, or 12 for k = 2, 4, or 6, respectively, has been critically commented on in [3]. The inapplicability of such parameterization has recently been admitted [21]. However, the paper [21] adds more to the controversies concerning the low symmetry properties of CF Hamiltonians, which will be discussed separately.

References

[1] Rudowicz C and Qin J 2003 *Phys. Rev.* B **67** 174420

[2] Rudowicz C and Qin J 2004 *J. Alloys Comp.* **385** 238

[3] Rudowicz C and Qin J 2004 *J. Lumin.* **110** 39

[4] Rudowicz C and Qin J 2005 *J. Alloys Comp.* **389** 256

[5] Rudowicz C and Qin J 2005 *Physica* B **364** 239

[6] Rudowicz C 1987 *Magn. Reson. Rev.* **13** 1
Erratum, *Magn. Reson. Rev.*1988 **13** 335

[7] Rudowicz C and Misra S K 2001 *Appl. Spectrosc. Rev.* **36** 11

[8] Rudowicz C 1985 *Chem. Phys.* **97** 43

[9] Rudowicz C 1986 *Chem. Phys.* **102** 437

[10] Rudowicz C 1986 *J. Chem. Phys.* **84** 5045

[11] Rudowicz C and Bramley R 1985 *J. Chem. Phys.* **83** 5192

[12] Rudowicz C 1985 *J. Phys.* **C18** 1415; *Erratum, J. Phys* **C18** 3837

[13] Rudowicz C in Newman D J and Ng B (Eds) 2000 *Crystal Field Handbook* (Cambridge: Cambridge Univ. Press) Appendix 4

[14] Rudowicz C and Qin J 2002 *Computers and Chemistry* **26** 149

[15] Rudowicz C and Sung H W F 2001 *Physica B* **300** 1

[16] Rudowicz C, Chua M and Reid M F 2000 *Physica* B **291** 327

[17] Wildner M, Andrut M and Rudowicz C in Beran A and Libowitzky E (Eds) 2004 *Spectroscopic Methods in Mineralogy - European Mineralogical Union Notes in Mineralogy* Vol 6 (Budapest: Eötvös University Press) ch 3

[18] Andrut M, Wildner M and Rudowicz C in Beran A and Libowitzky E (Eds) 2004 *Spectroscopic Methods in Mineralogy - European Mineralogical Union Notes in Mineralogy* Vol 6 (Budapest: Eötvös University Press) ch 4

[19] Rudowicz C and Chung C Y 2004 *J. Phys. Cond. Matter* **16** 5825

[20] Mulak J 2003 *Physica* B **337** 173

[21] Mulak J and Mulak M 2005 *J. Phys.* A*: Math. Gen.* **38** 6081

A microscopic model of oxygen vacancies in Ca-doped YIG

A Lehmann-Szweykowska*, A Szamer*, R J Wojciechowski*, R Micnas* and T Lulek**

* A. Mickiewicz University, Institute of Physics, Poznań, Poland

** Rzeszów University of Technology, Dept. of Physics, Rzeszów, Poland

Abstract. Annealing experiments carried out on calcium doped yttrium iron garnet (Ca:YIG) in an oxygen-reducing atmosphere, show dramatic decreases in the magnetisation over the entire temperature range. Simultaneously, the electric resistivity and the electric activation energy significantly increase. Within the framework of the iron-oxygen cluster model, we analyse and compare with each other two situations, in which compensating holes can be found. The first is their occupying hybridised eigen-energy levels of the cluster. In this case, it can be shown at which ions the holes tend to remain for a given energy level. It turns out that the active holes remain localised most at the oxygen ions. The other situation means the hole's localisation in an attractive trap centre. The latter is created at the missing oxygen ion between the tetrahedral and octahedral iron-oxygen ionic clusters in the proximity of a doping Ca^{2+} ion. For a sufficiently shallow traps, their localised states can be considered within the framework of hydrogen-like model. The ground state energy of the vacancy can thus be estimated, considering the dielectric constant of the host material and the hole's effective mass as variational parameters. Communication between the tetrahedral and octahedral iron-oxygen clusters with missing oxygen between them, is assumed to be maintained via the $s-d$ hybridisation.

1. Introduction

There has recently been observed a great interest in creating semiconducting materials with permanent magnetic moment by an appropriate modification of those which already have magnetic long range order but do not necessarily conduct the electric current. Yttrium iron garnet (YIG) seems to be a good candidate for such a role as it is a ferrimagnet with high Curie temperature (550 K). Pure (YIG) is an almost ideal insulator, however, doped with valence uncompensated ions it becomes a semiconductor. In the last decades, many experiments have been carried out indicating that both the electric and magnetic properties of yttrium-iron garnets depend critically on the type and number of impurities included into the host structure. Our recent analysis is aimed at an understanding of a role of oxygen vacancies, which are present in all garnet structures. The empirical data concern thin films of yttrium-iron garnets doped with divalent calcium ions Ca^{2+}, which substitute trivalent yttrium [1-3]. Calcium is as diamagnetic as yttrium,so it basically leaves the magnetic moment untouched. Nevertheless, it can have significant effects on the magnetic moment of the system. Also, the electric properties of YIG are dramatically altered by the doping. Ca:YIG is a p-type semiconductor. Calcium ion introduces a local charge imbalance at the dopant site which is minimised by the creation of secondary induced charge centres of opposite sign. In our model, monovalent or completely missing oxygen ions immediately adjacent to the included Ca^{2+} are considered as the principal charge compensating mechanism. The role of O^- centres has been already discussed in a great detail elsewhere [4,5]. In the present paper, we concentrate on the question of oxygen vacancies which are oxygen sites with no oxygen ions but containing one ($V_{\dot{O}}$ or two $V_{\ddot{O}}$ orbital compensating holes.In the first case, the vacancy compensates one adjacent Ca^{2+}, whereas the other vacancy compensates two calcium ions globally. It has been known that at temperatures below 100 K the magnetisation in Ca:YIG exhibits a anomalous decrease relative to that expected for pure YIG. Within the framework of our cluster model, we propose monovalent oxygen as an key instigator of this magnetisation anomaly. The lowering in magnetisation can be explained by a change of the local super-exchange interaction for ferromagnetic and a reduction of the tetrahedral cluster spin from $S = 5/2$ to $S = 2$ [4]. It has turned out that annealing in reducing atmosphere gives rise to even more dramatic decreases in the magnetisation of Ca:YIG over the entire temperature range. Furthermore, the observed shift in magnetisation for the N_2 annealed samples was linearly proportional to the change in crystal lattice constant[1,2]. The empirical results show conclusively that the magnetisation is related to the number of missing oxygen ions, which implicitly supports one of our most important previous conclusions [4] that actually compensating holes remain on the oxygen sites and only occasionally can hop at the iron sites, forming thereby Fe^{4+} ions, the latter action being highly unfavourable from the energetic point of view as it involves the highest energy level of the iron-oxygen cluster.The organisation of the paper is as follows. In the next section, we briefly analyse a microscopic model of a pair of the octahedral (a) and tetrahedral (d) ionic clusters with the emphasis put on a key role which is played by the oxygen mutual site in determination of both the magnetic properties and the charge transport (see: for details, [4-6]). The analysis is aimed at a better understanding of a situation which occurs when the mutual oxygen vanishes. To achieve this objective, it seems to be necessary and sufficient to consider a pair of a and d ionic clusters, while only inducing influence of the remaining system in a form of certain mean-field parameters, such as the dielectric constant, or the effective mass of the compensating hole. In the third section, we focus on the properties of the vacancy with one or two compensating holes trapped in it as the topic requires obviously a deeper viewing. The oxygen vacancy is considered to be shallow and, consequently, treated as an hydrogen atom with the Coulomb attraction formed between a positive compensating hole and a resultant negative charge of Ca^{2+} substituted for Y^{3+}. Assuming localisation of the hole at the empty oxygen site, we thereby determine the Bohr radius as being equal to the distance between the empty oxygen site and an Ca^{2+} ion in its proximity, which means that its ground state energy is known, as well.

2. A microscopic model of a pair of the a and d iron-oxygen clusters with the oxygen vacancy

Electronic properties of the single tetrahedral (d) and octahedral (a) ionic clusters are determined by the five 3d electrons localised at the central iron ions and the six 2p electrons at each neighbouring oxygen site. Considering systems with their charge transport maintained by positive compensating holes it seems quite natural to analyse their electronic structure in terms of orbital holes rather than electrons. We thereby start with the Anderson Hamiltonian in the atomic limit rewritten for the pair of the different ionic clusters, in terms of the orbital3d and 2p holes [4]:

$$\hat{H} = \hat{H}_{3d} + \hat{H}_{2p} + \hat{H}_{p-d} \tag{1}$$

where

$$\hat{H}_{3d} = E_a \sum_{i,m} \hat{n}_{i,m} + E_d \sum_{j,m} \hat{n}_{j,m} + \frac{U}{2} \sum_{i,m} \hat{n}_{i,m}\hat{n}_{i,-m} + 1/2U \sum_{j,m} \hat{n}_{j,m}\hat{n}_{j,-m} \tag{2}$$

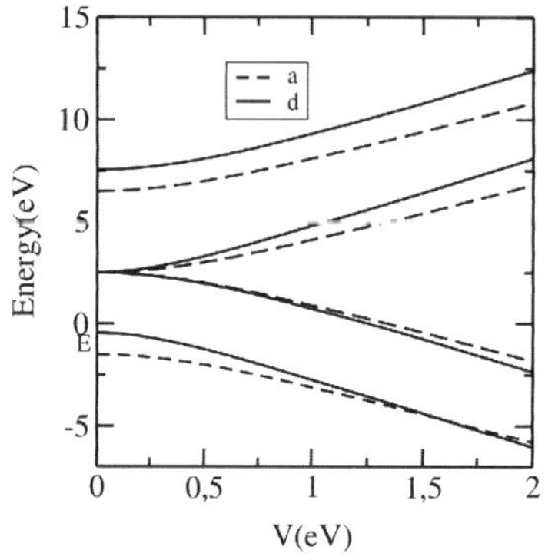

Figure 1. The hybridised single-hole energy levels of the octahedral(E^a) (dashed line) and tetrahedral (E^d)(solid line) clusters vs the $p - d$ hybridisation parameter (V) for the following values of the on-site energies:$E_a = -1.5eV, E_d = -0.44eV$, $E_p = 2.5eV$ and $U = 8eV$.

and

$$\hat{H}_{2p} = E_p \sum_{\sigma,m} \hat{n}_{\sigma,m} \tag{3}$$

The communication between the two systems of the orbital holes is maintained by the $p - d$ hybridisation which is assumed to take the following simple form:

$$\hat{H}_{p-d} = \sum_{l,\sigma} (V(l,m|\sigma,m)\hat{d}^{+}_{l,m}\hat{c}_{\sigma,m} + h.c.) \tag{4}$$

where E_a and E_d are the energy levels of all five holes localised either at the octahedral or tetrahedral iron site, and U is the on-site Coulomb repulsion parameter. E_a and E_d, in general, can differ from each other. The summations i, j, l run over five orbital d eigenstates, and m is the spin state.E_p is the energy of all six spin orbitals at the oxygen ions. $\hat{d}^+$ and $\hat{c}^+$ ($\hat{d}$, $\hat{c}$) are the fermion creation (annihilation) operators. The hybridisation parameters are reduced to one, being equal to V_{pd}.

In the first step of our procedure, the original Hamiltonian is projected onto each ionic cluster separately and diagonalised with only the inter-cluster hybridisation considered, which enables us to determine properly original single-hole eigenstates of the system within the framework of the LCMO approximation. The eigenstates of the octahedral cluster are liner combinations of appropriate 3d atomic orbitals, localised at the central iron site, with 2p atomic orbitals, localised at each of the six oxygen sites. Analogously, the eigenstates of the tetrahedral cluster are linear combinations of iron $3d$ orbitals with $2p$ states at each of its four oxygen neighbours. In the first case, the orbital eigenstates of the cluster are labelled by irreducible representations of the point group S_6(octahedral) [7], whereas in the second case, the eigen-states are determined by irreducible representations of the point group S_4 (tetrahedral). The clusters' eigenstates with mutually opposite spins are separated because of the on-site Coulomb repulsion at the iron sites. For our further discussion, it is sufficient to consider only a set of four eigen-states per a cluster. Of course, such a set must be labelled by the same irreducible representation. We select linear combinations of the $3d$ and $2p$ states with the largest value of the matrix element of $\hat{H}_{pd}$ between them. An illustration of our results is provided in Fig.1, where the hybridised energy levels of the clusters are given as functions of the hybridisation parameter. For different eigenstates of the clusters, there can be also determined probabilities of localisation both at the iron and oxygen sites. They are determined by the $3d$ and $2p$ contributions to the appropriate eigenfunctions.

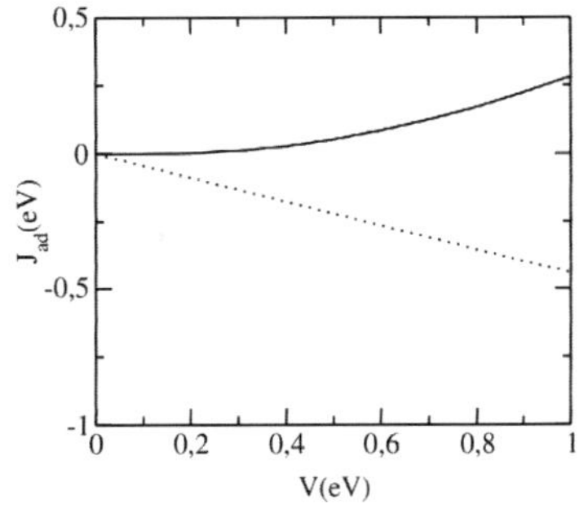

Figure 2. The superexchange integral J_{ad} between spins attached to the ground orbital eigenstates of the a and d clusters vs the hybridisation parameter V, for pure (the solid line) and doped (the dotted curve) system.The relevant model parameters take the following values: $U = 8eV$, $E_a = -1.5eV$, $E_d = -0.44eV$ and $E_p = 2.5eV$.

In the next step of the perturbation procedure, the inter-cluster p-d hybridisation is considered to yield single-hole states of the a-d pair of the clusters. For each pair their mutual oxygen ion is crucial as a bridge of the inter-cluster p-d hybridisation. By taking into account non-zero matrix elements of the hybridisation Hamiltonian between the tetrahedral $3d$ states and the octahedral $2p$ states of the mutual oxygen ion as well as between the octahedral $3d$ states and the tetrahedral $2p$ states at the oxygen, we can obtain a 8×8 transfer matrix for the $a - d$ pair of the clusters. Its matrix elements take the following form:

$$(\hat{H}_{pd})^{ad}_{i,j} = (\Psi^a_i, \hat{H}_{pd}\Psi^d_j)\delta(m, m') \tag{5}$$

where Ψ^a_i and Ψ^d_j with $i, j = 1, 2, 3, 4$ are the single-particle eigenstates of the clusters a and d, corresponding to the energy levels E^a_i and E^d_j, and the spins m, m', respectively. In the inter-cluster $2p - 3d$ hybridisation, the specific oxygen site, shared by the two clusters, plays an essential role and, consequently, the oxygen parts of the eigenstates of the two clusters (Ψ^a_i, Ψ^d_j),are reduced to their respective components, which are localised at the mutual site. Since the hybridisation is spin conserving, its matrix elements are non-zero only between the clusters' orbital states with the same spin, the transfer matrix is different for the case of parallel and anti-parallel orientations of the spins attached to the ground orbital states of the clusters. Consequently two different sets of eight energy levels of the pair are obtained. A parameter of superexchange interaction between spins of the clusters can be estimated as a difference in the magnetic energy of the pair for the parallel and anti-parallel orientation of their spins.

Within the framework of the simplified cluster model of only four hybridised eigenstates per a cluster, pure and doped yttrium-iron garnets can be distinguished from one another by a number of the holes localised at each cluster. In the case of pure YIG the number of the holes per a cluster is 1 and per a pair of the a and d clusters it is 2. In the case of $Ca : YIG$ the number of holes per a cluster can increase to 2 for certain clusters, which means also that for certain $a - d$ pairs the number of the holes is 3. Assuming that each doping Ca^{2+} introduces one extra hole we can roughly determine a fraction of the clusters, whose population gets doubled in the process. On the basis of a straightforward analysis of the low-temperature anomaly in the behaviour of the magnetisation of $Ca : YIG$, it seems to be sensible to attribute the anomaly to a presence of the holes localised at certain oxygen ions and thereby increasing the latter's valency from O^{-2} to O^-. As already mentioned, the low-temperature anomaly in magnetisation of $Ca : YIG$ can be interpreted as an effect of the hole's localisation at the oxygen ions. Since the inter-cluster antiferromagnetic superexchange interaction is then changed for ferromagnetic, it must give rise to a canting of the certain tetrahedral spins and consequently, to a moderate reduction of the resultant magnetic moment. The effect is more pronounced in the low-temperature range since at higher temperatures, the holes are prone to move. In Fig.2, the inter-cluster exchange interaction is presented vs the hybridisation parameter. The interaction change its sign for certain magnitude of the parameter. An occurrence of an oxygen vacancy at the site shared by the octahedral and tetrahedral clusters requires a revision and a modification of our original approach. As mentioned before, the oxygen vacancy with a trapped hole is described in terms of an hydrogen atom. Only the ground state of the atom is taken into account. To maintain a contact of the clusters, hybridisation between the $1s$ states of the vacancy and the $3d$ states at the octahedral and tetrahedral iron ions must be assumed. Possible hybridisation of the vacancy with the $2p$ states at the remaining oxygen ions of the clusters, is neglected as the oxygen-oxygen distance is considerably larger than that between oxygen and iron ions. Let us then reconsider a pair of the a and d clusters with a vacancy at their mutual corner. It is obvious that the inter-cluster transfer matrix must be completely changed. However, since there is only one missing oxygen neighbour out of the octahedral six and tetrahedral four oxygen ions, we find it justifiable to start our analysis of the problem with a modified version of our simplified cluster model. It is assumed that the $p-d$ hybridisation is maintained between the $3d$ states of at the central iron ions and the $2p$ states at all their nearest neighbours, save the one, which is missing. It suggests that in order to adapt the former LCMO hybridised single-cluster eigenstates to the present physical conditions, they should only have their oxygen contributions to multiplied by factor of $\sqrt{6/5}$ in the case of the octahedral cluster and by $\sqrt{4/3}$ in the case of the tetrahedral one. For the pair of the clusters, the original basis of its Hilbert space of the eigenstates consists of ten following wave-functions $\Psi^a_{i,m}, \Psi^d_{j,m'}, \phi_{1s,m}, \phi_{1s,-m}$, where $i, j = 1, 2, 3, 4$ and m, m' are the spin states.

Within this basis a new 10×10 transfer matrix $\hat{H}_{ad}$ can be obtained. It corresponds to the original Hamiltonian $\hat{H}$ extended by the the $1s-3d$ hybridisation term $\hat{H}_{sd}$, and projected onto the 10-dimensional Hilbert space of eight selected eigenstates of the clusters, and two eigenstates of the vacancy.From the ten diagonal matrix elements of the transfer matrix, four are equal to the energies of the octahedral cluster(E^a_i), and four to the energies of the tetrahedral cluster (E^d_j), and two to the energy (E_{1s},corresponding to the ground state of the vacancy, being a spin-orbital doublet.

Non-zero off-diagonal matrix elements of $\hat{H}_{ad}$ take either the form:

$$(\hat{H}_{ad})_{k,l} = (\Psi^a_{i,m}, \hat{H}_{sd}\phi_{1s,m'})\delta(m, m') \tag{6}$$

where for $k = 1, .., 4$ i takes the same values, and $l = 5, 6$ for $m', -m'$, respectively, or, the following analogous form:

$$(\hat{H}_{sd})_{k,l} = (\Psi^d_{j,m},\ hatH_{sd}\phi_{1s,m'})\delta(m, m') \tag{7}$$

where for $k = 7, .., 10$ j is equal to $j = 1, .., 4.m$ and m' are the respective spin states of the clusters (m) and the vacancy (m').The approach, presented here, has one more advantage. It provides information about localisation of the hole either at one of the central irons or at their respective oxygen neighbours. The transfer matrix elements take larger values than those of the second option.

Likewise, in the case of the $p-d$ hybridisation, all possible parameters of the $s-d$ hybridisation are reduced to only one, V_s. Diagonalisation of the transfer matrix yields ten energy levels of the pair of the clusters with the oxygen vacancy localised at their mutual site.

Again we can assume either a parallel or an anti-parallel mutual orientation of spins attached to the orbital ground states of clusters a and d. It turns out, however, that the orientation of the spins has no effect what-so-far on the form of the transfer matrix. Both ferromagnetic and

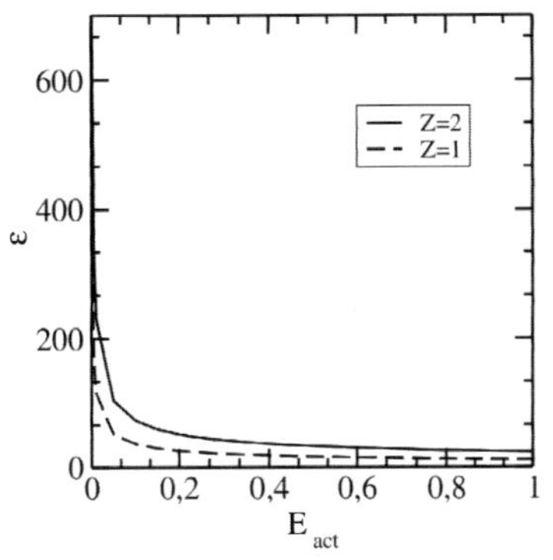

Figure 3. The relative dielectric constant vs the relative activation anergy for $Z = 1$ (dashed line) and $Z = 2$ (solid line). The effective mass is assumed to be 0.1 m_e.

anti-ferromagnetic transfer matrices are identical which further means that possibly the oxygen vacancy can totally destroy super-exchange interaction between spins localised at the iron sites. This conclusion remains in remarkable agreement with the empirical data on the YIG-derivative samples annealed in oxygen-reducing atmosphere and containing oxygen vacancies. Magnetic measurements indicate dramatic changes in the temperature behaviour of their magnetisation [1,2]. All the results we have obtained are expressed in terms of several characteristics of the system, which are the on-site energies E_a, E_d, E_p and U, the $p-d$ hybridisation parameter, V, as well as the parameters, which describe the vacancy itself, such as the ground state energy E_{1s}, and the parameter of the $1s-3d$ hybridisation, V_s. A deeper insight into the problem of an $a-d$ pair of the clusters with an oxygen vacancy requires more information about mutual relation between the two types of the system's parameters. Thus, in order to determine more precisely the characteristics of the vacancy, we need a closer look at the microscopic model of it.

3. Microscopic model of the oxygen vacancy

The simplest microscopic approach to a shallow point defect in semiconductors consists in an application of the hydrogen model [8]. The oxygen vacancy is a point defect, the question whether it is also shallow, however, requires further analysis.

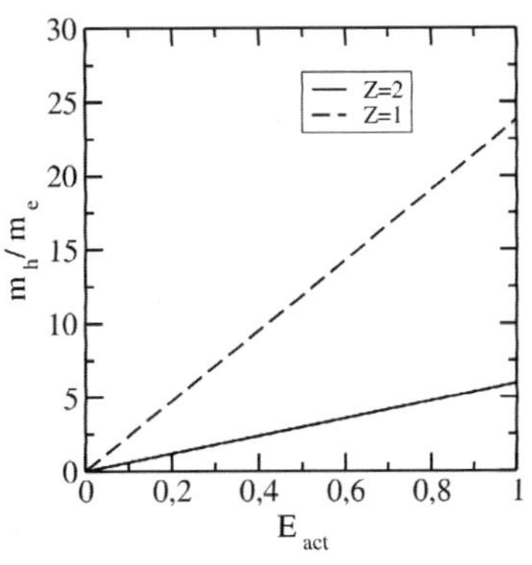

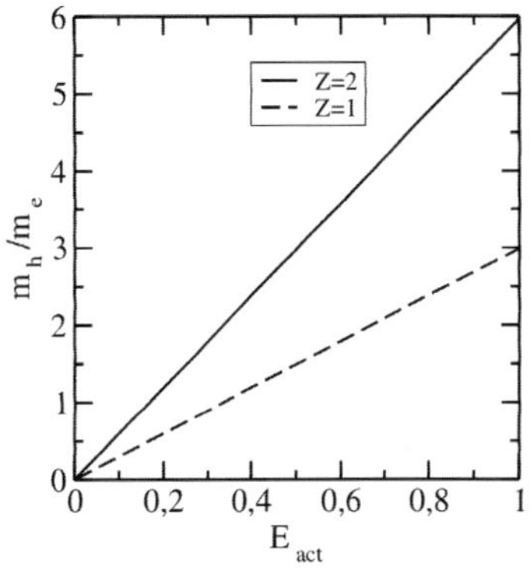

Figure 4. The relative hole's mass vs the relative activation energy for two different values of two relative dielectric constants (a)$\varepsilon = 18$, (b)$\varepsilon = 2.61$.

At the present moment, the hydrogen model is merely assumed. And, we have to realise that the specific geometry of garnets gets lost as the accepted model is characterised by spherical symmetry. Nevertheless, it serves well as a point of departure for our dealing with the problem. It is hoped that only the ground state of the vacancy is necessary in the process. To this objective, the text-book formula for the ground state $1s$ energy of the original hydrogen atom must be re-written in the following way:

$$E_{1s} = m_h Z^2 e^4 / (2\pi\hbar^2 (4\epsilon)^2) \tag{8}$$

the effective Bohr radius can be introduced, as well:

$$a^* = 4\pi\hbar^2\epsilon/(e^2 m_h) \tag{9}$$

where m_h is the effective mass of the hole and ϵ is a ratio of the dielectric constant of the host material to that of vacuum (ϵ_0). These two parameters are especially important since they characterise the host material, in which the impaired pair of the clusters is embedded. Within the framework of our cluster model, they can be varied to account for the actual experimental results.

Since we are especially interested in compensating holes localised at the oxygen vacancies, the length of the effective Bohr radius should be fixed to remain comparable with the distance between the vacancy and its dodecahedral neighbour, which is occupied by the doping Ca^{2+} ion. Consequently, possible delocalisation of the hole can be interpreted as its excitation to one of the non-occupied a or d cluster energy levels. The excited hole gains an access to all the oxygen sites of the hybridised cluster and can continue its motion farther on. Thus, within the framework of the cluster model, the difference in energy between the highest occupied localised level of the vacancy and the lowest non-occupied level of either of the clusters should respond to the activation energy, E_{act}. The latter was measured, in the annealed samples of $Ca:YIG$, at room temperature, and found to be in the range of $0.2 - 0.4eV$ [1-3]. A position of E_{1s} is thus automatically determined relative to the energy spectra of the clusters.

In order to maintain a proper relation between the ground state energy of the vacancy and its fixed distance from the nearest dodecahedral neighbour, we still have two variational parameters for further manipulation. Those are the effective relative dielectric constant ϵ/ϵ_0, and the effective relative hole mass m_h/m_e, where m_e is the electron mass.

In Fig.3, the relative dielectric constant vs ratio of the activation energy to the ionisation energy of the hydrogen atom ($13.6eV$), is presented for the vacancy with one hole ($Z = 1$) and that with two holes ($Z = 2$). The hole's effective mass was assumed to be equal to $0.1m_e$.The same value of the mass was used in experimental analysis. The measurements on $Ca:YIG$ thin films suggest the dielectric constant between 16 and 18. However, at this point, we cannot attain desired compatibility with the empirical data as ϵ turns out to be much lower and equal to 2.61. By assuming larger values for the effective mass such as the relative mass of the order of 10, one can obtain the microscopic results which remain in reasonably good agreement with the experimental data.

For the sake of this discussion, the following formula can be useful:

$$\epsilon/\epsilon_0 = Z\sqrt{E_{1s}/E_{act}}\sqrt{m_h/m_e} \tag{10}$$

For two different values of the relative dielectric constant, we determine the effective hole's mass as a function of the activation energy. The results are gathered and presented in Fig. 4. Again two cases are considered with one hole or two holes trapped in the vacancy. Our assumption of heavy mass of the hole corresponds with the assumption of its fixed position and thus it finds an experimental confirmation .

In order to summarise the results presented in this paper, we wanted to include an occurrence of the oxygen vacancies in the ionic cluster model of garnets, to re-interpret the empirical data in terms of the model quantities, and so achieve compatibility of the both aspects of the problem. It turns out that the magnetic properties of different derivatives of YIG, and in particular, those of $Ca : YIG$, as well as charge-transfer processes in these systems, interpreted within the framework of the ionic cluster model, result from an interplay of the two mutually opposite tendencies. Doping ions, such as Ca^{2+} in $Ca : YIG$ introduce compensating holes, which are positive charge carriers in the system and, when localised, they can change magnetic interaction between spins. On the other hand, oxygen vacancies tend to trap the holes, destroying thereby the antiferromagnetic inter-cluster coupling and simultaneously reducing effectively the number of the mobile charge carriers. Moreover, localisation of the hole results in a reduction of the resultant spin in the system. The mechanism of the charge transfer, as based on the hybridisation of the orbital states of the clusters, is simultaneously spin-polarised.

Acknowledgment
A partial support from the Foundation for Polish Science (FNP) is gratefully acknowledged.

References

[1] Turpin G B, Song Y J and Wigen P E 1995 *IEEE Transactions in Magnetism* **31** 3835
[2] Song Y J, Turpin G B, Bornfreund R E, Aoyama H and Wigen P E 1996 *Journal of Magnetism and Magnetic Materials* **154** 37
[3] Bornfreud R E and Wigen P E 1996 *J.Appl.Physics* **79** 5191
[4] Wojciechowski R J, Lehmann-Szweykowska A, Micnas R, Gehring G A and Wigen P E 2004 *Rev. B* **69** 214434-1-11
[5] Gehring G A, Lehmann-Szweykowska A, Wojciechowski R J, Wigen P E and Micnas R 2006 *phys. stat. sol. (b)* **243** 255
[6] Koper A, Lehmann-Szweykowska A, Wojciechowski R and Mucha M 1995 *J.Phys.: Condens.Matter* **7** 1391
[7] Lehmann-Szweykowska A, Lulek T and Kaczmarek M M 2001 *J.Phys.: Condens.Matter* **13** 3607
[8] Yu P Y and Cardona M 1996 *Fundamentals of Semiconductors* (Physics and Materials Properties) (Springer-Verlag Berlin Heidelberg New York)

Magnetic translation group and classification of states of an itinerant electron

Andrzej Wal

Institute of Physics, University of Rzeszow, Rejtana 16A, 35-959 Rzeszow, Poland

E-mail: wal@univ.rzeszow.pl

Abstract. We consider an itinerant electron on two-dimensional finite square lattice in a magnetic field. A magnetic translation group (MTG) for this system with the periodic Born-Karman conditions has been introduced. The irreducible representation of MTG is used for classification of energy levels of electron states for this model.

1. Introduction

The motion of a free electron in a plane perpendicular to a magnetic field is periodic. The corresponding states are quantized according to the Landau rule [1]. For an electron in both magnetic and periodic electric potential, an additional quantum effect appears - the Bohm-Aharanov effect. It implies that a jump from one node of the lattice to another one is accompanied by a phase. In this way the translation group of the planar lattice is no longer a symmetry group of the Hamiltonian of the electron. Brown [2] and Zak [3] introduced operators which commute with the Hamiltonian. They can form a group - the magnetic translation group (MTG) - which provides a restriction on the magnitude of the magnetic field. The magnetic flux through the cells of the planar lattice is quantaized and the number η of quanta through the elementary cell should be the rational number p/q, where p and q are coprime integers [2, 3, 4, 5].

The important part in the description of electron dynamics is played by gauge symmetry C (Weyl-Heisenberg group), according to which the vector potential of magnetic field is changing along translation. This group, together with the translational symmetry T, builds the magnetic translation group, which noncommutativity is in connection with quantum character of magnetic interaction. The group theoretical analysis of the MTG has been done by Lulek [6], Kuzma [7], Florek [8], and Geyler [9].

The Hamiltonian of an electron in a magnetic field can be reduced to the tight-binding Hamiltonian which depends on the phase factors associated with each edge of the planar lattice [10]. Such a system was the subject of studies during last years in context of integer and fractional Hall efects, high-T_C superconductors and even quantum computation [11, 12, 13].

2. The hopping Hamiltonian

The Hamiltonian of an electron in a periodic electric potential and homogeneous magnetic field is given by

$$\mathcal{H} = \frac{1}{2m}\left(\mathbf{p} + \frac{e}{c}\mathbf{A}\right)^2 + V(\mathbf{r}), \tag{1}$$

where the vector potential can be chosen in the form $\mathbf{A} = \frac{1}{2}\mathbf{H} \times \mathbf{r}$. In the case of a two-dimensional finite $N \times N$ square lattice Λ, for the tight-binding approximation this Hamiltonian can be reduced to the form:

$$H_l = \sum_{\lambda,\mu\in\Lambda} \exp(2\pi i \vartheta_{\lambda\mu}) c_\mu^+ c_\lambda \tag{2}$$

where λ, μ are vertices of the lattice Λ and c_μ^+ is the creation operator on the vertex μ. The summation includes all the nearest-neighbor sites (4 in case of a square lattice). Periodic boundary conditions are applied to the system.

The Hamiltonian is expressed by the phase factors $\vartheta_{\lambda\mu}$ which are defined for each edge. They fulfill the condition:

$$\vartheta_{\lambda\mu} = -\vartheta_{\mu\lambda} \tag{3}$$

and are determined mod 1. The sum of these factors along the closed sequence of points (circuit) $\lambda_1, \lambda_2, \lambda_3, \ldots, \lambda_n, \lambda_1$ gives the magnetic flux through the area bounded by this circuit:

$$\phi = \sum_{j=1}^{n} \vartheta_{\lambda_j, \lambda_{j+1}}, \quad \text{where } \lambda_{n+1} = \lambda_1. \tag{4}$$

The action of the Hamiltonian on the nodes of the lattice, numbered by two indices (j_1, j_2) can be described as follows:

$$\mathcal{H}|j_1, j_2\rangle = e^{2\pi i \vartheta_1}|j_1+1, j_2\rangle + e^{2\pi i \vartheta_2}|j_1-1, j_2\rangle + e^{2\pi i \vartheta_3}|j_1, j_2+1\rangle + e^{2\pi i \vartheta_4}|j_1, j_2-1\rangle \tag{5}$$

This action should commute with the action of the magnetic translation group.

3. Magnetic translation group

The elements of the MTG group, denoted hereafter as G, have the form (t, ξ), where $t \in T$ is a translation by a vector t and ξ is an element of the gauge group C:

$$C = <\xi_0> = \{\xi_0^n | n \in Z\}, \quad \xi_0 = e^{i\pi\eta}. \tag{6}$$

The multiplication rule for the group G is given by

$$(t, \xi)(t', \xi') = (t + t', \xi\xi' e^{i\pi\eta(t_x t'_y - t_y t'_x)}), \tag{7}$$

the exponent in the right hand side of the equation denotes the magnetic flux through the rectangle spanned by vectors t and t'. The group is generated by two elements $a = (10, 1)$ and $b = (01, 1)$, which correspond to translation, accompanied by the phase "1", in x and y-direction of the planar lattice, respectively (the magnetic field is applied in z-direction).

Let us denote by $T(a)$ and by $T(b)$ the oprators describing the action of the elements a and b on the nodes of the lattice respectively. These actions are given by:

$$T(a)|j_1, j_2\rangle = e^{i\pi\eta j_2}|j_1 - 1, j_2\rangle, \quad T(b)|j_1, j_2\rangle = e^{-i\pi\eta j_1}|j_1, j_2 - 1\rangle. \tag{8}$$

These operators commute with the Hamiltonian

$$[\mathcal{H}, T(a)] = [\mathcal{H}, T(b)] = 0. \tag{9}$$

The last equation provides, together with the property that the magnetic flux through an elementary cell is equal to η, the conditions for phase factors asociated with each edge. Their distribution is presented in figure 1.

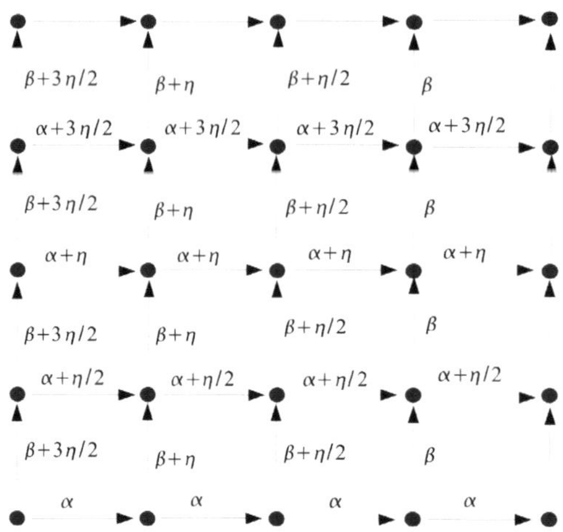

Figure 1. The distribution of phase factors $\vartheta_{\lambda\mu}$ which fulfill the equations (9). The symbols α and β denote arbitrary phases. This picture can be treated as a part of the larger lattice.

4. Classification of states of an electron

Let the eigenvalues of the Hamiltonian be denoted by w_i and corresponding eigenvectors by f_{ij}, the index j is member of the set $\{1, 2, \ldots d\}$ where d is the degeneration of the energy level. The action of the magnetic group on the nodes is described as follows

$$T(g)|j_1, j_2\rangle = T(t_1 t_2, \xi)|j_1, j_2\rangle = \xi e^{i\pi\eta(t_1 j_2 - t_2 j_1)}|j_1 - t_1, j_2 - t_2\rangle \tag{10}$$

where $g \in G$. To classify the states, the following matrix should be determined for a given eigenvalue w_i:

$$< f_{ij'}|T(g)|f_{ij} > \tag{11}$$

where $j, j' \in \{1, 2, \ldots d\}$. These matrices form the representation Γ' which can be decomposed into irreducible representations Γ of G.

5. Example: $N = 4$, $\eta = \frac{1}{2}$

The magnetic translation group for this case has been described in previous author's paper [14]. The irreducible representations of this group were also determined. There are four "physical" [9] representations and they are two-dimensional. Their characters are presented in table 1.

Table 1. The characters of the irreducible representations of group G for $N = 4$ and $\eta = 1/2$

Γ	(00,1)	(00,-1)	(20,1)	(20,-1)	(02,1)	(02,-1)	(22,1)	(22,-1)
Γ_6	2	-2	2	-2	2	-2	2	-2
Γ_8	2	-2	2	-2	-2	2	-2	2
Γ_{14}	2	-2	-2	2	2	-2	-2	2
Γ_{16}	2	-2	-2	2	-2	2	2	-2

The eigenvalues of the Hamiltonian (2) for this case are the following: $2\sqrt{2}$ (2), $-2\sqrt{2}$ (2), -2 (4), 2 (4) and 0 (4). The numbers in parentheses denote the degeneracy of the level. The representations Γ' calculated according to the equation (11) provide the classification of the energy levels of the electron. This classification is presented in table 2.

Table 2. The classification of energy levels of an electron. d is the degeneration of the state, Γ_i is the representation of MTG. For all other elements $g \in G$, the characters are $\chi(\Gamma'(g)) = 0$

energy	d	the character of a representation								decomposition
		(00,1)	(00,-1)	(20,1)	(20,-1)	(02,1)	(02,-1)	(22,1)	(22,-1)	
$2\sqrt{2}$	2	2	-2	2	-2	-2	2	-2	2	Γ_8
$-2\sqrt{2}$	2	2	-2	2	-2	-2	2	-2	2	Γ_8
-2	4	4	-4	0	0	0	0	4	-4	$\Gamma_{16} + \Gamma_6$
2	4	4	-4	0	0	0	0	4	-4	$\Gamma_{16} + \Gamma_6$
0	4	4	-4	-4	4	4	-4	-4	4	$\Gamma_{14} + \Gamma_{14}$

6. Conclusion

For an electron in a constant magnetic field and periodic electric potential the eigenvalue problem on the 4×4 lattice was solved. The energy states are determined by the phase factors $\vartheta_{\lambda\mu}$. The distribution of these factors on the planar lattice, which guarantees the commutation relation for G and $\mathcal{H}$ was determined. It should be noticed that periodic boundary conditions restrict the number of allowed size N of the lattice. Only those sizes are admissible for which the values of phase factors on the corresponding border are equivalent (mod 1). For the case $\eta = \frac{1}{2}$ and $N = 4$ the electron states have been classified using the irreducible representations Γ of the magnetic translation group G.

Acknowledgments

The author would like to thank Prof. T. Lulek for a helpful discussion.

References

[1] Landau L 1930 *Z. Phys.* **64** 629
[2] Brown E 1964 *Phys. Rev.* **133** A1038
[3] Zak J 1964 *Phys. Rev.* **134** A1602; A1607
[4] Brown E (1968) *Phys. Rev.* **166** 626
[5] Brown E and Meisel L V 1976 *Phys. Rev.* **B 13** 5271
[6] Lulek T 1994 *Rep. Math. Phys.* **34** 71
[7] Kuzma M, Mortensen K, Lulek B and Lulek T 1997 *Proc. Int. Conf. Symmetry and Structural Properties of Condensed Matter (Zajaczkowo)* ed T Lulek W Florek *at al* (Singapore: World Scientific) p 484
[8] Florek W 1994 *Rep. Math. Phys.* **34** 81
[9] Geyler V A and Popov I Yu 1995 *Phys. Lett.* **A 201** 359
[10] Lieb E H 1994 *Phys. Rev. Lett.* **73** 2158
[11] Mikhailov S A 2001 *Physica* B **299** 6
[12] Vourdas A 1996 *Proc. Int. Conf. Symmetry and Structural Properties of Condensed Matter (Myczkowce)* ed T Lulek W Florek *at al* (Singapore:World Scientific) p 310
[13] Vourdas A 2002 *Proc. Int. Conf. Symmetry and Structural Properties of Condensed Matter (Myczkowce)* ed T Lulek B Lulek *at al* (Singapore:World Scientific) p 55
[14] Wal A 2005 *Phys. Stat. Sol.* (b) **242** 291

Effect of time reversal symmetry on states of quasi – particles in crystals

H W Kunert

Department of Physics, University of Pretoria, 0001 Pretoria, South Africa.

Email: hkunert@nsnper1.up.ac.za

Abstract. The problem of additional time reversal (*TR*) degeneracy and the transformation properties of "time reversed' wave functions are investigated. In crystals the one electron Hamiltonian is invariant with respect to a space group of a crystal. The Bloch functions are the basis for irreducible representations (irrp's) of the wave vector groups **k**. When an irrp is complex the *TR* symmetry must be considered. Using Herring's criterion adopted to space groups we have investigated several hexagonal materials for optoelectronic devices such as ZnO, GaN, 6H-SiC. We have found many complex irrps. We discuss the effect of the *TR* on vibrational and electronic states together with optical selection rules in ZnO and GaN. In addition we list all complex irrps of the 32 crystallographic double point groups. Spin is taken into account throughout the whole discussion.

1. Introduction

States of quasiparticles in crystals such as electrons in conduction bands, holes in valence bands, excitons, phonons, plasmons, polaritons, etc., in short: "ons", are classified according to irrps of factor wave vector groups $G^{\mathbf{k}}/T$. Here the vector **k** runs over the entire first BZ and denotes three quantum numbers k_x, k_y and k_z . For hexagonal BZ the factor groups are determined by high symmetry points Γ = *GM*, *A*, *M*, *K*, *L*, *H*, and lines *R*, *Q*, *S*, Δ = *LD*, Σ = *SM*, Λ = *LD*, *T*, *U*, and *P*. The generators of irrps and the characters of the corresponding factor groups $G^{\mathbf{k}\Gamma}/T$,....., $G^{\mathbf{k}U}/T$, have been tabulated (CDML [1]).When an irrp *D* is complex, an extra *TR* degeneracy may occur. In such case any state of "ons" will be classified according the "joint" reps $D \oplus D^*$. This will affect many phenomena. For example, it will increase the dimension of the dynamical matrices .It will change the selection rules for optical transitions. It will also influence the scattering tensors and many other physical processes taking place in a crystal. It is therefore of importance to find out which irrps of a crystal are complex before analysis of experimental data.

In this contribution we use Herring [2,3] criterion for investigation of complex irrps modified by inclusion of spin and effect of *TR* operator on behaviour of wave vectors of a star $\{^*\mathbf{k}\}$ of the factor space group C_{6v}^4 – $P6_3/mc$ for ZnO, GaN, 6H-SiC, BeO, CdS, ZnS and many others.

The scope of the paper is following: In section 2 we discuss some theoretical aspects of the *TR* operator and Herring's criterion. In section 3 we analyze the experimental dispersion curves obtained by neutron scattering technique [4,5]. In section 3 we give some examples for complex irrps calculation. The section 4 is devoted to vibrational modes in wurtzite crystals due to *TR* symmetry. In section 5 we discuss the electronic band structure and optical selection rules in presence of *TR*. Complex irrps of 32 double crystallographic point groups and of C_{6v}^4 space group are listed explicitly.

2. Theoretical background

2.1. Time reversal operator for spinless particles. T = K

Consider a quantum mechanical system described by the time-dependent Schroedinger equation

$$\frac{ih}{2\pi}\frac{\partial\Psi(\mathbf{r},t)}{\partial t}=\left\{\left(\frac{-h}{2\pi}\right)^2\frac{1}{2m}\nabla^2+V(\mathbf{r})\right\}\Psi(\mathbf{r},t) \quad (1)$$

Replacing t by –t and taking the complex conjugate of both sides of the above equation, we have

$$\frac{ih}{2\pi}\frac{\partial\Psi^*(\mathbf{r},-t)}{\partial t}=\left\{\left(\frac{-h}{2\pi}\right)^2\frac{1}{2m}\nabla^2+V(\mathbf{r})\right\}\Psi^*(\mathbf{r},-t) \quad (2)$$

This shows that $\Psi^*(\mathbf{r}, -t)$ is also a solution of Schroedinger equation if $\Psi(\mathbf{r}, t)$ is.

Clearly, for spinless particles the K has the effect of reversing the direction of propagation of time and it is just the complex conjugation operator apart from a phase factor, usually denoted as $\exp\{-iEt/(h/2\pi)\}$.The matrices for **r** and **p** are real and purely imaginary so that the complex conjugation operator K has the following effect on **r** and **p**, and the operation of T on a wave function gives its complex conjugate:

$$K\mathbf{r}K^+=\mathbf{r} \quad (3a)$$

$$K\mathbf{p}K^+=-\mathbf{p} \quad (3b)$$

$$T\Psi(\mathbf{r})=\Psi^*(\mathbf{r}) \quad (3c)$$

$$T=K \quad (3d)$$

In the equation (3d) we have chosen the phase factor to be unity.

2.2. Time reversal operator for spin –1/2 particles, $s_y = 1/2(h/2\pi)\sigma_y$. T = $-i\sigma_y$K

In order to define the *TR* operator for particles with spin we follow the transformation properties of dynamical variables **r, p, s** and **j** [5] .It is clear that

$$T\mathbf{r}T^+=\mathbf{r} \quad (4a)$$

$$T\mathbf{p}T^+=-\mathbf{p} \quad (4b)$$

$$T(\mathbf{r}\times\mathbf{p})T^+=-(\mathbf{r}\times\mathbf{p}) \quad (4c)$$

$$T\mathbf{s}T^+=-\mathbf{s} \quad (4d)$$

$$T\mathbf{j}T^+=-\mathbf{j} \quad (4e)$$

$$Ks_xK^+=s_x \quad (5a)$$

$$Ks_yK^+=s_y \quad (5b)$$

$$Ks_zK^+=s_z \quad (5c)$$

$$T=UK \quad (5d)$$

$$K^2=1 \quad (5e)$$

$$TK=U \quad (5f)$$

for the particle with spin. The last equation (4) is in the standard ($\mathbf{r}$,s_z) representation in which s_x and s_z are real whereas the s_y matrix is imaginary. The *U* is to be determined. The *T* and *K* are antiunitary operators. Using equations (3a, b; 4a, b, d, e; 5a, b, c) we obtain

$$U = \exp[-i\pi s_y /(h/2\pi)] \quad (6a)$$

$$T = \exp[-i\pi s_y /(h/2\pi)]K \quad (6b)$$

$$T = -i\sigma_y K \quad (6c)$$

The equation (6c) represents the time reversal operator T for a spin –1/2 particle, $s_y = 1/2(h/2\pi)\sigma_y$ which follows from theory of the SO(2) group [6,7]. The equation(6c) can be easily extended to a system of n particles having arbitrary spin angular momenta. The extension leads to the Kramers' theorem which will not be discussed here.

2.3. Criteria for real and complex irreducible representations. Point groups

Frobenius and Shur [8] showed that it is sufficient knowing only the characters of an irrp to determine whether the rep is real or complex. If the sum of characters of squares of the group elements is equal to the order of the group $|g|$, then the reps is real: if the sum is $-|g|$, the rep is equivalent to its conjugate: and if the sum vanishes the reps D and D^* are complex and inequivalent. For single and double valued (spin included) irrps of 32 crystallographic point groups we write

$$(1/|g|)\sum_{g\in G}\chi(g^2) = \begin{cases} 1, \text{ case (a) } D \text{ is real} & (7a) \\ 0, \text{ case (b) } D \text{ and } D^* \text{ are complex and inequivalent} & (7b) \\ -1, \text{ case (c) } D \text{ and } D^* \text{ are complex and equivalent} & (7c) \end{cases}$$

In terms of basis functions Ψ of D and Ψ^* of D^* for case (a) the functions Ψ and Ψ^* are linearly dependent and no extra degeneracy occurs, for cases (b) and (c) Ψ and Ψ^* are linearly independent and an extra degeneracy occurs. Using the criteria we have tested all single and double valued (spin included – spinor) irrps of the 32 point group. The results are listed in table 1.

2.4. Criteria for real and complex irreducible representations. Space groups

The basis of irrps of space groups are Bloch functions $\Psi_{\mathbf{k}}(\mathbf{r}) = u(\mathbf{r})\exp(i\mathbf{kt})$, where the **k** runs over the entire first BZ and **t** are translations. As stated in Chapter 2, when spin excluded, the *TR* operator is just complex conjugation action on function. That means: $T\Psi_{\mathbf{k}}(\mathbf{r}) = T\{u(\mathbf{r})\exp(i\mathbf{kt})\} = u^*(\mathbf{r})\exp[i(-\mathbf{k})\mathbf{t}]$. Clearly, the *TR* operator transforms vector **k** into – **k.** That is an important effect which leads us to an essential reformulation of the equation 7. We deal now with characters of squares of space group elements $\chi[\{g|\tau_g + \mathbf{t}\}^2]$. The τ_g is a non-primitive translation associated with operator g which essentially belongs to the point group $G_0(\mathbf{k})$, **t** is primitive translation, and summation is over these g's which transform **k** into – **k.** The **k** is the first wave vector of the star $\{^*\mathbf{k}\}$.The total space group G contains all groups of all members of the star and it can be decomposed in terms of $G(\mathbf{k})$:

$$G = G(\mathbf{k}) + \{\varphi_2 | \tau_2\}G(\mathbf{k}) + \{\varphi_2 | \tau_2\}G(\mathbf{k}) + \ldots + \{\varphi_\sigma | \tau_\sigma\}G(\mathbf{k}) \quad (8)$$

We reserve the subscript σ for the coset representatives $\{\varphi_\sigma|\tau_\sigma\}$. And the members of a star is obtained by symmetry operators φ of the point group $G_0(\mathbf{k})$.

$$\{^*\mathbf{k}\} = \mathbf{k}_1 = \mathbf{k}_2 = \varphi_2\mathbf{k}\ldots\mathbf{k}_s = \varphi_s\mathbf{k} \quad (9)$$

The **k** is the first wave vector of the star in the fundamental domain of the BZ.
The characters of 230 space groups of these first wave vectors are tabulated [1].
Evaluating the character of the squared operators we obtain:

Table1. Real and complex irreducible representations of 32 crystallographic double point groups. Kronecker Product Tables. CDML [1]

Group Representation Γ_j	case (a) j	case (b) j	case (c) j
O_h	1+,2+,3+,4+,5+ 1-, 2-, 3-, 4-, 5-	6+,7+,8+,7-,8-,9-	
O	1,2,3,4,5		6,7,8
T_d	1,2,3,4,5		6,7,8
T_h	1+,4+,1-,4-	2+,3+,2-,3-	5+ ,6+,7+5-,4-,5-
T	1,4	2,3	5,6,7
D_{4h}	1	2+,3+,4+,5+,2-,3-,4-,5-	6+,7+,6-,7-
C_{4h}	1+,2+,1-,2-	3+,4+,3-,4-	5+,6+,7+,8+,5-,6,-7-,8-
C_{4v}	1,2,3,4		6,7
D_4	1,2,3,4,5		6,7
C_4	1,2	3,4,5,6,7,8	
S_4	1,2	3,4,5,6,7,8	
D_{2d}	1,2,3,4,5		6,7
D_{2h}	1+,2+,3+,4+,1-,2-,3-,4-,		5+,5-
C_{2h}	1+,2+,3+,4+,1-,2-,3-,4-		5+,5-
C_{2v}	1,2,3,4		5-
D_2	1,2,3,4		5
C_2	1,2	3,4	
D_{6h}	1+,2+,3+,4+,5+,6+, 1-,2-, 3-, 4-, 5-, 6-		7+,8+,9+,7-,8-,9-
C_{6h}	1+,2+,3+,4+,5+,6+, 1-, 2-, 3-, 4-, 5-, 6-,	7+,8+,9+,10+,11+,12 7-, 8-, 9-, 10-, 11-, 12-	
C_{6v}	1,2,3,4,5,6		7,8,9
D_6	1,2,3,4,5,6		7,8,9
D_{3h}	1,2,3,4,5,6		7,8,9
D_{3d}	1+,2+,3+,1-,2-,3-	4+,5+,4-,5-,	6+,6-
C_6	1,2	3,4,5,6,7,8,9,10,11,12	
C_{3h}	1,2	3,4,5,6,7,8,9,10,11,12	
C_{3v}	1,2,3	4,5	6
D_3	1,2,3	4,5	6
C_i	1+.2+,1-,2-		
C_3	1,5	2,3,4,6	
C_2	1,2	3,4	
C_s	1,2	3,4	
C_1	1		

$$\chi\left[\{g \mid \tau_g\}^2\right] = \chi\{g^2 \mid \tau_g + g\tau_g\} = \chi\{g^2 \mid \tau'_g + \tau_g + g\tau_g - \tau'_g\}$$

$$= \chi\{\in \mid \mathbf{t}_0\}\chi\{g^2 \mid \tau'_g\} = \exp(i\mathbf{k}\mathbf{t}_0)\chi\{g^2 \mid \tau'_g\}, \quad (10a)$$

$$\mathbf{t}_0 \equiv \tau_g + g\tau_g - \tau'_g, \quad (10b)$$

where τ_g' is known non-primitive translation associated with the operator g , $\in$ is the identity operator and $\mathbf{t}_0$ is to be calculated from equation (10b). The criterion for real and complex irrps of space groups becomes:

$$(|G|/|G(\mathbf{k})|)\sum_{g \subset G(\mathbf{k})} \exp(i\mathbf{k}\mathbf{t}_0)\chi\{g^2 \mid \tau'_g\}\delta_{\mathbf{k},-g\mathbf{k}} = 1 \text{ or } 0 \ -1, \quad (11)$$

where the summation is performed over $g^2 \subset G(\mathbf{k})$ such that $g\mathbf{k} = -\mathbf{k}$ and the $|G|/|G(\mathbf{k})$ is the number of symmetry elements in the small wave vector group $G(\mathbf{k})$.The formula differs from Bir and Picus [9 page 166 equation (18.32)]. We have utilized the CDML tables where the characters $\chi\{g|\tau_g\}$ of the irrps of 230 space groups are explicitly listed and therefore the characters of $\{g^2|\tau_g'\}$ arc chiefly available.

Using equation (11) and CDML tables we have tested all reps of the high symmetry points and lines of the ZnO space group C_{6v}^4. The results are:

(a) $\Gamma_{1,2,3,4,5,6}$, $M_{1,2,3,4}$, $K_{1,2,3}$, $\Sigma_{1,2}$, $T_{1,2}$, H_3. Real reps. No extra degeneracy
(b) $A_{1,2,3,4,5,6}$, $\Delta_{1,2,3,4,5,6}$, $H_{1,2}$ $L_{1,2,3,4}$, $U_{1,2,3,4}$, $P_{1,2,3}$, $S_{1,2}$. Complex reps.
(c) $R_{1,2}$. Complex reps.

In cases (b) and (c) the extra *TR* degeneracy arises and the "ons" are classified according to the "joint" or "stick" reps: $D \oplus D^*$.

Table 2. Vector representation of hexagonal crystallographic point groups.

1	E	h_1	x, y, z	1 0 0 0 1 0 0 0 1	13	I	h_{13}	-x,-y, -z	-1 0 0 0 -1 0 0 0 -1
2	C_6^+	h_2	x-y, x, y	1 -1 0 1 0 0 0 0 1	14	S_3^-	h_{14}	-x+y,-x,-z	-1 1 0 -1 0 0 0 0 -1
3	C_3^+	h_3	-y, x-y,z	0 -1 0 1 -1 0 0 0 1	15	S_5^-	h_{15}	y,-x+y,-z	0 1 0 -1 1 0 0 0 -1
4	C_2	h_4	-x, -y, z	-1 0 0 0 -1 0 0 0 1	16	σ_h		x, y, -z	1 0 0 0 1 0 0 0 -1
5	C_3^-	h_5	-x+y,-x,z	-1 1 0 -1 0 0 0 0 1	17	S_6^+		x-y,x,-z	1 -1 0 1 0 0 0 0 -1
6	C_6^-	h_6	y,-x+y,z	0 1 0 -1 1 0 0 0 1	18	S_3^+		-y,x-y,-z	0 -1 0 1 -1 0 0 0 -1
7	C_{21}''	h_9	x-y,-y,-z	1 -1 0 0 -1 0 0 0 -1	19	σ_{v1}		-x+y,y,z	-1 1 0 0 1 0 0 0 1
8	C_{22}'	h_{10}	x, x-y,-z	1 0 0 1 -1 0 0 0 -1	20	σ_{d2}			-1 0 0 -1 1 0 0 0 1
9	C_{23}''	h_{11}	y, x, -z	0 1 0 1 0 0 0 0 -1	21	σ_{v3}			0 -1 0 -1 0 0 0 0 1
10	C_{21}'	h_{12}	-x+y,y,-z	-1 1 0 0 1 0 0 0 -1	22	σ_{d1}			1 -1 0 0 -1 0 0 0 1
11	C_{22}''	h_7	-x,-x+y,-z	-1 0 0 -1 1 0 0 0 -1	23	σ_{v2}			1 0 0 1 -1 0 0 0 1
12	C_{23}'	h_8	-y, -x, -z	0 -1 0 -1 0 0 0 0 -1	24	σ_{d3}			0 1 0 1 0 0 0 0 1

Up to now we have considered only single valued reps of space groups (spin excluded). The inclusion of spin leads to the double valued reps. The criterion (equation 11) is also valid for spinors. In general, in order to perform calculation using equation (11) we must consider systems with (i) integral and (ii) half-odd-integral spin. Combination of the three cases (a), (b), and (c) with (i) and (ii) result in additional six possibilities those can expressed as follows:

- No extra degeneracy in case a(i) and c(ii),
- Doublet degeneracy in cases b(i), b(ii), a(ii) and c(i),

- In table 3, we discuss points A, Δ, M, L, H, K, and lines Λ, Σ.

In here we have considered spinor reps, case (ii)-1/2 spin, of points Γ, A, H, K, M, L, and Δ line. The characters of spinor reps of the double groups have been taken from CDML tables. Using the equation (11) we have obtained following results:

- Case a(ii) no *TR* degeneracy: $\Gamma_{6,7,8}$
- Case b(ii) doublet degeneracy: $\Delta_{7,8,9}$, $H_{4,5}$, $K_{4,5}$. Classification of states:

$$\Delta_7 \oplus \Delta_7^*, \Delta_8 \oplus \Delta_8^*, \Delta_9 \oplus \Delta_9^*, H_4 \oplus H_4^*, K_4 \oplus K_4^*$$

- Case c(ii) no *TR* degeneracy: $A_{7,8,9}$, H_6, K_6.

Table 3. Characters of the irreducible A, Δ, M, L, H, K, Σ and Λ representations.

Space Group C_{6v}^4(P6$_3$/mc). CDML symmetry elements labelling: $\{g|\tau_g\}$ of $G^{\mathbf{k}_A}/T$

$\{g\|\tau_g\}$	1	2.1	3	4.1	5	6.1	19	20.1	21	22.1	23	24.1
A_1	1	i	1	i	1	i	1	i	1	i	1	i
A_2	1	i	1	i	1	i	−1	$-i$	−1	$-i$	−1	$-i$
A_3	1	$-i$	1	$-i$	1	$-i$	−1	i	−1	i	−1	i
A_4	1	$-i$	1	$-i$	1	$-i$	1	$-i$	1	$-i$	1	$-i$
A_5	1 0	iw^* 0	w 0	i 0	w^* 0	iw 0	0 $-i$	0 $-w^*$	0 $-iw$	0 $-i$	0 $-iw$	0 $-w$
	0 1	0 iw	0 w^*	0 i	0 w	0 iw^*	i 0	w 0	iw^* 0	i 0	iw 0	w^* 0
$A_6 = A_5^*$, $A_4 = A_1^*$, $A_2 = A_3^*$												
$\{g\|\tau_g\}^2$	1	3t$_0$	5	1t$_0$	3	5t$_0$	1	1t$_0$	1	1t$_0$	1	1t
Characters of $\{g\|\tau_g\}^2$, t$_0$ ={E|001}, exp($i\mathbf{k}_A\mathbf{t_0}$) = exp[$i$(0,0,1/2)(0,0,1)] 2π = e$^{i\pi}$ = −1												
$A_{1,2,3,4}$	1	1(−1)	1	1(−1)	1	1(−1)	1	1(−1)	1	1(−1)	1	1(−1)
A_5	2	(w+w*)(-1)	(w*+w)	2(−1)	(w+w*)	(w*+w)(−1)	2	2(−1)	2	2(−1)	2	2(−1)
A_6	2	(w*+w)(-1)	(w+w*)	2(−1)	(w*+w)	(w+w*)(-1)	2	2(−1)	2	2(−1)	2	2(−1)
$A_{1,2,3,4,5,6}$ case (b)												

$G^{\mathbf{k}_\Delta}/T$

$\{g\|\tau_g\}$	1	2.1	3	4.1	5	6.1	19	20.1	21	22.1	23	24.1
Δ_1	1	1.T	1	1.T	1	1.T	1	1.T	1	1.T	1	1.T
Δ_2	1	1.T	1	1.T	1	1.T	−1	−1.T	−1	−1.T	−1	−1.T
Δ_3	1	−1.T	1	−1.T	1	−1.T	−1	1.T	−1	1.T	−1	1.T
Δ_4	1	−1.T	1	−1.T	1	−1.T	1	−1.T	1	−1.T	1	−1.T
Δ_5	1 0	Tw^* 0	w 0	T 0	w^* 0	Tw 0	0 T^2	0 Tw^*	0 $-wT^2$	0 T	0 $-w^*T^2$	0 wT
	0 1	0 Tw	0 w^*	0 T	0 w	0 Tw^*	T^2 0	Tw 0	w^*T^2 0	T 0	wT^2 0	w^*T 0

$\Delta_6 = \Delta_5^*$, $\Delta_4 = \Delta_1^*$, $\Delta_2 = \Delta_3^*$ $w = \exp(2\pi i/3)$, $T = \exp(i\pi\alpha)$

For Δ there is no point group operation $g \subset C_{6v}$ which transforms $\mathbf{k}_\Delta$ into $-\mathbf{k}_\Delta$. $\Delta_{1,2,3,4,5,6}$ case (b)

$G^{\mathbf{k}_M}/T$

$\{g\|\tau_g\}$	1	4.1	20.1	23
M_1	1	1	1	1
M_2	1	1	−1	−1
M_3	1	−1	1	−1
M_4	1	−1	−1	1
$\{g\|\tau_g\}^2$	1	{1|0,0,1}	{1|0,0,1}	1
exp($i\mathbf{k}_M\mathbf{t}$) = exp[i(1/2,0,0)(0,0,1)]2π = 1				
Characters of $\{g\|\tau_g\}^2$				
$M_{1,2,3,4}$	1	1	1	1
case (a)				

$G^{\mathbf{k}_L}/T$

$\{g\|\tau_g\}$	1	4.1	20.1	23
L_1	1	i	i	1
L_2	1	i	$-i$	−1
L_3	1	$-i$	i	−1
L_4	1	$-i$	$-i$	1
$\{g\|\tau_g\}^2$	1	{1|0,0,1}	{1|0,0,1}	1
exp[i($\mathbf{k}_L\mathbf{t}$)] = exp[i(1/2,0,1/2)(0,0,1)] = −1				
$L_{1,2,3,4}$	1	1(−1)	1(−1)	1
case (b)				

$G^{\mathbf{k}_H}/T$

$\{g\vert\tau_g\}$	1	3	5	20.1	22.1	24.1
H_1	1	1	1	i	i	i
H_2	1	1	1	$-i$	$-i$	$-i$
H_3	2	1	1	0	0	0
$\{g\vert\tau_g\}^2$	1	5	3	$\{1\vert 0,0,1\}$	$\{1\vert 0,0,1\}$	$\{1\vert 0,0,1\}$
$\exp(i\mathbf{k}_H\,\mathbf{t}) = \exp[i(1/3,1/3,1/2)(001)]2\pi = -1$						
Characters of $\{g\vert\tau_g\}^2$						
$H_{1,2}$	1	1	1	1(−1)	1(−1)	1(−1)
H_3	2	−1	−1	2(−1)	2(−1)	2(−1)
$H_{1,2}$ case (b), H_3 case (c)						

$G^{\mathbf{k}_K}/T$

$\{g\vert\tau_g\}$	1	3	5	20.1	22.1	24.1
K_1	1	1	1	1	1	1
K_2	1	1	1	−1	−1	−1
K_3	2	−1	−1	0	0	0
$\{g\vert\tau_g\}^2$	1	5	3	$\{1\vert 0,0,1\}$	$\{1\vert 0,0,1\}$	$\{1\vert 0,0,1\}$
$\exp(i\mathbf{k}_K\,\mathbf{t}) = \exp[i(1/3,1/3,0)(001)]2\pi = 1$						
$K_{1,2}$	1	1	1	1	1	1
K_3	2	−1	−1	2	2	2
$K_{1,2,3}$ case (a)						

$G^{\mathbf{k}_\Lambda}/T$

$\{g\vert\tau_g\}$	1	24.1
Λ_1	1	1
Λ_2	1	−1
$\{g\vert\tau_g\}^2$	1	$\{1\vert 0,0,1\}$
$\exp(i\mathbf{k}_\Lambda\mathbf{t}) = \exp[i(\alpha,\alpha,0)(0,0,1)]2\pi = 1$		
Characters of $\{g\vert\tau_g\}^2$		
$\Lambda_{1,2}$	1	1
case (a)		

$G^{\mathbf{k}_\Sigma}/T$

$\{g\vert\tau_g\}$	1	23
Σ_1	1	1
Σ_2	1	−1
$\{g\vert\tau_g\}^2$	1	1
$\exp(i\mathbf{k}_\Sigma\mathbf{t}) = \exp[i(\alpha,0,0)(0,0,1)]2\pi = 1$		
$\Sigma_{1,2}$	1	1
case (a)		

$1 = E$, $2.1 = \{C_6^+\vert 0,0,1/2\}$, $3 = \{C_3^+\}$, $4.1 = \{C_2\vert 0,0,1/2\}$, $5 = \{C_3^-\}$, $6.1 = \{C_6^-\vert 0,0,1/2\}$

$19 = \{\sigma_{v1}\}$, $20.1 = \{\sigma_{d2}\vert 0,0,1/2\}$, $21 = \{\sigma_{v3}\}$, $22.1 = \{\sigma_{d1}\vert 0,0,1/2\}$, $23 = \{\sigma_{v2}\}$, $24.1 = \{\sigma_{d3}\vert 0,0,1/2\}$

3. Examples. Points *M* and *L*

In here we test the irrps of ZnO of the high symmetry points *L*, $G(\mathbf{k}_L)$ and *M*, $G(\mathbf{k}_M)$.
Coordinates are; $\mathbf{k}_M \approx (1/2,0,0)$ and $\mathbf{k}_L \approx (1/2,0,0)$. The symmetry operators of the $G(\mathbf{k}_M)$/T and $G(\mathbf{k}_L)/T$ are $1 = E$, $4.1 = \{C_2\vert 001/2\}$, $20.1 = \{\sigma_{d2}\vert 001/2\}$, $23 = \{\sigma_{v2}\}$

Let's calculate the square of symmetry element 4.1 using the equations (10a,b)

$$\{4.1\}^2 = \{C_2^2\vert\, 001/2 + C_2(001/2)\},\ C_2(001/2) = \begin{matrix} 1 & -1 & 0 \\ 1 & 0 & 0 \\ 0 & 0 & 1 \end{matrix}\ (001/2) = (0,0,1/2).$$

Therefore: $\{C_2^2\vert\, 001/2 + 001/2\} = \{E\vert 001\}$, and similarly for $\{20.1\}^2$ and $\{23\}^2$. Thus the characters of the squared operators for point *M* and *L* are

$$\chi\{E^2\} = \chi\{E\},\ \chi\{(4.1)^2\} = \chi\{E\vert 001\},\ \chi\{(20.1)^2\} = \chi\{E\vert 001\},\ \chi\{(23)^2\} = \chi\{E\}$$

For matrices of the vector representation for hexagonal groups see table 2. Since the $\mathbf{k}_M$ and $\mathbf{k}_L$ have different coordinates the factor $\exp(i\mathbf{kt})$ will differ. For point *M*, we have

$$\exp(i\mathbf{k}_M\mathbf{t})\text{: for } \chi\{(4.1)^2\} \text{ and } \chi\{(20.1)^2\} \text{ is: } \exp[i(1/2,0,0)(0,0,1)2\pi] = \exp(0) = 1.$$

Therefore, the characters of the squared operators of point *M* are:

$$\chi\{E^2\} = \chi\{E\},\ \chi\{(4.1)^2\} = \chi\{E\},\ \chi\{(20.1)^2\} = \chi\{E\},\ \chi\{(23)^2\} = \chi\{E\}.$$

For point *L*, we have

$$\exp(i\mathbf{k}_L\mathbf{t})\text{: for } \chi\{(4.1)^2\} \text{ and } \chi\{(20.1)^2\} \text{ is: } \exp[i(1/2,01/2)(0,0,1)2\pi] = \exp(i\pi) = -1.$$

Therefore, the characters of the squared operators of point L are:

$$\chi\{E^2\} = \chi\{E\},\ \chi\{(4.1)^2\} = -1\chi\{E\},\ \chi\{(20.1)^2\} = -1\chi\{E\},\ \chi\{(23)^2\} = \chi\{E\}.$$

See table 2 for $G(\mathbf{k}_M)/T$ and $G(\mathbf{k}_L)/T$.

From equation (8) we obtain the star for $\mathbf{k}_M$ and $\mathbf{k}_L$

$$G = \{1\}G(\mathbf{k}_M) + \{2.1\}G(\mathbf{k}_M) + \{3\}G(\mathbf{k}_M),$$

where $\{1\} = \{E\}$, $\{2.1\} = \{C_6^+|0,0,1/2\}$, $\{3\} = \{C_3^+\}$ are the coset representatives by means of those the stars $\{^*\mathbf{k}_M\}$ and $\{^*\mathbf{k}_L\}$ are obtained: $E\mathbf{k}_M = 1\mathbf{k}_M$ (1/2,0,0), $(C_6^+)\mathbf{k}_M = 2\mathbf{k}_M$ (1/2,1/2,0), $C_3^+\mathbf{k}_M = 3\mathbf{k}_M$ (0,1/2,0) and for $\{^*\mathbf{k}_L\}$ we have $E\mathbf{k}_L = 1\mathbf{k}_L$ (1/2,0,1/2), $C_6^+\mathbf{k}_L = 2\mathbf{k}_L$ (1/2,1/2,1/2), $C_3^+\mathbf{k}_L = 3\mathbf{k}_L$ (0,1/2,1/2). The coset representatives are useful in the determination of the full induced irrps of $G\{^*\mathbf{k}\}$ [10] and in the performing the summation in equation (11). Using the calculated coset representatives we have proceeded (for M and L points) both ways and our results are in agreement with those obtained in here using characters of the small irrps tabulated.

4. The effect of time reversal symmetry on vibrational states in wurtzite structure crystals with C6v4 space group

Before going into discussion of experimental results we recall general utilization of group theory related to phonons in any crystal with well establish space group [11]. The Lattice Mode Representation (LMR) [12] of a crystal provides an exact number of primary (first order – non-interacting phonons) originating from high symmetry point and lines of a BZ and also their symmetries (irrps, degeneracies). For example, for ZnO we have following symmetry allowed vibrational modes: ($2\Gamma_1$, $2\Gamma_4$, $2\Gamma_5$, $2\Gamma_6$), ($2A_1$, $2A_4$, $2A_5$, $2A_6$), ($2\Delta_1$, $2\Delta_4$, $2\Delta_5$, $2\Delta_6$), ($2H_1$, $2H_2$, $2H_3$), ($2P_1$, $2P_2$, $2P_3$), ($2K_1$, $2K_2$, $2K_3$), ($4L_1$, $2L_2$, $2L_3$, $4L_4$), ($4M_1$, $2M_2$, $2M_3$, $4M_4$), ($4U_1$, $2U_2$, $2U_3$, $4U_4$), ($8R_1$, $4R_2$), ($8\Sigma_1$, $4\Sigma_2$), ($6Q_1$, $6Q_2$), ($6S_1$, $6S_2$), ($6\Lambda_1$, $6\Lambda_2$), and ($6T_1$, $6T_2$) [11]. Knowing these symmetries and using compatibility relations along high symmetry lines and points the group theoretical assignment of phonons follows. Phonons assignment in ZnO, GaN, and related compounds with the C_{6v}^4 space group is presented on figure 1 in [11].

4.1. Experimental phonon dispersion curves in ZnO. Assignment

Thoma and Hewat [4,5] considered lattice dynamics of ZnO by means of inelastic neutron scattering. Figure 1 in [4,5] shows the calculated and measured phonon dispersion curves for ZnO. Our phonon assignment for GaN (presented on figure 1 in [4,5]) is valid also for phonons in ZnO. According to our calculations, the Δ and A phonons are influenced by the *TR* symmetry. However, on both figures 1 in [4,5] there is no group theoretical assignment of vabrational modes. We have shown that at point A and axis Δ the representations spanned by the modes are $2A_1 \oplus 2A_4 \oplus 2A_5 \oplus 2A_6$ and $2\Delta_1 \oplus 2\Delta_4 \oplus 2\Delta_5 \oplus 2\Delta_6$, respectively. In other words, there are eight A's and eight Δ's modes. We also know that at point Γ there are also eight modes of $2\Gamma_1$, $2\Gamma_4$, $2\Gamma_5$, $2\Gamma_6$ symmetries. The presence of *TR* symmetry (at A and Δ) requires classification of modes according to $A \oplus A^*$, $\Delta \oplus \Delta^*$ reps. Using the compatibility relations the resulting mode's assignment is:

(i) Point Γ, from the bottom to the top: $\Gamma_1 \oplus \Gamma_5$, Γ_6, Γ_4, Γ_5, Γ_6, Γ_4, and Γ_1.

(ii) Point A, from the bottom to the top: $A_5 \oplus A_5^*$, $A_1 \oplus A_1^*$, $A_5 \oplus A_5^*$, $A_1 \oplus A_1^*$.

(iii) Line Δ: the dispersion curves connect the points Γ and A when going from the bottom to the top on A axis side: Δ_5, Δ_5^*, Δ_1, Δ_1^*, Δ_5, Δ_5^*, Δ_1^*, and Δ_1.

From reps provided in here follows that $A_1^* = A_4$, and $A_5^* = A_6$, and $\Delta_1^* = \Delta_4$, $\Delta_5^* = \Delta_6$, which is consistence with the number of modes at Γ, A, and Δ and with their symmetries obtained independently from Lattice Mode Representation. Figure 1 displays the schematic dispersion curves of Γ–Δ–A region of the BZ subjected to *TR* symmetry for ZnO in terms of "joint' reps. For simplicity, we used straight lines for connectivity. Generally, the frequencies of the modes at Γ, Δ and A point may shift while going from compound to compound (for instance from ZnS to BeO, all with the wurtzite structutre, C_{6v}^4/T group), but the connecting dispersion curves between Γ and A point will be kept by

Δ's experimental phonon data. The shifting may cause more accidental degeneracy. The existence of experimentally measured modes, those "generate" dispersion curves Δ_5, Δ_6, Δ_1, Δ_4 and four phonons on A axis $2(A_5 \oplus A_6)$, $2(A_1 \oplus A_4)$ evidently proves the presence of the Time Reversal symmetry in wurtzite crystals. Figure 1 displays the schematic dispersion curves $\Gamma - \Delta - A$ region of BZ subjected to *TR* symmetry in wurtzite crystal in term of "joint" reps.

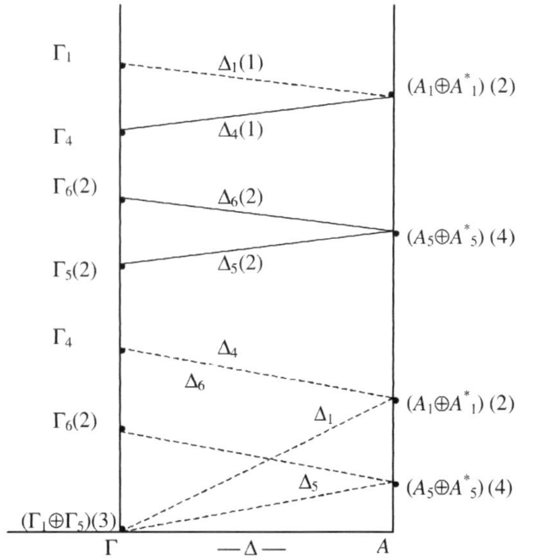

Figure 1. Assignment of phonons in ZnO, GaN, CdS, BeO, ZnS, CdSe influenced by time reversal symmetry. Numbers in brackets indicate the degree of degeneracy.

5. Effect of time reversal symmetry on electronic band structure and optical transitions in wurtzite crystals with C6v4 space group. Selection rules GaN band structure

Since the states of electrons in conduction and holes in valence bands are classified according to the irrps of the factor group of the wave vectors in the fundamental BZ domain (first vectors of a stars) the effect of *TR* must be taken into account. The wurtzite BZ and the calculated LDA energy bands of GaN are given, (see figure 2 and figure 4 pages 308 and 310 in [13]). On the figure 4 the classification of states is only in terms of single valued reps of the space group C_{6v}^4. It means that spin of electron and holes has been excluded. However, in GaN there is an appreciable spin orbit interaction [13.p.307] that splits the top of the valence $\Gamma_{1,6}$ $(\Gamma_1 \oplus \Gamma_6) \otimes D_{1/2}$ band onto $\Gamma_9 + \Gamma_7$ and $\oplus \Gamma_7$ the later being split by crystalline field. Therefore, the energy band structure presented on figure 4 [13] may not correspond to the true one. Nevertheless, *TR* symmetry has been taken into account. For example, the top of the valence band at point *A* (figure 4) is described by (5, 6) numbers, which means $A_5 \oplus A_5^*$ state *TR* degenerate. One of the lower valence band state at K point is classified by single valued irrp K_3 (two-dimensional). Inclusion of spin results in Kronecker Product $(KP)\, K_3 \otimes D_{1/2}$ sixfold degenerate spinor state, which must be decomposed onto double valued $(K_{4,5,6})$ irrps of *K*. The $D_{1/2}$ is the 2x2 spin matrix. Generally, all states of holes and electrons will be classified according to their spinor reps when spin included. To our best knowledge the electronic band structure of GaN and many other novel compounds do not include spin. Using compatibility relations and spin matrix we have described the GaN energy band (Fg.4) in terms of spinor states those are frequently subjected to *TR* effect, as showed in section 2d. Our results will be discussed elsewhere.

5.1. Selection rules

In here we briefly discuss additional conditions that invariance under *TR* imposes on the matrix elements of an operator V. Our discussion will be related to selection rules of some optical transitions between *TR* influenced states in GaN. In general we have:

$$\left\langle \varphi_f V \varphi_i \right\rangle = \int \varphi_f^* V \varphi_i d\mathbf{r}\,, \tag{12a}$$

$$\left\langle K\varphi_f \mid K\varphi_i \right\rangle = \left\langle \varphi_f \varphi_i \right\rangle^*, \tag{12b}$$

$$\left\langle \varphi_f V \varphi_i \right\rangle^* = \left\langle K\varphi_f, KV\varphi_i \right\rangle = \left\langle K\varphi_f, KVK^{-1}K\varphi_i \right\rangle, \tag{12c}$$

where the φ_i , φ_f V and K are initial and, final states, perturbation and Time Reversal operators, respectively. The $K\varphi_i$ and $K\varphi_f$ yield complex irrps D_i^* and D_f^* between those transition takes place. It is sufficient to consider only possible *KP*s: $D_i^* \otimes D_f^*$ and find out whether in their decomposition is a rep according to which the perturbation V transforms. The Kronecker Products of all irrps of 230 space groups are tabulated, CDML [1]. However, the *TR* symmetry has not been included. Therefore, for all irrps satisfying cases (b) and (c) of Herring's criterion the selection rules must be computed. The formal theory involved in such cases will be considered elsewhere. In here we discuss only practical transitions involved in *UV* reflectivity of GaN [14]. There are two cases concerning the effect of the *TR* on selection rules regarding the CDML tables. If a complex irrps satisfying (b) or (c) cases are in the set of irrps of a space group tabulated then selection rules can be still extracted from the CDML tables. For instance, at point A in ZnO we have states $A_1 \oplus A_1^*$, $A_2 \oplus A_2^*$, $A_3 \oplus A_3^*$, $A_4 \oplus A_4^*$, $A_5 \oplus A_5^*$ and $A_6 \oplus A_6^*$. However, from inspection of the tables we conclude that: $A_1^* = A_4$, $A_2^* = A_3$, $A_5^* = A_6$. For transition between levels $A_{5,6}$ and $A_{1,3}$ (5.94eV), [14. Table III p. 13 529]) one reads

$$(A_5 \oplus A_6) \otimes (A_1 \oplus A_3) = (A_5 \oplus A_1) \otimes (A_5 \oplus A_3) \oplus (A_6 \oplus A_1) \otimes (A_6 \oplus A_3)\,.$$

in Rashba's labelling [15]. The Rashba's A_2 corresponds to A_3 CDML and so A_3 – A_4, A_4 – A_2 correspondingly, (see table 4, this paper). The above *KP*s are tabulated (vol p). Therefore, in such cases no implications are involved. In this spirit, the selection rules for tabulated transitions (see Table III in [14]) between *A*, Δ, *L*, *U*, *P'* levels can still be found in the CDML tables. Nevertheless, considering transitions between *TR* degenerate K_5 and K_6 levels (spin included) we have

$$(K_5 \oplus K_5^*) \otimes (K_6 \oplus K_6^*) = K_5 \otimes K_6 \oplus K_5 \otimes K_6^* \oplus K_5^* \otimes K_6 \oplus K_5^* \otimes K_6^*\,.$$

The rep K_5 is one-dimensional, while K_6 is two-dimensional. Neither K_5^* nor K_6^* is in the set of *K*'s reps of the *G*(**K**) group. Only the product $K_5 \otimes K_6$ can be found in the CDML tables. All the other *KP*s have to be calculated. Clearly, *TR* symmetry results in additional *KP*s of complex reps those have to be determined.

Table 4 Labels of irreducible group C_{6v}^4 representations of the space group ($P6_3/mc$)

CDML															Rashba															Lyle et al.			
Γ	Δ	*A*	*L*	*M*	*U*	*K*	*H*	*P*	Σ	Λ	*Q*	*S*	*T*	*R*	Γ	Δ	*A*	*L*	*M*	*U*	*K*	*H*	*P*	Σ	*T*	*S*	*S'*	*T'*	*R*	Γ	Δ	*A*	*K*
1	1	1	1	1	1	1	1	1	1	1	1	1	1	1	1	1	1	1	1	1	1	1	1	1	1	1	1	1	1	1	1	1	1
2	2	2	2	2	2	2	2	2	2	2	2	2	2	2	2	2	2	4	4	4	2	2	2	2	2	2	2	2	2	2	2	2	2
3	3	3	3	3	3	3	3	3							4	4	4	2	2	2	3	3	3							3	3	3	3
4	4	4	4	4	4										3	3	3	3	3	3										4	4	4	4
5	5	5													5	5	5													6	6	6	6
6	6	6													6	6	6													5	5	5	5

CDML see [1], Rashba see [15], Lyle et al. see [16].

6. Conclusions

In this paper we discussed the effect of *TR* symmetry on quasi-particles states. We have shown that a presence of *TR* in crystals results in modified classification of states, phonon dynamical matrices, energy bands, optical selection rules, and others.

In conclusion: a comprehensive tables of complex irreducible representations of 230 double valued space group are needed. In table 5 we list the most common compounds and their symmetries. The table can be quite useful in making a choose of further investigations of *TR* effect in these compounds.

Table 5. Compounds and their symmetries

D^1_{2h} (Pmmm)	$YBa_2Cu_3O_7$ High Temperature Superconductor (HTS)
D^5_{2h} (Pmma)	Si_2G_2 Semiconductor Superlattices (SS)
D^{12}_h (Pnnm)	Marcasite structure. FeS_2 and crystals isomorphic to FeS_2: $FeAs_2$, FeP_2, $FeSb_2$, $FeSe_2$, $FeTe_2$, $CoTe_2$, $NiAs_2$, etc.
D^{16}_{2h} (Pnma)	MnP
D^{18}_{2h} (Cmca)	Si_2Ge_4 (SS)
D^{28}_{2h} (Imma)	La_2CuO_4
D^5_{2d} (C42m)	$(Si)_m(Ge)_n$: (m + n = 4k: m,n odd), Si_1Ge_3, $(GaAs)_1(AlAs)_1$, $(GaAs)_1(AlAs)_3$, $(GaAs)_2(AlAs)_2$, SS
D^9_{2d} (F42m)	$(Si)_m(Ge)_n$: (m + n = 4k + 2; m, n odd), Si_3Ge_3, $(GaAs)_1(AlAs)_2$, (SS)
D^1_{4h} (P4/mmm)	$YBa_2Cu_3O_6$ host material for the HTS: $Yba_2Cu_3O_{6+\delta}$ Tl Ba_2 $Ca_{n-1}Ca_nO_{2n+3}$ HTS
D^{14}_{4h} ($P4_2$/mnm)	Rutile type:TiO_2, and CoF_2, MnO_2, PhO_2, WO_2
D^{17}_{4h} (I4/mmm)	$Bi(Tl)_2Sr(Ba)_2Ca_{n-1}Cu_n,O_{2n+4}$ HTS
D^{19}_{4h} ($I4_1$/amd)	TiO_2 – anatase. $(Si)_m(Ge)_n$: (m + n = 2k + 1), Si_1Ge_2, (SS)
T^6_h (Pa3)	CoS_2, MnS_2, NiS_2
T^7_h (Ia3)	Tl_2O_3
T^2_d (F43m)	GaAs, ZnS, AgJ, AlP, CuBr, CuCl, HgS, ZnSe, ZnTe, CdTe
O^1_h (Pm3m)	Cesium chloride structure CsCl, CsBr, CsI, RbCl, TlCl, TlBr, Intermetallic binary compounds LiAg, AlNd, FeAl, FeTi, Compounds of magnesium MgAg, -Sr,-La, -Ce,-Pr, -Au, -Hg, -Tl, of beryllium BeCo, -Cu, -Pd, of copper CuZn and CuPd of zinc ZnAg, -La, -Ce, -Pr, -Au, of thallium TlCa, -b,-I, -Bi, perovskite structure $NaNbO_3$, $NaTaO_3$, $NaWO_3$, $CaTiO_3$, $KTaO_3$, $SrTiO_3$, $BaTiO_3$, $CaZrO_3$, $CaSnO_3$, $SrZrO_3$, $SrSnO_3$, $SrCeO_3$, $PbTiO_3$, $BaZrO_3$, $PbZrO_3$, $LaAlO_3$, $LaKO_3$, $LaCrO_3$, $LaMnO_3$, $LaFeO_3$, $KMnF_3$, $KMgF_3$, $KNiF_3$, $KCdF_3$, $RbCaF_3$, $RbMnF_3$, $CsCaF_3$, $CsCdCl_3$, $CsCdBr_3$, $CsHgCl_3$, $CsHgBr_3$,also Mn_3SnN
O^3_h (Pm3n)	The A – 15 or β - wolfram structure with the formula A_3B. Materials with the highest transition temperature to the super. Conducting state, Nb_3Ge, Nb_3Sn, Nb_3Al, Vi_3Si, V_3Ga, etc.
O^4_h (Pn3m)	Cuprite Cu_2O
O^5_h (Fm3m)	NaCl, AgBr, AgCl, BaO, BaS, CaO, CaS, CdO, KBr, KCl, MgO, PbS. Cuprite type Cu, Ag, Au, Al, Ce, Ir, La, Ni, Pd, Pb, Sr, Th etc. Fluoride type CaF_2, BaF_2, UO_2, ZrO_2, $SrCl_2$, K_2O, K_2S, Li_2O, Li_2S, Na_2O.
O^6_h (Fm3c)	The rare compounds of the $KTlBr_4$ x $2H_2O$ type. Alloys: $NaZn_{13}$, KZn_{13}, KCd_{13}, $CaZn_{13}$, $ZrBe_{13}$, Ube_{13}.
O^7_h (Fd3m)	Si, Ge, diamond, β - form SiO_2, spinel Al_2MgO_4
O^8_h (Fd3c)	Voltaits, $(NH_4)_2Fe^{II}_5Fe^{III}_4(SO_4)_{12}\cdot 18H_2O$ $K_2(Fe^{III}, Al)_4Zn_5(SO_4)_{12}\cdot 18H_2O$ $Rb_2Fe^{II}_5Fe^{III}_4(SO_4)_{12}\cdot 18H_2O$, $Tl_2(Fe^{III},Al)_4Zn_5(SO_4)_{12}\cdot 18H_2O$
O^9_h (Im3m)	Metallic elements Li, Na, K, Ba, V, Nb, Ta, α - Cr, Mo, α - W, α - Fe, and Eu also C_2Cl_6
O^{10}_h (Ia3d)	Garnets. Calcium aluminium orthosilicate $Ca_3Al_2(SiO_4)_3$, HoGaG, ErGaG, TmGaG, YbGaG, YbAlG, LuAlG, DyAlG, DyGaG, $R_3Al_5O_{12}$, $Tm_3Ga_2(GaO_4)_3$, TbAlG, YIG, TbIG, GdIG, LuGaG, DyIG, ErIG
D^4_3 ($P3_121$)	Sulfurdioxide SiO_2, α - quartz structure, selenium
D^3_{3d} (P3m1)	CdI_2, La_2O_3
D^6_{3d} (R3c)	$CaCO_3$, $CdCO_3$, $FeCO_3$, $MgCO_3$, $MnCO_3$, $NaNO_3$, $ZnCO_3$. Corundum: α - Al_2O_3 (sapphire)
D^4_{6h} ($P6_3$/mmc)	β - ice, NiAs – structure, CoS, FeS, CrS, CuSn, NiSb, Mg, Ba, Cd, La, Nd, Y, Zn
C^4_{6v} ($P6_3mc$)	ZnS – wurtzite, BeO, CdS, ZnO, GaN, 2H-SiC,4H-SiC, 6H-SiC, 8H-SiC

References

[1] Cracknell A P, Davies B L, Miller S C and Love W F 1979, *Kronecker Product Tables vol.1-4 IFI* (New York, Washington, London: Plenum Press)

[2] Conyers and Herring 1937 *Phys. Rev.* **52** 361

[3] Wigner E P 1959 *Group Theory and its Application to the Quantum, Mechanics of Atomic Spectra* (New York, London: Academic Press)

[4] Thoma K, Dorner B, Duesing G and Wegener W 1974 *Solid State Commun.* **15** 1111

[5] Hewat A W 1974 *Solid State Commun.* **8** 187

[6] Schiff L I 1968 *Quantum Mechanics* (New York: McGraw-Hill)

[7] Normand J-M A 1947 *Lie Group: Rotation in Quantum Mechanics* (North Holland Publishing Company)

[8] Frobenius and Schur 1906 *Berl. Ber.* 186

[9] Bir G L and Picus G E 1974 *Symmetry and Strain Induced Effects in Semiconductors* (John Wiley & Sons)

[10] Kunert H W 1982 *Physica* **114** A 600-603

[11] Kunert H W 2004 *Eur. Phys. J. Appl. Phys.* **27** 252

[12] Kunert H W 2003 *Applied Surface Science* **212-213** 890

[13] Lambrecht W R L and Segall B 1994 *Properties of Group III Nitrides* ed Edgar J H (London: INSPEC)

[14] Lambrecht W R L, Segall B, Rife J, Hunter W R and Wickenden D K 1995 *Phys. Rev.* B **51** 13516

[15] Rashba E I 1959 *Fiz. Tverd. Tel.* **1** 407 (English transl.: 1959 *Soviet Phys. Solid State* **1** 368)

[16] Lyle P, Hamilton D R and Choyke W J 1966 *Phys. Rev.* **143** 526

Symmetric and asymmetric dispersion relation in graphene

Magdalena Margańska[1], Marek Szopa[1] and Elżbieta Zipper[1]

[1] Department of Theoretical Physics, University of Silesia, ul. Uniwersytecka 4, 40 007 Katowice, Poland

E-mail: magda@phys.us.edu.pl

Abstract. The standard tight-binding dispersion relation for graphene and carbon nanotubes has an electron-hole symmetry. This symmetry has not been observed experimentally until recently. We discuss here the effect of the overlap between π-orbitals at neighbouring sites, belonging to different sublattices, on the dispersion relation. The e-h asymmetry increases with the degree of overlap, although this effect is visible only for energy bands far from the Fermi level. We estimate the observable effects of this asymmetry on the DOS which could be observed in conductivity measurements and explain the symmetry found in the experiment. The degree of the overlap determines also the magnitude of the magnetic moment induced in the nanotube by the Aharonov-Bohm magnetic flux.

1. Introduction

Carbon nanotubes belong to the most promising materials for nanotechnology. It is therefore worthwhile to investigate their properties in detail. In calculations involving the dispersion relation of nanotubes it is commonly assumed that they inherit the standard tight-binding dispersion of graphene, and the periodic conditions around the nanotube quantize the momenta, forming energy bands [1]. In calculations concerning small nanotubes their curvature must also be taken into account. Its main consequences are a change in the angle between neighbouring π-bonds, in their length, and in the degree of $\sigma - \pi$ hybridization. It has been theoretically predicted and confirmed experimentally [2] that the curvature opens a band gap at the Fermi level. This gap decreases with increasing nanotube radius and almost vanishes at $d \approx 1.5$nm.
In the present paper we will investigate another contribution to the tight-binding model which is important also in large nanotubes. It is usually assumed that the π orbitals on the two sublattices of graphene do not intersect and the energy spectrum has the electron-hole symmetry. However it was found that in graphene the neighbouring π orbitals overlap, which removes the electron-hole symmetry of the spectrum. We present below a short study of the effect which this modification of the standard tight-binding Hamiltonian may have on transport and magnetic properties of carbon nanotubes.

2. Dispersion relation

On graphene plane we work in the basis $\mathbf{T}_1 = \sqrt{3}e_x$, $\mathbf{T}_2 = (\sqrt{3}/2)e_x + (3/2)e_y$. In this basis the dispersion relation without overlap on the adjacent carbon sites is

$$E_S(\mathbf{k}) = \pm\gamma\sqrt{1 + 4\cos^2\left(\frac{\sqrt{3}}{2}k_x\right) + 4\cos\left(\frac{\sqrt{3}}{2}k_x\right)\cos\left(\frac{3}{2}k_y\right)}. \tag{1}$$

The subscript S stands for "symmetric" and γ is the hopping integral for graphene, which is most often assumed to have the value 2.5, 2.7 or 3 eV, depending on the authors.
In real graphene the wavefunctions on neighbouring atoms always overlap, although this overlap may be small. In order to take this effect into account, a mixing between electrons from the two sublattices must be introduced. It may be expected that bonding states will not be as profitable in energy as they were without overlap – and vice versa, the antibonding states will have to overcome the natural tendency of the wavefunctions to overlap. Indeed, calculations confirm this intuition, yielding [1]

$$E_A(\mathbf{k}) = \frac{\epsilon_{2p} \mp \gamma|E_S(\mathbf{k})|}{1 \pm s|E_S(\mathbf{k})|}, \tag{2}$$

where s is the value of the overlap, ϵ_{2p} is the electron's on-site energy and the A subscript stands for "asymmetric". Following [1], we assume $s = 0.129$. The difference between the symmetric and asymmetric dispersion is shown in Fig. 1. In both cases the momentum spectrum remains

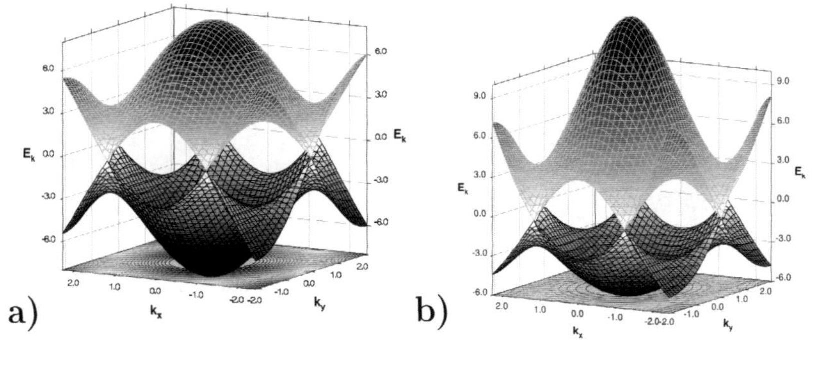

Figure 1. The dispersion relation of graphene. a) symmetric, without considering the overlap, and b) asymmetric, with overlap $s = 0.129$ as in [1]. Note the flattening of the valence band, not so favourable energetically anymore, and the elongation of the conduction band.

the same, determined by the transverse and longitudinal boundary conditions.

$$\mathbf{k} \cdot \mathbf{C}_h = k_C C_h = 2\pi l_c, \qquad \mathbf{k} \cdot \mathbf{L} = k_L L = 2\pi l_l, \qquad l_c, l_l \in \mathbb{Z}, \tag{3}$$

where $\mathbf{C}_h$ is the circumference (or the chiral vector) of the nanotube, and $\mathbf{L}$ is its length. The energy bands and the density of states are different in the two cases. Close to the Fermi surface, where the denominator in Eq. (2) is close to 1, the overlap does not result in a visible difference (see Fig. 2), but far from the Fermi level it is clearly different whether the dispersion takes into account the overlap or not.

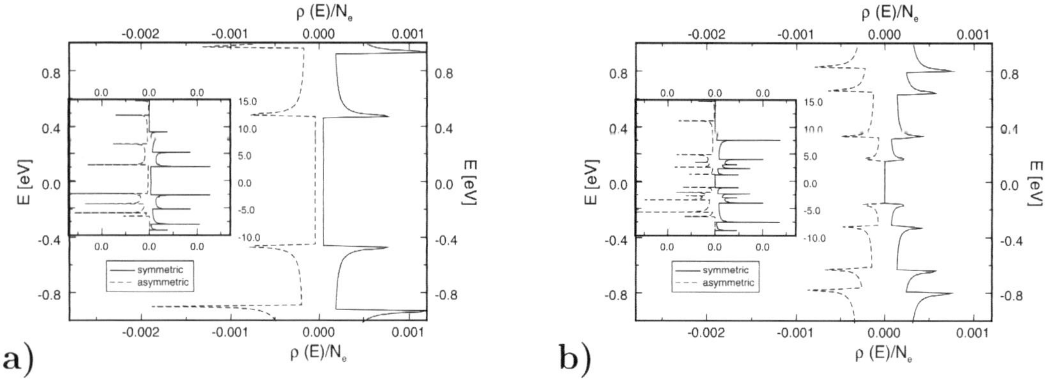

Figure 2. The density of states in a) metallic and b) semiconducting carbon nanotube, both in symmetric and asymmetric case. The top plots are in the full range of energies for small (3,3) metallic and (3,2) semiconducting tubes, for the sake of clarity. The bottom plots show the density of states of realistic, (20,20) and (20,19) nanotubes close to the Fermi level, in the range (-1eV, 1eV).

3. π overlap in transport experiments

In the CN's, as in graphene, there are only two Fermi points. Moreover, due to the small size of nanotubes, the energy gaps between individual one-dimensional energy subbands are large. These two features justify the treatment of nanotubes as one-dimensional wires with two conducting channels. In the Landauer-Büttiker formalism the (undoped) nanotube conductance of thermally activated electrons is [3]

$$G_{act}(T) = \frac{2e}{h} \sum_{i=1,2} |t_i|^2 \frac{2}{exp(\Delta E_i/kT) + 1}, \tag{4}$$

where $|t_i|^2$ are the transmission probability in the i-th channel and ΔE_i is the energy gap of the i-th subband at the Fermi level. In 2004 the group from Delft found a perfect symmetry between hole and electron energy levels in a semiconducting nanotube [4]. The experiment was performed in a standard quantum dot device setup, with the nanotube lying across two electrodes. The quantum dot differential conductance plots showed the electron-hole symmetry for the doping with up to 20 electrons/holes. This remarkable result, obtained in a transport experiment, probes the effective doping range of $\approx (-0.15, 0.15)$eV. The density of states, obtained numerically, of a CN whose size corresponds to that of the CN used in the experiment is plotted in the bottom part of Fig. 2b, in the range (-1eV,+1eV). It is clear that the visible deviation from symmetric DOS appears only when the doping exceeds $\sim \pm 0.5$eV. Thus it follows from our calculations that e-h symmetry is preserved for small doping, in agreement with the Delft experiment [4]. The asymmetry of the spectrum due to the overlap can be revealed at larger doping and, consequently, the overlap of the π-orbitals at adjacent carbon sites can be determined.

4. π overlap and the orbital magnetic moments

Due to their cylindrical topology carbon nanotubes in magnetic fields parallel to the nanotube axis display the Aharonov-Bohm effect. One of its manifestations are the persistent currents [5]. The change of phase of the electron running clockwise around the tube and of the electron running anticlockwise produces a shift in their momenta, resulting in a nonequilibrium between currents carried by electrons with k_C and $-k_C$. The net current carried by all the electrons in the system does not decay provided that $\Delta E_g > kT$, where ΔE_g is the energy gap at the Fermi level, therefore it is called 'persistent'.

Persistent currents run perpendicular to the axis of the system, and they are not transport currents – their origin is purely topological. As they run around a closed loop, they induce an orbital magnetic moment in the nanotube. This magnetic moment determines the magnetic response of a nanotube in parallel magnetic field. This magnetic moment depends on the value of the flux penetrating the nanotube and it is given by [6]

$$\mu_{orb}(\phi,T) = -\pi R^2 \frac{\partial F(\phi,T)}{\partial \phi} = -\pi R^2 \sum_{\mathbf{k}} \frac{\partial E_{\mathbf{k}}(\phi)}{\partial \phi} f_{FD}(E,\mu_{chem},T), \quad (5)$$

where F is the free energy, μ is the chemical potential, ϕ is the magnetic flux penetrating the nanotube, and $f_{FD}(E,\mu_{chem},T)$ is the Fermi-Dirac distribution function.

Our theoretical calculations [6] show also that in nanotubes with shifted Fermi level (due to electron or hole-doping) the response varies depending on the amount of doping (cf. Fig. 3). This variation changes when the overlap is taken into account. The boundaries between dark and bright regions correspond to paramagnetic jumps in the persistent current, present when there are occupied electron states at the Fermi level. The values of doping and magnetic flux at which these jumps occur reflect the geometry of the Brillouin zone of the nanotube and the features of the dispersion relation. However, the response at slight doping is again nearly e-h symmetric, with differences seen only at larger values of doping, especially in the hole-doping (lower) region of the plot. The fundamental difference between transport phenomena and persistent currents

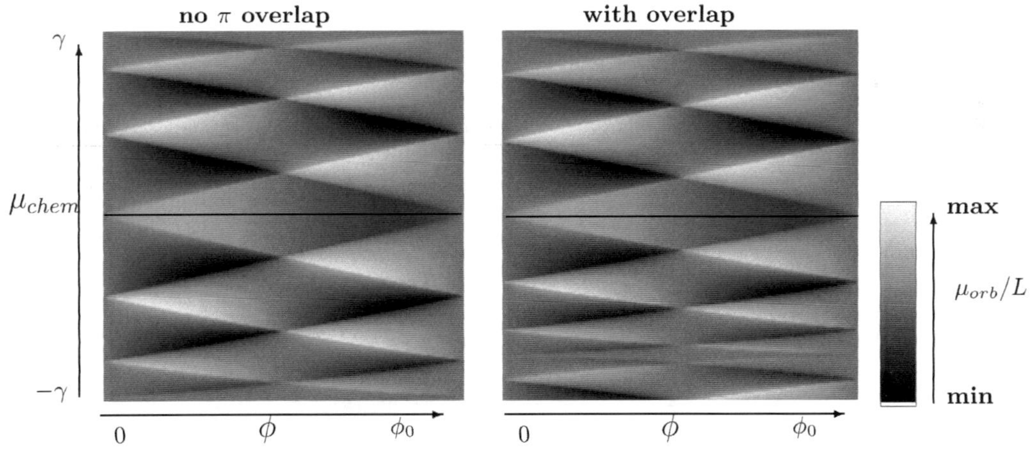

Figure 3. The magnetic moments (greyscale) dependence on ϕ and μ_{chem} in an armchair (7,7) nanotube at $T = 0$K and the doping in the range $(-\gamma, +\gamma)$.

is that the latter contain contributions from all states in the Brillouin zone, also those deep below the Fermi level. Thus, even though the pattern of the magnetic response does not yield information about the size of the overlap, its amplitude does. The amplitude of the magnetic moment for a metallic (20,20) and semiconducting (20,19) nanotube with and without overlap are presented in Table 1. It is plain to see that since $\partial E/\partial k_C$ is smaller in the asymmetric case, the individual currents are diminished and the amplitude of the magnetic moment can drop by a factor of 3 as compared to the symmetric case. As till now there is no conclusive

Table 1. Amplitude of the magnetic moments in undoped metallic (20,20) and semiconducting (20,19) nanotubes ($L = 0.1\mu m$) with symmetric and asymmetric dispersion relation. $T = 0$K.

CN type	μ_0 - symmetric	μ_0 - asymmetric
metallic	$85\mu_B$	$28\mu_B$
semiconducting	$52\mu_B$	$15\mu_B$

experiment measuring the magnetic moment, we cannot state which relation is better suited for CN's. However, the calculated differences of μ_{orb} for the two models are remarkable and when they are measured they could give us information about the precise value of the π orbital overlap in carbon nanotubes.

References

[1] Saito R, Dresselhaus G and Dresselhaus M S 1998 *Physical Properties of Carbon Nanotubes*, (London : Imperial College Press)

[2] Kleiner A, Eggert S 2001 *Phys. Rev.* **B64** 113402
Ding J W, Yan X H, Cao J X, Wang D L, Tang Y, Yang Q B 2003 *J. Phys: Condens. Matter* **15** 439

[3] Maiti A, Svizhenko A, Anantram M P 2002 *Phys. Rev. Lett* **88** 126805
Minot E D, Yaish Y, Sazonova V and McEuen P L 2004 *Nature* **428** 536

[4] Jarillo-Herrero P, Sapmaz S, Dekker C, Kouwenhoven L P, van der Zant H S J 2004 *Nature* **429** 389

[5] Büttiker M, Imry Y, Landauer R *Phys. Lett* 1983 **A 96** 365
Cheung H -F, Gefen Y, Riedel E K and Shih W -H 1988 *Phys. Rev.* **B 37** 6050

[6] Margańska M, Szopa M, Zipper E 2005 *Phys. Rev. B* **72** 115406

Charge transport in thin layers of organic crystals

Marian Kuzma

Institute of Physics, University of Rzeszow, 35-310 Rzeszow, Rejtana 16a

E-mail: kuzma@univ.rzeszow.pl

Abstract. Recently, the organic thin layers have been studied widely due to their good properties for application in field–efect transistors. In this paper we survey recent studies on carrier transport properties in such molecular layers which are promising for applying in transistors. First we present the main electronic properties of organic molecules forming layers or crystals and the relation of this properties on specific features of band structure of organic crystals. The layers of pentacene are studied particularly as an example. The crystal structure of this material is presented. The experimental methods of carrier mobility measurements in high resistivty layers studied are described. The experimental results are in agreement with Visenberg model of charge transport in non-ordered materials. We discuss the charge transport properties measured for single crystal transistors. The mobility in these crystals is significantly greater in comparison with layers particularly at low temperatures and shows an interesting anisotropy in respect to crystallographic directions.

1. Introduction

Organic conductors have been studied for a long time (e.g. anthracene since 1950). Organic crystals are formed by organic molecules bonded by Van-der-Waals forces. Therefore, the electronic properties of organic crystals are different form those in inorganic crystals, where bonds are of the ionic or covalent origin. The number of trapped states and strong electron-phonon coupling in organic crystals, and resulting from these the electronic, molecular and lattice polarisations determine charge transport properties via small polaron formation and a kind of many body interaction between them [1]. Thin layers of organic materials have crystal structure far different from single crystal one (epitaxial). There is a weak order, as well as a great number of defects. Thus, the electrical conductivity is hopping one instead of band-like. Nevertheless, even non-perfect organic layers are promising for field effect transistors production and the investigations of transport properties in such layers are carried out widely.

The desirable goal of material research in this field is increasing the charge mobility in p- tape as well as in n-type channel organic crystals. This can be attained not only by careful purifying or searching new materials but also by the crystal structure improvement of the layers for these materials. The pentacene as a p-type material shows the mobility in thin layers exceeding 1 cm^2/Vs. In single ambipolar pentacene transistors the electron mobility also exceeds 1cm^2/Vs. In other n-type channel organic materials the n mobility is restricted to 0.1 cm^2/Vs for naphthalenetetracarbolixylic materials and C60 and to 0,01 cm^2/Vs for fluorinated copperphtalocyanine [2]. However, in ambipolar perylene single crystal layers, the electron mobility achieves the values 5.5 cm^2/Vs at room temperature and 120 cm^2/Vs at 50K [3]. In the paper [4] the Hall mobility 4 cm^2/Vs in single crystals of bis (1,2,5-thiadiazolo)-p-quinobis(1,3-dithiole) (BTQBT) was reported, whereas in channel layers the field-effect mobility was 0.044 cm^2/Vs [5]. The advantage of this material over pentacene is that the BTQBT planar molecules lie flat on substrates during organic molecular–beam deposition process [6].

The present paper is oriented mainly towards to the structure and charge transport properties of pentacene which seems to be the best candidate for channel organic material in field–effect–transistors [7, 8].

2. Electronic properties of free organic molecules and electronic structure of organic crystals

Isolated molecule is either in neutral or negative or positive ion state. The affinity energy A (the energy liberated by the attachment of an electron by neutral molecule) and the ionisation energy I are vertical or adiabatic A_v, A_a I_v, I_a.

The terms of ionisation and affinity energy can be considered in the frame of the molecular orbitals theory [9]. The position of molecular energy levels of ionised molecule are shifted in respect to the levels in neutral molecule. This changing is particularly visible in aromatic hydrocarbons [10, 11]. Moreover, the intrinsic or extrinsing (medium) polarisation effects influence also the position of these levels e.g. for naphthalene I_{gv} = 9.30 eV [11], whereas the experimental value for positive ionised molecule this energy is 8.15 eV. The difference arises from polarisation of the naphthalene molecule by positive space charge.

In molecular crystals the ionic molecular states form narrow bands due to forces forming the crystal (figure 1). Both I_g and A_g energies are released upon the taking the electron from the infinity and placing it in valence orbital of positive ion or on molecule respectively. Otherwise these energies should be inserting into system in order to the entire removal of an electron from valence orbital or from molecule. The additional electron on a molecule in a crystal gains the additional energy Pc (polarisation energy) forming lowest conduction level. Likewise, the molecule with a missing electron profits the additional polarisation energy P_h in a crystal as well as forming the valence level. These two levels form energy gap E_g. However, it should be noticed here that the conduction and valence levels do not mean the delocalized carriers as they are in inorganic semiconductors. The carriers are strongly localised in a simple homomolecular organic crystals. There are two main reasons for strong localisation of carriers in those crystals: the high value of polarisation energies in aromatic hydrocarbons and the small nearest–neighbour interaction energy. In narrow-band semiconductors there is also high polarisation energy, but the inter-site energy is relatively much greater making the carriers mobile. Contrary, in wide–gap materials (Ge, Si) the polarisation energy is very small comparing to inter–site energy, therefore the carriers are free to.

It should be noticed however that not all organic crystals exhibit the same degree of localisation. The carriers on higher conducting levels show greater mobility.

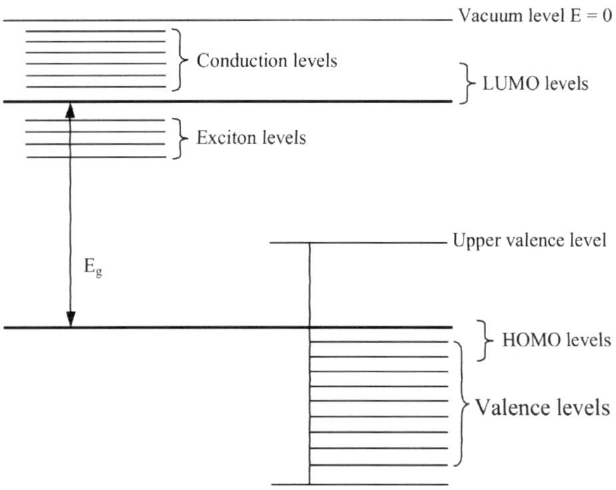

Figure 1. Scheme of energy levels of molecular crystals.

The excitation charge–transfer levels (CT) in figure 1 represent the energy required for removing of the electron from one molecule and placing it on another one being nearest or next- nearest neighbour. These states were previously called ionic states [13] and are similar in the origin to the ordinary excitons, but here the hole-electron distance is now extended to one or more distances of lattice constant. The charge- transfer of neutral molecules forms electron- hole pairs in which electron and hole is more or less correlated depending on the distance r between them. For r sufficient small i.e. smaller from value $r_c = kT$, the Coulomb interaction energy W_e correlates the electron-hole interaction energy providing the CT energy:

$$E_{CT}(r) = I_g - A_g - P_{eh}(r) - W_e(r) \quad (1)$$

where $P_{eh}(r)$ is the polarisation energy of the electron–hole correlated pair. For $r > r_c$ the CT energy is equal E_g. Thus the energy gap is the minimal energy which forms uncorrelated electron-hole pairs in the crystal. The CT energies for aromatic hydrocarbon crystals decrease considerably with the increase of the number of benzene rings in linear molecule [14, 15, 16] (figure 2), e.g. this excitation is 4.38 eV in naphthalene [15] whereas in tetracene this energy is about 2.8 eV [14]. It should be expected that the lowest CT level in pentacene approaches the lowest conduction band level (Fig 2).

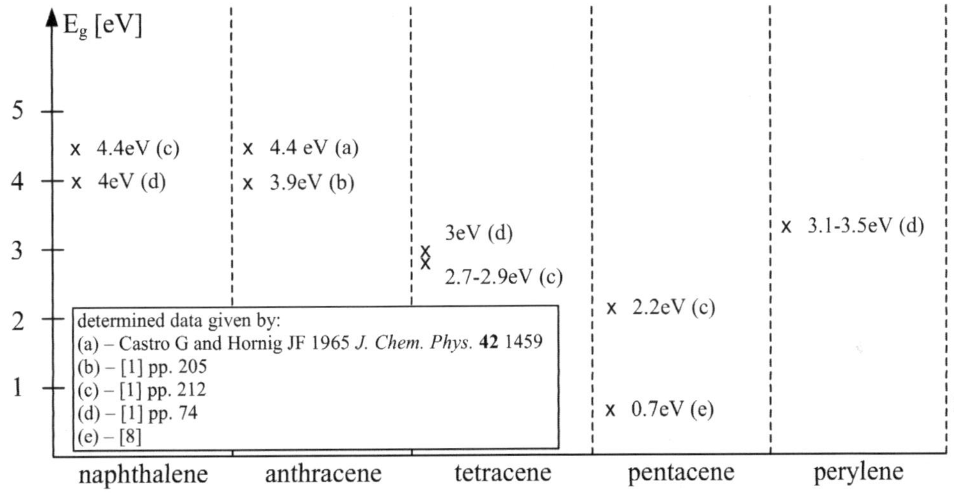

Figure 2. Energy band gap in molecular crystals defined as a minimal energy necessary to induce electrical conductivity in a crystal ($E_g = I_c - A_c$).

The strong electronic polarisation effects in molecular crystals are bar for the one-electron approximation calculations of the band structure of excess carriers usually applied in inorganic crystals. The electronic polaron formalism which is rather many-electron approach should be applied for crystals in which the hopping time of carriers from one site to neighbouring one is slow in respect to response time of electron system (10^{-15} s). The one-electron band structures for naphthalene, anthracene, tetracene and pentacene were calculated by Katz, Rice et al. [18, 19]. The one-electron crystal states were constructed from molecular orbitals of isolated molecules. The crystal potential was formed as a sum of neutral molecular potentials placed in the nodes of crystal lattice. As an example the band structure of electron and holes in anthracene crystal is presented in figure 3 [18]. The bandwidth depends clearly on the direction in k space; this feature is typical for all aromatic hydrocarbons. The effective mass is usually 5–20 times the effective mass of electron and holes in the bottom of the conduction band and the top of the valence band respectively.

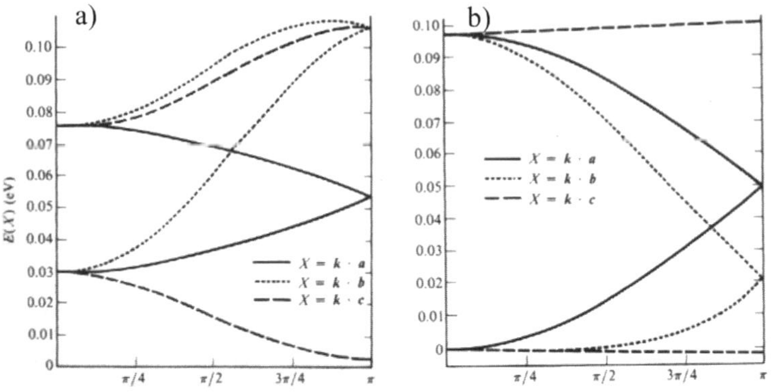

Figure 3. Energy band structure of anthracene crystal [18], (a), (b) – hole and electron bands respectively.

3. Single crystal structure and anisotropy of carrier mobility

The kind of molecular packing in molecular crystals influences strongly on intrinsic transport properties. The overlap between electronic orbitals of nearest molecules in crystals seems to be very important factor in this matter. The overlap C–C would be the great for flat molecules packed as parallel stacking as in violanthrene crystals [20]. Such layered structure of a crystal should have good electronic properties at least in the plane of the layers. However the realization of such layered structure in organic crystals of acenes is not observed [21]. These grown in herringbone structure (see figure 4), forming monoclinic (naphthalene and anthracene [1]) and triclinic (tetracene [22, 23] and pentacene [8]) crystallographic systems. The inversion centre of molecules is placed in the corners of unit cell as well as in (1/2, 1/2, 0) point. The angle δ between molecular planes of non-equivalent molecules is in the range of 51.1 – 52.3° for different acenes. The angle between molecular axis and c axes is different for non-equivalent molecules as well.

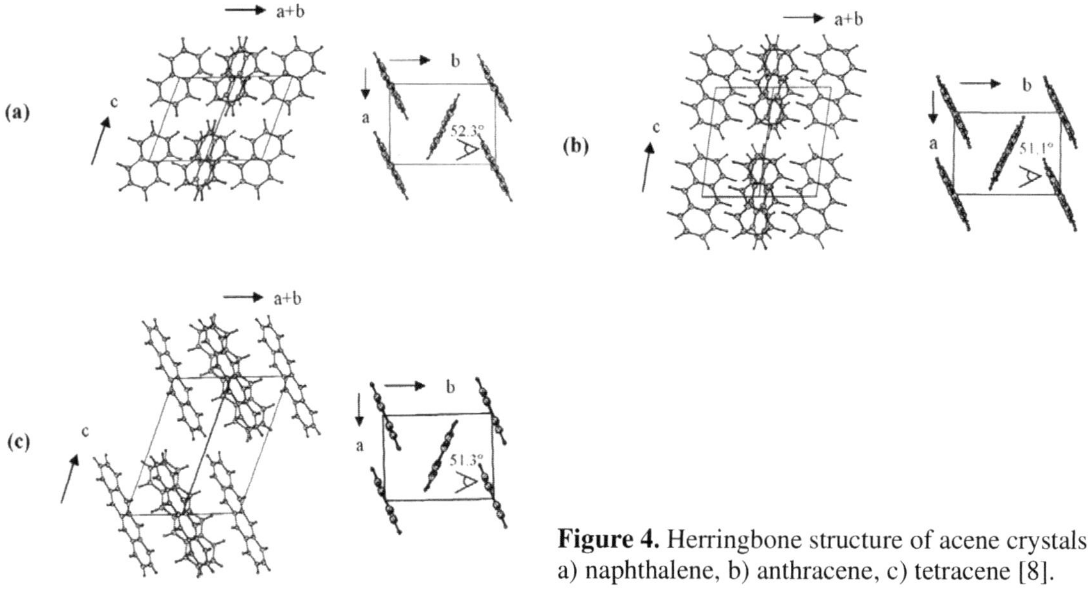

Figure 4. Herringbone structure of acene crystals a) naphthalene, b) anthracene, c) tetracene [8].

The crystal structure of pentacene single crystals (figure 5) and layers was studied by Campbell [24, 25], Dmitrakopoulos [26], Luo [27] and Matteus [28, 8]. Matteus [8] obtained four different polymorphic structures of pentacene layers. These structures differ by d (001) values which were 1.41

nm, 1.4 nm, 1.50 nm and 1.54 nm. The interlayer space d (001) for single crystals is 1.41 nm independently on a method of crystal growth.

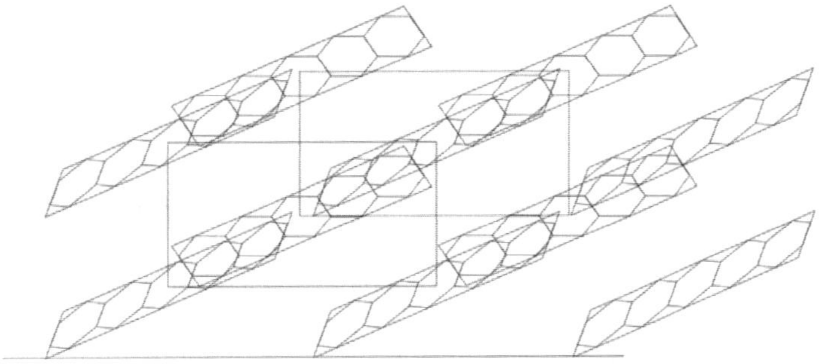

Figure 5. Scheme of pentacene crystal structure.

Herringbone structure of aromatic hydrocarbon crystals is in tight relation to hydrogen–hydrogen and hydrogen–carbon electronic state interactions. The overlap of these states from neighbouring molecules depends on the structure. Thus the anisotropy of charge carrrier properties should be expected. For pentacene single crystals the anisotropy of hole mobility was investigated by Schon, Kloc and Battlog [7]. Single crystals of pentacene were grown by vapour transport in hydrogen stream. Electrical properties were studied by current–voltage characteristics at wide range of temperature 2–500 K. The conductivity was greater in the plane parallel to the growth surface compared to that in the direction perpendicular to the a–b layers of the crystals. In the plane a–b the mobility of holes was approximately 2.9, 3.2, 2.3 2.20 cm^2/Vs for directions forming angle 0, 30, 60 and 90° respectively with b axis. The anisotropy is visibly easy. The higher mobility was in the direction 30° in respect to b axis.

A strong mobility anisotropy on the growth surface of rubrene single crystals was studied by researchers in Ref. [29]. Here the PDMS stamp technique [30, 31] was used to fabrication FET structure on the surface of the rubrene crystals with various orientation of the channel in the plane a-b. The mobility was determined from I-V FET characteristics. The highest mobility almost 3 cm^2/Vs coincides with **b** direction whereas in **a** direction this value is 4 times lower. Studies of the anisotropy of mobility in perfect single crystals are crucial for observations of the nature of electrical conductivity in organic crystals and layers. The measurements on thin-film organic field effect transistors never show such anisotropy due to a number of structural defects as grain boundaries. Rubrene is conjugated chain with non-planar side molecules. The single crystal structure is herringbone to. The high mobility of this crystals may be attributed to the better overlap of the electronic orbitals of adjacent molecules.

4. Theory of hole mobility in FET with non–single crystal layer

Two basic theoretical models of charge transport in organic layers are usually proposed: of hopping and band–like transport. In such crystals the transport properties depend strongly on a nature, number and distribution of traps. The traps originate in a point and extended crystal structure defects. The self–trapping of charges also contributes to the conductivity of the organic crystals. This type of trapping occurs usually in crystals with narrow conductivity band. Self trapping can occur by optical modes of phonons or by acoustic modes.

The dependence of the field effect mobility on the temperature and gate bias for good ordered organic layers was proposed by Horowitz et al. [32]. The model bases on multiple trapping and release of charge carriers which transit to the extended levels in transport band i.e. to the valence band in the case of hole carriers. Thus the band model of conductivity takes place. The concentration of the band carriers depends on the energy levels of localised states, the temperature and the applied voltage.

In non ordered, amorphic or pury ordered layers there are only strongly localized states. There, the charge transport is expected to be hopping one only. Vissenberg and Matters [33] calculated field effect mobility for such case using variable–range hopping idea [34]. The main assumption of the model is that the transport properties are determined by the tail (for sufficiently low temperatures T< T_o and low carrier densities) of the density g(E) of states

$$g(\varepsilon)=\frac{N_t}{k_B T_0}\exp\left(\frac{\varepsilon}{k_B T_0}\right),\ (-\infty<\varepsilon\leq 0), \tag{2}$$

where N_t is concentration of states.

Taken into account the Fermi–Dirac energy distribution f (ε, ε_F) of the carriers as well as the percolation theory [35] the authors of ref [33] obtained the Arrhenius–like behaviour of the conductivity versus temperature $\sigma \propto \exp(-E_a / k_B T)$:

$$\sigma(\delta,T)=\sigma_0\left(\frac{\pi N_t \delta (T_0/T)^3}{(2\alpha)^3 B_c \Gamma(1-T/T_0)\Gamma(1+T/T_0)}\right)^{T/T_0} \tag{3}$$

where δ is a fraction of states occupied by carriers, B_c is percolation criterion [36, 37] of the system and Γ function

$$\Gamma(z)=\int_0^\infty dy \exp(-y) y^{z-1} \tag{4}$$

is in relation to the occupation number δ

$$\delta=\frac{1}{N_t}\int d\varepsilon\, g(\varepsilon) f(\varepsilon,\varepsilon_F)\cong \exp\left(\frac{\varepsilon_F}{k_B T_0}\right)\Gamma(1-T/T_0)\Gamma(1+T/T_0), \tag{5}$$

The formulae (3) points on the nonlinear dependence of conductivity on carriers density $N_t\delta$ (the linearity gives models with Boltzmann carrier distribution)

$$\sigma \propto (\delta N_t)^{T_0/T} \tag{6}$$

It should be notice that temperature dependence of conductivity (3) is in contrast with Mott behaviour $\sigma \propto \exp\left[-(T_1/T)^{1/4}\right]$ obtained also for variable–range hopping but with constant DOS [34].
Basing on the model of conductivity presented above, Vissenberg and Matters [33] gave the temperature and voltage dependence of mobility of carriers measured by FET method. The gate voltage V_g induces in semiconductor layer the potential V(x) in the distance x from interface insulator–semiconductor modifying carrier distribution δ(x) in semiconductor

$$\delta(x)=\delta_0 \exp\left(\frac{eV(x)}{k_B T_0}\right) \tag{7}$$

The field–effect mobility depends on gate bias as [38]

$$\mu=\mu(V_g)=\left(\frac{L}{ZC_i V_d}\right)\frac{\partial I_{sd}}{\partial V_g}\Big|_{V_d\to 0} \tag{8}$$

where I_{sd} is source -drain current and gate biases are negative and greater then –2V.
The resultant source-drain current for the case of distance dependence charge distribution δ(x) is

$$I_{sd}=\frac{WV_D}{L}\int_0^t dx\,\sigma[\delta(x),T] \tag{9}$$

where L is the length, W is the width and t is thickness of the channel.

From (7), (8) and (9) the field effect mobility has been obtained [33]:

$$\mu_{FE} = \frac{\sigma_0}{e}\left(\frac{\pi(T_0/T)^3}{(2\alpha)^3 B_c \Gamma(1-T/T_0)\Gamma(1+T/T_0)}\right)^{T_0/T}\left[\frac{(C_i V_G)^2}{2k_B T_0 \varepsilon_s}\right]^{T_0/T-1} \tag{10}$$

where α^{-1} is overlap parameter.

5. Methods of mobility measurements

Three basic methods of charge carriers mobility measurements are implemented for materials with high resistance: 1) time of flight (TOF) spectroscopy, 2) current voltage characteristics and 3) field effect transistor characteristics.

Time of flight spectroscopy bases on measurements of electrical currents caused by photoexcitation of electron-hole carriers near the surface of the crystal placed in applied electrical field. The short pulse (nanoseconds) laser light is used for excitation of carriers. The wavelength of light should be adjusted to its absorption in thin, near the surface layer of the sample. Polarity of the voltage applied selects the type of carriers. The carriers pass the width L of crystal in transient time τ, which can be determined from the shape of decay signal. The mobility is calculated for trap free crystals as $\mu = L/E\tau$ where E is field applied. For the case with shallow traps having energy E_t^s and density N_t^s the measured mobility μ_m is in the following relation to the ideal free mobility μ:

$$\mu_m(T) = \frac{\mu(T)}{1+(N_t^s/N_v)[\exp(E_t^s/k_B T)-1]} \tag{11}$$

This method can be used for bulk materials of very good crystalinity, nevertheless it has some advantages in respect to classical methods: the results of TOF method are free from contact influences [39].

The second method; the space charge limited I–V characteristics is based on general equation for current J in unipolar (hole) sample caused by electric field

$$J = \mu\rho_f F - D\frac{d\rho_f}{dx} \tag{12}$$

where D = μkt/E is the Einstein diffusion coefficient and μ is the mobility of the carriers.
The electric field have to fulfill Poisson's equation

$$\frac{dF}{dx} = \frac{\rho}{\varepsilon\varepsilon_0} \tag{13}$$

where ρ is total charge density i.e. the sum of free carriers ρ_f and trapped one $\rho_t = \rho_f + \rho_t$. The free carriers density depends on total density as well as on the position x: $\rho_f = \rho_f(\rho, x)$.

The simplest solution of equation (12) is given by Mott and Gurney [40] for the case without traps and ignoring diffusion

$$J = \frac{9}{8}\mu\varepsilon\varepsilon_0 \tag{14}$$

This solution is a Child Low and it determines the space charge limited current (SCLC).

The solutions for various cases of trapping were given by Bonham and Jarvis [41]. These cases are listed in Table 1. For shallow traps with exponentially distributed energy the current density is the square dependence on the applied voltage U as well. However the constant trapping factor Θ plays significant role :

$$J = \frac{9}{8}\mu\varepsilon\varepsilon_0\Theta\frac{U^2}{L^3} \quad (15)$$

$$\Theta = \frac{N_V}{H_t}\exp\left(-\frac{E_t}{kt}\right) - \theta \quad (16)$$

where N_v and H_t is total density of states and traps respectively
At the sufficiently large voltage above trap- filled voltage U_{TLF} the Fermi level is shifted towards valence band over-passing trap levels and is making charge carrier free. In result the rapid increase of the current is observed. The symmetry of models is presented in table 1 and pictorial presentation of I–V characteristics is shown in figure 6.

Table 1. Currents in I–V characteristics.

Model	Assumptions	I(V) dependence	Descriptions
Matt and Gurnay [40]	- trap–free - lack of diffusion - ohmic contacts	$J = \frac{9}{8}\mu\varepsilon\varepsilon_0\frac{U^2}{L^3}$	$U = \frac{W}{e} - V$
Bonham and Jarvis [41]	- traps are shallow - traps energy are discrete $\rho_f << \rho_t$ - ohmic electrodes	$J = \frac{9}{8}\mu\varepsilon\varepsilon_0\vartheta\frac{U^2}{L^3}$, where $\vartheta = \frac{N_V}{H_t}\exp\left(-\frac{E_t}{kT}\right)$	$\theta = \frac{\rho_f}{\rho} = \vartheta = const$ H_t - total density of traps uniformly distributed in a crystal
Helfrich and Mark [58, 59, 41]	- traps density depends exponentially on the trap energy	$H(E_t) = \frac{H_t}{kTl}\exp\left(-\frac{E_t}{kTl}\right)$, $l > 1$ $J = N_V\mu e\left[\frac{\varepsilon\varepsilon_0 l\sin(\pi/l)}{eH_t(\pi/l)(l+1)}\right]^l\left(\frac{2l+1}{l+1}\right)^{l+1}\frac{U^{l+1}}{L^{2l+1}}$	E_t - energy of trap calculated from the top of valence band (E = 0) l – experimental parameter

Thirst's method allows to determine the mobility of carriers from electrical characteristics of field effect transistor in which the organic material studied plays the role of active layer. figure 7 shows the geometry of the thin film field effect transistors (figure 7a) and field effect transistor made on single crystal (figure 7b). A typical current-voltage characteristics for field effect transistors are presented in figure 8, where figure 8a shows the dependence of current drain–source I_{ds} versus voltage drain-source V_{ds} for different gate voltage V_g. In figure 8b the dependence of saturated current I_{ds} on V_g is shown as well. The field effect mobility μFE can be determined from the slope of a plot in figure 8b. According to the formulae [42]:

$$I_{ds,s} = W/2LC_i\mu_{FE}(V_gV_{th})^2 \quad (17)$$

where W and L is width and length of channel and C_i is a capacitance of insulator (per unit area).

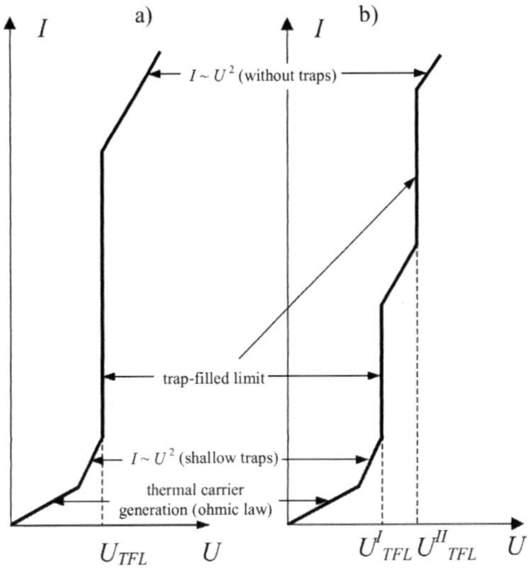

Figure 6. One carriers I–V characteristics for a) one trap level and b) two trap levels.

6. Charge transport properties of pentacene films

Pentacene is promising material for organic electronic devices [43, 44, 45, 46].

The structure and ordering of pentacene thin films depend highly on a growth method. The crystal defects, grain boundaries and disorder dominate charge transport properties in layer obtained. Therefore the structural quality determines the type of transport and its temperature behaviour. In ultra pure single crystals the band like transport is observed at low temperatures whereas at high temperatures it crosses over to hopping mechanism [7, 47]. The same behaviour would be expected in epitaxial layers of pentacene. However, single crystalline layers of pentacene has not grown so far. Brown et al. [38] obtained layers using precursor method (figure 9). Films deposited were composed of microcrystallites. The transport properties depends strongly on conversion temperature and the best mobility of 0.03 $cm^2V^{-1}s^{-1}$ was observed for conversion temperature of 180° C.

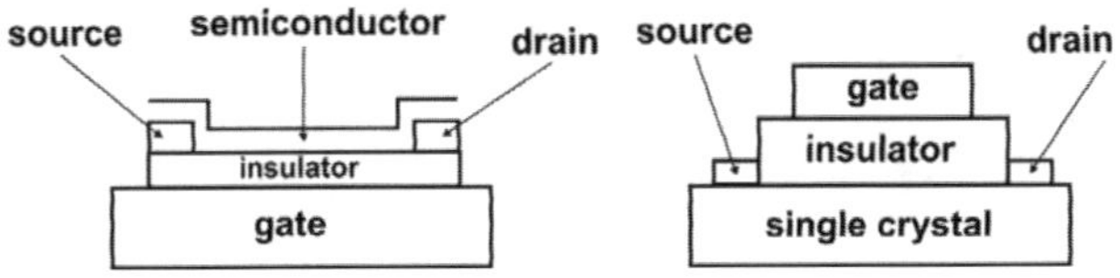

Figure 7. Scheme of field effect transistor.

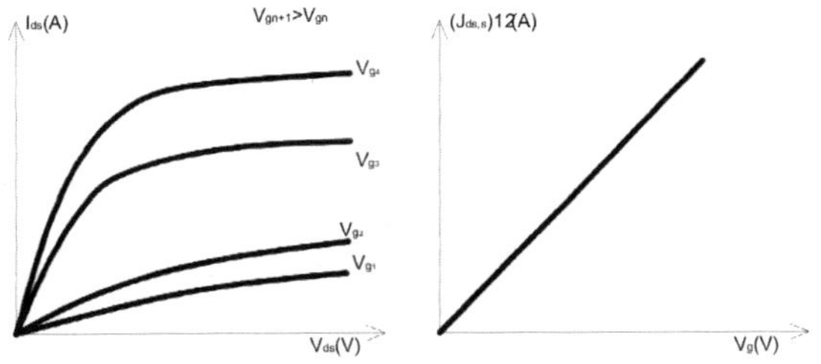

Figure 8. Current-voltage characteristics of field effect transistor.

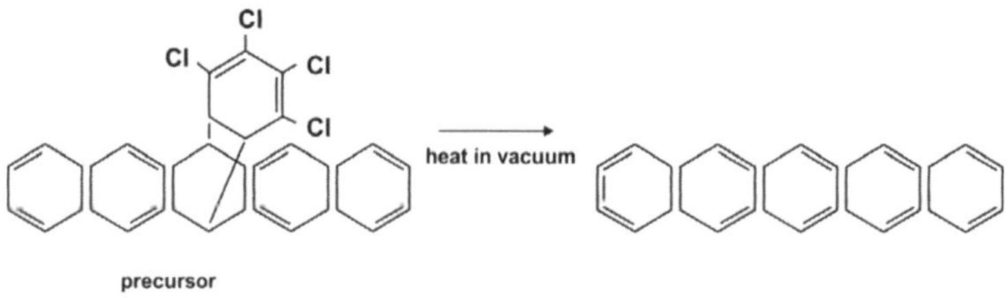

Figure 9. Growth of pentacene films from precursor.

The temperature dependence of the field–effect mobility μFE on temperature in these layers is justified very well [33] (figure 10a) by formulae (10) describing theoretically hopping mechanism of conductivity presented in previous section. The experimentally determined activation energy as a function of gate voltage is also in agreement with prediction by formulae (10) [33] (see figure 10b). The parameters of fitting shown in figure 10 are $\sigma_0 = 1.6$ x 1010 S/m, $\alpha^{-1} = 0.22$ nm, $T_0 = 385$ K [33].

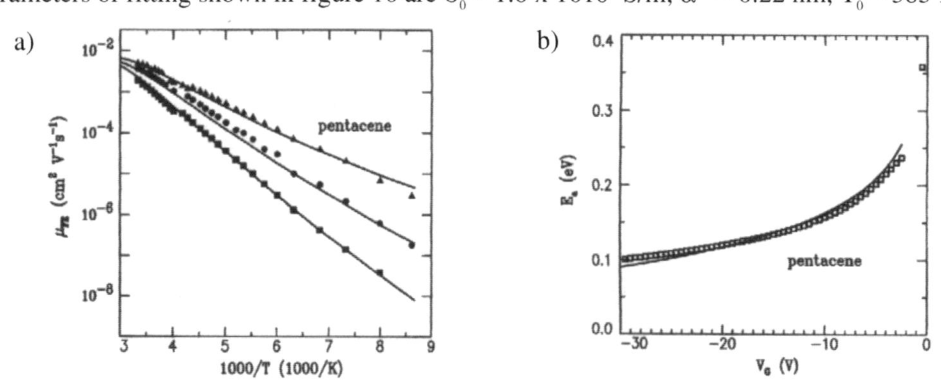

Figure 10. Fitting of experimental data [38] of hole mobility (a) and activation energy, (b) in pentacene thin films by hopping conductivity theory [33].

Mattheus [8] investigated pentacene layers formed by high vacuum evaporation on highly doped silicon (gate) wafers with 1μm SiO_2 layer as an insulator (see figure 7). The 1.54 nm polymorphic pentacene was studied by I–V characteristics of FET. The charge transport properties were determined from linear part of the characteristic [42]. The mobility was calculated from channel conductance g_C

$$g_C = \frac{\partial I_{sd}}{\partial V_{sd}}\Big|_{Vg} = \frac{Z}{L}\mu C_i (V_g - V_t) \qquad (18)$$

The temperature dependence of mobility in the range of temperature 145–300 K was measured as well. These results point on the almost independence of mobility on the temperature. The value of mobility was in the range $1.3 \pm 0.1 \times 10^{-2}$ cm²/Vs.

Stadlober et al. [48] studied the influence of polymorphism, grain size and film thickness on electrical properties of pentacene layers deposited in high–vacuum conditions in the standard FET geometry device with organic PVP or PMMA dielectric layers. The mobility in layers composed in grains with size below $1\mu m$ was 10^{-2} cm²/Vs whereas for 1μm grains this mobility increased steeply and stayed stable (0.35–0.4 cm²/Vs) for layers with larger grains. This interesting result would be explained by assumption that the mean free path of carriers is about 1μm. Then the grain boundary scattering is greater for layers composed of smaller grains. It was also noticed that the mobility depends on the layer thickness. It was the same for layers having thickness up to 150 nm. For thicker films the mobility decreases probably due to higher resistance of contacts or to the presence of the bulk phase inclusions in the layer. From this the authors concluded that the bulk phase had smaller intrinsic mobility.

Table 2. Semiconductor polymer materials applied in organic FET.

Polymer	Symbol	Molecule	Mobility μ (300K) $[cm^2/Vs]$	Ref.
	TVB	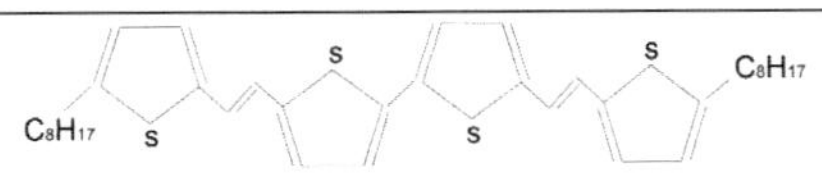		
Thienylene-vinylene oligemers	TVC	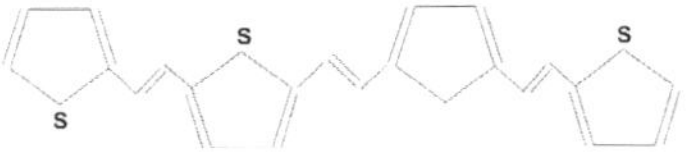	10^{-5} – – 10^{-1}	[52]
	TVD	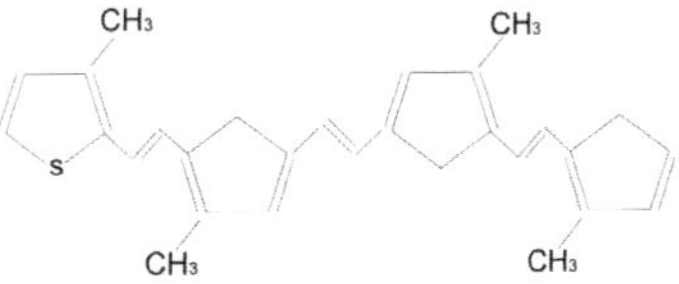		
Thienylene-vinylene-thienylene	TVT	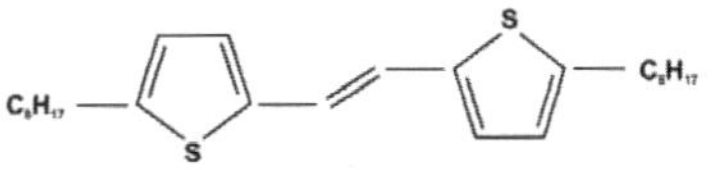	$1.1-5.1\cdot$ $\cdot 10^{-5}$	[53]
Polythienylene-vinylene	PTV	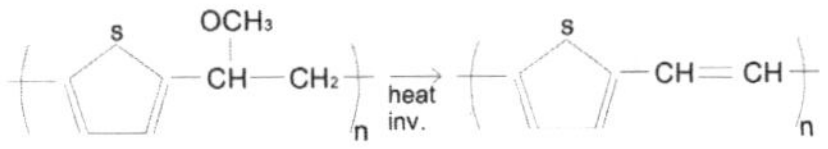	10^{-4} – – 10^{-3}	[38]
Poly (β'-dodecyloxy (-α, α'–α, α") tertienyl (poly DOT_3)	DOT_3	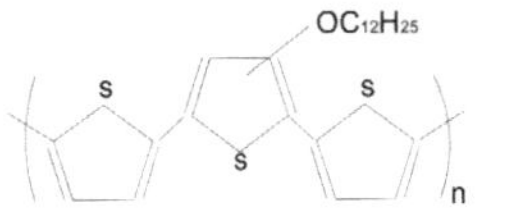	10^{-6} – – 10^{-4}	[54], [55]
Rubrene		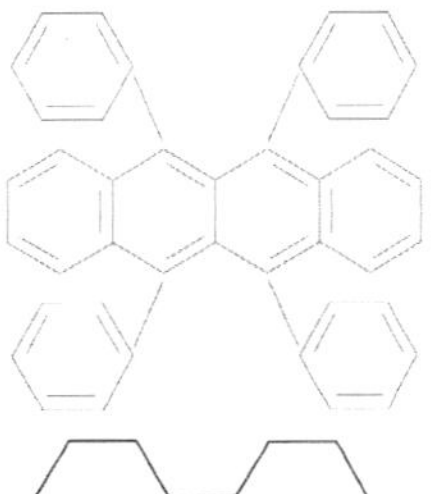	15 (single crystal 300K)	[29], [31]
Perylene bulk		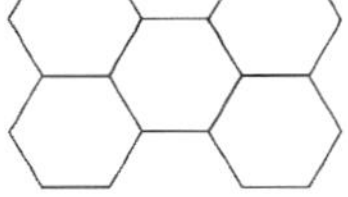	hole 0.4 electr. 5.5	[3]
Pentacene film			0.7	[49]
		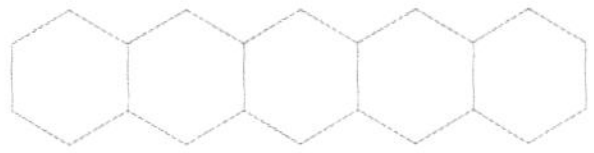	2, 0.1	[56], [57]
Pentacene bulk			3.2 40000 (5K)	[7]
Bis (1,2,5-thiadiazolo)-*p*-quinobis (1,3-dithiole)	BTQBT	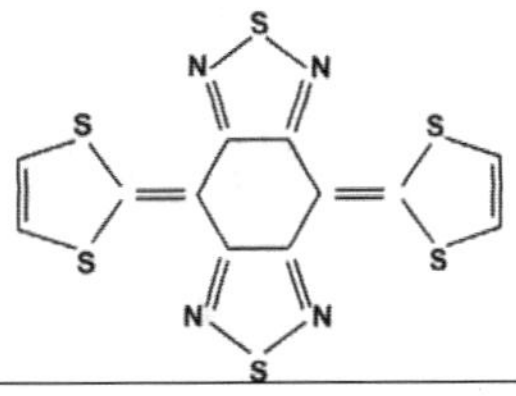	0.044	[5]

7. Conclusions

The recent interest of the progress in organic electronic is focused towards thin film transistors with active semiconductor layer formed from flat oligomeric molecules.

The great progress was achieved in increasing of charge carrier mobility in layers: up to the value about 1 cm^2/Vs in pentacene [48, 49, 50]. Paralelly, the studies of single crystals of organic materials with high mobility are developed strongly in the aim to research the mechanisms of the electronic transport not limited by structural defects. Several works reported optimistic results. In tetracene single crystals the mobility of 1 cm^2/Vs was measured and in paper [7] the mobility in single pentacene crystals reaches the value 3.2–4 cm^2/Vs at room temperature and 40000 cm^2/Vs at 5K. Schon, Kloc and Batlogg [3] reported 5.5 cm^2/V cm in perylene at room temperature. In rubrene single crystals the mobility was measured 7.54 cm^2/Vs at room temperature [51]. However, the carrier mobility in layers of organic materials is still far less then these for single crystals (see table 2). The reasons are defects, disorder, grain boundaries. The main mechanism of charge transport is hopping. Some cases of crossover to band like model are observed only for single crystals [7]. The charge transport in such structures is far from full understanding. The dominated molecular orbital overlap mechanism needs further studies. The temperature dependence of mobility in layers usually varies for samples even forming in the same growth conditions with strongly technological parameters imposed. Generally it seems that the transport properties in layers is not bulk effect [48]. The interface semiconductor–insulator and the first layer are particularly important in conduction. Therefore the study of crystal structure of layers and the initial stage of layer forming should be investigated carefully in the aim to understand better conductivity properties and to reach effort in the increase of mobility. Therefore, the searching of new organic materials consisting of planar molecules, which would lie flat on the substrate in the deposition process, seems to be also promising direction of research [5].

Acknowledgement

This work was supported by the grant no. 3 T11B 036 27 of the Polish Committee for Scientific Research (KBN).

References

[1] Pope M and Swenberg C E 1982 *Electronic Processes in Organic Crystals* (Oxford:Clarendon Press)
[2] Bao Z, Lovinger A J and Brown J 1998 *J. Am. Chem. Soc.* **120** 207
[3] Schon J H, Kloc Ch and Batlogg B 2000 *Appl. Phys. Lett.* **77** No 23
[4] Imaeda K, Yamashita Y, Li Y, Mori T, Inokuchi H and Sano M 1992 *J. Mater. Chem.* **2** 115
[5] Xue J and Forrest S R 2001 *Appl. Phys Lett.* **79** No 22
[6] Forrest S R 1997 *Chem. Rev.* **97** 1793
[7] Schon J H, Kloc Ch and Batlogg B 2001 *Phys. Rev.* B **63** 245201
[8] Mattheus C C 2002 *Polymorphism and electronic properties of pentacene* Dissertation Groningen
[9] Koopmans T 1934 *Physica* **1** 104
[10] Hoyland R and Goodman L 1962 *J. Chem. Phys.* **36** 12
[11] Buenker R J and Peyerimhoff S D 1969 *Chem. Phys. Lett.* **3** 37
[12] Sworakowski J 1974 *Sci. Pap. Instit. Org. Phys. Chem. Wroclaw Tech. Univ.*, Monograph No 2
[13] Lyons L E 1967 *J. Chem.* **10** 365
[14] Pope M, Burgos J and Giachino J 1965 *J. Chem. Phys. Lett.* **12** 140
[15] Pope M and Burgos J 1966 *Mol. Cryst.* **1** 395
[16] Silinsh E A, Jurgis A, Gailis A K and Taure L F 1974 *Sci. Pap. Instit. Org. Phys. Chem. Wroclaw Tech. Univ.* **740**
[17] Karl N 1974 *Festkorperprobleme* **14** 261
[18] Katz J L, Rice S A, Choi S and Jortner J 1963 *J. Chem. Phys.* **39** 1683

[19] Silbey R, Jortner J, Rice S A and Vala M J 1965a *J. Chem. Phys.* **42** 733, 1965b *J. Chem. Phys.* **43** 2959
[20] Oonishi I, Fujisawa S, Aoki J, Ohashi Y and Sasada Y 1986 *Bull. Chem. Soc. Jpn.* **59** 2233
[21] Desiraju G R and Gavezzotti A 1989 *Acta Cryst.* B **45** 473
[22] Holmes D, Kumaraswamy S, Matzger A J and Vollhardt K P C 1999 *Chem. Eur. J.* **5** 3399
[23] Campbell R B, Monteath Robertson J and Trotter J 1962 *Acta Cryst.* B **45** 473
[24] Campbell R B, Monteath Robertson J and Trotter J 1961 *Acta Cryst.* **14** 705
[25] Campbell R B, Monteath Robertson J and Trotter J 1962 *Acta Cryst.* B **15** 289
[26] Dimitrakopoulos C D, Brown A R and Pomp A 1996 J. *Appl. Phys.* **80** 2501
[27] Luo Y, Wang G, Theobald J A and Beton P H 2003 *Surface Science* **537** 241-245
[28] Mattheus C C, Dros A B, Baas J, Meetsma A, de Boer J L and Palstra T T M 2001 *Acta Cryst.* C **57** 939
[29] Sundar V C, Zaumseil J, Podzorov V, Menard E, Willett R L, Someya T, Gershenson M E and Rogers J A 2004 *Science* **303** 1644
[30] Zaumseil J, Someya T, Bao Z N, Loo Y L, Cirelli R and Rogers J A 2003 *Appl. Phys. Lett.* **82** 793
[31] de Boer R W I, Gershenson M E, Morpurgo A F and Porzorov V 2004 *Phys. Stat. Sol.* (a) **201** No 6 1302-13331
[32] Horowitz G, Hajlaoui R and Delannoy P 1995 *J. Phys.* III **5** 355
[33] Vissenberg M C J M and Matters M 1998 *Phys. Rev.* B **57** No 20
[34] Mott N F and Davis E A 1979 *Electronic Processes in Noncrystalline Materials* (Oxford:Clarendon Press)
[35] Sahimi M 1994 *Applications of Percolation Theory* (London:Taylor & Francis)
[36] Grunewald M, Thomas P and Wurtz D 1979 *Phys. Status Solidi* B **94** K1
[37] Shapiro F R and Adler D 1985 *J. Non. Cryst. Solids* **74** 189
[38] Brown A R, Jarrett C P, de Leeuw D M and Matters M 1997 *Synthetic Metals* **88** 37-55
[39] de Boer R W I, Jochemsen M, Klapwijk T M, Morpurgo A F, Niemax J, Tripathi A K and Pflaum J 2004 *Journal of Applied Physics* **95** No 3
[40] Mott N F and Gurney R W 1948 *Electronic Processes in Ionic Crystats* (2nd edn.) (Oxford:Oxford Univ. Press)
[41] Bonham J S and Jarvis D H 1977 *Aust. J. Chem.* **30** 705
[42] Sze S M 1981 *Physsies of Semiconductor Devices* (New York: Wiley)
[43] Kim S H, Hwang D-H, Park H, Chu H Y, Lee J-I, Do L-M and Zyung T 2000
[44] Schon J H 2001 *Appl. Phys. Lett.* **79** No 25
[45] Park J H, Kang C H, Kim Y J, Lee Y S and Choi J S 2004 *Mat. Sci. Eng.* C **24** 27-29
[46] Kymissis I, Dimitrakopoulos C D and Purushothaman S 2000 *J. Vac. Sci. Tech.* B **20** 956-959
[47] Schon J H, Kloc Ch and Batlogg B 2000 *Phys. Rev. Lett.* **86** No 17
[48] Stadlober B, Satzinger V, Maresch H, Somitsch D, Haase A, Pichler H, Rom W and Jakopic G 2003 Structural and Electrical Properties of Polymorphic Pentacene Thin Films *Proceedings of SPIE* Vol. 5217 *(Organic Field Effect Transistors II)* ed C D Dimitrakopoulos and A Dodabalapur (WA:Bellingham) pp 112–23
[49] Gundlach D J, Lin Y Y, Jackson T N, Nelson S F and Schlom D G 1997 *IEEE Electron Device Lett.* **18** 87
[50] Lin Y Y, Gundlach D J, Nelson S F and Jackson T N 1997 *IEEE Elec. Dev. Lett.* **18** 606
[51] Podzorov V, Sysoev S E, Loginowa E, Pudalov V M and Gershenson M E 2003 *Appl. Phys. Lett.* **83** 3504
[52] Reynolds S, Shepherd J T, Main C, Marshall J M and Maud J M 2000 *Journal of Non-Crystalline Solids* **266-269** 994-998
[53] Salih A J, Haynes D M and Hepburn A R 1995 *Synthetic Metals* **71** 2257-2258
[54] Brown A R, de Leeuw D M, Havinga E E and Pomp A 1994 *Synth. Met.* **68** 65
[55] Jarrett C P, Brown A R, Friend R H and de Leeuw D M 1995 *J. Appl. Phys.* **77** 6289
[56] Nelson S F, Lin Y Y, Gundlach D J and Jackson T N 1998 *Appl. Phys. Lett.* **72** 1854

[57] Matters M, de Leeuw D M, Vissenberg M J C M, Hart C M, Herwing P T, Geuns T, Mutsaers C M J and Drury C J 1999 *Optical Materials* vol 12 **2–3** pp 189–97
[58] Helfrich W and Mark P 1962 *Z. Physik.* **168** 495
[59] Helfrich W and Mark P 1962 *Z. Physik.* **166** 370

Noisy dynamics of magnetic flux in mesoscopic cylinders

J Dajka[1], J Łuczka[1], M Mierzejewski[1] and P Hänggi[2]

[1] Institute of Physics, University of Silesia, 40-007 Katowice, Poland
[2] Institute of Physics, University of Augsburg, D-86135 Augsburg, Germany

E-mail: `dajka@server.phys.us.edu.pl`

Abstract. We study magnetic fluxes and currents in mesoscopic systems of cylindrical symmetry like rings, toroids and cylinders. We analyze the time evolution of the magnetic flux and the characteristic time of a formation of the ordered state. We investigate how, starting from some symmetric initial state, the magnetic flux or the current approach their corresponding asymptotic state.

1. Introduction

Quantum phenomena manifested at the mesoscopic level have attracted much theoretical and experimental attention [1]. Phase coherence and persistent currents in mesoscopic systems of cylindrical symmetry can be mentioned as examples. Persistent currents predicted as early as in 1938 [2] have been observed experimentally only since 1990 [3]. In this paper we study the dynamics of magnetic fluxes and currents in mesoscopic cylinder under conditions when dissipation and fluctuations can play an important role [4].

2. Evolution equation for the magnetic flux

We consider a collection of rings composed of individual current channels, which form a cylinder. There are N_z channels in the direction of the cylinder axis and N_r in the direction of the cylinder radius. We assume that the thickness of the cylinder wall is small as compared with the radius. Because of the mutual inductance between rings, the current in one ring induces a flux in the other rings. In turn, the flux induces a current, and so on. We assume that the rings are not contacted. So, there is no tunneling of electrons among the channels and the charge carriers moving in the different rings are independent. It has been shown [5] that the effective interaction between the ring currents, when taken in the selfconsistent mean field approximation, results in the magnetic flux $\phi_{ind} = LI_{tot}$ felt by all electrons, where L is the cylinder inductance and I_{tot} is the total current in the cylinder. At temperature $T > 0$, the total magnetic flux ϕ consists of a sum of the external flux ϕ_{ext} and the flux ϕ_{ind} stemming from the total current, which in turn is a sum of the dissipative 'normal' Ohmic current and the non-dissipative persistent current resulting from the presence of the 'phase-coherent' electrons in the system [4], i.e. it assumes the form:

$$\phi = \phi_{ext} + LI_{tot} = \phi_{ext} + L[I_{coh}(\phi, T) + I_{nor}(\phi, T)]. \qquad (1)$$

Note that ϕ_{ext} is induced by an external magnetic field and can either take a fixed value or it can be a random function. Taking into account an explicit form of the "normal" current, as it follows from the Lenz's and Ohm's rules complemented with the Johnson-Nyquist noise term [6], Eq. (1) takes the form [4]

$$\frac{1}{R}\frac{d\phi}{dt} = -\frac{1}{L}(\phi - \phi_{ext}) + I_{coh}(\phi, T) + \sqrt{\frac{2k_BT}{R}}\,\Gamma(t), \tag{2}$$

where R is the resistance of the cylinder, k_B is the Boltzmann constant and $\Gamma(t)$ denotes a zero-mean Gaussian delta-correlated white noise modeling Nyquist equilibrium current noise. This equation takes the form of a classical Langevin equation and it constitutes our basic evolution equation.

The dimensionless form of (2) reads [4]

$$\dot{x} = -V'(x) + \sqrt{2D}\,\tilde{\Gamma}(\tilde{t}), \tag{3}$$

where the dot denotes a derivative with respect to the rescaled time $\tilde{t} = t/\tau_0$ with $\tau_0 = L/R$ being the relaxation time of the averaged normal current. The prime denotes a derivative with respect to the dimensionless flux $x = \phi/\phi_0$, where the flux quantum $\phi_0 = h/e$ is the ratio of the Planck constant h and the elementary charge e. The generalized potential reads $V(x) \equiv V(x, \lambda, i_0, p, T) = \frac{1}{2}x^2 - \lambda x - i_0 F(x, p, T)$, where $\lambda = \phi_{ext}/\phi_0$ is the rescaled external flux. The prefactor is $i_0 = NLI_0/\phi_0$ with $N = N_z N_r$. The function $F(x) \equiv F(x, p, T) = \int f(x, p, T)dx$ characterizes the coherent current and $f(x, p, T) = pf_e(x, T) + (1 - p)f_o(x, T)$, where $f_e(x, T) = \sum_{n=1}^{\infty} A_n(T)\sin(2n\pi x) = f_o(x - \frac{1}{2}, T)$. The functions f_e and f_o describe the coherent current flowing in the channel with an even or odd number of electrons, respectively [7]. The amplitudes $A_n(T)$ are decreasing functions of temperature. Their explicit forms are given in [4, 7]. The quantity $p \in [0, 1]$ denotes the probability of the occurrence of the single current channel with an even number of electrons. In the following we consider the symmetric case, i.e. $p = 1/2$.

The dimensionless intensity D of rescaled Gaussian white noise $\tilde{\Gamma}(\tilde{t}) \equiv \sqrt{\tau_0}\,\Gamma(\tau_0\tilde{t})$ is $D = k_BT/2\epsilon_0$, where the characteristic magnetic energy $\epsilon_0 = \phi_0^2/2L$. Let us notice that the resistance R does not enter into the rescaled equation (3). Moreover, the prefactor i_0 depends on the geometry and material of the sample [8]. Although formally the above equations can also be applied to a single mesoscopic ring or toroids, we consider a cylinder because for such a system the prefactor i_0 in the effective potential can take sufficiently large value (because $N = N_z N_r$). We choose the parameters of the system in such a way that the diffusion coefficient $D \sim 0.001T/T^*$ and $i_0 = 1$. The characteristic temperature T^* is defined by the relation $k_BT^* = \Delta_F/2\pi^2$, where Δ_F marks the energy gap at the Fermi surface. The values of parameters which occur in the amplitudes $A_n(T)$ are the same as in [4, 8].

As it follows from (1), the total current I_{tot} is linearly related to the magnetic flux ϕ (and to the rescaled flux x). As a consequence, the properties and behavior of the current are identical to the properties and behavior of the magnetic flux. Therefore, below we use equivalently these two characteristics of the system. ¿From now on, we will use only the dimensionless variables and omit the 'tilde' for the rescaled time, $\tilde{t} \equiv t$. Temperature will be measured in units of T^*.

3. Dynamics of the magnetic flux

First, let us consider the deterministic case of the Langevin stochastic equation (3) formally neglecting the Nyquist noise term $\tilde{\Gamma}(\tilde{t})$, i.e.,

$$\dot{x} = -V'(x). \tag{4}$$

The stationary solutions x_s of (4), for which $\dot{x}_s = 0$, correspond to extrema of the generalized potential $V(x)$. The solutions x_s of the gradient differential equation (4) are stable provided they correspond to a minimum of the generalized potential and they are unstable in the case of a maximum. For the case when the externally induced flux is zero we have $\lambda = 0$ and when $x_s \neq 0$, the stable stationary states of the dynamical system (4) correspond to the so called *self-sustaining currents* which can flow without any external driving [9].

The Langevin equation (3) defines a Markov diffusion process. Its probability density $p(x,t)$ obeys the Fokker-Planck equation [10]

$$\frac{\partial}{\partial t}p(x,t) = \frac{\partial}{\partial x}V'(x)p(x,t) + D\frac{\partial^2}{\partial x^2}p(x,t) \tag{5}$$

with the natural boundary condition, i.e. $\lim_{|x|\to\infty} p(x,t) = 0$.

The stationary solution $p_s(x)$ is asymptotically stable and ergodic, i.e. it does not depend on the initial condition. Its explicit form is

$$p_s(x) = N_0 \mathrm{e}^{-V(x)/D} \tag{6}$$

with the normalization constant N_0. The properties of steady-states were investigated in [4, 8]. In the following we study the time evolution of the probability density $p(x,t)$ in the absence of the external flux, $\lambda = 0$, and for the symmetric initial conditions. As an example, we assume that the initial flux probability distribution is Gaussian, i.e. $p(x,0) = \exp(-x^2/2\sigma)/\sqrt{2\pi\sigma}$ with $\sigma = 0.005$. The numerical solutions of the Fokker-Planck equation (5) are depicted below.

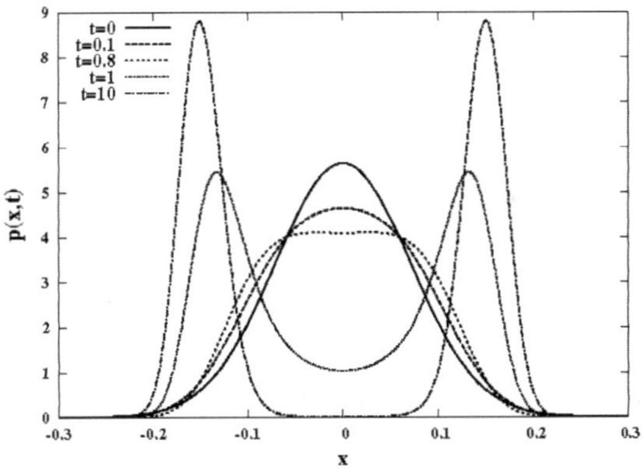

Figure 1. The probability density of the magnetic flux at several instants of time measured in units of τ_0. The temperature is set at $T = 1$. The graph at t=10 and the asymptotic probability density are indistinguishable.

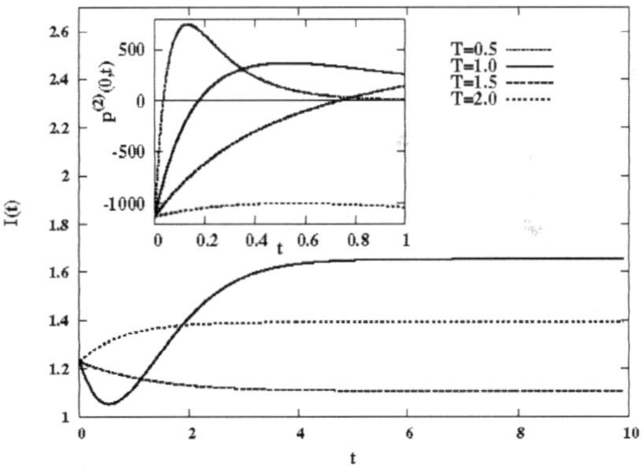

Figure 2. The Shannon information defined by the equation (7)(main graph) and the second derivative of the probability density at zero (inset) as a function of time in units of τ_0 and different temperatures. The change in sign of $p^{(2)}(0,t)$ indicates the formation of the bistability.

Because the initial condition as well as the generalized potential are symmetric with respect to the magnetic flux x, the averaged magnetic flux becomes $< x(t) >= 0$. ¿From the practical point of view, the experimental results are accumulated around the most probable values which correspond to maxima of the probability distribution $p(x,t)$. In the stationary state, they correspond to the minima of the generalized potential $V(x)$. If the maximum x_M of $p(x,t)$ is non-zero, it corresponds to the case of the self-sustaining current and an ordered state. In the stationary case and for the fixed values of parameters that we have assumed, the ordered states exist only below some critical temperature T_c, for details see [8]. For the assumed values of the parameters, $T_c \approx 1.66$. In figure 1 we depict the time evolution of $p(x,t)$ in the bistability case when in the stationary regime two symmetric ordered states coexist, i.e. when $p_s(x)$ has two symmetric maxima. One can observe the time-transition from the monostable state to the bistable state. We can introduce some characteristic time t_f of the formation of the ordered states. It can be related to the time when the maximum of $p(x,t)$ at $x = 0$ changes to the non-zero value, $x_M \neq 0$. It leads to the definition of this characteristic time t_f as the time when $p^{(2)}(0,t_f) = 0$, i.e. when the second derivative with respect to x at zero changes its sign (note that for this definition the time t_f depends on the initial condition). The initially 'concave-up' function becomes 'concave-down' at zero. We see that this time t_f is an increasing function of temperature, cf. figure 2. It tends to infinity when temperature tends to its critical value, $T \to T_c$. Above the critical temperature [4], there are no values of parameters which can guarantee the change of sign of $p(x,t)$ at $x = 0$. Let us notice that for vanishing temperature the curve representing $p^{(2)}(0,t)$ approaches asymptotically zero. It is clear since at $T = 0$ the asymptotic probability density of the flux is $p_s(x) = \delta(x - x_s)/2 + \delta(x + x_s)/2$.

The quantity of interest is the entropy or the information gained or loosed during the evolution. We investigate the time-evolution of the Shannon information defined by the relation

$$I[p](t) = \int_{-\infty}^{\infty} p(x,t)\ln(p(x,t))dx. \tag{7}$$

The time evolution of the Shannon information belongs to one of three qualitatively different classes. At larger temperatures, the information can be either an increasing or a decreasing function of time, depending on the relation between the initial and asymptotic states. The most interesting non-monotonic behavior appears at low temperatures. The initial information lost connected with the formation of bistability is followed by the information gain originating from the quasi-localization of the flux x in maxima of $p(x,t)$. This localization is caused by a relatively long escape time at low temperatures [4].

In conclusion, we have investigated the time evolution of the probability distribution of the magnetic flux in a mesoscopic cylinder. For our chosen parameters, there is a critical temperature below which ordered states can exist in the system. This bistability phenomenon corresponds to the self-sustaining currents flowing in the system. For this case, we have defined the characteristic time of formation of the ordered states. It is an increasing function of temperature and it diverges as temperature approaches its critical value.

Acknowledgments

The authors acknowledge the support of the Foundation for Polish Science (JD, grant SN-302-934/2005), the ESF Program *Stochastic Dynamics: fundamentals and applications* (JŁ and PH) and the KBN grant PBZ-MIN-008/P03/2003 (JD and J Ł).

References

[1] Imry Y 1997 *Introduction to Mesoscopic Physics* (New York: Oxford University Press)
[2] Hund F 1938*Ann. Phys. (Leipzig)* **32** 102
[3] Levy L P, Dolan G, Dunsmuir J and Bouchiat H 1990 *Phys. Rev. Lett.* **64** 2074
[4] Dajka J, Łuczka J, Szopa M and Zipper E 2003*Phys. Rev. B* **67** 073305
[5] Lisowski M, Zipper E and Stebelski M 1999 *Phys. Rev. B* **59** 8305
[6] Dajka J, Łuczka J, Szopa M and Hänggi P 2005 *phys. stat. sol.* **b242** 196
[7] Cheung H F, Gefen Y, Riedel E K and Shih W H 1988 *Phys. Rev. B* **37**,6050
[8] Dajka J, Kostur M, Łuczka J, Szopa M and Zipper E 2003 *Acta Phys. Pol.* B **34** 3793
[9] Wohlleben D, Esser M, Freche P, Zipper E and Szopa M 1991 *Phys. Rev. Lett.* **66** 3191
[10] Hänggi P and Thomas H 1982 *Phys. Rep.* **88** 207

Symmetry and classification of electron states in polyacenes

M Labuz[1], A Wal, G Wisz and M Kuzma

University of Rzeszow, Institute of Physics, Rejtana 16a, 35-310 Rzeszow, Poland

E-mail: [1]labuz@univ.rzeszow.pl

Abstract. Organic materials have been recently the object of the intense studies due to their opto-electronic properties which follow from the states of the outer-shell electrons, taking part in molecule bonding. The symmetry of molecules provides the classification of quantum states according to the group theory method. We apply molecular state space factorizaton scheme for the states structure classification of selected polyacenes.

1. Introduction

Organic semiconductors are strongly studied from many years and a great progress was made in the performance of organic devices particularly in organic field-effect transistors (FET). The key element in FET transistors is thin semiconductor organic film. From many new organic compounds such as thienylene – vinylene oligomers (TVB, TVC, TVD) [1], rubrene [2], metal phthalocyanines (MP) [3-5], which manifest interesting semiconducting parameters, the pentacene was selected as the most promising for FET. Such organic electronics will result in many low-cost and large area applications. The low-temperature processes of devices in relation to the nonorganic semiconductors industry result in applications of extraordinary substrates as flexible, large area, low-weight polymeric or even biological plates for integrated electronic structures. The possibility of submicron scale of integrated electronic circuits is the next feature of organic electronics, which overcome the possibilities of standard microelectronics.

Usefulness of organic crystals for FET transistors production depends on good parameters that characterise the electric conduction in such crystals, first of all large mobility of charge. The mobility is in strong relation with electron-phonon coupling, because the mechanism of conductivity can be hopping or polaron one. Microscopic description of the model being discussed is still unclear and theoretical investigations are nowadays up-to-date. To work the problem out correctly it is necessary to create classification of both electron and phonon states for the organic crystals being investigated. The properties of charge carriers depend on the electronic properties of molecules as well as the kind of interactions between them. As regards to aromatic hydrogencarbon molecules the π orbitals were studied mainly, which determine not only bonding between carbon atoms, but they also control other electronic properties of the molecule. However, in organic crystals the σ carbon atomic orbitals as well as the hydrogen 1s, 2s and 2p orbitals contribute in overlap effects of adjacent molecules. Such calculations for overlap of H orbitals with π system of carbon atoms of neighbour molecule were done recently by Mattheus [6] for herringbone arrangement in pentacene crystals. It seems to us that precise insight into symmetry of orbital studied should give more information about the conductivity mechanism in these crystals as well as charge carriers mobility anisotropy measured e.g. in paper [7].

In our paper we propose classification of the electron states in cyclic molecules of aromatic hydrocarbons and derivatives because of its major application as semiconducting layers in FET transistors. In particular, we focus our attention on the classification of molecular electron states in benzene and naphthalene, which play the role of the model investigations of pentacene widely applying in FET transistors [8,9]. The classification introduced in the paper bases on factorization of molecular orbital representation, the algorithm related to that, proposed earlier by Lulek [10] and applied for classification of normal vibrations [11].

2. Molecular orbitals and LCAO approximation

The LCAO method [12-14] introduces molecular orbitals Ψ_m expressed by the formula

$$\Psi_m = \sum_i c_{im}\varphi_i \, . \tag{1}$$

Introducing symmetry orbitals ψ_k

$$\psi_k = \sum_i a_{ik}\varphi_i \, , \tag{2}$$

we obtain the series (1) as

$$\Psi_m = \sum_k b_{km}\psi_k \, , \tag{3}$$

where φ_i denotes i-th atomic orbital. Coefficients c_{im} can be obtained from the eigenvalue problem simplified by group-theory, whereas a_{ik} coefficients can be found directly from group-theory calculations.

The eigenvalue problem of the system

$$\left(\hat{H} - E_m\right)\Psi_m = 0 \tag{4}$$

is reduced to the secular equation:

$$\begin{vmatrix} H_{11} - E_m & \dots & H_{1k} - E_m S_{1k} \\ \vdots & \ddots & \vdots \\ H_{k1} - E_m S_{k1} & \dots & H_{kk} - E_m \end{vmatrix} = 0 \, , \tag{5}$$

where $H_{ii} = \int \varphi_i \hat{H} \varphi_i d\tau$, $H_{ij} = \int \varphi_i \hat{H} \varphi_j d\tau$, $S_{ij} = \int \varphi_i \varphi_j d\tau$ are Coulomb, interaction and overlap integrals respectively and can be evaluated, e.g. also using some approximation method (e.g. Hückel method). The solution of this eigenvalue problem results in determining of energies E_m and c_{im} coefficients and thereby the strict form of orbital molecules.

Let the set of molecular orbitals forms Hilbert space M of quantum electronic states of the aromatic hydrocarbons. Such a space is a direct product of two subspaces: P and A spanned by positions of C-H atom groups in the molecule and by atomic orbitals of carbon atoms, respectively:

$$M = P \otimes A \, . \tag{6}$$

Furthermore the subspace A is a sum of A_π and A_σ subspaces of π and σ atomic orbitals:

$$A = A_\pi \oplus A_\sigma \, . \tag{7}$$

Thus:

$$M = (P \otimes A_\pi) \oplus (P \otimes A_\sigma) \, . \tag{8}$$

The group G of symmetry of the molecule acts in the space $\mathtt{M}$ forming representation M, whereas representations P and A act in subspaces $\mathtt{P}$ and $\mathtt{A}$, respectively.

Let vectors $|\Gamma w\gamma\rangle$, $|\Lambda\lambda\rangle$ and $|\Delta\delta\rangle$ form basis of the Hilbert space $\mathtt{M}$, $\mathtt{P}$, and $\mathtt{A}$ respectively. The positional representation P and orbital representation A decomposes into irreducible representations Λ and Δ of the group G:

$$\mathsf{P} = \sum_{\Lambda} \oplus n(\mathsf{P},\Lambda)\Lambda \tag{9}$$

$$\mathsf{A} = \sum_{\Delta} \oplus n(\mathsf{A},\Delta)\Delta \tag{10}$$

Taking into account the decomposition

$$\Lambda \otimes \Delta = \sum_{\Gamma} \oplus c(\Lambda,\Delta,\Gamma)\Gamma\,, \tag{11}$$

the basis vectors of the $\mathtt{M}$ (molecular) space are determined using Clebsch-Gordan procedure as:

$$|\Lambda t,\Delta d,\Gamma w\gamma\rangle = \sum_{\lambda\delta} \begin{bmatrix} \Lambda & \Delta & \Gamma w \\ \lambda & \delta & \gamma \end{bmatrix} |\Lambda t\lambda\rangle |\Delta d\delta\rangle\,, \tag{12}$$

where $t = 1,2,\ldots,n(\mathsf{P},\Lambda)$, $d = 1,2,\ldots,n(\mathsf{A},\Delta)$, $w = 1,2,\ldots,c(\Lambda,\Delta,\Gamma)$ are repetition indices of the first (t and d) and second range (w). The symbol $\begin{bmatrix} \Lambda & \Delta & \Gamma w \\ \lambda & \delta & \gamma \end{bmatrix}$ is Clebsch-Gordan coefficient. The vectors on the right side of the equation (12) are

$$|\Lambda t\lambda\rangle = \sum_{i=1}^{N} b^{i}_{\Lambda t\lambda} |i\rangle \tag{13}$$

$$|\Delta d\delta\rangle = \sum_{\alpha} a^{\alpha}_{\Delta d\delta}\varphi^{\alpha} \tag{14}$$

3. Electron states of benzene

In the case of benzene molecular orbitals, the electron configuration of each carbon atom is $1s^2 2s^2 2p^2$. In the language of chemical bonding the sp^2 hybridisation is a set of three (so called σ) atomic orbitals $\varphi_s, \varphi_{p_x}, \varphi_{p_y}$ (figure 1b). However in the symmetry with three-fold axis in the carbon position (as in the case of benzene ring) it is convenient to introduce orbitals $\varphi_{\sigma_a}, \varphi_{\sigma_b}, \varphi_{\sigma_c}$ (figure 1c) [15]

$$\begin{aligned}
\varphi_{\sigma_a} &= \frac{1}{\sqrt{3}}\left(\varphi_s + \sqrt{2}\varphi_{p_x}\right) \\
\varphi_{\sigma_b} &= \frac{1}{\sqrt{3}}\left(\varphi_s + \sqrt{\frac{3}{2}}\varphi_{p_x} - \frac{1}{\sqrt{2}}\varphi_{p_y}\right) \\
\varphi_{\sigma_c} &= \frac{1}{\sqrt{3}}\left(\varphi_s - \sqrt{\frac{3}{2}}\varphi_{p_x} - \frac{1}{\sqrt{2}}\varphi_{p_y}\right) \\
\varphi_{\pi} &= \varphi_{p_z}
\end{aligned} \tag{15}$$

The geometry of φ_σ, $\sigma = (\sigma_a, \sigma_b, \sigma_c)$ vectors is introduced in figure 1c. They are situated along a, b, c axis, which form 120° angle with each other. The φ_{p_z} atomic orbital remains unchanged and is perpendicular to the plane of benzene molecule (we denote this orbital as φ_π). The set of atomic orbitals φ_{σ_i} and φ_π spans the Hilbert space, which we call the atomic orbital space A with basis vectors φ_{σ_i}. The action of symmetry group D_{6h} of benzene molecule on the set $\{\varphi_{\sigma_i}\}$ forms the atomic representation A. Vector φ_π does not mix with the vectors φ_σ, therefore the representation is one-dimensional and φ_π changes its sign by operations C_2', C_2'', I, S_3, S_6, σ_h (table 1). The characters of the representation A and its decomposition into irreducible representations of the group D_{6h} are collected in table 1. In table 1 the structure of A_σ and A_π representations is presented too. The positions (i = 1,...,6) of carbon atoms in benzene (figure 1a) constitute basis vectors $|i\rangle$ of positional space P. The elements of the group D_{6h} acting in this space form positional representation P with characters included in table 1.

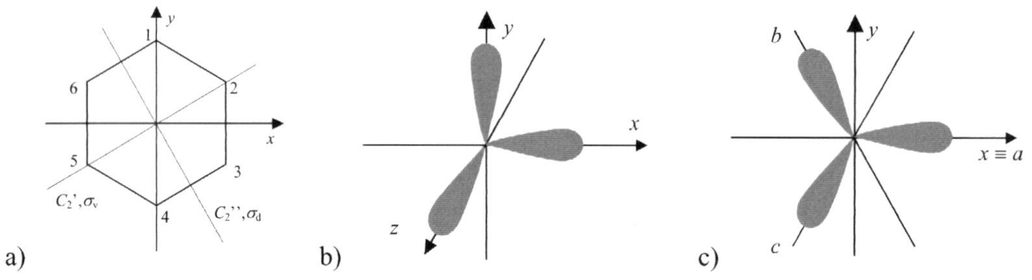

Figure 1. Benzene. a) symmetry elements of benzene ring, b) orbitals $\varphi_{p_x}, \varphi_{p_y}, \varphi_{p_z}$ and c) $\varphi_{\sigma_a}, \varphi_{\sigma_b}, \varphi_{\sigma_c}$ of carbon atoms.

Table 1. Benzene. The characters of positional P and atomic representation A with decomposition of P, A, A_σ and A_π.

D_{6h}	E	$2C_6$	$2C_3$	C_2	$3C_2$'	$3C_2$''	I	$2S_3$	$2S_6$	σ_h	$3\sigma_d$	$3\sigma_v$	Decomposition
P	6	0	0	0	2	0	0	0	0	6	0	2	$A_{1g} + E_{2g} + B_{1u} + E_{1u}$
A	4	1	1	-2	0	-2	-4	-1	-1	2	0	2	$A_{2u} + B_{1u} + E_{1u}$
A_σ	3	0	0	-3	1	-1	-3	0	0	3	-1	1	$B_{1u} + E_{1u}$
A_π	1	1	1	1	-1	-1	-1	-1	-1	-1	1	1	A_{2u}

The positional representation P decomposes into irreducible ones as follows (see table 1):

$$\mathsf{P} = A_{1g} + E_{2g} + B_{1u} + E_{1u}, \tag{16}$$

whereas atomic representation decomposes as (see table 1):

$$\mathsf{A} = \mathsf{A}_\sigma + \mathsf{A}_\pi = A_{2u} + B_{1u} + E_{1u}, \tag{17}$$

where $A_\pi = A_{2u}$. The irreducible basis vectors of the positional subspace for benzene are:

$$|A_1 a_1\rangle = \frac{1}{\sqrt{6}}\left(|1\rangle + |2\rangle + |3\rangle + |4\rangle + |5\rangle + |6\rangle\right)$$

$$|B_{1u}b_1\rangle = \frac{1}{\sqrt{6}}\left(|1\rangle - |2\rangle + |3\rangle - |4\rangle + |5\rangle - |6\rangle\right) \quad (18)$$

$$|E_1e_1\rangle = \frac{1}{\sqrt{12}}\left(2|1\rangle + |2\rangle - |3\rangle - 2|4\rangle - |5\rangle + |6\rangle\right), \qquad |E_1e_2\rangle = \frac{1}{2}\left(|2\rangle + |3\rangle - |5\rangle - |6\rangle\right)$$

$$|E_2e_1\rangle = \frac{1}{\sqrt{12}}\left(2|1\rangle - |2\rangle - |3\rangle + 2|4\rangle - |5\rangle - |6\rangle\right), \qquad |E_2e_2\rangle = \frac{1}{2}\left(|2\rangle - |3\rangle + |5\rangle - |6\rangle\right).$$

4. Electron states of naphthalene and polyacenes of higher order

Naphthalene, anthracene, tetracene and pentacene consists of two, three, four and five benzene rings respectively. Due to the adjoin of k benzene rings together, these molecules show symmetry D_{2h} which is subgroup of the C_{6h} group symmetry of benzene. The symmetry of all polyacenes is the same independently on the number k of benzene rings (figure 2). The symmetry group decomposes all carbon positions i into orbits of action of group $G=D_{2h}$ on a set $\{i\}$ (table 2) for all $g \in D_{2h}$:

$$O_r = \{gr\}, \quad (19)$$

where $r \in \{i\}$ is an generator of orbit O_r. There are k four-element orbits and one two-element orbit for each molecule (table 4).

Characters of atomic A and positional representations P_k are collected in table 3, where symbol k=II concerns naphthalene, k=III – anthracene, k=IV – tetracene and k=V – pentacene. The decomposition of these representations into irreducible ones are presented in this table as well.

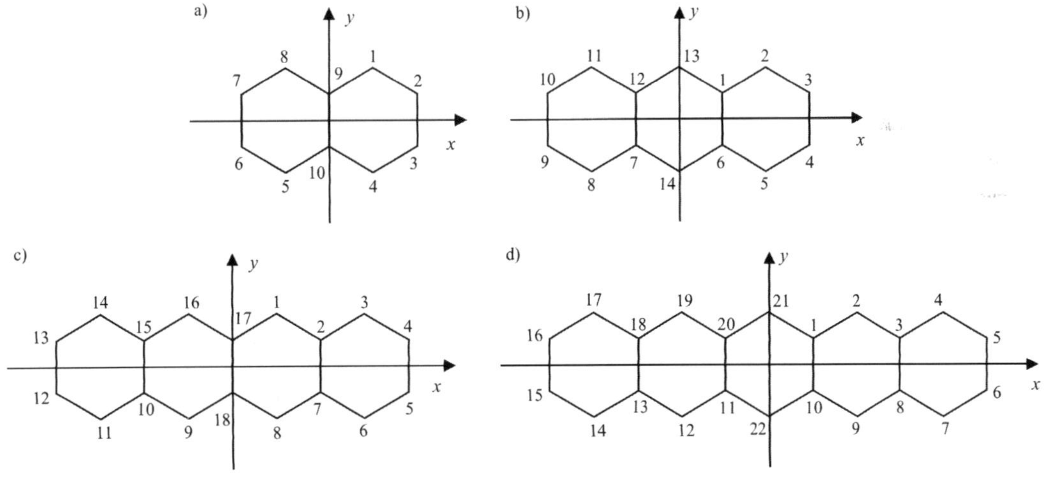

Figure 2. Enumeration of carbon atoms in: a) naphthalene, b) anthracene, c) tetracene, d) pentacene.

Table 2. Enumeration of orbits in polyacenes.

Polyacene	Orbits
Naphthalene	$O_1 = \{1,4,5,8\}$, $O_2 = \{2,3,6,7\}$, $O_9 = \{9,10\}$
Anthracene	$O_1 = \{1,6,7,12\}$, $O_2 = \{2,5,8,11\}$, $O_3 = \{3,4,9,10\}$, $O_{13} = \{13,14\}$
Tetracene	$O_1 = \{1,8,9,16\}$, $O_2 = \{2,7,10,15\}$, $O_3 = \{3,6,11,14\}$, $O_4 = \{4,5,12,13\}$ $O_{17} = \{17,18\}$

Pentacene	$O_1 = \{1,10,11,20\}$, $O_2 = \{2,9,12,19\}$, $O_3 = \{3,8,13,18\}$, $O_4 = \{4,7,14,17\}$ $O_5 = \{5,6,15,16\}$, $O_{21} = \{21,22\}$

Table 3. The characters of atomic representation A and positional representations P_{II}, P_{III}, P_{IV}, P_V of higher order polyacenes

D_{2h}	E	$C_2(z)$	$C_2(y)$	$C_2(x)$	I	$\sigma(xy)$	$\sigma(xz)$	$\sigma(yz)$	Decomposition
A	4	-2	-2	0	-4	2	2	0	$B_{1u} + B_{2u} + 2B_{3u}$
A_σ	3	-3	-1	1	-3	3	1	-1	$B_{2u} + 2B_{3u}$
A_π	1	1	-1	-1	-1	-1	1	1	B_{1u}
P_{II}	10	0	2	0	0	10	0	2	$3A_g + 2B_{1g} + 3B_{2u} + 2B_{3u}$
P_{III}	14	0	2	0	0	14	0	2	$4A_g + 3B_{1g} + 4B_{2u} + 3B_{3u}$
P_{IV}	18	0	2	0	0	18	0	2	$5A_g + 4B_{1g} + 5B_{2u} + 4B_{3u}$
P_V	22	0	2	0	0	22	0	2	$6A_g + 5B_{1g} + 6B_{2u} + 5B_{3u}$

According to enumeration of carbon atoms as in figure 2 the irreducible basis vectors $|\Lambda\lambda\rangle = N\sum_{i=1} b^i_{\Lambda\lambda}|i\rangle$ (where N is normalization factor) of positional representation are collected in table 4.

Table 4. Irreducible basis $|\Lambda\lambda\rangle$ of the positional representation for polyacenes.

			Representation			
Molecule	Orbit	N	$\vert A_g a_g >$	$\vert B_{1g} b_{1g} >$	$\vert B_{2u} b_{2u} >$	$\vert B_{3u} b_{3u} >$
Naphthalene	O_1	1/2	$\vert 1\rangle+\vert 4\rangle+\vert 5\rangle+\vert 8\rangle$	$\vert 1\rangle-\vert 4\rangle+\vert 5\rangle-\vert 8\rangle$	$\vert 1\rangle-\vert 4\rangle-\vert 5\rangle+\vert 8\rangle$	$\vert 1\rangle+\vert 4\rangle-\vert 5\rangle-\vert 8\rangle$
	O_2	1/2	$\vert 2\rangle+\vert 3\rangle+\vert 6\rangle+\vert 7\rangle$	$\vert 2\rangle-\vert 3\rangle+\vert 6\rangle-\vert 7\rangle$	$\vert 2\rangle-\vert 3\rangle-\vert 6\rangle+\vert 7\rangle$	$\vert 2\rangle+\vert 3\rangle-\vert 6\rangle-\vert 7\rangle$
	O_3	$1/\sqrt{2}$	$\vert 9\rangle+\vert 10\rangle$		$\vert 9\rangle-\vert 10\rangle$	
Anthracene	O_1	1/2	$\vert 1\rangle+\vert 6\rangle+\vert 7\rangle+\vert 12\rangle$	$\vert 1\rangle-\vert 6\rangle+\vert 7\rangle-\vert 12\rangle$	$\vert 1\rangle-\vert 6\rangle-\vert 7\rangle+\vert 12\rangle$	$\vert 1\rangle+\vert 6\rangle-\vert 7\rangle-\vert 12\rangle$
	O_2	1/2	$\vert 2\rangle+\vert 5\rangle+\vert 8\rangle+\vert 11\rangle$	$\vert 2\rangle-\vert 5\rangle+\vert 8\rangle-\vert 11\rangle$	$\vert 2\rangle-\vert 5\rangle-\vert 8\rangle+\vert 11\rangle$	$\vert 2\rangle+\vert 5\rangle-\vert 8\rangle-\vert 11\rangle$
	O_3	1/2	$\vert 3\rangle+\vert 4\rangle+\vert 9\rangle+\vert 10\rangle$	$\vert 3\rangle-\vert 4\rangle+\vert 9\rangle-\vert 10\rangle$	$\vert 3\rangle-\vert 4\rangle-\vert 9\rangle+\vert 10\rangle$	$\vert 3\rangle+\vert 4\rangle-\vert 9\rangle-\vert 10\rangle$
	O_4	$1/\sqrt{2}$	$\vert 13\rangle+\vert 14\rangle$		$\vert 13\rangle-\vert 14\rangle$	
Tetracene	O_1	1/2	$\vert 1\rangle+\vert 8\rangle+\vert 9\rangle+\vert 16\rangle$	$\vert 1\rangle-\vert 8\rangle+\vert 9\rangle-\vert 16\rangle$	$\vert 1\rangle-\vert 8\rangle-\vert 9\rangle+\vert 16\rangle$	$\vert 1\rangle+\vert 8\rangle-\vert 9\rangle-\vert 16\rangle$
	O_2	1/2	$\vert 2\rangle+\vert 7\rangle+\vert 10\rangle+\vert 15\rangle$	$\vert 2\rangle-\vert 7\rangle+\vert 10\rangle-\vert 15\rangle$	$\vert 2\rangle-\vert 7\rangle-\vert 10\rangle+\vert 15\rangle$	$\vert 2\rangle+\vert 7\rangle-\vert 10\rangle-\vert 15\rangle$
	O_3	1/2	$\vert 3\rangle+\vert 6\rangle+\vert 11\rangle+\vert 14\rangle$	$\vert 3\rangle-\vert 6\rangle+\vert 11\rangle-\vert 14\rangle$	$\vert 3\rangle-\vert 6\rangle-\vert 11\rangle+\vert 14\rangle$	$\vert 3\rangle+\vert 6\rangle-\vert 11\rangle-\vert 14\rangle$
	O_4	1/2	$\vert 4\rangle+\vert 5\rangle+\vert 12\rangle+\vert 13\rangle$	$\vert 4\rangle-\vert 5\rangle+\vert 12\rangle-\vert 13\rangle$	$\vert 4\rangle-\vert 5\rangle-\vert 12\rangle+\vert 13\rangle$	$\vert 4\rangle+\vert 5\rangle-\vert 12\rangle-\vert 13\rangle$
	O_5	$1/\sqrt{2}$	$\vert 17\rangle+\vert 18\rangle$		$\vert 17\rangle-\vert 18\rangle$	
Pentacene	O_1	1/2	$\vert 1\rangle+\vert 10\rangle+\vert 11\rangle+\vert 20\rangle$	$\vert 1\rangle-\vert 10\rangle+\vert 11\rangle-\vert 20\rangle$	$\vert 1\rangle-\vert 10\rangle-\vert 11\rangle+\vert 20\rangle$	$\vert 1\rangle+\vert 10\rangle-\vert 11\rangle-\vert 20\rangle$
	O_2	1/2	$\vert 2\rangle+\vert 9\rangle+\vert 12\rangle+\vert 19\rangle$	$\vert 2\rangle-\vert 9\rangle+\vert 12\rangle-\vert 19\rangle$	$\vert 2\rangle-\vert 9\rangle-\vert 12\rangle+\vert 19\rangle$	$\vert 2\rangle+\vert 9\rangle-\vert 12\rangle-\vert 19\rangle$
	O_3	1/2	$\vert 3\rangle+\vert 8\rangle+\vert 13\rangle+\vert 18\rangle$	$\vert 3\rangle-\vert 8\rangle+\vert 13\rangle-\vert 18\rangle$	$\vert 3\rangle-\vert 8\rangle-\vert 13\rangle+\vert 18\rangle$	$\vert 3\rangle+\vert 8\rangle-\vert 13\rangle-\vert 18\rangle$
	O_4	1/2	$\vert 4\rangle+\vert 7\rangle+\vert 14\rangle+\vert 17\rangle$	$\vert 4\rangle-\vert 7\rangle+\vert 14\rangle-\vert 17\rangle$	$\vert 4\rangle-\vert 7\rangle-\vert 14\rangle+\vert 17\rangle$	$\vert 4\rangle+\vert 7\rangle-\vert 14\rangle-\vert 17\rangle$
	O_5	1/2	$\vert 5\rangle+\vert 6\rangle+\vert 15\rangle+\vert 16\rangle$	$\vert 5\rangle-\vert 6\rangle+\vert 15\rangle-\vert 16\rangle$	$\vert 5\rangle-\vert 6\rangle-\vert 15\rangle+\vert 16\rangle$	$\vert 5\rangle+\vert 6\rangle-\vert 15\rangle-\vert 16\rangle$
	O_6	$1/\sqrt{2}$	$\vert 21\rangle+\vert 22\rangle$		$\vert 21\rangle-\vert 22\rangle$	

The basis vectors of atomic representations for all polyacenes are as follows:

$$|B_{2u}b_2\rangle = \frac{1}{\sqrt{2}}(\varphi_{\sigma_b} - \varphi_{\sigma_c}) \equiv \frac{1}{\sqrt{2}}(|b> - |c>)$$

$$|B_{3u}b_3^{(1)}\rangle = \varphi_{\sigma_a} \equiv |a> \tag{20}$$

$$|B_{3u}b_3^{(2)}\rangle = \frac{1}{\sqrt{2}}(\varphi_{\sigma_b} + \varphi_{\sigma_c}) \equiv \frac{1}{\sqrt{2}}(|b> + |c>),$$

where for abbreviation we denote $\varphi_{\sigma_a} \equiv |a>$, $\varphi_{\sigma_b} \equiv |b>$, $\varphi_{\sigma_c} \equiv |c>$.

Using formula (12) and taking into account the vectors from the table 4 as well as the ones expressed in (20) we obtain the symmetry orbitals $|\Lambda t, \Delta d, \Gamma w\gamma\rangle$. Some of the symmetry orbitals calculated for pentacene are presented in table 5, where $|il\rangle$ $(l = a, b, c)$ denotes product $|i\rangle|l\rangle$.

Table 5. The examples of σ–type symmetry orbitals $|\Lambda t, \Delta d, \Gamma w\gamma\rangle$ of pentacene.

Λt	Δd	$\Gamma\gamma$	N	$\|\Lambda t, \Delta d, \Gamma w\gamma>$
A_g1	B_{2u}	$B_{2u}b_{2u}$	$1/2\sqrt{2}$	$\|1b>-\|1c>+\|10b>-\|10c>+\|11b>-\|11c>+\|20c>-\|20c>$
	$B_{3u}1$	$B_{3u}b_{3u}$	$1/2$	$\|1a>+\|10a>+\|11a>+\|20a>$
	$B_{3u}2$	$B_{3u}b_{3u}$	$1/2\sqrt{2}$	$\|1b>+\|1c>+\|10b>+\|10c>+\|11b>+\|11c>+\|20c>+\|20c>$
A_g2	B_{2u}	$B_{2u}b_{2u}$	$1/2\sqrt{2}$	$\|1b>-\|1c>+\|10b>-\|10c>+\|11b>-\|11c>+\|20c>-\|20c>$
	$B_{3u}1$	$B_{3u}b_{3u}$	$1/2$	$\|1a>+\|10a>+\|11a>+\|20a>$
	$B_{3u}2$	$B_{3u}b_{3u}$	$1/2\sqrt{2}$	$\|1b>+\|1c>+\|10b>+\|10c>+\|11b>+\|11c>+\|20c>+\|20c>$
⋮				
$B_{3u}4$	B_{2u}	$B_{1g}b_{1g}$	$1/2\sqrt{2}$	$\|4b>-\|4c>+\|7b>-\|7c>+\|14b>-\|14c>+\|17c>-\|17c>$
	$B_{3u}1$	A_ga_g	$1/2$	$\|4a>+\|7a>+\|14a>+\|17a>$
	$B_{3u}2$	A_ga_g	$1/2\sqrt{2}$	$\|4b>+\|4c>+\|7b>+\|7c>+\|14b>+\|14c>+\|17c>+\|17c>$
$B_{3u}5$	B_{2u}	$B_{1g}b_{1g}$	$1/2\sqrt{2}$	$\|5b>-\|5c>+\|6b>-\|6c>+\|15b>-\|15c>+\|16c>-\|16c>$
	$B_{3u}1$	A_ga_g	$1/2$	$\|5a>+\|6a>+\|15a>+\|16a>$
	$B_{3u}2$	A_ga_g	$1/2\sqrt{2}$	$\|5b>+\|5c>+\|6b>+\|6c>+\|15b>+\|15c>+\|16c>+\|16c>$

5. Conclusions

The factorization of the symmetry orbitals subspace into positional and orbital subspaces is a useful tool for the group-theory classification of such orbitals. However, in the case of molecules consisted of large number of atoms and having a low symmetry, this procedure does not allow to classify orbits entirely because of the large number of irreducible representations included in the reducible ones (e.g. in the case of pentacene the index of multiplicity $t = 5$ or 6). Nevertheless, for aromatic hydrocarbons such index can be satisfactorily connected with the orbit numbers of carbon atoms. This provides almost complete distinction of symmetry orbitals. Only the indices of multiplicity d for atomic representation have no group-theory grounds.

Acknowledgement

This work was supported by the grant no. 3 T11B 036 27 of the Polish Committee for Scientific Research (KBN).

References

[1] Reynolds S, Shepherd J T, Main C, Marshall J M and Maud J M 2000 *J. Non-Cryst. Sol.* 266-269

[2] Sundar V C, Zaumseil J, Podzorov V, Menard E, Willet R L, Someya T, Gershenson M E and Rogers J A 2004 *Science* **303** 1644

[3] Liao M S and Scheiner S 2001 *J. Chem. Phys.* **114** 9780

[4] Awaga K and Maruyama Y 1991 *Phys. Rev.* B **44** 2589

[5] Evangelisti M, Bartolome J, de Jongh L J and Filoti G 2002 *Phys. Rev.* B **66** 144410

[6] Mattheus C C 2002 *Polymorphism and electronic properties of pentacene* Dissertation Groningen

[7] Schon J H, Kloc C and Batlogg B 2001 *Phys. Rev.* B **63** 245201

[8] Park J H, Kang Ch H, Kim Y J, Lee Y S and Cho J S 2004 *Mat. Sci. Eng.* C **24** 27-29

[9] Klauk H, Gundlach D J, Nichols J A and Jackson T N 1999 *IEEE Transactions on Electron Devices* **46**, 1258 - 1263

[10] Lulek T 1980 *Acta Physica Polonica* A **57** 407

[11] Kuzma M, Kuplowski I and Lulek T 1980 *Acta Physica Polonica* A **57** 415-28

[12] Streitweiser A Jr 1961 *Molecular Orbital Theory for Organic Chemists* (John Willey & Sons:New York).

[13] Cotton F A 1973 *Chemical Applications of Group Theory* (in Polish, Warsaw: PWN)

[14] Harris D C and Bertolucci M D 1978 *Symmetry and Spectroscopy. An Introduction to Vibrational and Electronic Spectroscopy* (New York : Dover Publications, Inc)

[15] Haken H and Wolf H C 1998 *Molecular Physics and Elements of Quantum Chemistry* (in Polish, Warsaw: PWN)

Longitudinal normals for piezoelectric elastic media

A Duda[1], T Paszkiewicz[2], S Wolski[2]

[1]Institute of Theoretical Physics, University of Wrocław, pl. Maxa Borna 9, PL-50-204, Wrocław, Poland

[2]Chair of Physics, Rzeszów University of Technology, ul. W. Pola 1, PL-35-959 Rzeszów, Poland

E-mail: ardud@ift.uni.wroc.pl

Abstract. We show that the problem of finding components of longitudinal normals for all piezoelectric classes is equivalent to the problem of solving two sets of two polynomial equations up to 6th degree. It is proved that the number of longitudinal normals for triclinic media cannot exceed 36. If the number of longitudinal normals contained in any plane is finite, then it cannot be larger than 6. Directions of longitudinal normals are found for quartz, $LiGaO_2$, and $Ba_2NaNb_5O_{15}$. For these compounds we numerically studied how piezoelectric terms of the propagation matrix influence the number and directions of the longitudinal normals.

1. Introduction

The goal of this paper is to study the problem of one of kinds of special directions of propagation of acoustic waves in piezoelectric elastic media. Special directions of propagation of sound could generally be divided into longitudinal normals, transverse normals, and degenerate directions. Longitudinal normals are directions along which a pure longitudinal wave could propagate. Transverse normals are directions along which a pure transverse wave could propagate. Degenerate directions, called also the acoustic axes [1]-[3], are directions for which phase velocity of at least two modes are equal.

The propagation matrices for acoustic waves propagating in a longitudinal or transverse direction have comparatively simple form, making it possible to use results of measurement of phase velocities of sound propagating in these directions to determine elastic constants of a medium. Moreover, Helbig [4] has shown that ratios of components of polarization vectors propagating in directions close to longitudinal normals are related in a simple way to ratios of appropriate combinations of elastic constants of a medium. This fact could be used to obtain information about elastic properties of media. For these reasons, it is important to know the maximal number as well as components of longitudinal normals for each acoustic symmetry class. This general problem has been the subject of research of many authors [1]-[9], who investigated properties of longitudinal normals for non-piezoelectric media. However, Bogdanov [10] considered longitudinal normals contained in coordinate planes of piezoelectric media belonging to the symmetry class $mm2$.

In our previous papers [11, 12], with the help of the method of eliminants, we studied general properties of longitudinal axes of non-piezoelectric elastic media, and calculated their directions for a number of media of different symmetries. In the present paper, we shall extend this method to piezoelectric elastic media belonging to *all* symmetry classes.

2. Longitudinal normals for triclinic media

For piezoelectric media, the polarization vectors of the elastic wave which propagates in the direction $\mathbf{n}$ are eigenvectors of the propagation matrix $\mathbf{\Gamma}^{(p)}(\mathbf{n}) = [\mathbf{\Gamma}^{(np)}(\mathbf{n}) + \mathbf{e}\otimes\mathbf{e}/\rho\varepsilon]$, with the elements $\Gamma_{il}^{(p)}(\mathbf{n})$

$$\Gamma_{il}^{(p)}(\mathbf{n}) = \rho^{-1}\sum_{j,m=1}^{3}\left[C_{ijlm}^{E} + \frac{e_{ij}(\mathbf{n})e_{lm}(\mathbf{n})}{\varepsilon_S(\mathbf{n})}\right]n_j n_m, \tag{1}$$

where C_{ijmk} are elements of tensor of elastic stiffness constants, while

$$\Gamma_{il}^{(np)}(\mathbf{n}) = \rho^{-1}\sum_{j,m=1}^{3} C_{ijlm}^{E} n_j n_m, \tag{2}$$

$$e_{ij}(\mathbf{n}) = \sum_{k=1}^{3} e_{ij,k}n_k, \varepsilon_S(\mathbf{n}) = \sum_{l,r=1}^{3} n_l\varepsilon_{lr}^{S}n_r, \tag{3}$$

where ρ is the density, $e_{lm,r}$ are elements of the tensor of electro-acoustical constants, and ε_{lm}^{S} $(l, m, r = 1, 2, 3)$ are elements of the tensor of dielectric permitivity constants characterizing elastic piezoelectric media. Squares of phase velocities are eigenvalues of matrices $\mathbf{\Gamma}^{(u)}(\mathbf{n})$, $u = p, np$.

A pure longitudinal wave could propagate in the direction $\mathbf{n}_u$, if, and only if, the following equation is fulfilled

$$\mathbf{\Gamma}_u(\mathbf{n}_u)\mathbf{n}_u = [c_u(\mathbf{n}_u)]^2\,\mathbf{n}_u \tag{4}$$

where $c(\mathbf{n})$ is the phase velocity and $\mathbf{\Gamma}_u$ is the propagation matrix for a given medium (piezoelectric or non-piezoelectric). We shall denote unit vectors fulfilling this condition by $\mathbf{n}_{ul}$.

Eq. (4) implies that

$$\left[n_{ul}^{(1)}, n_{ul}^{(2)}, n_{ul}^{(3)}\right] = \left[\Omega_{ul}^{(1)}, \Omega_{ul}^{(2)}, \Omega_{ul}^{(3)}\right], \tag{5}$$

where

$$\Omega_{ul}^{(i)} \equiv \lambda_u^{-1}\sum_{v=1}^{3}\Gamma_{iv}^{(u)}(\mathbf{n}_{ul})n_{ul}^{(v)}, \tag{6}$$

and $\lambda_u = c_u^2$. Further, we shall omit the index u.

2.1. Longitudinal normals lying out of OXY plane

The z-axis of the Cartesian coordinate system is directed in accord with general conventions (cf. ref. [13]). Therefore, one may consider longitudinal axes lying in plane perpendicular to z-axis as well as longitudinal axes lying out this plane.

Two vectors $\mathbf{n}$ and Ω are equal if

$$\Omega_l^{(1)}/\Omega_l^{(3)} = n_l^{(1)}/n_l^{(3)},\ \Omega_l^{(2)}/\Omega_l^{(3)} = n_l^{(2)}/n_l^{(3)},\ (n_l^3 \neq 0). \tag{7}$$

By making substitutions

$$x = n_l^{(1)}/n_l^{(3)},\ y = n_l^{(1)}/n_l^{(3)}, \tag{8}$$

the set of equations (7) is transformed into a set of two polynomial equations for x and y. Unlike for non-piezoelectric media [11, 12], for piezoelectric media these equations are not of fourth but of sixth degree.

For each solution of the aforementioned set, one can find exactly one longitudinal normal, because for each pair (x, y) one can find exactly one triple of numbers $[n_l^{(1)}, n_l^{(2)}, n_l^{(3)}]$ which fulfill the condition

$$\left[n_l^{(1)}\right]^2 + \left[n_l^{(2)}\right]^2 + \left[n_l^{(3)}\right]^2 = 1. \tag{9}$$

Namely,

$$\left[\frac{\pm x}{\sqrt{1+x^2+y^2}}, \frac{\pm y}{\sqrt{1+x^2+y^2}}, \frac{\pm 1}{\sqrt{1+x^2+y^2}}\right], \tag{10}$$

On the other hand, each longitudinal normal crosses the sphere in two points.

According to the Bezout theorem [14], if the number of solutions of the set of polynomial equations is finite, then it cannot be larger then the product of degrees of equations which form this set. Because the degree of each equation belonging to this set is 6, the maximal number of longitudinal normals cannot be larger than 36. Of course, there can exist longitudinal axes lying in the chosen plane, but generally we can choose the z-axis in such a way that longitudinal axes lying in plane perpendicular to z-axis are absent. The choice of z-axis cannot influence the number of solutions of considered set of equations.

By solving the set of Eqs. (7), one can determine all longitudinal normals for a given medium. This set could be written in the following form

$$\begin{aligned} &A_0(y) + A_1(y)x + A_2(y)x^2 + A_3(y)x^3 + A_4(y)x^4 \\ &+A_5(y)x^5 + A_6(y)x^6 = 0, \end{aligned} \tag{11}$$

$$\begin{aligned} &B_0(y) + B_1(y)x + B_2(y)x^2 + B_3(y)x^3 + B_4(y)x^4 \\ &+B_5(y)x^5 + B_6(y)x^6 = 0, \end{aligned} \tag{12}$$

where

$$A_\sigma(y), B_\sigma(y) \ (\sigma = 0, 1, 2, 3) \tag{13}$$

are univariate polynomials in the variable y. They depend on characteristics of electro-elastic properties of piezoelectric media, namely, on components $C_{ij,kl}$, ε_{ij} and e_{ijk} $(i, j, k, l = 1, ..., 3)$.

With the use of MAPLE system we obtained the explicit form of Eqs. (11), (12) for each symmetry class. Generally, the explicit form of Eqs. (11), (12) for piezoelectric media different than cubic, are too involved to be displayed here. Let us note that in the case of cubic nonpiezoelectric media these equations are very simple and their solutions are vectors defining the symmetry axes [11, 12]. For cubic piezoelectric media belonging to the $\bar{4}3m$ symmetry class, these equations read

$$\begin{aligned} &2C_{44}\,\epsilon\, n_1^2\, n_2\, n_3^3 + 12e_{14}^2\, n_1^2\, n_2\, n_3^3 + C_{11}\,\epsilon\, n_1^2\, n_2^3\, n_3 \\ &+2C_{44}\,\epsilon\, n_2\, n_3^5 + C_{11}\,\epsilon\, n_3\, n_2^5 + C_{12}\,\epsilon\, n_2\, n_3^5 \\ &+C_{12}\,\epsilon\, n_1^2\, n_2\, n_3^3 - 2C_{44}\,\epsilon\, n_2^5\, n_3 - 2C_{44}\,\epsilon\, n_1^2\, n_2^3\, n_3 \\ &-12e_{14}^2\, n_1^2\, n_2^3\, n_3 - C_{12}\,\epsilon\, n_2^5\, n_3 - C_{11}\,\epsilon\, n_1^2\, n_2\, n_3^3 \\ &-C_{11}\,\epsilon\, n_2\, n_3^5 - C_{12}\,\epsilon\, n_1^2\, n_2^3\, n_3 = 0, \end{aligned} \tag{14}$$

$$\begin{aligned} &C_{12}\,\epsilon\, n_1\, n_2^2\, n_3^3 + 2C_{44}\,\epsilon\, n_1\, n_3^3\, n_2^2 - 2\, C_{44}\,\epsilon\, n_1^5\, n_3 \\ &+C_{11}\,\epsilon\, n_1^3\, n_2^2\, n_3 + C_{12}\,\epsilon\, n_3^5\, n_1 + 12\, e_{14}^2\, n_1\, n_2^2\, n_3^3 \\ &+2\, C_{44}\,\epsilon\, n_1\, n_3^5 + C_{11}\,\epsilon\, n_1^5\, n_3 - 12\, e_{14}^2\, n_1^3\, n_2^2\, n_3 \\ &-C_{11}\,\epsilon\, n_1\, n_2^2 n_3^3 - 2\, C_{44}\,\epsilon\, n_1^3\, n_2^2\, n_3 - C_{11}\,\epsilon\, n_1\, n_3^5 \\ &-C_{12}\,\epsilon\, n_1^5\, n_3 - C_{12}\,\epsilon\, n_1^3\, n_2^2\, n_3 = 0. \end{aligned} \tag{15}$$

The set of Eqs. (11), (12) is solved by the method of eliminants [11, 12]. In Sect. 3, we show their solutions for several piezoelectric elastic media of low and middle symmetry. We also considered also two piezoelectric cubic media, namely, CuCl belonging to the symmetry class $\bar{4}3m$ and $Bi_{12}GeO_{20}$ (belonging to the class 23). For these media, results of calculations show that all their longitudinal axes are directed along the fourfold, threefold, and twofold symmetry axes.

2.2. Longitudinal normals inside OXY plane

For directions inside OXY plane, $n_l^{(3)} = 0$. Therefore, Eqs. (5) are equivalent to

$$\sum_{j=1}^{3} \Gamma_{3j} n_l^{(j)} = 0, \quad \left(\sum_{j=1}^{3} \Gamma_{2j} n_l^{(j)} \right) n_1^{(j)} = \left(\sum_{j=1}^{3} \Gamma_{1j} n_j^{(j)} \right) n_2^{(j)}. \tag{16}$$

Table 1. Components of the direction vectors of longitudinal normals for three piezoelectric compounds. Their symmetry classes are indicated in brackets. The corresponding phase velocities (m/s) are also indicated.

No.	SiO_2 (32)	$LiGaO_2$ (mm2)	$BaNaNb_5O_{15}$ (mm2)
(1)	[1,0,0]	[1,0,0]	[1,0,0]
	3245 5120 5746	3381 4060 5782	3531 3787 6715
(2)	[0, 0, 1]	[0, 1, 0]	[0,1,0]
	4727 4727 6360	3751 4060 5354	3503 3787 6827
(3)	[0.5,0.886,0]	[0, 0, 1]	[0,0,1]
	3245 5120 5746	3365 3693 6005	3502 3529 5763
(4)	[0.5,-0.866, 0]	[0.458, 0.610, 0.646]	[0.588, 0.809,0]
	3245 5120 5746	3531 3788 5737	3513 3637 6870
(5)	[0,0.755,0.656]	[-0.458, 0.610, 0.646]	[0.588, -0.809, 0]
	3362 3904 7061	3531 3788 5737	3513 3637 6870
(6)	[0.654,-0.377,0.656]	[0.458, -0.610, 0.646]	
	3362 3904 7061	3531 3788 5737	
(7)	[-0.654,-0.377,0.656]	[-0.458,-0.610,0.646]	
	3362 3904 7061	3531 3788 5737	
(8)	[0.284,0.164,0.945]	[0.853,0.522,0]	
	4080 5067 6414	3482 3794 5850	
(9)	[-0.284,0.164,0.945]	[0.853,-0.522,0]	
	4080 5067 6414	3482 3794 5850	
(10)	[0, -0.328, 0.945]	[0.717,0,0.697]	
	4080 5067 6414	3765 3886 5603	
(11)	[0.827, 0.478, 0.295]	[-0.717,0,0.697]	
	4367 4847 5334	3765 3886 5603	
(12)	[-0.827,0.478,0.295]	[0.5, -0.866, 0]	
	4367 4847 5334	3483 4222 6572	
(13)	[0,-0.955,0.295]	[0.5, 0.866, 0]	
	4367 4847 5334	3483 4222 6572	

Again, for a general case, the set of polynomial equations obtained from Eqs. (5) is too complicated to be displayed. The degrees of the corresponding polynomials equal 6, and they depend on the components $n_l^{(1)}$, $n_l^{(2)}$. Therefore, they cannot have more than six solutions. For cubic media belonging to the symmetry class $\bar{4}3m$, the considered equations are simple and identical with the suitable equation for non-piezoelectric media. One of them is a trivial identity $0 = 0$, and the second one reads

$$n_l^{(1)} (n_l^{(2)})^5 = (n_l^{(1)})^5 n_l^{(2)}. \tag{17}$$

We see that, similarly as in the case of non-piezoelectric media, in the case of piezoelectric media belonging to both $\bar{4}3m$ and 23 symmetry classes, the only longitudinal axes lying in the planes perpendicular to the four-fold symmetry axes are the symmetry axes.

3. Numerical results

3.1. Numerical results — longitudinal axes for selected compounds

We solved Eqs. (11) and (12) for three of compounds considered by Koos and Wolfe [15] and McCurdy and Every [16], namely, one trigonal - quartz SiO_2, belonging to the 32 symmetry class, and for two orthorhombic compounds - $LiGaO_2$ and $BaNaNb_5O_{15}$ belonging to the $mm2$ symmetry class. The results are shown in Table 1. Even in the case of two compounds belonging to the same class of the orthorhombic symmetry, the number of longitudinal normals is quite different. This indicates the strong dependence of the number and directions of the longitudinal axes on numerical values of their elasto-electric characteristics. The number of longitudinal normals of media of *different* crystalline symmetries, e.g. trigonal and orthorhombic considered by us, could be the same and much smaller than their maximal number.

The obtained vectors $\mathbf{n}_l$ were used for numerical calculation of corresponding phase velocities. Results of calculations are collected in Table 1.

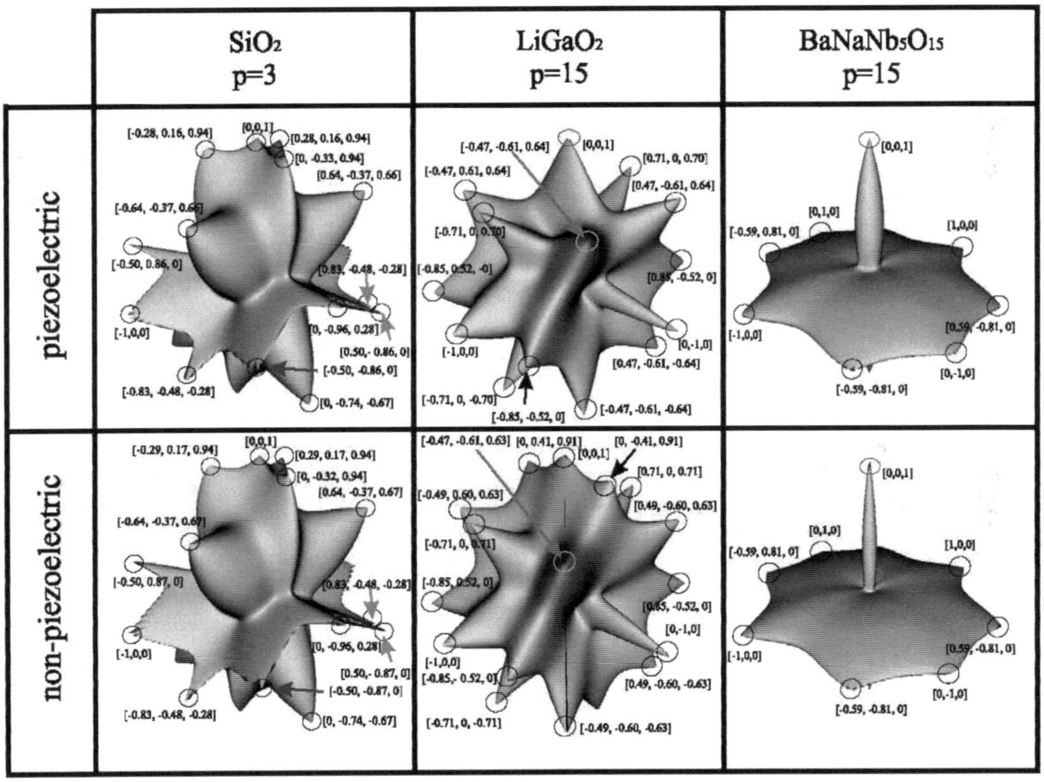

Figure 1. The plots of the surfaces $\Sigma(\mathbf{r})$, defined in the text, for three considered compounds and for propagation matrices with and without piezoelectric terms. The power p and direction vectors of longitudinal axes (cf. Table 1) are indicated.

3.2. Influence of piezoelectric term on longitudinal axes

In this section we shall study the influence of the term of $\Gamma^{(p)}(\mathbf{n})$ matrix related to piezoelectricity on the number of longitudinal normals and their directions. The numerical solutions of Eqs. (11) and (12) with the use of matrix elements of complete $\mathbf{\Gamma}^{(p)}$ are presented in Table 1 for the chosen compounds.

To verify our numerical results and study the influence of the piezoelectric term of the propagation matrix, for each of chosen compounds we draw the surface $\Sigma(\mathbf{r})$ being the geometrical locus of points $\mathbf{r} = \mathcal{P}(\mathbf{n})\mathbf{n}$, where $\mathcal{P}(\mathbf{n}) = [F(\mathbf{n})]^{\mathrm{p}}$. The function $F(\mathbf{n}) = \frac{2}{\pi} \arccos[\mathbf{n} \cdot e_{\mathbf{ql}}(\mathbf{n})]$ is normalized to the unity. For each propagation direction $\mathbf{n}$, and for both Γ_u $(u = p, np)$ matrices the polarization vector $\mathbf{e}_{ql}(\mathbf{n})$ of the quasi-longitudinal mode is numerically calculated. Then, the arcos of the angle between $\mathbf{n}$ and the polarization vector $\mathbf{e}_{ql}(\mathbf{n})$ is calculated. The surface $\Sigma(\mathbf{r})$ has maxima for pure longitudinal modes. To make them more pronounced, we plot the p-th power $(p = 3$ or $15)$ of $F(\mathbf{n})$.

Checking Fig 1, we conclude that the piezoelectric term of the propagation matrix generally influences the directions of longitudinal axes and changes their number. For example, "non-piezoelectric" $LiGaO_2$ crystal has two longitudinal axes more than "piezoelectric" one. In the case of quartz and $BaNaNb_5O_{15}$ crystals, the piezoelectric term does not change either number or directions of longitudinal axes. Numerical calculations performed with the use of MAPLE system confirm these results.

References

[1] Khatkevich A G 1963 *Sov. Phys. Crystallogr.* **7** 601
[2] Khatkevich A G 1965 *Sov. Phys. Crystallogr.* **9** 579
[3] Khatkevich A G 1977 *Sov. Phys. Crystallogr.* **22** 701
[4] Helbig K 1993 *Geophysics* **58** 680
[5] Sakadi Z 1941 *Proc. Physico-Mathematical Soc. Jpn.* **23** 539
[6] Borgnis F E 1955 *Phys. Rev.* **98** 1000
[7] Truesdell C 1966 *J. Acoust. Soc. Am.* **40** 729
[8] Kolodner I 1966 *J. Acoust. Soc. Am.* **40** 729
[9] Fedorov F I 1968 *Theory of Elastic Waves in Crystals* (New York: Plenum Press)
[10] Bogdanov S V 1995 *Akust. Zhurnal* **41** 27
[11] Duda A and Paszkiewicz T 2003 *Eur. Phys. J.* **B 31** 327
[12] Duda A and Paszkiewicz T 2003 *Proc. Int. Conf. on Symmetry and Structural Properties of Condensed Matter (Myczkowce)* ed T. Lulek, B. Lulek, A. Wal (Singapore: World Scientific) p 441
[13] Sirotin Yu I and Shaskolskaya M P 1982 *Fundamentals of Crystal Physics* (Moscow: Mir Publishers)
[14] Bix R 1998 *Conics and Cubics: Concrete Introduction to Algebraic Curves* (Heidelberg: Springer)
[15] Koos G L and Wolfe J P 1984 *Phys. Rev.* **B 30** 3470
[16] Every A G and McCurdy A K 1987 *Phys. Rev.* **B 36** 1432
[17] *Acoustic Crystals*, ed M. P. Shaskolskaya (Moscow: Nauka, 1982)

EPR studies of chromium ions in CdTe single crystals

I Stefaniuk[1], M Bester[1] and I E Lipiński[2]

[1]Institute of Physics, University of Rzeszow, Rejtana 16 A, 35-959 Rzeszow, Poland
[2]Institute of Physics, Szczecin University of Technology, Al. Piastów 48, 70-310 Szczecin, Poland.

Abstract. We have studied single CdTe crystals doped with relatively high concentration of chromium. The EPR measurements were carried out at room temperature and the angular dependence of the EPR spectrum was obtained. The spectra detected at liquid nitrogen temperature indicate two types of EPR centers. Simulations of the spectra at room temperature confirm the Cr^{2+} charge state of the dopand ion at tetrahedral symmetry sites.

1. Introduction

Diluted Magnetic Semiconductors (DMSs) are of interest for study mainly due to the exchange interaction between the localised magnetic moments and the band electrons [1,2,3]. This property of DMS has potential applications in spin-electronics [4]. Most studied DMS is $Cd_{1-x}Mn_xTe$ [2]. Blinowski et al. [5] pointed out very interesting properties of Cr based DMS, namely prediction of ferromagnetism in CdTe:Cr. Moreover, CdTe:Cr shows interesting laser properties in the range from 2 nm to 3 nm [6]. In this paper we investigate CdTe:Cr bulk crystals grown from the vapour phase. Ludwig and Lorentz [7] studied Cr^{1+} ($3d^5$) ions substituted into the cadmium sites in CdTe using electron Nuclear Double Resonance (ENDOR) and EPR. Note that usually chromium ions substituting cadmium ions are expected in the Cr^{2+} valence state i.e. in the $3d^4$ configuration. However, the authors [7] determined the total spin S=5/2, g=1.9997±0.0003 as well the parameters of the cubic zero-field splitting and the hyperfine interaction with second neighbours Cd^{111} and Cd^{113}. Vallin and Watkins [8] investigated EPR spectra of Cr^{2+} (S=2) in CdTe. The Jahn-Teller coupling coefficients were determined from the changes of the fine structure of EPR spectra under uni-axial stress. The spectra observed at 4.2 K and frequency ν_0=20 GHz consist of many lines in a wide range of magnetic field. The spectra show a complex angular dependence. The spin Hamiltonian (SH) parameters D, a and F were determined from the theoretical simulations of the fine structure of EPR spectrum of Cr^{2+} in CdTe. The anisotropy of the spectra was attributed to the static Jahn-Teller distortion around Cr^{2+} ion [8]. The temperature broadening of the EPR lines was also studied. More recently the contributions from the spin-orbit coupling of Cr^{2+} ions and Te ligands were calculated using the molecular-orbital approach in [9]. The studies [9] explained the difference in the zero-field splitting parameter for Cr^{2+} in CdTe ($b_2^0>0$) and in CdS ($b_2^0<0$). The Cr^{3+} ions were studied by EPR in GaAs crystals [10] concluding that the trivalent state was predominant if the chromium concentration significantly exceeded the concentration of shallow donors in the crystal. Crystals doped during the growth process should have a higher concentration of dopant ions than those doped by diffusion. The samples used in the studies [7, 8] were bulk CdTe crystals into which the chromium ions were introduced by diffusion. In our previous paper [11] we have studies single crystals CdTe:Cr with low concentration of Cr << 1%. The angular dependence of fine structure in room temperature was simulated by EPR-NMR programme [12] In the present paper we investigate the samples doped by chromium during the growth process with the concentration of Cr = 5% in alloy obtained in first stage of crystal growth. Here the properties of one very broad line in the spectrum are studied.

2. Experimental

Two types of CdTe:Cr material were investigated: (i) one obtained by simple melting of the components Cd, Te, Cr (alloy), and (ii) a single crystal grown by vapour method from this alloy. The EPR spectra of CdTe:Cr samples were recorded using the X-band Bruker ESP-300 spectrometer and ELEXSYS Bruker E 500 one at room (293 K) and liquid nitrogen (77 K) temperature. The sample No. 1 was cut from the alloy containing 5% of chromium. The sample No. 2 was cleaved from the single crystals. The (110) plane of sample No. 2 was identified as the cleaved plane. The samples have the form of a rectangle of the size 3 x 2 x 1.5 mm^3.

3. Cr^{2+} ion in crystal field

The electronic configuration of free Cr atom is $1s^2\ 2s^2\ 2p^6\ 3s^2\ 3p^6\ 3d^5\ 4s^1$. Cr atom incorporated into $A^{II}B^{VI}$ crystal lattice occupy cation sites while two of the outer electrons delocalize into the conduction and valence bands. The Cr^{2+} ion in $A^{II}B^{VI}$ semiconductor creates the deep impurity levels. The ground state for free Cr^{2+} ion is ^5D with the orbital quantum number L=2 and the total spin S=2 yielding 25-fold degeneracy. In the crystal field of D_{2d} symmetry the degeneracy is partially removed: ^5D term splits into fifteen-fold degenerate orbital triplet $^5T_2$ and tenfold degenerate orbital doublet ^5E. The Jahn-Teller distortion leads to splitting of the $^5T_2$ ground state into fivefold degenerate $^5B_2$ orbital singlet and the ^5E orbital doublet. The spin-orbit coupling leads to further splitting [13, 14, 15, 16, 17]. The arising energy level scheme is presented in figure 1. In external magnetic field the degeneracy of each electron spin level is removed (not presented in figure 1).

For description of the EPR spectra we employ the spin Hamiltonian with $S = 2$ [18]:

$$\hat{H} = \mu_B(\mathbf{B}\cdot\mathbf{g}\cdot\mathbf{S}) + \sum_{k,q} B_k^q O_k^q(S_x, S_y, S_z) \qquad (1)$$

where g is the gyromagnetic tensor, μ_B is the Bohr magneton, **B** is a magnetic field intensity, (S_x, S_y, S_z) are the spin operators, O_k^q are the Stevens operators [19, 20] and B_k^q are the fine structure parameters.

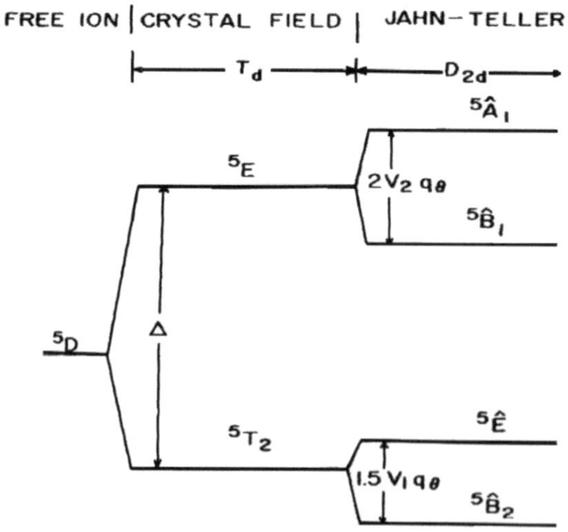

Fig. 1. Energy level splitting for Cr^{2+} ions in the crystal field of D_{2d} symmetry in the $A^{II}B^{VI}$ crystal (adapted from [21]).

4. Results

Selected EPR spectra of CdTe:Cr crystals are presented in figures 2a-2d. The spectrum consists of one broad and strong component on which a few weak fine structure lines are superimposed. In this paper we study only the major intense and broad component. The effective spectroscopic g-factor and the peak-to-peak linewidth H_{pp} of this line H_{pp} (we denote them as g_E and H_{ppE}) were readout from the spectrum. The way in which these parameters have been determined is sketched in figure 2a. Since the broad EPR line is strongly asymmetric the accuracy of parameters measured directly from the experimental spectrum is limited. Therefore, additionally the experimental line was fitted using the Lorenzian type curve. We were using this procedure because this type of curve describes satisfactory, in high temperature range, an experimental EPR lines of DMSs with manganium (see [22] and references therein). From the fitting we determined the corresponding parameters g_F and H_{ppF}. The obtained values g_E, H_{ppE} and g_F, H_{ppF} are listed in Table 1.

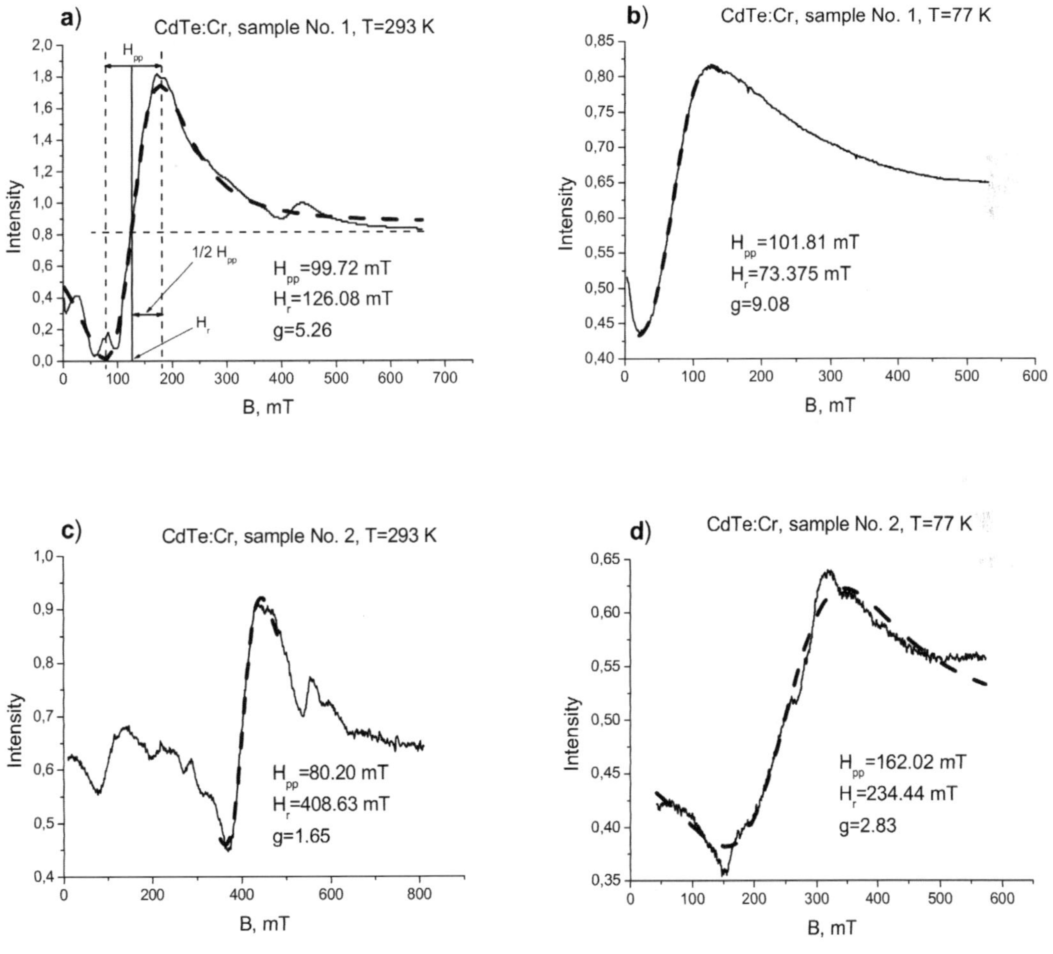

Fig. 2. Representative EPR spectra for CdTe:Cr crystals: a) alloy sample No.1, at T = 293 K, b) sample No.1, at T = 77 K, c) single crystal sample No.2 , at T = 293 K, d) sample No.2, at T = 77 K; solid lines represent experimental data, dashed lines represent Lorentzian fits to the experimental data

Table 1. The spectroscopic g parameters and the peak-to-peak linewidths for CdTe: Cr.

	Sample			
	No. 1		No. 2	
	T=293 (K)	T=77 (K)	T=293 (K)	T=77 (K)
g_E	5.26±0.21	9.08±1.28	1.65±0.02	2.83±0.08
g_F	5.22±0.02	9.46±0.02	1.67±0.01	2.66±0,01
H_{ppE}, (mT)	99.72±10.26	101.81±20.63	80.20±9.47	162.02±12.70
H_{ppF}, (mT)	101.49±0.50	107.68±0.24	80.99±1.18	190.04±0.62

5. Discussion and Conclusions

Our study reveals that the position of the EPR lines is independent of the orientation of the samples, whereas it depends on the crystallisation level and temperature. For the samples cut from the alloy (sample No. 1, figure 2a,b), the gyromagnetic factor g is 5.26 at T=293 K and 9.08 at T=77 K, whereas for the single crystal sample the value of g is reduced to g=1.65 and 2.83 at T=293 K and T=77 K, respectively. Such large differences in the g values are not common and may likely arise from the exchange interactions between magnetic ions for large Cr concentration. The linewidth increases with decreasing temperature. Particularly, in the case of the single crystal, the change of linewidth is significant; from 80 mT up to 162 mT. However, the same EPR behaviour was observed by Furdyna et al. [2, 22] in $Cd_{1-x}Mn_xTe$. The strong temperature dependence of the large linewidth was interpreted in terms of the Dzyaloshinski-Moriya exchange interactions[23]. In paper [22] Lorenzian shape was attributed to that EPR measurements particularly for samples with low concentration of magnetic ions. In our case the concentration of chromium is rather high. Therefore the deviation of EPR line in figure 2 from the Lorenzian shape may be expected. This may explain the discrepancy between the experimental and fitted values in table 1. The EPR spectra of both samples at room temperature show a clear fine structure; so more pronounced for the single crystal. The angular dependence for the sample No. 2 at room temperature figure 2c [24] as well as simulation of the fine structure of the spectrum indicate the Cr^{2+} charge of the dopant ion. Importantly, the fine structure of both samples disappears at the liquid nitrogen temperature. Our previous EPR studies [11] on CdTe:Cr samples with a minor concentration of Cr ions yield results differing considerably from those obtained in the present paper. The two-component spectrum has also been observed, however, the Cr ions occurred in the Cr^{3+} state instead of Cr^{2+}.

Acknowledgments

This work was partially supported by the grant No 1 PO3B 021 27 from The Polish Committee for Scientific Research (KBN). We are grateful to Prof. J. Kuriata and Dr M. Kuźma for help in EPR measurements.

References

[1] Dobrowolski W, Kossut J and Story T 2003 *II-VI and IV-VI Diluted Magnetic Semiconductors– new bulk materials and low dimensional quantum structures* (Handbook of Magnetic Materials vol. 15) ed. Bushow K H J (Elsevier)

[2] Furdyna J K and Kossut J 1988 *Eds. Semiconductors and Semimetals Diluted Magnetic Semiconductors* vol. 25 (Academic Press)

[3] Kacman P 2001 *Semicond. Sci. Technol* **16** R25

[4] Awshalom D D, Loss D and Samarth N 2002 *Eds. Semiconductors Spintronics and Quantum Computation* (Springer)

[5] Blinowski J, Kacman P and Majewski J A 1995 *Acta Physica Polonica* A **88** 683

[6] Bluiett A G, Hommerich U, Shah R T, Trivedi S B, Kutcher S W and Wang C C 2002 *J. of Electr. Mat.* **31** 806
[7] Ludwig G W and Lorenz M R 1963 *Phys. Rev.* **131** 601
[8] Vallin J T and Watkins D G 1974 *Phys. Rev.* B **9** 2051
[9] D-H Li, F-Z Li and Y-Y Zhou 1998 *Solid State Communications* **105** 59
[10] Krebs J J and Stauss G H 1977 *Phys. Rev.* B **15** 17
[11] Stefaniuk I, Bester M, Virt I S and Kuźma M 2005 *Acta Physica Polonica* A **105** 413
[12] Mc Gavin D G, Mombourquette M Y and J.A. Weil, the computer program "EPR version 5.13", Department of Chemistry, University of Saskatchewan, Canada 1993
[13] Boonman M E J, Mac W, Twardowski A, Wittlin A, van Bentum P J M, Maan J C and Demianiuk M 2000 *Phys. Rev.* B **61** 5358
[14] Rudowicz C and Konkiewicz B 1975 *Acta Phys. Polon.* **A47** 291
[15] Rudowicz C 1975 *Acta Phys. Polon.* **A47** 305
[16] Zhou Y and Rudowicz C 1996 *J. Phys. Chem. Solids* **57** 1191
[17] Rudowicz C and Zhou YY 1999 *J. Phys. Chem. Solids* **60**, 17
[18] Abragam A and Bleaney A 1970 *Electron Paramagnetic Resonance of Transition Ions* (Clarendon Press Oxford)
[19] Rudowicz C 1985 *J. Physics* **C18** 1415
Erratum: Rudowicz C 1985 **C18** 3837
[20] Rudowicz C 1987 *Magn. Res. Rev.* **13** 1
Rudowicz C 1988 **13** 335
[21] Vallin J T, Slack G A and Roberts S 1970 *Phys. Rev.* B **2** 4313
[22] Samarth N and Furdyna J K 1988 *Phys. Rev.* B **37** 9227
[23] Larson B E, Hass K C Ehrenreich H and Carlsson A E 1985 *Solid State Commun.* **56** 347
[24] Stefaniuk I, Bester M, Kuriata J, Kuzma M, to be published (2006)

The $D^1\Delta \rightarrow X^1\Sigma^+$ Transition in $^{13}C^{16}O$ Molecule

J Domin[1], U Domin[2], K Gałan[1] and R Kępa[2]

Rzeszów University of Technology, W.Pola 2, 35-959 Rzeszów, Poland
Atomic and Molecular Physics Laboratory, Institute of Physics, Rzeszów University, Rejtana 16A, 35-310 Rzeszów, Poland

Abstract. $D^1\Delta \rightarrow X^1\Sigma^+$ forbidden transition in $^{13}C^{16}O$ molecule was observed here. The resolved structure of 10 - 17 band was analyzed and new spectroscopic data were received for $D^1\Delta$ state. These data are in a good agreement with other authors referring this upper state of $^{13}C^{16}O$ molecule. Some perturbations in the rotational structure were observed also

1. Introduction

Several rovibronic states of CO molecule were investigated by many researches [1]-[5]. Most of the accuring band systems in this molecule were refered in Herzberg's monography [1] and also by Simmons and Tilford [3]. The first report on $D^1\Delta \rightarrow X^1\Sigma^+$ transition in absorption was published by Tiford and Simmons [6], as a forbidden band system. More detailed investigations were realised lately by Kittrell *et al* [7]-[9]. Now we just mention an observation of 10 - 17 band in $^{13}C^{16}O$ molecule. The registration of this band took a place during the realisation of a research project of IV-th Positive System [10] for this molecule.

2. Experimental

An emission spectrum in the near ultraviolet region was excited in a water-cooled Geissler's tube filled up with $^{13}C^{16}O$ molecule containing 95% of ^{13}C isotope. The spectrum was ohotographes with PGS-2 spectrograph at X^{th} spectral order. To separate the overlapping spectral orders, a standard monochromator was used for this spectrograph. The thorium lines excited in a water–colled lamp known as the hollow–cathode type lampe were used as the standard spectrum [11].

Spectral analysis

Registration of a single band belonging to $D^1\Delta \rightarrow X^1\Sigma^+$ transitions give a chance us for more detailed quantum description of the upper state. Up to now, this description is not satisfactory becouse the optical transitions from this state are not too probable. Additionaley the perturbations of the neighbouring states are some source of data [9] for $D^1\Delta$ state.

An analysis of the registered band classified as 10 - 17 band of $D^1\Delta \rightarrow X^1\Sigma^+$ elctronic transition was rather difficult. Just one cuircumstance which makes easier this analysis is a fact that $X^1\Sigma^+$ ground state is very well known and spectroscopic parameters for this state have an excellent precision. For these reasons a reconctruction of the vibrational levels in the ground state is possible with very high accuracy. So combination relations can be used for ground state for a rotational analysis of our registered band. Unfortunately some strong perturbation is observed in this band. However the combination differences $\Delta_2^{''}F(J)$ are not influenced by the perturbations in $X^1\Sigma^+$ ground state. The knowledge of these values for $^{13}C^{16}O$ molecule [12] allows to perform the rotational analysis. The rotational structure of our band is listed in Table 1. This structure consists of three branches: P, Q and R, and it seems to be suppising cuircumstance, because $D^1\Delta \rightarrow X^1\Sigma^+$ transition is attributed to electric–quadrupole transition. However such rotational assignment corresponds quite well with discoussion proposed by Tilford in his first report about [6].

Table 1. Observed wavenumbers (in cm^{-1}) and rotational assignments for the 10-17 band of the $D^1\Delta \rightarrow X^1\Sigma^+$ system of the $^{13}C^{16}O$ molecule

J	$R(J)$		$Q(J)$		$P(J)$	
0						
1		43135.590				
2		43134.905				
3		43133.978	43125.904			
4		43132.096	43121.226		43114.234	
5	43125.904	43130.392	43115.128		43106.620	
6	43124.700		43111.694		43098.804	
7	43123.100		43106.620		43090.564	43087.177
8	43121.200		43099.939			43079.177
9		43121.200	43093.333			43070.544
10		43114.239		43087.177		43062.764
11		43109.064		43079.177	43056.333	
12		43104.092		43070.540	43043.960	
13		43098.116		43061.004	43032.740	
14		43096.769			43021.990	
15		43083.779			43009.953	
16					42997.503	
17					42984.658	
18					42971.370	
19					42957.664	

Having the rotational assignment for our band we could determine the band origin by an extrapolation of Q branch, which seems to be more regular than P and R branches. This value can by estimated also from Deslandres table:

$\sigma_0 = 43128.65cm^{-1}$ (Q - branch)
$\sigma_0 = 43128.94cm^{-1}$ (Deslandres table)

These two values have a sufficient convergence. The first one from an extrapolation of Q branche seems to be slightly shifled by perturbation. An estimation on the second value was performed on the basis of the completed vibrational constants (of the Kittrell [8] and also our estimation) for $D^1\Delta$ state:

$T_e = 65975.84cm^{-1}$
$\omega_e = 1038.007cm^{-1}$
$\omega_e x_e = 9.4275cm^{-1}$
$\omega_e y_e = 0.5075cm^{-1}$

Taking into accaunt the divergence beetwen Tilford [6] and Kittrell [8] the next investigations are necessary to receive an accuracy comporable with other electronic states in $^{13}C^{16}O$ molecule.

Also the identification of the accuring perturbations actually is not possible. Some existing data [10] concerning $^{13}C^{16}O$ molecule could be a basis for the first trial of the identification. All these aspects will be undertaken in next our research efforts.

3. Conclusion

Observation of the 10 -17 band belonging to $D^1\Delta \rightarrow X^1\Sigma^+$ system permits to receive additional vibrational constants, that is $\omega_e y_e$ for $D^1\Delta$ state. Now, this molecular set of parameters allows to reconstruct the vibronic terms in $D^1\Delta$ state. In this way we can try to complete $D^1\Delta \rightarrow X^1\Sigma^+$ system and in the consequence to achieve a typical spectroscopic data set for $D^1\Delta$ state. Moreover an evidence of the accruing perturbations caused by $D^1\Delta$ state [9] will be completed.

References

[1] Herzberg G 1961 *Spectra of Diatomic Molecules* Van Nostrand Princeton
[2] Tilford S G Simmons J D 1972 *J.Phys.Chem Ref. Data* **1** 147
[3] Simmons J D Tilford S G 1971 *J.Res.Nat.B.S.* **75A** 455
[4] Krupenie P H 1964 *Nat.Bur.Stand.Ref.Data* NSRDS-NBS 5
[5] Kępa R Domin U and Porada K 2003 *Acta Phys.Polon.* **A103** 441
[6] Simmons J D Tilford S G 1966 *J.Chem.Phys.* **45** 2965
[7] Kittrell C Comeron S Butler L Field R W and Barrow R F 1963 *J.Chem.Phys.* **78** 3623
[8] Kittrell C Garetz B A 1989 *Spectrochim. Acta* **45A** 31
[9] Bruce Garetz A Kittrell C Le Floch A C 1990 *J.Chem.Phys.* **94** 843
[10] Domin J 1986 *Acta Phys.Hung.* **60** 43
[11] Giacchetti A 1970 *J.Opt.Soc.Am.* **60** 477
[12] Dale R M Herman M Johns J W C McKellar A R W Nagler S and Strathy I K M 1979 *Canad. J. Phys.* **57** 677

AUTHOR INDEX

AUTHOR INDEX